1 ref.

GROUNDWATER CHEMICALS DESK REFERENCE

JOHN H. MONTGOMERY
LINDA M. WELKOM

LEWIS PUBLISHERS

Library of Congress Cataloging-in-Publication Data

Montgomery, John H. (John Harold), 1955-
  Groundwater chemicals desk reference/John H. Montgomery,
Linda M. Welkom.
    p. cm.
  Includes bibliographical references.
  ISBN 0-87371-286-2
  1. Water, Underground—Pollution—Handbooks, manuals, etc.
2. Pollutants—handbooks, manuals, etc. I. Welkom, Linda M.
II. Title.
TD426.M66   1989   89-13382
628.1'61—dc20   CIP

Second Printing 1990

LEWIS PUBLISHERS, INC.
121 South Main Street, Chelsea, Michigan 48118

PRINTED IN THE UNITED STATES OF AMERICA

To my mother, Liane Malinofsky, for her encouragement; my wife, Patricia Elizabeth for her patience and support, and my daughter, Kelly Elizabeth.

John H. Montgomery

To my parents, Lydia and Mike, for their understanding; my brother Scott and my friend Toby for their support.

Linda M. Welkom

Preface

Protection of groundwater, a most vital natural resource, requires that regulators and public and private interests cooperate. To this end, professionals from environmental consultants to the local fire officers need reliable, accurate and readily accessible data to accomplish these tasks. Unfortunately, the necessary data are scattered throughout many reference books, papers and journals. This book is designed to include, in one reference, all the information needed by those involved in the protection and remediation of the groundwater environment. Its format is easy to use by a wide spectrum of professionals. The data fields have been selected to fulfill the minimum technical requirements of the user based on the extensive experience of the authors, their colleagues, and others.

This book should be useful to government agencies, environmental scientists, emergency response teams and cleanup contractors (for its physicochemical properties, exposure data, uses category and analytical test methods); environmental personnel from consulting and industrial firms (for its exposure data and physicochemical data); chemical engineers, scientists and industrial hygienists (for its synonym index, uses category, exposure and physicochemical data); real estate developers, insurance underwriters and environmental attorneys (for its uses category and synonym index).

A unique feature is the listing of competing values for most of the physicochemical properties. For example, all reported solubilities found in the documented literature have been cited for each compound enabling the reader to determine the reliability or uncertainty of these values. This feature enhances the value of all the physicochemical data.

The book is based on more than 425 references. It is broad enough and comprehensive enough to serve its purpose as a desk reference, but small enough to be taken out into the field. The publisher and authors would appreciate hearing from readers regarding corrections, suggestions, or comments for revisions in future editions.

Acknowledgments

The authors are indebted to I.G. "Butch" Grossman for his editorial review, technical support and suggestions. He provided much encouragement and sound advice. We are also grateful to the following: Maria Baratta and Angelo Papa of the New Jersey Department of Environmental Protection for their technical assistance; the New Jersey State Library for the use of its resources; and Brian Lewis and Janet Tarolli of Lewis Publishers for their indispensable support and encouragement.

Introduction

The compounds profiled include all the Priority Pollutants promulgated by the U.S. Environmental Protection Agency (U.S. EPA) under the Clean Water Act of 1977 [1]. Many of these priority pollutants were included among the Target Compounds promulgated by the U.S. EPA under the Comprehensive Environmental Response, Compensation and Liability Act (CERCLA) in 1980 and the Superfund Amendments and Reauthorization Act (SARA) of 1986. All chemicals described in the book are classified as priority pollutants and/or target compounds.

The compound headings are those commonly used by the U.S. EPA and are arranged alphabetically. Positional and/or structural prefixes set in italic type are not an integral part of the chemical name and are disregarded in alphabetizing. These include *asym-*, *sym-*, *n-*, *sec-*, *cis-*, *trans-*, α-, β-, *o-*, *m-*, *p-*, *n-*, etc.

Synonyms: These are listed alphabetically following the convention used for the compound headings. Compounds in boldface type are the Chemical Abstracts Service (CAS) names listed in the ninth Collective Index. If no synonym appears in boldface type, then the compound heading is the CAS assigned name. Synonyms include chemical names, common or generic names, trade names, registered trademarks, government codes and acronyms. All synonyms found in the literature are listed except foreign, slang and obsolete names.

Although synonyms were retrieved from several references, most of them were retrieved from the Registry of Toxic Effects of Chemical Substances [2].

Structural Formula: This is given for every compound regardless of its complexity. The structural formula is a graphic representation of atoms or group(s) of atoms relative to each other. Clearly, the limitation of structural formulas is that they depict these relationships in two dimensions.

CHEMICAL DESIGNATIONS

CAS Registry Number: This is a unique identifier assigned by the American Chemical Society to chemicals recorded in the CAS Registry System. This number is used to access various chemical databases such as the Hazardous Substances Data Bank (HSDB), CAS Online, Chemical Substances Information Network and many others. This entry is also useful to conclusively identify a substance regardless of the assigned name.

DOT Designation: This is a four-digit number assigned by the U.S. Department of Transportation (DOT) for hazardous materials and is identical to the United Nations identification number (which is preceded by the letters UN). This number is required on shipping papers, on placards or orange panels on tanks, and on a label or package containing the material. These numbers are widely used for personnel responding to emergency situations, e.g., overturned tractor trailers, in which the identification of the transported material is quickly and easily determined. Additional information may be obtained through the U.S. Department of Transportation, Research and Special Programs Administration, Materials Transportation Bureau, Washington, D.C. 20590.

Empirical Formula: This is arranged by carbon, hydrogen and remaining elements in alphabetical order in accordance with the Hill system [3]. Empirical formulas are useful in identifying isomers (i.e., compounds with identical empirical formulas) and are required if one wishes to calculate the formula weight of a substance.

Formula Weight: This is calculated to the nearest hundredth using the empirical formula and the 1981 Table of Standard Atomic Weights as reported in Weast [4]. Formula weights are required for many calculations, such as converting weight/volume units, e.g., mg/L or g/L, to molar units (mol/L); with density for calculating molar volumes; and for estimating Henry's law constants.

RTECS Number: Many compounds are assigned a unique accession number consisting of two letters followed by seven numerals. This number is needed to quickly and easily locate additional toxicity and health-based data which are cross-referenced in the Registry of Toxic Effects of Chemical Substances (RTECS) [2]. Contact the National Institute for Occupational Safety and Health (NIOSH), U.S. Department of Health and Human Services, 4676 Columbia Pkwy., Cincinnati, Ohio 45226 for additional information.

PHYSICAL AND CHEMICAL PROPERTIES

Appearance and Odor: The appearance, including the physical state (solid, liquid, or gas) of a chemical at room temperature (20-25 °C) is provided. If the compound can be detected by the olfactory sense, the odor is noted. Unless noted otherwise, the information provided in this category is for the pure substance and was obtained from many sources [5,6,7,8,9,10].

Boiling Point: This is defined as the temperature at which the vapor pressure of a liquid equals the atmospheric pressure. Unless otherwise noted, all boiling points are reported at one atmosphere pressure (760 mm Hg). Although not used in environmental assessments, boiling points are useful in assessing entry of toxic substances into the body. Body contact with high-boiling liquids is the most common means of entry into the body whereas the inhalation route is the most common for low-boiling liquids [11].

Dissociation Constant: In an aqueous solution, an acid (HA) will dissociate into the carboxylate anion (A^-) and hydrogen ion (H^+) and may be represented by the general equation:

$$HA_{(aq)} \rightleftharpoons H^+ + A^-$$

At equilibrium, the ratio of the products (ions) to the reactant (non-ionized electrolyte) is related by the equation:

$$K_a = \frac{[H^+][A^-]}{[HA]}$$

where K_a is the dissociation constant. This expression shows that K_a increases if there is increased ionization and vice versa. A strong acid (weak base) such as hydrochloric acid ionizes readily and has a large K_a, whereas a weak acid (or stronger base) such as benzoic acid ionizes to a lesser extent and has a lower K_a. The dissociation constants for weak acids are sometimes expressed as K_b, the dissociation constant for the base, and both are related to the dissociation constant for water by the expression:

$$K_w = K_a + K_b$$

where

K_w = dissociation constant for water (10^{-14} at 25 °C)
K_a = acid dissociation constant
K_b = base dissociation constant

The dissociation constant is usually expressed as $pK_a = -\log_{10} K_a$. The above equation becomes:

$$pK_w = pK_a + pK_b$$

When the pH of the solution and the pK_a are equal, 50% of the acid will have dissociated into ions. The percent dissociation of an acid or base can be calculated if the pH of the solution and the pK_a of the compound are known [12]:

For organic acids:
$$\alpha_a = \frac{100}{1 + 10^{(pH-pKa)}}$$

For organic bases:
$$\alpha_b = \frac{100}{1 + 10^{(pKw-pKb-pH)}}$$

where

α_a = percent of the organic acid that is non-dissociated
α_b = percent of the organic base that is non-dissociated
pK_a = log dissociation constant for acid
pK_w = log dissociation constant for water (14.00 at 25 °C)
pK_b = log dissociation constant for base ($pK_b = pK_w - pK_a$).
pH = hydrogen ion activity (concentration) of the solution

Since ions tend to remain in solution, the degree of dissociation will affect processes such as volatilization, photolysis, adsorption and bioconcentration [13].

Henry's Law Constant: Sometimes referred to as the air-water partition coefficient, the Henry's law constant is defined as the ratio of the partial pressure of a compound in air to the concentration of the compound in water at a given temperature under equilibrium conditions. If the vapor pressure and solubility of a compound are known, this parameter can be calculated at 1 atm (760 mm Hg) as follows:

$$H = \frac{P \times FW}{760 \times S}$$

where

H = Henry's law constant (atm·m^3/mol)
P = pressure (mm Hg)

S = solubility (mg/L)
FW = gram formula weight

Henry's law constant can also be expressed in dimensionless form and may be calculated using one of the following equations:

$$H' = \frac{H}{R \times K} \quad \text{or} \quad H' = \frac{S_a}{S}$$

where

H = Henry's law constant (atm·m³/mol)
H' = Henry's law constant (dimensionless)
R = ideal gas constant (8.20575 x 10⁻⁵ atm·m³/mol·K)
K = temperature of water (degrees Kelvin)
S_a = solute concentration in air (mol/L)
S = aqueous solute concentration (mol/L)

It should be noted that estimating Henry's law constant assumes that the gas obeys the ideal gas law and the aqueous solution behaves as an ideally dilute solution. The solubility and vapor pressure data inputted into the equations are valid only for the pure compound and must be in the same standard state at the same temperature.

Henry's law constants provided an indication of the relative volatility of a substance. According to Lyman and others [14], if H < 10⁻⁷ atm·m³/mol, the substance has a low volatility. If H is greater than 10⁻⁷ but less than 10⁻⁵ atm·m³/mol, the substance will volatilize slowly. Volatilization becomes an important transfer mechanism in the range 10⁻⁵ < H < 10⁻³ atm·m³/mol. Values of H exceeding 10⁻³ atm·m³/mol indicate volatilization will proceed rapidly.

Ionization Potential: The ionization potential of a compound is defined as the energy required to remove a given electron from the molecule's atomic orbit (outermost shell) and is expressed in electron volts (eV). One electron volt is equivalent to 23,053 cal/mol.

Knowing the ionization potential of a contaminant is required in determining the appropriate photoionization lamp for detecting that contaminant or family of contaminants. Photoionization instruments are equipped with a radiation source (ultraviolet lamp), pump, ionization chamber, an amplifier and a recorder (either digital or meter). Generally, compounds with ionization potentials smaller than the radiation source (UV lamp rating) being used will readily ionize and will be detected by the instrument. Conversely, compounds with ionization potentials higher than the lamp rating will not ionize and will not be detected by the instrument.

Log K_{oc}: The soil/sediment partition or sorption coefficient is defined as the ratio of adsorbed chemical per unit weight of organic carbon to the aqueous solute concentration. This value provides an indication of the tendency of a chemical to partition between particles containing organic carbon and water. Compounds that bind strongly to organic carbon have characteristically low solubilities, whereas compounds with low tendencies to adsorb onto organic particles have high solubilities.

Chemicals that sorb onto organic materials in an aquifer (i.e., organic carbon), are retarded in their movement in groundwater. The sorbing solute travels at linear velocity that is lower than the groundwater flow velocity by a factor of R_d, the retardation factor. If the K_{oc} of a compound is known, the retardation factor may be calculated using the following equation from Freeze and Cherry [15]:

$$R_d = \frac{V_w}{V_c} = 1 + \frac{BK_d}{n_e}$$

where

R_d = retardation factor (unitless)
V_w = average linear velocity of groundwater (e.g., ft/day)
V_c = average linear velocity of contaminant (e.g., ft/day)
B = average soil bulk density (g/cm^3)
n_e = effective porosity (unitless)
K_d = distribution (sorption) coefficient (cm^3/g)

By definition, K_d is related to K_{oc} by the equation

$$K_{oc} = \frac{K_d}{f_{oc}}$$

where f_{oc} is the fraction of naturally occurring organic carbon in soil.

Correlations between K_{oc} and bioconcentration factors in fish and beef have shown a log-log linear relationship [16] as well as solubility of organic compounds in water [17,18]. Moreover, the log K_{oc} has been shown to be related to molecular connectivity indices [19] and high performance liquid chromatography (HPLC) capacity factors [20].

In instances where experimentally determined K_{oc} values are not available, they can be estimated using recommended regression equations as cited in Lyman and others [14]. All the K_{oc} estimations are based on regression equations in which the solubility or the K_{ow} of the substance is known.

Log K_{ow}: The K_{ow} of a substance is the *n*-octanol/water partition coefficient and is defined as the ratio of the solute concentration in the water-saturated *n*-octanol phase to the solute concentration in the *n*-octanol-saturated water phase. Values of K_{ow} are therefore unitless.

The partition coefficient has been recognized as a key parameter in predicting the environmental fate of organic compounds. The K_{ow} has been shown to be linearly correlated with bioconcentration factors (BCF) in aquatic organisms [21]; in fish [16,21,22,23]; soil/sediment partition coefficients (K_{oc}) [24,25]; solubility of organic compounds in water [21,26,27,28,29]; molecular surface area [30,31]; molar refraction [32]; HPLC capacity factors [33,34]; and HPLC retention times [35,36,37].

For ionizable compounds such as acids, amines and phenols, K_{ow} values are a function of pH. Unfortunately, many investigators have neglected to report the pH of the solution at which the K_{ow} was determined. If a K_{ow} value is used for an ionizable compound for which the pH is known, both values should be noted.

Melting Point: The melting point of a substance is defined as the temperature at which a solid substance undergoes a phase change to a liquid. The reverse process, the temperature at which a liquid freezes to a solid, is called the freezing point. For a given substance, the melting point is identical to the freezing point.

Unless noted otherwise, all melting points are reported at the standard pressure of 1 atmosphere. Although the melting point of a substance is not directly used in predicting its behavior in the environment, it is useful in determining the phase in which the substance would be found under typical conditions.

Solubility in Organics: The presence of solvents other than water can alter a compound's solubility. Consequently, its fate and transport in soils, sediments and groundwater will be changed due to the presence of these cosolvents. For example, soils contaminated with compounds having low water solubilities tend to remain bound to the soil by adsorbing onto organic carbon and/or by interfacial tension with water. A solvent introduced to an unsaturated soil environment (e.g., a surface spill, leaking aboveground tank, etc.) may come in contact with existing soil contaminants. As the solvent interacts with the existing contamination, it may mobilize it, thereby facilitating its migration. Consequently, the organic solvent can facilitate the leaching of contaminants from the soil to the water table. Therefore, the presence of cosolutes must be considered when predicting the fate and transport of contaminants in the unsaturated zone and the water table.

Solubility in Water: The water solubility of a compound is defined as the saturated concentration of the compound in water at a given temperature and pressure. This parameter is perhaps the most important factor in estimating a chemical's fate and transport in the aquatic environment. Compounds with high water solubilities tend to desorb from soils and sediments (i.e., they have low K_{oc} values), are less likely to volatilize from water, and are susceptible to biodegradation. Conversely, compounds with low solubilities tend to adsorb onto soils and sediments (have high K_{oc}), volatilize more readily from water, and bioconcentrate in aquatic organisms. The more soluble compounds commonly enter the water table more readily than their less soluble counterparts.

The water solubility of a compound varies with temperature, pH (particularly, ionizable compounds such as acids and bases), and other dissolved constituents, e.g., inorganic salts (electrolytes) and organic chemicals including naturally occurring organic carbon, such as humic and fulvic acids. At a given temperature, the variability/discrepancy of water-solubility measurements documented by investigators may be attributed to one or more of the following: (1) purity of the compound, (2) analytical method employed, (3) particle size (for solid solubility determinations only), (4) adsorption onto the container and/or suspended solids, (5) time allowed for equilibrium conditions to be reached, (6) losses due to volatilization, and (7) chemical transformations (e.g., hydrolysis).

The water solubility of chemical substances has been related to bioconcentration factors, soil/sediment partition coefficients (K_{oc}), [17,18], *n*-octanol/water partition coefficients (K_{ow}) [27,28,29], and soil organic matter [38]. Regression equations generated from these relationships have demonstrated a log-log linear relationship for these properties. The reported regression equations are useful in estimating the solubility of a compound in water if experimental values are not available.

Unless otherwise noted, all reported solubilities were determined using distilled water. For some compounds, solubilities were determined using well water, surface water, natural seawater or artificial seawater.

Specific Density: The specific density, also known as relative density, is defined as:

$$\text{Specific density} = \frac{D_s}{D_w}$$

where

D_s = density of a substance (g/mL or g/cm^3)
D_w = density of distilled water (g/mL or g/cm^3)

Values of specific density are unitless and are reported in the form:

$$D \text{ at } T_s/T_w$$

where

D = specific density of the substance
T_s = temperature of the substance at the time of measurement (°C, Celsius)
T_w = water temperature (°C).

For example, the value 1.1750 at 20/4 °C indicates a specific density of 1.1750 for the substance at 20 °C with respect to water at 4 °C. At 4 °C, the density of water is exactly 1.0000 g/mL (g/cm^3). Therefore, the specific density of a substance is equivalent to the density of the substance relative to the density of water at 4 °C.

The density of a hydrophobic substance enables it to sink or float in water. Density values are especially important for liquids migrating through the unsaturated zone and encountering the water table as "free product." Generally, liquids that are less dense than water "float" on the water table. Conversely, organic liquids that are more dense than water commonly "sink" through the water table, e.g., dense non-aqueous phase liquids such as chloroform, dichloroethane and tetrachloroethylene.

Hydrophilic substances, on the other hand, behave differently. Acetone, which is less dense than water, does not float on water because it is freely miscible with it in all proportions. Therefore, the solubility of a substance must be considered in assessing its behavior in the subsurface.

Transformation Products: Chemicals released in the environment are susceptible to several degradation pathways. These include hydrolysis, photolysis and biodegradation. Compounds transformed by one or more of these processes may result in the formation of more toxic or less toxic substances. In addition, the transformed product(s) will behave differently than the parent compound due to changes in their physicochemical properties. Many researchers focus their attention on transformation rates rather than the transformation products. Consequently, only limited data exist on the resultant end products. Where available, compounds that are transformed into identified products are listed.

Vapor Density: The vapor density of a substance is defined as the ratio of the mass of vapor per unit volume. An equation for estimating vapor density is readily derived from a varied form of the ideal gas law:

$$P \times V = \frac{M \times R \times K}{FW}$$

where

P = pressure (atm)
V = volume (L)
M = mass (g)
R = ideal gas constant (8.20575×10^{-5} atm·m^3/mol·K)
K = temperature (Kelvin)
FW = gram formula weight (g/mol)

Recognizing that the density of a substance is defined as:

$$D = \frac{M}{V}$$

Substituting this equation into the first equation, rearranging, and simplifying results in an expression to determine the specific vapor density which is unitless:

$$p_v = \frac{P \times FW}{R \times T}$$

The specific vapor density, p_v, is simply the ratio of the vapor density of the substance to that of air under the same pressure and temperature. According to Weast [4], the vapor density of dry air at 20 °C and 760 mm Hg is 1.204 g/L. At 25 °C, the vapor density of air decreases slightly to 1.184 g/L. Calculated specific vapor densities are reported relative to air (set equal to 1) only for compounds which are liquids at room temperature (i.e., 25 °C). These are reported in addition to the calculated vapor densities.

Vapor Pressure: The vapor pressure of a substance is defined as the pressure exerted by the vapor (gas) of a substance when it is under equilibrium conditions. It provides a semi-quantitative rate at which it will volatilize from soil and/or water. The vapor pressure of a substance is a required input parameter for calculating the air-water partition coefficient (see **Henry's Law Constant**), which in turn is used in volatilization rate constant calculations.

FIRE HAZARDS

Flash Point: The flash point is defined as the minimum temperature at which a substance releases ignitable flammable vapors in the presence of an ignition source. e.g., spark or flame. Flash points may be determined by two methods - Tag closed cup (ASTM method D56) or Cleveland open cup (ASTM method D93). Unless otherwise noted, all flash point values represent closed cup method determinations. Flash point values determined by the open cup method are slightly higher than those determined by the closed cup method; however, the open cup method is more representative of actual conditions.

According to Sax [5], a material with a flash point of 100 °F or less is considered dangerous, whereas a material having a flash point greater than 200 °F is considered to have a low flammability. Substances with flash points within this temperature range are considered to have moderate flammabilities.

Lower Explosive Limit: The minimum concentration (vol% in air) of a flammable gas or vapor required for ignition or explosion to occur in the presence of an ignition source (see also **Flash Point**).

Upper Explosive Limit: The maximum concentration (vol% in air) of a flammable gas or vapor required for ignition or explosion to occur in the presence of an ignition source (see also **Flash Point**).

HEALTH HAZARD DATA

Immediately Dangerous to Life or Health: According to the National

Institute of Occupational Safety and Health [39], the IDLH level ". . . for the purpose of respirator selection represents a maximum concentration from which, in the event of respirator failure, one could escape within 30 minutes without experiencing any escape-impairing or irreversible health effects." Concentrations are reported in parts per million (ppm) or milligrams per cubic meter (mg/m^3).

Permissible Exposure Limits (PEL) in Air: The permissible exposure limits in air, set by the Occupational Health and Safety Administration (OSHA), can be found in the Code of Federal Regulations [40]. Unless noted otherwise, the PEL are 8-hour time-weighted average (TWA) concentrations. If NIOSH [39] and/or the American Conference of Governmental Industrial Hygienists (ACGIH) has published recommended exposure limits, these are also included. The ACGIH's recommended exposure limits, commonly known as threshold limit values (TLV), are subdivided into three exposure classes [41]. These are defined as follows:

Threshold Limit Value-Time Weighted Average (TLV-TWA) - the time-weighted average concentration for a normal 8-hour workday and a 40-hour workweek, to which nearly all workers may be repeatedly exposed, day after day, without adverse effect.

Threshold Limit Value-Short Term Exposure Limit (TLV-STEL) - the concentration to which workers can be exposed continuously for a short period of time without suffering from 1) irritation, 2) chronic or irreversible tissue damage, or 3) narcosis of sufficient degree to increase the likelihood of accidental injury, impair self-rescue or materially reduce work efficiency, and provided that the daily TLV-TWA is not exceeded. It is not a separate independent exposure limit; rather, it supplements the time-weighted average (TWA) limit where there are recognized acute toxic effects from a substance whose toxic effects are primarily of a chronic nature. STELs are recommended only where toxic effects have been reported from high short-term exposures in either humans or animals.

A STEL is defined as a 15-minute time-weighted average exposure which should not be exceeded at any time during a workday even if the eight-hour time-weighted average is within the TLV. Exposures at the STEL should not be longer than 15 minutes and should not be repeated more than four times per day. There should be at least 60 minutes between successive exposures at the STEL.

An averaging period other than 15 minutes may be recommended when this is warranted by observed biological effects.

Threshold Limit Value-Ceiling (TLV-C) - the concentration that should not be exceeded during any part of the working exposure.

For additional information from OSHA, write to Technical Data Center, U.S. Department of Labor, Washington, D.C. 20210. The ACGIH's address is 6500 Glenway Ave., Bldg. D-7, Cincinnati, OH 45211-7881.

MANUFACTURING

Selected Manufacturers: The names of selected companies have been provided if additional information on a chemical is required. Manufacturers are an excellent source of Material Safety Data Sheets, which list physical, chemical, health-hazard, and other data. Most of the manufacturers names have been obtained from the CHRIS (Chemical Hazard Response Information System) Manual [9] and the Hazardous Substance Data Bank [8]. Some of the less prominent companies may have subsequently changed their names and/or addresses, may have been purchased by another company or may even have closed their operations. Accordingly, at least one but not more than three companies are reported for each compound.

Uses: Descriptions of specific uses are based on one or more of the following sources - HSDB [8], CHRIS Manual [9] and Verschueren [42]. This information is useful in attempting to identify potential sources of the industrial and environmental contamination.

REFERENCES

1.   "Guidelines Establishing Test Procedures for the Analysis of Pollutants," U.S. Code of Federal Regulations, 40 CFR 136, 44(233):69464-69575.

2.   "Registry of Toxic Effects of Chemical Substances," U.S. Department of Health and Human Services, National Institute for Occupational Safety and Health (1985), 2050 p.

3.   Hill, E.A. "On a System of Indexing Chemical Literature; Adopted by the Classification Division of the U.S. Patent Office," *J. Am. Chem. Soc.*, 22(8):478-494 (1900).

4. Weast, R.C., Ed. *CRC Handbook of Chemistry and Physics*, 67th ed. (Boca Raton, FL: CRC Press, Inc., 1986), 2406 p.
5. Sax, N.I. *Dangerous Properties of Industrial Materials* (New York: Van Nostrand Reinhold Co., 1984), 3124 p.
6. Windholz, M., Budavari, S., Blumetti, R.F., and E.S. Otterbein, Eds., *The Merck Index*, 10th ed., (Rahway, NJ: Merck and Co., 1983), 1463 p.
7. Hawley, G.G. *The Condensed Chemical Dictionary* (New York: Van Nostrand Reinhold Co., 1981), 1135 p.
8. Hazardous Substances Data Bank. National Library of Medicine, Toxicology Information Program (1989).
9. "CHRIS Hazardous Chemical Data" U.S. Department of Transportation, U.S. Coast Guard, U.S. Government Printing Office (October, 1978).
10. *Toxic and Hazardous Industrial Chemicals Safety Manual for Handling and Disposal with Toxicity and Hazard Data* (Tokyo, Japan: International Technical Information Institute, 1986), 700 p.
11. Shafer, D. *Hazardous Materials Training Handbook* (Madison, CT: Bureau of Law and Business, Inc.), 195 p. *Hazardous Materials Training Handbook* (Madison, CT: Business and Legal Reports, 1987), 206 p.
12. Guswa, J.H., Lyman, W.J., Donigan, A.S. Jr., Lo, T.Y.R., Lo and E.W. Shanahan. *Groundwater Contamination and Emergency Response Guide* (Park Ridge, NJ: Noyes Publications, 1984), 490 p.
13. Howard, P.H. *Handbook of Environmental Fate and Exposure Data for Organic Chemicals* (Chelsea, MI: Lewis Publishers, Inc., 1989), 574 p.
14. Lyman, W.J., W.F. Reehl, and D.H. Rosenblatt. *Handbook of Chemical Property Estimation Methods: Environmental Behavior of Organic Compounds* (New York: McGraw-Hill, Inc., 1982).
15. Freeze, R.A. and J.A. Cherry. *Groundwater* (Englewood Cliffs, NJ: Prentice-Hall, Inc., 1974), 604 p.
16. Kenaga, E.E. "Correlation of Bioconcentration Factors of Chemicals in Aquatic and Terrestrial Organisms with Their Physical and Chemical Properties," *Environ. Sci. Technol.*, 14(5):553-556 (1980).
17. Means, J.C., S.G. Wood, J.J. Hassett, and W.L. Banwart. "Sorption of Polynuclear Aromatic Hydrocarbons by Sediments and Soils," *Environ. Sci. Technol.*, 14(2):1524-1528 (1980).
18. Abdul, S.A., Gibson, T.L., and D.N. Rai. "Statistical Correlations for Predicting the Partition Coefficient for Nonpolar Organic

Contaminants between Aquifer Organic Carbon and Water," *Haz. Waste Haz. Mater.*, 4(3):211-222 (1987).

19. Sabljić, A., and M. Protić. "Relationship Between Molecular Connectivity Indices and Soil Sorption Coefficients of Polycyclic Aromatic Hydrocarbons," *Bull. Environ. Contam. Toxicol.*, 28(2):162-165 (1982).

20. Hodson, J., and N.A. Williams. "The Estimation of the Adsorption Coefficient (K_{oc}) for Soils by High Performance Liquid Chromatography," *Chemosphere*, 19(1):67-77 (1988).

21. Isnard, S., and S. Lambert. "Estimating Bioconcentration Factors from Octanol-Water Partition Coefficient and Aqueous Solubility," *Chemosphere*, 17(1):21-34 (1988).

22. Neely, W.B., Branson, D.R., and G.E. Blau. "Partition Coefficient to Measure Bioconcentration Potential of Organic Chemicals in Fish," *Environ. Sci. Technol.*, 8(13):1113-1115 (1974).

23. Oliver, B.G., and A.J. Niimi. "Bioconcentration Factors of Some Halogenated Organics for Rainbow Trout: Limitations in Their Use for Prediction of Environmental Residues," *Environ. Sci. Technol.*, 19(9):842-849 (1985).

24. Kenaga, E.E., and C.A.I. Goring. "Relationship between Water Solubility, Soil Sorption, Octanol-Water Partitioning and Concentration of Chemicals in Biota," in *Aquatic Toxicology, ASTM STP 707*, Eaton, J.G., Parrish, P.R., and A.C. Hendricks, Eds. (Philadelphia: American Society for Testing and Materials, 1980), pp 78-115.

25. Chiou, C.T., L.J. Peters, and V.H. Freed. "A Physical Concept of Soil-Water Equilibria for Nonionic Organic Compounds," *Science*, 206:831-832 (1979).

26. Banerjee, S., S.H. Yalkowsky, and S.C. Valvani. "Water Solubility and Octanol/Water Partition Coefficients of Organics. Limitations of the Solubility-Partition Coefficient Correlation," *Environ. Sci. Technol.*, 14(10):1227-1229 (1980).

27. Chiou, C.T., V.H. Freed, D.W. Schmedding, and R.L. Kohnert. "Partition Coefficients and Bioaccumulation of Selected Organic Chemicals," *Environ. Sci. Technol.*, 11(5):475-478 (1977).

28. Chiou, C.T., D.W. Schmedding, and M. Manes. "Partitioning of Organic Compounds in Octanol-Water Systems," *Environ. Sci. Technol.*, 16(1):4-10 (1982).

29. Miller, M.M., Wasik, S.P., Huang, G.-L., Shiu, W.-Y., and D. Mackay. "Relationships between Octanol-Water Partition Coefficient and Aqueous Solubility," *Environ. Sci. Technol.*, 19(6):522-529 (1985).

30. Amidon, G.L., S.H. Yalkowsky, S.T. Anik, and S.C. Valvani.

"Solubility of Nonelectrolytes in Polar Solvents. V. Estimation of the Solubility of Aliphatic Monofunctional Compounds in Water using a Molecular Surface Area Approach," *J. Phys. Chem.*, 79(21):2239-2246 (1975).

31.   Camilleri, P., S.A. Watts, and J.A. Boraston. "A Surface Area Approach to Determination of Partition Coefficients," *J. Chem. Soc. Perkin Trans. II*, (September 1988), pp 1699-1707.

32.   Yoshida, K., Tadayoshi, S., and F. Yamauchi. "Non-Steady-State Equilibrium Model for the Preliminary Prediction of the Fate of Chemicals in the Environment," *Ecotoxicol. Environ. Safety*, 7(2):179-190 (1983).

33.   Eadsforth, C.V. "Application of Reverse-Phase H.P.L.C. for the Determination of Partition Coefficients," *Pestic. Sci.*, 17(3):311-325 (1986).

34.   Brooke, D.N., A.J. Dobbs, and N. Williams. "Octanol:Water Partition Coefficients (P): Measurement, Estimation, and Interpretation, Particularly for Chemicals with $P > 10^5$," *Ecotoxicol. Environ. Safety*, 11(3):251-260 (1986).

35.   Sarna, L.P., P.E. Hodge, and G.R.B. Webster. "Octanol-Water Partition Coefficients of Chlorinated Dioxins and Dibenzofurans by Reversed-Phase HPLC Using Several C_{18} Columns," *Chemosphere*, 13(9):975-983 (1984).

36.   Burkhard, L.P., and D.W. Kuehl. "n-Octanol/Water Partition Coefficients by Reverse Phase Liquid Chromatography/Mass Spectrometry for Eight Tetrachlorinated Planar Molecules," *Chemosphere*, 15(2):163-167 (1986).

37.   Webster, G.R.B., Friesen, K.J., Sarna, L.P., and D.C.G. Muir. "Environmental Fate Modeling of Chlorodioxins: Determination of Physical Constants," *Chemosphere*, 14(6/7):609-622 (1985).

38.   Chiou, C.T., P.E. Porter, and D.W. Schmedding. "Partition Equilibria of Nonionic Organic Compounds between Organic Matter and Water," *Environ. Sci. Technol.*, 17(4):227-231 (1983).

39.   "NIOSH Pocket Guide to Chemical Hazards," U.S. Department of Health and Human Services, U.S. Government Printing Office (1987), 241 p.

40.   "General Industry Standards for Toxic and Hazardous Substances," U.S. Code of Federal Regulations, 29 CFR 1910.1000, Subpart Z (January 1977).

41.   *Threshold Limit Values and Biological Exposure Indices for 1986-1987* (Cincinnati, OH: American Conference of Governmental Hygienists, 1986), 111 p.

42.   Verschueren, K. *Handbook of Environmental Data on Organic Chemicals* (New York: Van Nostrand Reinhold Co., 1983), 1310 p.

Abbreviations and Symbols

α	Alpha
α_a	Percent of acid that is non-dissociated
α_b	Percent of base that is non-dissociated
ACGIH	American Conference of Governmental Industrial Hygienists
aq	Aqueous
ASTM	American Society for Testing and Materials
asym	Asymmetrical
atm	Atmosphere
β	Beta
BCF	Bioconcentration factor
°C	Degrees Centigrade (Celsius)
cal	Calorie
CAS	Chemical Abstracts Service
CERCLA	Comprehensive Environmental Response, Compensation a n d Liability Act
CHRIS	Chemical Hazard Response Information System
cm	Centimeter
D	Density
DOT	Department of Transportation (U.S.)
D_s	Density of a substance
D_w	Density of water
δ	Delta
eV	Electron volts
°F	Degrees Fahrenheit
FW	Gram formula weight
g	Gram
γ	Gamma
GC/MS	Gas chromatography/mass spectrometry
HPLC	High performance liquid chromatography
HSDB	Hazardous Substances Data Bank
K	Kelvin (°C + 273.15)
kg	Kilogram
K_{oc}	Soil/sediment partition coefficient
K_{ow}	*n*-Octanol/water partition coefficient
kPa	Kilopascal
L	Liter
m	Meter
m	Meta
M	Molarity (moles/liter)
mg	Milligram
Ma	Mass
mL	Milliliter

mm Hg	Millimeters of mercury
mmol	Millimole
mol	Mole
n, N	Normal
ng	Nanogram
NIOSH	National Institute for Occupational Safety and Health
nm	Nanometer
σ	Sigma
o	Ortho
OSHA	Occupational Safety and Health Administration
p	Para
P	Pressure
Pa	Pascal
PEL	Permissible exposure limits
pK_a	Log dissociation constant of an acid
pK_b	Log dissociation constant of a base
pK_w	Log dissociation constant of water
ppb	Parts per billion
ppm	Parts per million
p_r	Relative vapor density
ρ	Rho
R	Ideal gas constant (8.20575×10^{-5} atm·m^3/mol·K)
RCRA	Resource Conservation and Recovery Act
R_d	Retardation factor
RTECS	Registry of Toxic Effects of Chemical Substances
sec	secondary
sp.	Species
S	Solubility
SARA	Superfund Amendments and Reauthorization Act
STEL	Short-term exposure limit
sym	Symmetrical
T	Temperature
TLV	Threshold limit value
TWA	Time-weighted average
μ	Micro-
μg	Microgram
U.S. EPA	U.S. Environmental Protection Agency
V, vol	Volume
wt	Weight

Contents

xxvi   **Contents**

ACENAPHTHENE

Synonyms: **1,2-Dihydroacenaphthylene**; Ethylene naphthalene; 1,8-Ethylene naphthalene; 1,8-Hydroacenaphthylene; Periethylene naphthalene.

Structural Formula:

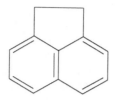

CHEMICAL DESIGNATIONS

CAS Registry Number: 83-32-9

DOT Designation: None assigned.

Empirical Formula: $C_{12}H_{10}$

Formula Weight: 154.21

RTECS Number: AB 1255500

PHYSICAL AND CHEMICAL PROPERTIES

Appearance: White crystalline solid.

Boiling Point: 279 °C [1].

Henry's Law Constant: 1.5 x 10^{-4} atm·m^3/mol [2]; 2.41 x 10^{-4} atm·m^3/mol [3]; 1.9 x 10^{-4} atm·m^3/mol [4]; 7.92 x 10^{-5} atm·m^3/mol at 25 °C [5].

Ionization Potential: No data found.

Log K_{oc}**:** 1.25 [6].

Log K_{ow}**:** 3.92 [7]; 4.33 [8].

Melting Point: 96.2 °C [1]; 95 °C [9].

Solubility in Organics: Soluble in ethanol, toluene, chloroform, benzene, and acetic acid [10].

Solubility in Water: 3.47 mg/L at 25 °C [11]; 3.93 mg/L at 25 °C [12]; 1.84 ppm in seawater at 25 °C, 2.42 ppm [13].

Specific Density: 1.0242 at 90/4 °C [1].

Transformation Products: No data found.

Vapor Pressure: 10 mm at 131.2 °C, 40 mm at 168.2 °C, 100 mm at 197.5 °C, 400 mm at 250 °C, 760 mm at 277.5 °C [1]; 0.001-0.01 mm at 20 °C [13]; 0.00155 mm at 25 °C [14]; 0.149 mm at 65 °C, 0.231 mm at 70 °C, 0.351 mm at 75 °C, 0.529 mm at 80 °C, 0.787 mm at 85 °C, 1.151 mm at 90 °C, 1.388 mm at 92.5 °C, 1.463 mm at 93.195 °C [15].

FIRE HAZARDS

Flash Point: No data found.

Lower Explosive Limit (LEL): No data found.

Upper Explosive Limit (UEL): No data found.

HEALTH HAZARD DATA

Immediately Dangerous to Life or Health (IDLH): No data found.

Permissible Exposure Limits (PEL) in Air: No standards set.

MANUFACTURING

Selected Manufacturers:

Fluka Chemical Corp.
980 South Second St.
Ronkonkoma, NY 11779

Pfaltz & Bauer, Inc.
172 East Aurora St.
Waterbury, CT 06708

Uses: Manufacturing of dye intermediates, pharmaceuticals, insecticides, fungicides, and plastics.

REFERENCES

1. Weast, R.C., Ed. *CRC Handbook of Chemistry and Physics*, 67th ed. (Boca Raton, FL: CRC Press, Inc., 1986), 2406 p.
2. Lyman, W.J., Reehl, W.F., and D.H. Rosenblatt. *Handbook of Chemical Property Estimation Methods: Environmental Behavior of Organic Compounds* (New York: McGraw-Hill, Inc., 1982).
3. Warner, H.P., Cohen, J.M., and J.C. Ireland. "Determination of Henry's Law Constants of Selected Priority Pollutants," Office of Science and Development, U.S. EPA Report-600/D-87/229 (1987), 14 p.
4. Petrasek, A.C., Kugelman, I.J., Austern, B.M., Pressley, T.A., Winslow, L.A., and R.H. Wise. "Fate of Toxic Organic Compounds in Wastewater Treatment Plants," *J. Water Poll. Control Fed.*, 55(10):1286-1296 (1983).
5. Hine, J., and P.K. Mookerjee. "The Intrinsic Hydrophilic Character of Organic Compounds. Correlations in Terms of Structural Contributions," *J. Org. Chem.*, 40(3):292-298 (1975).
6. Mihelcic, J.R., and R.G. Luthy. "Microbial Degradation of Acenaphthene and Naphthalene under Denitrification Conditions in Soil-Water Systems," *Appl. Environ. Microbiol.*, 54(5):1188-1198 (1988).
7. Banerjee, S., Yalkowsky, S.H., and S.C. Valvani. "Water Solubility and Octanol/Water Partition Coefficients of Organics. Limitations of the Solubility-Partition Coefficient Correlation," *Environ. Sci. Technol.*, 14(10):1227-1229 (1980).
8. Walton, W.C. *Practical Aspects of Ground Water Modeling* (Worthington, OH: National Water Well Association, 1985), 587 p.
9. Sax, N.I. *Dangerous Properties of Industrial Materials* (New York: Van Nostrand Reinhold Co., 1984), 3124 p.
10. "Chemical, Physical, and Biological Properties of Compounds Present at Hazardous Waste Sites," U.S. EPA Report-530/SW-89-010 (1985), 619 p.
11. Eganhouse, R.P., and J.A. Calder. "The Solubility of Medium Weight Aromatic Hydrocarbons and the Effect of Hydrocarbon Co-solutes and Salinity," *Geochim. Cosmochim. Acta*, 40(5):555-561 (1976).
12. Mackay, D., and W.Y. Shiu. "Aqueous Solubility of Polynuclear Aromatic Hydrocarbons," *J. Chem. Eng. Data*, 22(4):399-402 (1977).
13. Sax, N.I., Ed. *Dangerous Properties of Industrial Materials Report*

(New York: Van Nostrand Reinhold Co., 1984), 4(1): 112 p.

14. Mabey, W.R., Smith, J.H., Podoll, R.T., Johnson, H.L., Mill, T., Chou, T.-W., Gates, J., Partridge, I.W., Jaber, H., and D. Vandenberg. "Aquatic Fate Process Data for Organic Priority Pollutants - Final Report," Office of Regulations and Standards, U.S. EPA Report-440/4-81-014 (1982), 407 p.

15. Osborn, A.G., and D.R. Douslin. "Vapor Pressures and Derived Enthalpies of Vaporization of Some Condensed-Ring Hydrocarbons," *J. Chem. Eng. Data*, 20(3):229-231 (1975).

ACENAPHTHYLENE

Synonym: Cyclopenta[*de*]naphthalene.

Structural Formula:

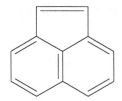

CHEMICAL DESIGNATIONS

CAS Registry Number: 208-96-8

DOT Designation: None assigned.

Empirical Formula: $C_{12}H_8$

Formula Weight: 152.20

RTECS Number: AB 1254000

PHYSICAL AND CHEMICAL PROPERTIES

Appearance: Solid.

Boiling Point: 280 °C [1].

Henry's Law Constant: 2.8×10^{-4} atm·m³/mol [2]; 1.14×10^{-4} atm·m³/mol [3].

Ionization Potential: 8.73 eV [4].

Log K_{oc}: 3.68 using method of Karickhoff and others [5].

Log K_{ow}: 4.07 [6].

Melting Point: 92-93 °C [7].

Solubility in Organics: Soluble in ethanol, ether, and benzene [8].

Solubility in Water: 3.93 mg/L at 25 °C [9].

Specific Density: 0.8988 at 16/2 °C [7].

Transformation Products: Based on data for structurally similar compounds, acenaphthylene may undergo photolysis to yield quinones [8].

Vapor Pressure: 0.0290 mm at 20 °C [10].

FIRE HAZARDS

Flash Point: No data found.

Lower Explosive Limit (LEL): No data found.

Upper Explosive Limit (UEL): No data found.

HEALTH HAZARD DATA

Immediately Dangerous to Life or Health (IDLH): No data found.

Permissible Exposure Limits (PEL) in Air: No individual standards have been set, however, as a constituent in coal tar pitch volatiles, the following exposure limits have been established: 0.2 mg/m^3 (benzene-soluble fraction) [11]; 0.1 mg/m^3 10-hour TWA (cyclohexane-extractable fraction) [12]; 0.2 mg/m^3 TWA (benzene solubles) [13].

MANUFACTURING

Uses: Research chemical. Derived from industrial and experimental coal gasification operations where maximum concentrations detected in gas, liquid, and coal tar streams were 28 mg/m^3, 4.1 mg/m^3, and 18 mg/m^3, respectively [14].

REFERENCES

1. *Catalog Handbook of Fine Chemicals* (Milwaukee, WI: Aldrich

Chemical Co., 1988), 2212 p.

2. Lyman, W.J., Reehl, W.F., and D.H. Rosenblatt. *Handbook of Chemical Property Estimation Methods: Environmental Behavior of Organic Compounds* (New York: McGraw-Hill, Inc., 1982).

3. Warner, H.P., Cohen, J.M., and J.C. Ireland. "Determination of Henry's Law Constants of Selected Priority Pollutants," Office of Science and Development, U.S. EPA Report-600/D-87/229 (1987), 14 p.

4. Franklin, J.L., Dillard, J.G., Rosenstock, H.M., Herron, J.T., Draxl K., and F.H. Field. "Ionization Potentials, Appearance Potentials and Heats of Formation of Gaseous Positive Ions," National Bureau of Standards Report NSRDS-NBS 26, U.S. Government Printing Office (1969), 289 p.

5. Karickhoff, S.W., Brown, D.S., and T.A. Scott. "Sorption of Hydrophobic Pollutants on Natural Sediments," *Water Res.*, 13:241-248 (1979).

6. Yoshida, K., Shigeoka, T., and F. Yamauchi. "Non-Steady State Equilibrium Model for the Preliminary Prediction of the Fate of Chemicals in the Environment," *Ecotoxicol. Environ. Safety*, 7(2):179-190 (1983).

7. Weast, R.C., Ed. *CRC Handbook of Chemistry and Physics*, 67th ed. (Boca Raton, FL: CRC Press, Inc., 1986), 2406 p.

8. "Chemical, Physical, and Biological Properties of Compounds Present at Hazardous Waste Sites," U.S. EPA Report-530/SW-89-010 (1985), 619 p.

9. Verschueren, K. *Handbook of Environmental Data on Organic Chemicals* (New York: Van Nostrand Reinhold Co., 1983), 1310 p.

10. Sims, R.C., Doucette, W.C., McLean, J.E., Grenney, W.J., and R.R. Dupont. "Treatment Potential for 56 EPA Listed Hazardous Chemicals in Soil," U.S. EPA Report-600/6-88-001 (1988), 105 p.

11. "General Industry Standards for Toxic and Hazardous Substances," U.S. Code of Federal Regulations 1910, Subpart Z Section 1910.1000 (July 1982).

12. "NIOSH Pocket Guide to Chemical Hazards," U.S. Department of Health and Human Services, U.S. Government Printing Office (1987), 241 p.

13. *Documentation of the Threshold Limit Values and Biological Exposure Indices* (Cincinnati, OH: American Conference of Governmental Industrial Hygienists, 1986), 744 p.

14. Cleland, J.G. "Project Summary - Environmental Hazard Rankings of Pollutants Generated in Coal Gasification Processes," Office of Research and Development, U.S. EPA Report-600/S7-81-101 (1981), 19 p.

ACETONE

Synonyms: Chevron acetone; Dimethylformaldehyde; Dimethylketal; Dimethyl ketone; DMK; Ketone propane; β-Ketopropane; Methyl ketone; Propanone; **2-Propanone**; Pyroacetic acid; Pyroacetic ether; RCRA waste number U002; UN 1090.

Structural Formula:

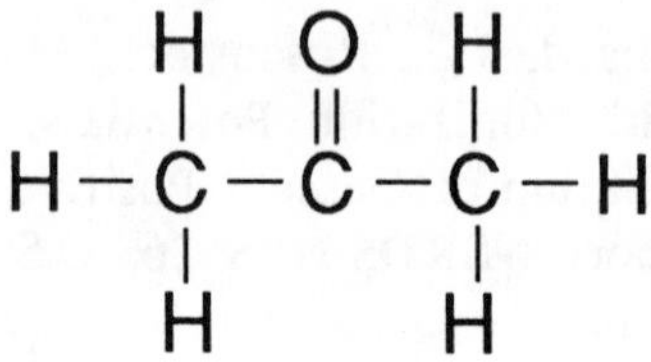

$$
\begin{array}{c}
\text{H} \quad\ \text{O} \quad\ \text{H} \\
| \quad\quad || \quad\quad | \\
\text{H} - \text{C} - \text{C} - \text{C} - \text{H} \\
| \quad\quad\quad\quad\ | \\
\text{H} \quad\quad\quad \text{H}
\end{array}
$$

CHEMICAL DESIGNATIONS

CAS Registry Number: 67-64-1

DOT Designation: 1090

Empirical Formula: C_3H_6O

Formula Weight: 58.08

RTECS Number: AL 3150000

PHYSICAL AND CHEMICAL PROPERTIES

Appearance and Odor: Colorless, volatile liquid with a sweet fragrant odor.

Boiling Point: 56.2 °C [1].

Henry's Law Constant: 3.97 x 10^{-5} atm·m^3/mol at 25 °C [2].

Ionization Potential: 9.69 eV [3].

Log K_{oc}: -0.43 using method of Rao and Davidson [4].

Log K_{ow}: -0.24 [5].

Melting Point: -95.35 °C [1].

Solubility in Organics: Soluble in ethanol, ether, benzene, and chloroform [6].

Solubility in Water: Miscible in all proportions.

Specific Density: 0.7899 at 20/4 °C [1]; 0.7912 at 20/4 °C [7].

Transformation Products: No data found.

Vapor Density: 2.37 g/L at 25 °C, 2.01 (air = 1).

Vapor Pressure: 266 mm at 25 °C [3]; 89 mm at 5 °C, 270 mm at 30 °C, 400 mm at 39.5 °C [8]; 760 mm at 56.5 °C, 1,520 mm at 78.6 °C, 3,800 mm at 113.0 °C, 7,600 mm at 144.5 °C, 15,200 mm at 181.0 °C, 30,400 mm at 214.5 °C [1]; 180 mm at 20 °C [9].

FIRE HAZARDS

Flash Point: -17 °C [3].

Lower Explosive Limit (LEL): 2.6% [3].

Upper Explosive Limit (UEL): 12.8% [3].

HEALTH HAZARD DATA

Immediately Dangerous to Life or Health (IDLH): 20,000 ppm [3].

Permissible Exposure Limits (PEL) in Air: 1,000 ppm ($\approx$2,350 mg/m^3) [10]; 250 ppm ($\approx$590 mg/m^3) 10-hour TWA [3]; 750 ppm ($\approx$1,780 mg/m^3) TWA, 1,000 ppm STEL [9].

MANUFACTURING

Selected Manufacturers:

Allied Chemical Corp.
Specialty Chemical Division
Wilmington Turnpike
Marcus Hook, PA 19061

Shell Chemical Co.
Industrial Chemical Division
Houston, TX 77001

Union Carbide Corp.
Chemicals and Plastics Division
New York, NY 10017

Uses: Intermediate for chemicals (methyl isobutyl ketone, methyl isobutyl carbinol, methyl methacrylate, bisphenol-A); paint, varnish, and lacquer solvent; spinning solvent for cellulose acetate; to clean and dry parts for precision equipment; solvent for potassium iodide, potassium permanganate, cellulose acetate, nitrocellulose; acetylene; delustrant for cellulose acetate fibers; specification testing for vulcanized rubber products; extraction of principals from animal and plant substances; ingredient in nail polish remover; sealants and adhesives; pharmaceutical manufacturing; organic synthesis.

REFERENCES

1. Weast, R.C., Ed. *CRC Handbook of Chemistry and Physics*, 67th ed. (Boca Raton, FL: CRC Press, Inc., 1986), 2406 p.
2. Hine, J., and P.K. Mookerjee. "The Intrinsic Hydrophilic Character of Organic Compounds. Correlations in Terms of Structural Contributions," *J. Org. Chem.*, 40(3):292-298 (1975).
3. "NIOSH Pocket Guide to Chemical Hazards," U.S. Department of Health and Human Services, U.S. Government Printing Office (1987), 241 p.
4. Rao, P.S.C., and J.M. Davidson. "Estimation of Pesticide Retention and Transformation Parameters Required in Nonpoint Source Pollution Models," in *Environmental Impact of Nonpoint Source Pollution*, Overcash, M.R., and J.M. Davidson, Eds. (Ann Arbor, MI: Ann Arbor Science Publishers, Inc., 1980), pp 23-67.
5. Leo, A., Hansch, C., and D. Elkins. "Partition Coefficients and Their Uses," *Chem. Rev.*, 71(6):525-616 (1971).
6. "Chemical, Physical, and Biological Properties of Compounds Present at Hazardous Waste Sites," U.S. EPA Report-530/SW-89-010 (1985), 619 p.
7. Huntress, E.H., and S.P. Mulliken. *Identification of Pure Organic Compounds - Tables of Data on Selected Compounds of Order I* (New York: John Wiley and Sons, Inc., 1941), 691 p.
8. Verschueren, K. *Handbook of Environmental Data on Organic*

Chemicals (New York: Van Nostrand Reinhold Co., 1983), 1310 p.

9. *Threshold Limit Values and Biological Exposure Indices for 1986-1987* (Cincinnati, OH: American Conference of Governmental Industrial Hygienists, 1986), 111 p.

10. "General Industry Standards for Toxic and Hazardous Substances," U.S. Code of Federal Regulations 1910, Subpart Z, Section 1910.1000 (July 1982).

ACROLEIN

Synonyms: Acraldehyde; Acrylaldehyde; Acrylic aldehyde; Allyl aldehyde; Aqualin; Aqualine; Biocide; Crolean; Ethylene aldehyde; Magnacide; NSC 8,819; Propenal; **2-Propenal**; Prop-2-en-1-al; 2-Propen-1-one; RCRA waste number P003; Slimicide; UN 1092.

Structural Formula:

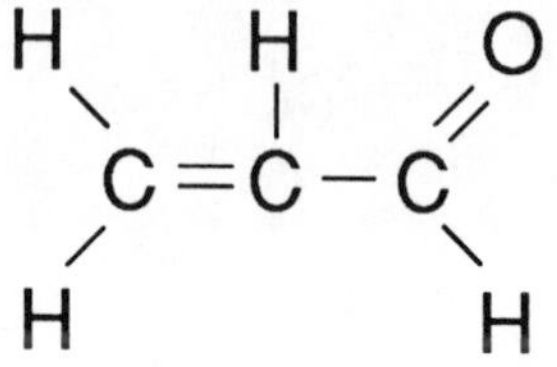

$$
\begin{array}{ccc}
\text{H} & \text{H} & \text{O} \\
\diagdown & | & \diagup\!\!\diagup \\
\text{C} = & \!\!\text{C} - & \!\!\text{C} \\
\diagup & & \diagdown \\
\text{H} & & \text{H}
\end{array}
$$

CHEMICAL DESIGNATIONS

CAS Registry Number: 107-02-8

DOT Designation: 1092 (inhibited), 2607 (stabilized dimer).

Empirical Formula: C_3H_4O

Formula Weight: 56.06

RTECS Number: AS 1050000

PHYSICAL AND CHEMICAL PROPERTIES

Appearance and Odor: Colorless to yellow watery liquid with a very sharp, disagreeable, pungent, irritating odor. Polymerizes readily if not inhibited.

Boiling Point: 52.7 °C [1].

Henry's Law Constant: 4.4 x 10^{-6} atm·m^3/mol at 25 °C [2].

Ionization Potential: 10.10 eV [3].

Log K_{oc}: -0.28 using method of Rao and Davidson [4].

Log K_{ow}: -0.10 [5]; -0.090 [6]; 0.90 [7].

Melting Point: -86.9 °C [8].

Solubility in Organics: Soluble in ethanol, ether, and acetone [9].

Solubility in Water: 20.8% at 20 °C [10]; 2-3 parts water [11]; 400,000 mg/L [12]; 200,000 ppm at 25 °C [13].

Specific Density: 0.8410 at 20/4 °C [8]; 0.8427 at 20/20 °C [1].

Transformation Product: Microbes in site water degraded acrolein to β-hydroxypropionaldehyde [14].

Vapor Density: 2.29 g/L at 25 °C, 1.94 (air = 1).

Vapor Pressure: 220 mm at 20 °C, 330 mm at 30 °C [10]; 1 mm at -64.5 °C, 10 mm at -36.7 °C, 40 mm at -15 °C, 100 mm at 2.5 °C, 400 mm at 34.5 °C, 760 mm at 52.5 °C [8]; 214 mm at 20 °C [15]; 215 mm at 20 °C [1]; 265 mm at 25 °C [2]; 135.71 mm at 10 °C, 210 mm at 20 °C [16].

FIRE HAZARDS

Flash Point: -25 °C [17]; -18 °C (open cup) [18].

Lower Explosive Limit (LEL): 2.8% [3].

Upper Explosive Limit (UEL): 31% [3].

HEALTH HAZARD DATA

Immediately Dangerous to Life or Health (IDLH): 5 ppm [3].

Permissible Exposure Limits (PEL) in Air: 0.1 ppm ($\approx$0.25 mg/m^3) [19]; 0.1 ppm TWA, 0.3 ppm ($\approx$0.8 mg/m^3) STEL [15].

MANUFACTURING

Selected Manufacturers:

Aldrich Chemical Co., Inc.
940 West Saint Paul Ave.
Milwaukee, WI 53233

Shell Chemical Co.
Industrial Chemicals Division
P.O. Box 2463
Houston, TX 77001

Uses: Intermediate for synthetic glycol, pharmaceuticals, polyurethane and polyester resins; herbicide; warning agent in gases.

REFERENCES

1. Standen, A., Ed. *Kirk-Othmer Encyclopedia of Chemical Technology, Volume 1*, 2nd ed. (New York: John Wiley and Sons, Inc., 1963), 990 p.
2. Howard, P.H. *Handbook of Environmental Fate and Exposure Data for Organic Chemicals* (Chelsea, MI: Lewis Publishers, Inc., 1989), 574 p.
3. "NIOSH Pocket Guide to Chemical Hazards," U.S. Department of Health and Human Services, U.S. Government Printing Office (1987), 241 p.
4. Rao, P.S.C., and J.M. Davidson. "Estimation of Pesticide Retention and Transformation Parameters Required in Nonpoint Source Pollution Models," in *Environmental Impact of Nonpoint Source Pollution*, Overcash, M.R., and J.M. Davidson, Eds. (Ann Arbor, MI: Ann Arbor Science Publishers, Inc., 1980), pp 23-67.
5. Mills, W.B., Porcella, D.B., Ungs, M.J., Gherini, S.A., Summers, K.V., Mok, L., Rupp, G.L., and G.L. Bowie. "Water Quality Assessment: A Screening Procedure for Toxic and Conventional Pollutants in Surface and Groundwater-Part I," Office of Research and Development, U.S. EPA Report-600/6-85-002a (1985), 638 p.
6. Walton, W.C. *Practical Aspects of Ground Water Modeling* (Worthington, OH: National Water Well Association, 1985), 587 p.
7. Veith, G.D., Macek, K.J., Petrocelli, S.R., and J. Carroll. "An Evaluation of Using Partition Coefficients and Water Solubility to Estimate Bioconcentration Factors for Organic Chemicals in Fish," in *Aquatic Toxicology, ASTM STP 707*, Eaton, J.G., Parrish, P.R., and A.C. Hendricks, Eds. (Philadelphia, PA: American Society for Testing and Materials, 1980), pp 116-129.
8. Weast, R.C., Ed. *CRC Handbook of Chemistry and Physics*, 67th ed. (Boca Raton, FL: CRC Press, Inc., 1986), 2406 p.
9. "Chemical, Physical, and Biological Properties of Compounds Present at Hazardous Waste Sites," U.S. EPA Report-530/SW-89-010 (1985), 619 p.

10. Verschueren, K. *Handbook of Environmental Data on Organic Chemicals* (New York: Van Nostrand Reinhold Co., 1983), 1310 p.
11. Windholz, M., Budavari, S., Blumetti, R.F., and E.S. Otterbein, Eds. *The Merck Index*, 10th ed., (Rahway, NJ: Merck and Co., 1983), 1463 p.
12. Bailey, G.W., and J.L. White. "Herbicides: A Compilation of Their Physical, Chemical, and Biological Properties," *Res. Rev.*, 10:97-122 (1965).
13. Amoore, J.E., and E. Hautala. "Odor as an Aide to Chemical Safety: Odor Thresholds Compared with Threshold Limit Values and Volatilities for 214 Industrial Chemicals in Air and Water Dilution," *J. Appl. Toxicol.*, 3(6):272-290 (1983).
14. Kobayashi, H., and B.E. Rittman. "Microbial Removal of Hazardous Organic Compounds," *Environ. Sci. Technol.*, 16(3):170A-183A (1982).
15. *Documentation of the Threshold Limit Values and Biological Exposure Indices for 1986-1987* (Cincinnati, OH: American Conference of Governmental Industrial Hygienists, 1986), 111 p.
16. "Extremely Hazardous Substances-Superfund Chemical Profiles. Volume 1," (Park Ridge, NJ: Noyes Data Corp., 1988), 932 p.
17. Weiss, G. *Hazardous Chemicals Data Book* (Park Ridge, NJ: Noyes Data Corp., 1986), 1069 p.
18. *Catalog Handbook of Fine Chemicals* (Milwaukee, WI: Aldrich Chemical Co., 1988), 2212 p.
19. "General Industry Standards for Toxic and Hazardous Substances," U.S. Code of Federal Regulations 1910, Subpart Z, Section 1910.1000 (July 1982).

ACRYLONITRILE

Synonyms: Acritet; Acrylon; Acrylonitrile monomer; An; Carbacryl; Cyanoethylene; ENT 54; Fumigrain; Miller's fumigrain; Nitrile; Propenenitrile; **2-Propenenitrile**; RCRA waste number U009; TL 314; UN 1093; VCN; Ventox; Vinyl cyanide.

Structural Formula:

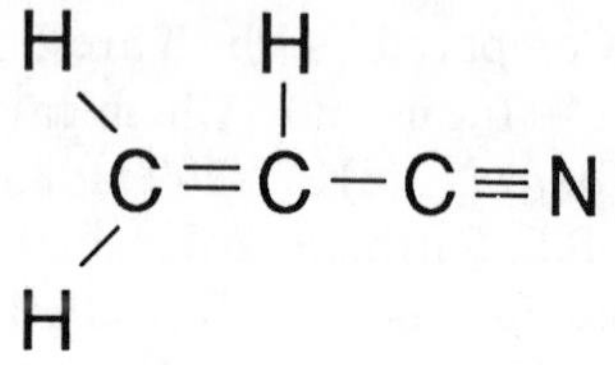

$$\underset{\underset{H}{/}}{\overset{\overset{H}{\diagdown}}{C}} = \overset{\overset{H}{|}}{C} - C \equiv N$$

CHEMICAL DESIGNATIONS

CAS Registry Number: 107-13-1

DOT Designation: 1093

Empirical Formula: C_3H_3N

Formula Weight: 53.06

RTECS Number: AT 5250000

PHYSICAL AND CHEMICAL PROPERTIES

Appearance and Odor: Clear, colorless, watery liquid with a sweet irritating odor resembling peach pits.

Boiling Point: 77.5-79 °C [1].

Henry's Law Constant: 1.10×10^{-4} atm·m^3/mol at 25 °C [2].

Ionization Potential: 10.91 eV [3].

Log K$_{oc}$: -1.13 using method of Rao and Davidson [4].

Log K$_{ow}$: -0.92 [5]; 0.25 [6]; 1.20 [7].

Melting Point: -83 °C [8].

Solubility in Organics: Soluble in ethanol, ether, acetone, and benzene [9]. Also soluble in carbon tetrachloride, toluene, and most other organic solvents [10].

Solubility in Water: 7.2% at 0 °C, 7.35% at 20 °C, 7.9% at 40 °C [11]; 80,000 mg/L at 25 °C [12]; 15.6 wt% at 20 °C [13]; 79,000 mg/L at 25 °C [6]; 9.1% at 60 °C [14]; 7.5% at 25 °C [15].

Specific Density: 0.8060 at 20/4 °C [1]; 0.8004 at 25/4 °C [14].

Transformation Product: In an aqueous solution at 50 °C, ultraviolet light photooxidized acrylonitrile to carbon dioxide. After 24 hours, the concentration of acrylonitrile was reduced 24.2% [8].

Vapor Density: 2.17 g/L at 25 °C, 1.83 (air = 1).

Vapor Pressure: 83 mm at 20 °C [3]; 137 mm at 30 °C [8]; 1 mm at -51.0 °C, 10 mm at -20.3 °C, 40 mm at 3.8 °C, 100 mm at 22.8 °C, 400 mm at 58.3 °C, 760 mm at 78.5 °C [1]; 110-115 mm at 25 °C [16]; 50 mm at 8.7 °C, 100 mm at 23.6 °C, 250 mm at 45.5 °C, 500 mm at 64.7 °C, 760 mm at 77.3 °C [11]; 107.8 mm at 25 °C [2].

FIRE HAZARDS

Flash Point: -1 °C [3].

Lower Explosive Limit (LEL): 3.05 ± 0.5% [14].

Upper Explosive Limit (UEL): 17.0 ± 0.5% [14].

HEALTH HAZARD DATA

Immediately Dangerous to Life or Health (IDLH): 4 ppm (carcinogen) [17].

Permissible Exposure Limits (PEL) in Air: 2 ppm ($\approx$4.5 mg/m^3), 10 ppm 15-minute ceiling [18]; 1 ppm TWA, 10 ppm 15-minute ceiling [3]; 2 ppm TWA [16].

MANUFACTURING

Selected Manufacturers:

E.I. duPont de Nemours and Co.
Electrochemicals Department
Wilmington, DE 19898

Monsanto Industrial Chemicals Co.
800 North Lindbergh Blvd.
St. Louis, MO 63166

Uses: Copolymerized with methylacrylate, methylmethacrylate, vinyl acetate, vinyl chloride, or 1,1-dichloroethylene to produce acrylic and modacrylic fibers and high-strength fibers; ABS (acrylonitrile-butadiene-styrene) and acrylonitrile-styrene copolymers; nitrile rubber; cyanoethylation of cotton; synthetic soil block (acrylonitrile polymerized in wood pulp); organic synthesis; grain fumigant; monomer for a semi-conductive polymer that can be used similar to inorganic oxide catalysts in dehydrogenation of *t*-butyl alcohol to isobutylene and water; pharmaceuticals; antioxidants; dyes and surfactants.

REFERENCES

1. Weast, R.C., Ed. *CRC Handbook of Chemistry and Physics*, 67th ed. (Boca Raton, FL: CRC Press, Inc., 1986), 2406 p.
2. Howard, P.H. *Handbook of Environmental Fate and Exposure Data for Organic Chemicals* (Chelsea, MI: Lewis Publishers, Inc., 1989), 574 p.
3. "NIOSH Pocket Guide to Chemical Hazards," U.S. Department of Health and Human Services, U.S. Government Printing Office (1987), 241 p.
4. Rao, P.S.C., and J.M. Davidson. "Estimation of Pesticide Retention and Transformation Parameters Required in Nonpoint Source Pollution Models," in *Environmental Impact of Nonpoint Source Pollution*, Overcash, M.R., and J.M. Davidson, Eds. (Ann Arbor, MI: Ann Arbor Science Publishers, Inc., 1980), pp 23-67.
5. Leo, A., Hansch, C., and D. Elkins. "Partition Coefficients and Their Uses," *Chem. Rev.*, 71(6):525-616 (1971).
6. Mabey, W.R., Smith, J.H., Podoll, R.T., Johnson, H.L., Mill, T., Chou, T.-W., Gates, J., Partridge, I.W., Jaber, H., and D. Vandenberg. "Aquatic Fate Process Data for Organic Priority Pollutants - Final

Report," Office of Regulations and Standards, U.S. EPA Report-440/4-81-014 (1982), 407 p.

7. Veith, G.D., Macek, K.J., Petrocelli, S.R., and J. Carroll. "An Evaluation of Using Partition Coefficients and Water Solubility to Estimate Bioconcentration Factors for Organic Chemicals in Fish," in *Aquatic Toxicology, ASTM STP 707*, Eaton, J.G., Parrish, P.R., and A.C. Hendricks, Eds. (Philadelphia, PA: American Society for Testing and Materials, 1980), pp 116-129.

8. Verschueren, K. *Handbook of Environmental Data on Organic Chemicals* (New York: Van Nostrand Reinhold Co., 1983), 1310 p.

9. "Chemical, Physical, and Biological Properties of Compounds Present at Hazardous Waste Sites," U.S. EPA Report-530/SW-89-010 (1985), 619 p.

10. Yoshida, K., Shigeoka T., and F. Yamauchi. "Non-Steady-State Equilibrium Model for the Preliminary Prediction of the Fate of Chemicals in the Environment," *Ecotoxicol. Environ. Safety*, 7(2):179-190 (1983).

11. *Environmental Health Criteria 28: Acrylonitrile* (Geneva: World Health Organization, 1983), 125 p.

12. Lyman, W.J., Reehl, W.F., and D.H. Rosenblatt. *Handbook of Chemical Property Estimation Methods: Environmental Behavior of Organic Compounds* (New York: McGraw-Hill, Inc., 1982).

13. Riddick, J.A., Bunger, W.B., and T.K. Sakano. *Organic Solvents - Physical Properties and Methods of Purification. Volume II.* (New York: John Wiley and Sons, Inc., 1986), 1325 p.

14. Standen, A., Ed. *Kirk-Othmer Encyclopedia of Chemical Technology, Volume 1*, 2nd ed. (New York: John Wiley and Sons, Inc., 1963), 990 p.

15. Gunther, F.A., Westlake, W.E., and P.S. Jaglan. "Reported Solubilities of 738 Pesticide Chemicals in Water," *Res. Rev.*, 20:1-148 (1968).

16. *Documentation of the Threshold Limit Values and Biological Exposure Indices for 1986-1987* (Cincinnati, OH: American Conference of Governmental Industrial Hygienists, 1986), 111 p.

17. Weiss, G. *Hazardous Chemicals Data Book* (Park Ridge, NJ: Noyes Data Corp., 1986), 1069 p.

18. "General Industry Standards for Toxic and Hazardous Substances," U.S. Code of Federal Regulations 1910, Subpart Z, Section 1910.1000 (July 1982).

ALDRIN

Synonyms: Aldrec; Aldrex; Aldrex 30; Aldrite; Aldrosol; Altox; Compound 118; Drinox; ENT 15,949; Hexachlorohexahydro-*endo,exo*-dimethanonaphthalene; **1,2,3,4,10,10-Hexachloro-1,4,4a,5,8,8a-hexahydro-1,4:5,8-dimethanonaphthalene**; 1,2,3,4,10,10-Hexachloro-1,4,4a,5,8,8a-hexahydro-1,4-*endo,exo*-5,8-dimethanonaphthalene; 1,2,3,4,10,10-Hexachloro-1,4,4a,5,8,8a-hexahydro-*exo*-1,4-*endo*-5,8-dimethanonaphthalene; 1,4,4a,5,8,8a-Hexahydro-1,4-*endo,exo*-5,8-dimethanonaphthalene; HHDN; NA 2,761; NA 2,762; NCI-C00044; Octalene; RCRA waste number P004; Seedrin; Seedrin liquid.

Structural Formula:

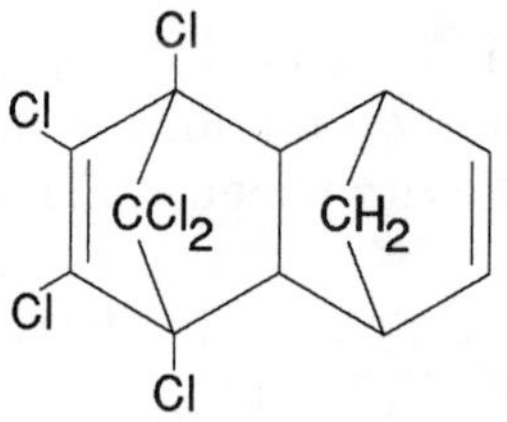

CHEMICAL DESIGNATIONS

CAS Registry Number: 309-00-2

DOT Designation: 2761 (additional numbers may exist and may be provided by the supplier).

Empirical Formula: $C_{12}H_8Cl_6$

Formula Weight: 364.92

RTECS Number: IO 2100000

PHYSICAL AND CHEMICAL PROPERTIES

Appearance and Odor: White, odorless crystals when pure; technical grades are tan to dark brown with a mild chemical odor.

Boiling Point: 145 °C at 2 mm [1].

Henry's Law Constant: 1.4×10^{-6} atm·m^3/mol [2]; 4.96×10^{-4} atm·m^3/mol [3].

Ionization Potential: No data found.

Log K$_{oc}$: 2.61 [4].

Log K$_{ow}$: 5.52 [5]; 5.66 [6]; 5.17 [7].

Melting Point: 104 °C [8]; 107 °C [9].

Solubility in Organics: Soluble in most organic solvents [10].

Solubility in Water: 1.01 mg/L [11]; 0.027 mg/L at 25-29 °C [12]; 17 μg/L at 25 °C [13]; 0.017-0.18 mg/L at 25 °C [14]; 0.011 mg/L at 25 °C [7]; 0.013 ppm [15]; 105 ppb at 15 °C, 180 ppb at 25 °C, 350 ppb at 35 °C, 600 ppb at 45 °C (particle sizes $\leq$ 5μ) [16].

Specific Density: 1.70 at 20/4 °C [1].

Transformation Products: Dieldrin is the major metabolite from the microbial degradation of aldrin by oxidation or epoxidation. Dieldrin may further degrade to photodieldrin [17]. Under oceanic conditions, aldrin may undergo dihydroxylation at the chlorine free double bond to produce aldrin diol [11].

Vapor Pressure: 2.31 x 10^{-5} mm at 20 °C [18]; 6 x 10^{-6} mm at 20 °C [19]; 7.5 x 10^{-5} mm at 20 °C [20]; 6 x 10^{-6} mm at 25 °C [21].

FIRE HAZARDS

Flash Point: Non-combustible [19].

Lower Explosive Limit (LEL): Not applicable [19].

Upper Explosive Limit (UEL): Not applicable [19].

HEALTH HAZARD DATA

Immediately Dangerous to Life or Health (IDLH): 100 mg/m^3 (carcinogen) [22].

Permissible Exposure Limits (PEL) in Air: 0.25 mg/m^3 [23]; lowest detectable limit (0.15 mg/m^3) TWA [19]; 0.25 mg/m^3 TWA [24].

MANUFACTURING

Selected Manufacturer:

Shell Chemical Co.
2401 Crow Canyon Rd.
San Ramon, CA 94583

Uses: Insecticide and fumigant.

REFERENCES

1. Hayes, W.J. *Pesticides Studied in Man* (Baltimore, MD: The Williams and Wilkens Co., 1982), pp 234-247.
2. Eisenreich, S.J., Looney, B.B., and J.D. Thornton. "Airborne Organic Contaminants in the Great Lakes Ecosystem," *Environ. Sci. Technol.*, 15(1):30-38 (1981).
3. Warner, H.P., Cohen, J.M., and J.C. Ireland. "Determination of Henry's Law Constants of Selected Priority Pollutants," Office of Science and Development, U.S. EPA Report-600/D-87/229 (1987), 14 p.
4. Kenaga, E.E. "Correlation of Bioconcentration Factors of Chemicals in Aquatic and Terrestrial Organisms with Their Physical and Chemical Properties," *Environ. Sci. Technol.*, 14(5):553-556 (1980).
5. Travis, C.C., and A.D. Arms. "Bioconcentration of Organics in Beef, Milk and Vegetation," *Environ. Sci. Technol.*, 22(3):271-274 (1988).
6. Isnard, S., and S. Lambert. "Estimating Bioconcentration Factors from Octanol-Water Partition Coefficient and Aqueous Solubility," *Chemosphere*, 17(1):21-34 (1988).
7. Walton, W.C. *Practical Aspects of Ground Water Modeling* (Worthington, OH: National Water Well Association, 1985), 587 p.
8. Weast, R.C., Ed. *CRC Handbook of Chemistry and Physics*, 67th ed. (Boca Raton, FL: CRC Press, Inc., 1986), 2406 p.
9. Sims, R.C., Doucette, W.C., McLean, J.E., Greeney, W.J., and R.R. Dupont. "Treatment Potential for 56 EPA Listed Hazardous Chemicals in Soil," National Technical Information Service, U.S. EPA Report-600/6-88-001 (1988), 105 p.
10. "Chemical, Physical, and Biological Properties of Compounds Present at Hazardous Waste Sites," U.S. EPA Report-530/SW-89-010 (1985), 619 p.
11. Verschueren, K. *Handbook of Environmental Data on Organic Chemicals* (New York: Van Nostrand Reinhold Co., 1983), 1310 p.

12. Park, K.S., and W.N. Bruce. "The Determination of the Water Solubility of Aldrin, Dieldrin, Heptachlor and Heptachlor Epoxide," *J. Econ. Entomol.*, 61(3):770-774 (1968).

13. Weil, L., Dure, G., and K.E. Quentin. "Solubility in Water of Insecticide Chlorinated Hydrocarbons and Polychlorinated Biphenyls in View of Water Pollution," *Z. Wasser Forsch.*, 7(6):169-175 (1974).

14. "Treatability Manual - Volume 1: Treatability Data," Office of Research and Development, U.S. EPA Report-600/8-80-042a (1980), 1035 p.

15. Kenaga, E.E., and C.A.I. Goring. "Relationship Between Water Solubility, Soil Sorption, Octanol-Water Partitioning and Concentration of Chemicals in Biota," in *Aquatic Toxicology, ASTM STP 707*, Eaton, J.G., Parrish, P.R., and A.C. Hendricks, Eds. (Philadelphia, PA: American Society for Testing and Materials, 1980), pp 78-115.

16. Biggar, J.W., and I.R. Riggs. "Apparent Solubility of Organochlorine Insecticides in Water at Various Temperatures," *Hilgardia*, 42(10):383-391 (1974).

17. Kearney, P.C., and D.D. Kaufman. *Herbicides: Chemistry, Degradation and Mode of Action* (New York: Marcel Dekker, Inc., 1976), 1036 p.

18. Martin, H., Ed. *Pesticide Manual*, 3rd ed. (Worcester, England: British Crop Protection Council, 1972).

19. "NIOSH Pocket Guide to Chemical Hazards," U.S. Department of Health and Human Services, U.S. Government Printing Office (1987), 241 p.

20. Windholz, M., Budavari, S., Blumetti, R.F., and E.S. Otterbein, Eds. *The Merck Index*, 10th ed., (Rahway, NJ: Merck and Co., 1983), 1463 p.

21. Gunther, F.A., and J.D. Gunther. "Residues of Pesticides and other Foreign Chemicals in Foods and Feeds," *Res. Rev.*, 36:69-77 (1971).

22. Weiss, G. *Hazardous Chemicals Data Book* (Park Ridge, NJ: Noyes Data Corp., 1986), 1069 p.

23. "General Industry Standards for Toxic and Hazardous Substances," U.S. Code of Federal Regulations 1910, Subpart Z, Section 1910.1000 (July 1982).

24. *Documentation of the Threshold Limit Values and Biological Exposure Indices for 1986-1987* (Cincinnati, OH: American Conference of Governmental Industrial Hygienists, 1986), 111 p.

ANTHRACENE

Synonyms: Anthracin; Green oil; Paranaphthalene; Tetra olive N2G.

Structural Formula:

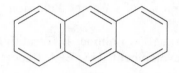

CHEMICAL DESIGNATIONS

CAS Registry Number: 120-12-7

DOT Designation: None assigned.

Empirical Formula: $C_{14}H_{10}$

Formula Weight: 178.24

RTECS Number: CA 9350000

PHYSICAL AND CHEMICAL PROPERTIES

Appearance and Odor: White to yellow crystals with a bluish or violet fluorescence and a weak aromatic odor.

Boiling Point: 339.9 °C [1].

Henry's Law Constant: 0.0014 atm·m^3/mol [2]; 6.51 x 10^{-5} atm·m^3/mol at 25 °C [3]; 1.77 x 10^{-5} atm·m^3/mol at 25 °C [4].

Ionization Potential: 7.55 eV [5]; 7.58 eV [6].

Log K_{oc}: 4.27 [7]; 4.41 [8]; 4.205 [9].

Log K_{ow}: 4.45 [10]; 4.54 [11]; 4.34 [12].

Melting Point: 216.2-216.4 °C [13].

Solubility in Organics: Soluble in acetone and benzene [14].

Solubility in Water: 0.075 mg/L at 15 °C, 1.29 mg/L at 25 °C in distilled water, 0.6 mg/L at 25 °C in salt water [15]; 0.045 mg/L at 25 °C [16]; 0.073 mg/L at 25 °C [17]; 0.030 mg/L at 25 °C [18]; 0.075 mg/L at 25 °C [19]; 0.1125 mg/L at 25 °C [20]; 12.7 μg/L at 5.2 °C, 17.5 μg/L at 10.0 °C, 22.2 μg/L at 14.1 °C, 29.1 μg/L at 18.3 °C, 37.2 μg/L at 22.4 °C, 43.4 μg/L at 24.6 °C, 55.7 μg/L at 28.7 °C [21]; 1.31×10^{-7} M at 8.6 °C, 1.37×10^{-7} M at 11.1 °C, 1.44×10^{-7} M at 12.2 °C, 1.54×10^{-7} M at 14 °C, 1.66×10^{-7} M at 15.5 °C, 1.81×10^{-7} M at 18.2 °C, 2.22×10^{-7} M at 20.3 °C, 2.34×10^{-7} M at 23.0 °C, 2.30×10^{-7} M at 25.0 °C, 2.67×10^{-7} M at 26.2 °C, 3.25×10^{-7} M at 28.5 °C, 3.90×10^{-7} M at 31.3 °C [22]; 0.041 mg/L at 20 °C [23].

Specific Density: 1.283 at 25/4 °C [24]; 1.24 at 20/4 °C [25].

Transformation Products: Catechol is the central metabolite in the bacterial degradation of anthracene. Intermediate byproducts included 3-hydroxy-2-naphthoic acid and salicylic acid [26].

Vapor Pressure: 1.95×10^{-4} mm at 25 °C [27]; 1 mm at 145 °C, 10 mm at 187.2 °C, 40 mm at 217.5 °C, 100 mm at 250 °C, 400 mm at 310.2 °C, 760 mm at 342 °C [24]; 1.95×10^{-4} mm at 20 °C [28]; 1.7×10^{-5} mm at 25 °C [29]; 0.00669 mm at 85.25 °C, 0.0102 mm at 90.15 °C, 0.0164 mm at 95.65 °C, 0.0249 mm at 100.70 °C, 0.0344 mm at 104.70 °C, 0.0603 mm at 111.90 °C, 0.0852 mm at 116.40 °C, 0.1100 mm at 119.95 °C [30]; 0.000086 mm at 65.7 °C, 0.000105 mm at 67.10 °C, 0.000118 mm at 68.75 °C, 0.000125 mm at 69.91 °C, 0.000156 mm at 71.25 °C, 0.000167 mm at 72.20 °C, 0.000194 mm at 73.55 °C, 0.000281 mm at 77.25 °C, 0.000367 mm at 79.95 °C, 0.000393 mm at 80.40 °C [31].

FIRE HAZARDS

Flash Point: 121.1 °C [25].

Lower Explosive Limit (LEL): 0.6% [25].

Upper Explosive Limit (UEL): No data found.

HEALTH HAZARD DATA

Immediately Dangerous to Life or Health (IDLH): Potential human carcinogen [32].

Permissible Exposure Limits (PEL) in Air: No individual standards have been set, however, as a constituent in coal tar pitch volatiles, the following exposure limits have been established: 0.2 mg/m^3 (benzene-soluble fraction) [33]; 0.1 mg/m^3 10-hour TWA (cyclohexane-extractable fraction) [32]; 0.2 mg/m^3 TWA (benzene solubles) [34].

MANUFACTURING

Selected Manufacturers:

Aldrich Chemical Co.
940 West Saint Paul Ave.
Milwaukee, WI 53233

Eastman Organic Chemicals
Rochester, NY 14650

Gallard-Schlesinger Chemical Mfg. Co.
584 Mineola Ave.
Carle Place, NY 11514

Uses: Dyes; starting material for the preparation of alizarin, phenanthrene, carbazole, anthraquinone, and insecticides; in calico printing; as component of smoke screens; scintillation counter crystals; organic semiconductor research; wood preservative.

REFERENCES

1. Dean, J.A., Ed., *Lange's Handbook of Chemistry*, 11th ed. (New York: McGraw-Hill, Inc., 1973), 1570 p.
2. Petrasek, A.C., Kugelman, I.J., Austern, B.M., Pressley, T.A., Winslow, L.A., and R.H. Wise. "Fate of Toxic Organic Compounds in Wastewater Treatment Plants," *J. Water Poll. Control Fed.*, 55(10):1286-1296 (1983).
3. Southworth, G.R. "The Role Volatilization in Removing Polycyclic Aromatic Hydrocarbons from Aquatic Environments," *Bull. Environ. Contam. Toxicol.*, 21(4/5):507-514 (1979).
4. Hine, J., and P.K. Mookerjee. "The Intrinsic Hydrophilic Character of Organic Compounds. Correlations in Terms of Structural Contributions," *J. Org. Chem.*, 40(3):292-298 (1975).
5. Franklin, J.L., Dillard, J.G., Rosenstock, H.M., Herron, J.T., Draxl

K., and F.H. Field. "Ionization Potentials, Appearance Potentials and Heats of Formation of Gaseous Positive Ions," National Bureau of Standards Report NSRDS-NBS 26, U.S. Government Printing Office (1969), 289 p.

6. Yoshida, K., Shigeoka, T., and F. Yamauchi. "Non-Steady State Equilibrium Model for the Preliminary Prediction of the Fate of Chemicals in the Environment," *Ecotoxicol. Environ. Safety*, 7(2):179-190 (1983).

7. Abdul, S.A., Gibson, T.L., and D.N. Rai. "Statistical Correlations for Predicting the Partition Coefficient for Nonpolar Organic Contaminants Between Aquifer Organic Carbon and Water," *Haz. Waste Haz. Mater.*, 4(3):211-222 (1987).

8. Karickhoff, S.W., Brown, D.S., and T.A. Scott. "Sorption of Hydrophobic Pollutants on Natural Sediments," *Water Res.*, 13:241-248 (1979).

9. Nkedi-Kizza, P., Rao, P.S.C., and A.G. Hornsby. "Influence of Organic Cosolvents on Sorption of Hydrophobic Organic Chemicals by Soils," *Environ. Sci. Technol.*, 19(10):975-979.

10. Hansch, C., and T. Fujita. "ρ-σ-π Analysis. A Method for the Correlation of Biological Activity and Chemical Structure," *J. Am. Chem. Soc.*, 86(8):1616-1617 (1964).

11. Miller, M.M., Wasik, S.P., Huang, G.-L., Shiu, W.-Y., and D. Mackay. "Relationships Between Octanol-Water Partition Coefficient and Aqueous Solubility," *Environ. Sci. Technol.*, 19(6):522-529 (1985).

12. Mackay, D. "Correlation of Bioconcentration Factors," *Environ. Sci. Technol.*, 16(5):274-278 (1982).

13. *Catalog Handbook of Fine Chemicals* (Milwaukee, WI: Aldrich Chemical Co., 1988), 2212 p.

14. "Chemical, Physical, and Biological Properties of Compounds Present at Hazardous Waste Sites," U.S. EPA Report-530/SW-89-010 (1985), 619 p.

15. Verschueren, K. *Handbook of Environmental Data on Organic Chemicals* (New York: Van Nostrand Reinhold Co., 1983), 1310 p.

16. May, W.E., Wasik, S.P., and D.H. Freeman. "Determination of the Aqueous Solubility of Polynuclear Aromatic Hydrocarbons by a Coupled Column Liquid Chromatographic Technique," *Anal. Chem.*, 50(1):175-179 (1978).

17. Mackay, D., and W.Y. Shiu. "Aqueous Solubility of Polynuclear Aromatic Hydrocarbons," *J. Chem. Eng. Data*, 22(4):399-402 (1977).

18. Schwarz, F.P., and S.P. Wasik. "Fluorescence Measurements of Benzene, Naphthalene, Anthracene, Pyrene, Fluoranthene, and Benzo[e]pyrene in Water," *Anal. Chem.*, 48(3):524-528 (1976).

19. Wauchope, R.D., and F.W. Getzen. "Temperature Dependence of

Solubilities in Water and Heats of Fusion of Solid Aromatic Hydrocarbons," *J. Chem. Eng. Data*, 17:38-41 (1972).

20. Sahyun, M.R.V. "Binding of Aromatic Compounds to Bovine Serum Albumin," *Nature*, 209(5023):613-614 (1966).

21. May, W.E., Wasik, S.P., and D.H. Freeman. "Determination of the Solubility Behavior of Some Polycyclic Aromatic Hydrocarbons in Water," *Anal. Chem.*, 50(7):997-1000 (1978).

22. Schwarz, F.P. "Determination of Temperature Dependence of Solubilities of Polycyclic Aromatic Hydrocarbons in Aqueous Solutions by a Fluorescence Method," *J. Chem. Eng. Data*, 22(3):273-277 (1977).

23. Kishi, H., and Y. Hashimoto. "Evaluation of the Procedures for the Measurement of Water Solubility and *n*-Octanol/Water Partition Coefficient of Chemicals Results of a Ring Test in Japan," *Chemosphere*, 18(9/10):1749-1749 (1989).

24. Weast, R.C., Ed. *CRC Handbook of Chemistry and Physics*, 67th ed. (Boca Raton, FL: CRC Press, Inc., 1986), 2406 p.

25. Weiss, G. *Hazardous Chemicals Data Book* (Park Ridge, NJ: Noyes Data Corp., 1986), 1069 p.

26. Chapman, P.J. "An Outline of Reaction Sequences Used for the Bacterial Degradation of Phenolic Compounds" in *Degradation of Synthetic Organic Molecules in the Biosphere: Natural, Pesticidal, and Various Other Man-Made Compounds* (Washington, DC: National Academy of Sciences, 1972), pp 17-55.

27. Radding, S.B., Mill, T., Gould, C.W., Lia, D.H., Johnson, H.L., Bomberger, D.S., and C.V. Fojo. "The Environmental Fate of Selected Polynuclear Aromatic Hydrocarbons," Office of Toxic Substances, U.S. EPA Report-560/5-75-009 (1976), 122 p.

28. "Treatability Manual - Volume 1: Treatability Data," Office of Research and Development, U.S. EPA Report-600/8-80-042a (1980), 1035 p.

29. Mabey, W.R., Smith, J.H., Podoll, R.T., Johnson, H.L., Mill, T., Chou, T.-W., Gates, J., Partridge, I.W., Jaber, H., and D. Vandenberg. "Aquatic Fate Process Data for Organic Priority Pollutants - Final Report," Office of Regulations and Standards, U.S. EPA Report-440/4-81-014 (1982), 407 p.

30. Macknick, A.B., and J.M. Prausnitz. "Vapor Pressures of High Molecular Weight Hydrocarbons," *J. Chem. Eng. Data*, 24(3):175-178 (1979).

31. Bradley, R.S., and T.G. Cleasby. "The Vapour Pressure and Lattice Energy of Some Aromatic Ring Compounds," *J. Chem. Soc. (London)*, pp 1690-1692 (1953).

32. "NIOSH Pocket Guide to Chemical Hazards," U.S. Department of

Health and Human Services, U.S. Government Printing Office (1987), 241 p.

33. "General Industry Standards for Toxic and Hazardous Substances," U.S. Code of Federal Regulations 1910, Subpart Z, Section 1910.1000 (July 1982).

34. *Documentation of the Threshold Limit Values and Biological Exposure Indices for 1986-1987* (Cincinnati, OH: American Conference of Governmental Industrial Hygienists, 1986), 111 p.

BENZENE

Synonyms: Annulene; Benxole; Benzol; Benzole; Benzolene; Bicarburet of hydrogen; Carbon oil; Coal naphtha; Coal tar naphtha; Cyclohexatriene; Mineral naphthalene; Motor benzol; NCI-C55276; Nitration benzene; Phene; Phenyl hydride; Pyrobenzol; Pyrobenzole; RCRA waste number U019; UN 1114.

Structural Formula:

CHEMICAL DESIGNATIONS

CAS Registry Number: 71-43-2

DOT Designation: 1114

Empirical Formula: C_6H_6

Formula Weight: 78.11

RTECS Number: CY 1400000

PHYSICAL AND CHEMICAL PROPERTIES

Appearance and Odor: Clear, colorless to light yellow watery-liquid with an aromatic or gasoline-like odor.

Boiling Point: 80.100 °C [1].

Henry's Law Constant: 0.00548 atm·m³/mol at 25 °C [2]; 0.00538 atm·m³/mol [3].

Ionization Potential: 9.25 eV [4]; 9.56 eV [5].

Log K_{oc}: 1.69 [6]; 1.92 [7]; 1.96, 2.00 [8].

Log K_{ow}: 2.13 [9]; 2.11 [10]; 1.56, 2.15 [11]; 2.12 [12]; 1.95 [13].

Melting Point: 5.533 °C [1].

Solubility in Organics: Freely miscible with ethanol, ether, glacial acetic acid, acetone, chloroform, and carbon tetrachloride [14].

Solubility in Water: 1,780 mg/L at 20 °C [15]; 820 mg/L at 22 °C [16]; 1,800 mg/L at 25 °C [17]; 0.093 vol% at 20 °C [18]; 1,790 mg/L at 25 °C [19]; 1,750 mg/L at 25 °C [20]; 1,850 mg/L at 30 °C [21]; 1,755 mg/L at 25 °C [22]; 1,740 mg/L at 25 °C [23]; 1,780 mg/L at 25 °C [24]; 1,791 mg/L at 25 °C [25]; 0.153 wt% at 0 °C, 0.163 wt% at 10 °C, 0.175 wt% at 20 °C, 0.180 wt% at 25 °C, 0.190 wt% at 30 °C, 0.206 wt% at 40 °C, 0.225 wt% at 50 °C, 0.250 wt% at 60 °C, 0.277 wt% at 70 °C, 0.344 wt% at 80 °C, 0.393 wt% at 90 °C, 0.504 wt% 107.4 °C [26]; 1,678 ppm at 0 °C, 1,755 ppm at 25 °C [27]; 1,740 mg/L at 25 °C, 1,391 mg/L in artificial seawater at 25 °C [28]; 1,710 mg/L at 20 °C [29]; 1,696 ppm at 25.00 °C [30]; 1,860 ppm at 25 °C [31]; 1,800 ppm at 25 °C [32]; 1,000 mg/L in fresh water at 25 °C, 1,030 mg/L in salt water at 25 °C [33]; 0.18775 wt% at 23.5 °C [34]; 0.0233 M at 25 °C [35].

Specific Density: 0.8765 at 20/4 °C [36]; 0.87895 at 20/4 °C, 0.87366 at 25/4 °C [37]; 0.8784 at 20/4 °C, 0.8680 at 30/4 °C, 0.8572 at 40/4 °C [38]; 0.87378 at 25/4 °C [39].

Transformation Products: A mutant of *Pseudomonas putida* dihydroxylyzed benzene into *cis*-benzene glycol accompanied by partial dehydrogenation yielding catechol [40]. Bacterial dioxygenases can cleave catechol at the *ortho*- and *meta*- positions to yield *cis,cis*-muconic acid and α-hydroxymuconic semialdehyde, respectively [41].

Vapor Density: 3.19 g/L at 25 °C, 2.70 (air = 1).

Vapor Pressure: 60 mm at 15 °C, 76 mm at 20 °C, 118 mm at 30 °C [15]; 95.2 mm at 25 °C [42]; 760 mm at 80.1 °C, 1,520 mm at 103.8 °C, 3,800 mm at 142.5 °C, 7,600 mm at 178.8 °C, 15,200 mm at 221.5 °C [36]; 100 mm at 26.075 °C [1]; 397 mm at 60.3 °C, 556 mm at 70.3 °C, 764 mm at 80.3 °C, 1,031 mm at 90.3 °C, 1,370 mm at 100.3 °C [43].

FIRE HAZARDS

Flash Point: -11 °C [4].

Lower Explosive Limit (LEL): 1.3% [4].

Upper Explosive Limit (UEL): 7.1% [4].

HEALTH HAZARD DATA

Immediately Dangerous to Life or Health (IDLH): 2,000 ppm (carcinogen) [44].

Permissible Exposure Limits (PEL) in Air: 10 ppm ($\approx$30 mg/m^3), 10-minute 50 ppm ceiling [45]; 0.1 ppm TWA, 1 ppm 15-minute ceiling [4]; 10 ppm TWA [46].

MANUFACTURING

Selected Manufacturers:

Commonwealth Oil Refining Co., Inc.
Penuelas, PR 00724

Phillips Petroleum Co.
Phillips Puerto Rico Core, Inc.
Banco Popular Center
Hato Rey, PR 00936

Shell Chemical Co.
Petrochemical Division
P.O. Box 2463
Houston, TX 77001

Uses: Manufacture of ethylbenzene (preparation of styrene monomer), dodecylbenzene (for detergents), cyclohexane (for nylon), nitrobenzene, aniline, maleic anhydride, diphenyl, benzene hexachloride, benzene sulfonic acid, phenol, dichlorobenzene, insecticides, pesticides, fumigants, explosives, aviation fuel, flavors, perfume, medicine, dyes, and other organic chemicals; paints, coatings, plastics and resins; food processing; photographic chemicals; nylon intermediates; paint removers; rubber cement; antiknock gasoline; solvent.

REFERENCES

1.   Standen, A., Ed. *Kirk-Othmer Encyclopedia of Chemical Technology,*

Volume 3, 2nd ed. (New York: John Wiley and Sons, Inc., 1964), 927 p.

2. Hine, J., and P.K. Mookerjee. "The Intrinsic Hydrophilic Character of Organic Compounds. Correlations in Terms of Structural Contributions," *J. Org. Chem.*, 40(3):292-298 (1975).

3. Jury, W.A., Spencer, W.F., and W.J. Farmer. "Behavior Assessment Model for Trace Organics in Soil: III. Application of Screening Model," *J. Environ. Qual.*, 13(4):573-579 (1984).

4. "NIOSH Pocket Guide to Chemical Hazards," U.S. Department of Health and Human Services, U.S. Government Printing Office (1987), 241 p.

5. Yoshida, K., Shigeoka, T., and F. Yamauchi. "Non-Steady State Equilibrium Model for the Preliminary Prediction of the Fate of Chemicals in the Environment," *Ecotoxicol. Environ. Safety*, 7(2):179-190 (1983).

6. Abdul, S.A., Gibson, T.L., and D.N. Rai. "Statistical Correlations for Predicting the Partition Coefficient for Nonpolar Organic Contaminants Between Aquifer Organic Carbon and Water," *Haz. Waste Haz. Mater.*, 4(3):211-222 (1987).

7. Schwarzenbach, R.P., and J. Westall. "Transport of Nonpolar Organic Compounds from Surface Water to Groundwater. Laboratory Sorption Studies," *Environ. Sci. Technol.*, 15(11):1360-1367 (1981).

8. Rogers, R.D., McFarlane, J.C., and A.J. Cross. "Adsorption and Desorption of Benzene in Two Soils and Montmorillonite Clay," *Environ. Sci. Technol.*, 14(4):457-461 (1980).

9. Hansch, C., and T. Fujita. "ρ-σ-π Analysis. A Method for the Correlation of Biological Activity and Chemical Structure," *J. Am. Chem. Soc.*, 86(8):1616-1617 (1964).

10. Mackay, D. "Correlation of Bioconcentration Factors," *Environ. Sci. Technol.*, 16(5):274-278 (1982).

11. Leo, A., Hansch, C., and D. Elkins. "Partition Coefficients and Their Uses," *Chem. Rev.*, 71(6):525-616 (1971).

12. Veith, G.D., Macek, K.J., Petrocelli, S.R., and J. Carroll. "An Evaluation of Using Partition Coefficients and Water Solubility to Estimate Bioconcentration Factors for Organic Chemicals in Fish," in *Aquatic Toxicology, ASTM STP 707*, Eaton, J.G., Parrish, P.R., and A.C. Hendricks, Eds. (Philadelphia, PA: American Society for Testing and Materials, 1980), pp 116-129.

13. Eadsforth, C.V. "Application of Reverse-Phase H.P.L.C. for the Determination of Partition Coefficients," *Pestic. Sci.*, 17(3):311-325 (1986).

14. "Chemical, Physical, and Biological Properties of Compounds

Present at Hazardous Waste Sites," U.S. EPA Report-530/SW-89-010 (1985), 619 p.

15. Verschueren, K. *Handbook of Environmental Data on Organic Chemicals* (New York: Van Nostrand Reinhold Co., 1983), 1310 p.

16. Chiou, C.T., Freed, V.H., Schmedding, D.W., and R.L. Kohnert. "Partition Coefficients and Bioaccumulation of Selected Organic Chemicals," *Environ. Sci. Technol.*, 11(5):475-478 (1977).

17. Howard, P.H., and P.R. Durkin. "Sources of Contamination, Ambient Levels, and Fate of Benzene in the Environment," Office of Toxic Substances, U.S. EPA Report-560/5-75-005 (1974), 73 p.

18. Meites, L., Ed. *Handbook of Analytical Chemistry*, 1st ed. (New York: McGraw-Hill, Inc., 1963), 1782 p.

19. Bohon, R.L., and W.F. Claussen. "The Solubility of Aromatic Hydrocarbons in Water," *J. Am. Chem. Soc.*, 73(4):1571-1578 (1951).

20. Banerjee, S., Yalkowsky, S.H., and S.C. Valvani. "Water Solubility and Octanol/Water Partition Coefficients of Organics. Limitations of the Solubility-Partition Coefficient Correlation," *Environ. Sci. Technol.*, 14(10):1227-1229 (1980).

21. Gross, P.M., and J.H. Saylor. "The Solubilities of Certain Slightly Soluble Organic Compounds in Water," *J. Am. Chem. Soc.*, 53(5):1744-1751 (1931).

22. McDevit, W.F., and F.A. Long. "The Activity Coefficient of Benzene in Aqueous Salt Solutions," *J. Am. Chem. Soc.*, 74:1773-1777 (1952).

23. Andrews, L.J., and R.M. Keefer. "Cation Complexes of Compounds Containing Carbon-Carbon Double Bonds. IV. The Argentation of Aromatic Hydrocarbons," *J. Am. Chem. Soc.*, 71(11):3644-3647 (1949).

24. McAuliffe, C. "Solubility in Water of Paraffin, Cycloparaffin, Olefin, Acetylene, Cycloolefin, and Aromatic Compounds," *J. Phys. Chem.*, 70(4):1267-1275 (1966).

25. May, W.E., Wasik, S.P., and D.H. Freeman. "Determination of the Solubility Behavior of Some Polycyclic Aromatic Hydrocarbons in Water," *Anal. Chem.*, 50(7):997-1000 (1978).

26. Stephen, H., and T. Stephen. *Solubilities of Inorganic and Organic Compounds - Part 1, Volume 1* (London: Pergamon Printing and Art Services, Ltd., 1963), 960 p.

27. Polak, J., and B.C.-Y. Lu. "Mutual Solubilities of Hydrocarbons and Water at 0 and 25 °C," *Can. J. Chem.*, 51(24):4018-4023 (1973).

28. Price, L.C. "Aqueous Solubility of Petroleum as Applied to its Origin and Primary Migration," *Am. Assoc. Pet. Geol. Bull.*, 60(2):213-244 (1976).

29. Freed, V.H., Chiou, C.T., and R. Haque. "Chemodynamics: Transport and Behavior of Chemicals in the Environment - A

Problem in Environmental Health," *Environ. Health Perspect.*, 20:55-70 (1977).

30. Keely, D.F., Hoffpauir, M.A., and J.R. Meriwether. "Solubility of Aromatic Hydrocarbons in Water and Sodium Chloride Solutions of Different Ionic Strengths: Benzene and Toluene," *J. Chem. Eng. Data*, 33(2):87-89 (1988).

31. Stearns, R.S., Oppenheimer, H., Simon, E., and W.D. Harkins. "Solubilization of Long-Chain Colloidal Electrolytes," *J. Chem. Phys.*, 15(7):496-507 (1947).

32. Alexander, D.M. "The Solubility of Benzene in Water," *J. Phys. Chem.*, 63(6):1021-1022 (1959).

33. Krasnoshchekova, R.Y., and M. Gubergrits. "Solubility of *n*-Alkylbenzene in Fresh and Salt Waters," [Chemical Abstracts 83(16):136583p]: *Vodn. Resur.*, 2:170-173 (1975).

34. Schwarz, F.P. "Measurement of the Solubilities of Slightly Soluble Organic Liquids in Water by Elution Chromatography," *Anal. Chem.*, 52(1):10-15 (1980).

35. Ben-Naim, A., and J. Wilf. "Solubilities and Hydrophobic Interactions in Aqueous Solutions of Monoalkylbenzene Molecules," *J. Phys. Chem.*, 84(6):583-586 (1980).

36. Weast, R.C., Ed. *CRC Handbook of Chemistry and Physics*, 67th ed. (Boca Raton, FL: CRC Press, Inc., 1986), 2406 p.

37. Huntress, E.H., and S.P. Mulliken. *Identification of Pure Organic Compounds - Tables of Data on Selected Compounds of Order I* (New York: John Wiley and Sons, Inc., 1941), 691 p.

38. Sumer, K.M., and A.R. Thompson. "Refraction, Dispersion, and Densities of Benzene, Toluene, and Xylene Mixtures," *J. Chem. Eng. Data*, 13(1):30-34 (1968).

39. Kirchnerová, J., and G.C.B. Cave. "The Solubility of Water in Low-Dielectric Solvents," *Can. J. Chem.*, 54(24):3909-3916 (1976).

40. Dagley, S. "Microbial Degradation of Stable Chemical Structures: General Features of Metabolic Pathways" in *Degradation of Synthetic Organic Molecules in the Biosphere: Natural, Pesticidal, and Various Other Man-Made Compounds* (Washington, DC: National Academy of Sciences, 1972), pp 1-16.

41. Chapman, P.J. "An Outline of Reaction Sequences Used for the Bacterial Degradation of Phenolic Compounds" in *Degradation of Synthetic Organic Molecules in the Biosphere: Natural, Pesticidal, and Various Other Man-Made Compounds* (Washington, DC: National Academy of Sciences, 1972), pp 17-55.

42. Mackay, D., and P.J. Leinonen. "Notes - Rate of Evaporation of Low-Solubility Contaminants from Water Bodies to Atmosphere," *Environ. Sci. Technol.*, 9(13):1178-1180 (1975).

43. Eon, C., Pommier, C., and G. Guiochon. "Vapor Pressures and Second Virial Coefficients of Some Five-membered Heterocyclic Derivatives," *J. Chem. Eng. Data*, 16(4):408-410 (1971).
44. Weiss, G. *Hazardous Chemicals Data Book* (Park Ridge, NJ: Noyes Data Corp., 1986), 1069 p.
45. "General Industry Standards for Toxic and Hazardous Substances," U.S. Code of Federal Regulations 1910, Subpart Z, Section 1910.1000 (July 1982).
46. *Documentation of the Threshold Limit Values and Biological Exposure Indices for 1986-1987* (Cincinnati, OH: American Conference of Governmental Industrial Hygienists, 1986), 111 p.

BENZIDINE

Synonyms: Azoic diazo component 112; Benzidine base; *p*-Benzidine; 4,4'-Bianiline; *p,p*'-Bianiline; **(1,1'-Biphenyl)-4,4'-diamine**; 4,4'-Biphenyldiamine; *p,p*'-Biphenyldiamine; 4,4'-Biphenylenediamine; *p,p*'-Biphenylenediamine; C.I. 37,225; C.I. azoic diazo component 112; 4,4'-Diaminobiphenyl; *p,p*'-Diaminobiphenyl; 4,4'-Diamino-1,1'-biphenyl; 4,4'-Diaminodiphenyl; *p*-Diaminodiphenyl; *p,p*'-Diamino-diphenyl; 4,4'-Dianiline; *p,p*'-Dianiline; 4,4'-Diphenylenediamine; *p,p*'-Diphenylenediamine; Fast corinth base B; NCI-C03361; RCRA waste number U021; UN 1885.

Structural Formula:

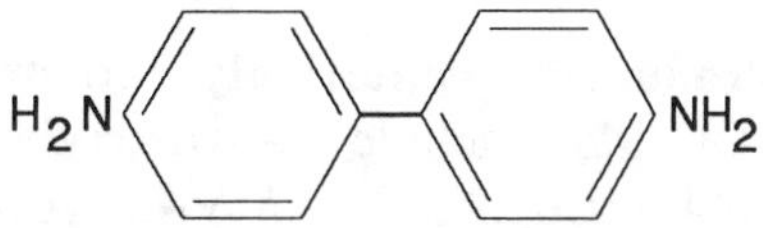

CHEMICAL DESIGNATIONS

CAS Registry Number: 92-87-5

DOT Designation: 1885

Empirical Formula: $C_{12}H_{12}N_2$

Formula Weight: 184.24

RTECS Number: DC 9625000

PHYSICAL AND CHEMICAL PROPERTIES

Appearance: Grayish-yellow powder or white to pale reddish crystals.

Boiling Point: 401.7 °C [1]; 400-401 °C at 740 mm [2].

Dissociation Constants: 3.63 (pK_1); 4.70 (pK_2) [3].

Henry's Law Constant: 3.88 x 10^{-11} atm·m^3/mol at 25 °C (estimated) [4].

Ionization Potential: No data found.

Log K_{oc}: 1.60 using method of Karickhoff and others [5].

Log K_{ow}: 1.81 [6]; 1.34 [7]; 1.63 [8].

Melting Point: 128 °C [9]; 117.2 °C [10].

Solubility in Organics: Soluble in ethanol and ether [11].

Solubility in Water: 400 mg/L at 12 °C, 9,400 mg/L at 100 °C [12]; 500 mg/L at 25 °C [13]; 400 mg/L at 25 °C [6]; 1 part in 2,447 parts water at 12 °C [2]; 360 mg/L at 24 °C [8]; 520 mg/L at 25 °C [14].

Specific Density: 1.250 at 20/4 °C [12].

Transformation Products: Tentatively identified biooxidation compounds using GC/MS include *n*-hydroxybenzidine, 3-hydroxybenzidine, 4-amino-4'-nitrophenyl, *N,N'*-dihydroxybenzidine, 3,3'-dihydroxybenzidine, and 4,4'-dinitrobiphenyl [12]. In the presence of hydrogen peroxide and acetylcholine at pH 11 and 20 °C, benzidine oxidized to 4-amino-4'-nitrobiphenyl [15].

Vapor Density: 7.50 g/L presumably at 20 °C [16].

Vapor Pressure: Using the reported specific vapor density value of 6.36, the vapor pressure was calculated to be 0.83 mm at 20 °C [16].

FIRE HAZARDS

Flash Point: Combustible solid [10].

Lower Explosive Limit (LEL): Unknown [10].

Upper Explosive Limit (UEL): Unknown [10].

HEALTH HAZARD DATA

Immediately Dangerous to Life or Health (IDLH): Human carcinogen [10].

Permissible Exposure Limits (PEL) in Air: No standards set.

MANUFACTURING

Selected Manufacturers:

Fluka Chemical Corp.
980 South Second St.
Ronkonkoma, NY 11779

Pfaltz & Bauer, Inc.
375 Fairfield Ave.
Stamford, CT 06902

Uses: Organic synthesis; manufacture of dyes, especially Congo Red; detection of blood stains; stain in microscopy; laboratory reagent; stiffening agent in rubber compounding.

REFERENCES

1. Sax, N.I. *Dangerous Properties of Industrial Materials* (New York: Van Nostrand Reinhold Co., 1984), 3124 p.
2. Standen, A., Ed. *Kirk-Othmer Encyclopedia of Chemical Technology, Volume 3*, 2nd ed. (New York: John Wiley and Sons, Inc., 1964), 927 p.
3. Dean, J.A., Ed., *Lange's Handbook of Chemistry*, 11th ed. (New York: McGraw-Hill, Inc., 1973), 1570 p.
4. Howard, P.H. *Handbook of Environmental Fate and Exposure Data for Organic Chemicals* (Chelsea, MI: Lewis Publishers, Inc., 1989), 574 p.
5. Karickhoff, S.W., Brown, D.S., and T.A. Scott. "Sorption of Hydrophobic Pollutants on Natural Sediments," *Water Res.*, 13:241-248 (1979).
6. Walton, W.C. *Practical Aspects of Ground Water Modeling* (Worthington, OH: National Water Well Association, 1985), 587 p.
7. Mabey, W.R., Smith, J.H., Podoll, R.T., Johnson, H.L., Mill, T., Chou, T.-W., Gates, J., Partridge, I.W., Jaber, H., and D. Vandenberg. "Aquatic Fate Process Data for Organic Priority Pollutants - Final Report," Office of Regulations and Standards, U.S. EPA Report-440/4-81-014 (1982), 407 p.
8. Hassett, J.J., Means, J.C., Banwart, W.L., and S.G. Wood. "Sorption Properties of Sediments and Energy-Related Pollutants," Office of Research and Development, U.S. EPA Report-600/3-80-041 (1980), 150 p.

9. Weast, R.C., Ed. *CRC Handbook of Chemistry and Physics*, 67th ed. (Boca Raton, FL: CRC Press, Inc., 1986), 2406 p.

10. *Documentation of the Threshold Limit Values and Biological Exposure Indices for 1986-1987* (Cincinnati, OH: American Conference of Governmental Industrial Hygienists, 1986), 111 p.

11. "Chemical, Physical, and Biological Properties of Compounds Present at Hazardous Waste Sites," U.S. EPA Report-530/SW-89-010 (1985), 619 p.

12. Verschueren, K. *Handbook of Environmental Data on Organic Chemicals* (New York: Van Nostrand Reinhold Co., 1983), 1310 p.

13. Morrison, R.T., and R.N. Boyd. *Organic Chemistry* (Boston: Allyn and Bacon, Inc., 1971), 1258 p.

14. Shriner, C.R., Drury, J.S., Hammons, A.S., Towill, L.E., Lewis, E.B., and D.M. Opresko. "Reviews of the Environmental Effects of Pollutants: II. Benzidine," Office of Research and Development, U.S. EPA Report-600/1-78-024 (1978), 157 p.

15. Aksnes, G., and K. Sandberg. "On the Oxidation of Benzidine and *o*-Dianisidine with Hydrogen Peroxide and Acetylcholine in Alkaline Solution," *Acta Chem. Scand.*, 11:876-880 (1957).

16. Sims, R.C., Doucette, W.C., McLean, J.E., Grenney, W.J., and R.R. Dupont. "Treatment Potential for 56 EPA Listed Hazardous Chemicals in Soil," U.S. EPA Report-600/6-88-001 (1988), 105 p.

BENZO[a]ANTHRACENE

Synonyms: BA; B(*a*)A; Benzanthracene; **Benz[*a*]anthracene**; 1,2-Benzanthracene; 1,2-Benz[*a*]anthracene; 2,3-Benzanthracene; Benzanthrene; 1,2-Benzanthrene; Benzoanthracene; 1,2-Benzoanthracene; Benzo[*a*]phenanthrene; Benzo[*b*]phenanthrene; 2,3-Benzophenanthrene; Naphthanthracene; RCRA waste number U018; Tetraphene.

Structural Formula:

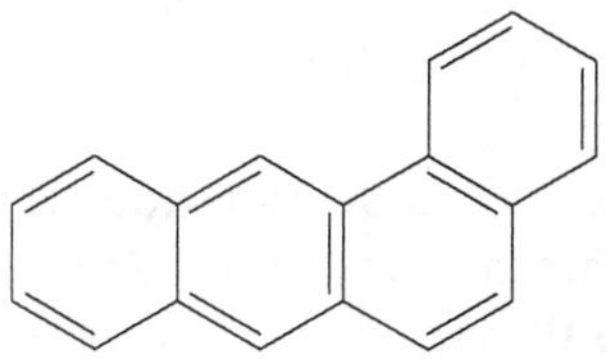

CHEMICAL DESIGNATIONS

CAS Registry Number: 56-55-3

DOT Designation: None assigned.

Empirical Formula: $C_{18}H_{12}$

Formula Weight: 228.30

RTECS Number: CV 9275000

PHYSICAL AND CHEMICAL PROPERTIES

Appearance: Colorless leaflets or plates with a greenish-yellow fluorescence.

Boiling Point: 437.6 °C [1]; 400 °C [2].

Henry's Law Constant: 6.6 x 10^{-7} atm·m^3/mol [3]; 8.0 x 10^{-6} atm·m^3/mol [4].

Ionization Potential: 8.01 eV [5]; 7.45 eV [6].

Log K_{oc}: 6.14 [3].

Log K_{ow}: 5.61 [7]; 5.91 [6]; 5.90 [8].

41

Melting Point: 162 °C [9]; 155-157 °C [10]; 158-160 °C [11].

Solubility in Organics: Soluble in ethanol, ether, acetone, and benzene [12].

Solubility in Water: 0.010 mg/L [13]; 0.014 mg/L at 25 °C [14]; 0.0094 mg/L at 25 °C, 0.0122 mg/L at 29 °C [15]; 0.011 mg/L [16]; 0.044 mg/L at 24 °C (practical grade) [17]; 5.7 ng/ml at 20 °C [18].

Specific Density: 1.274 at 20/4 °C [19].

Transformation Products: No biodegradation was observed in estuarine water nor during enrichment procedures [13]. In an enclosed marine ecosystem containing planktonic primary production and heterotrophic benthos, the major metabolites were water soluble and could not be extracted with organic solvents. The only degradation product identified was benz[*a*]anthracene-7,12-dione [20]. Under aerobic conditions, *Cunninghanella elegans* degraded benzo[*a*]anthracene to 3,4-, 8,9- and 10,11-dihydrols [21]. A strain of *Beijerinckia* was able to oxidize benzo[*a*]anthracene producing 1-hydroxy-2-anthranoic acid as the major product. Two other metabolites identified were 2-hydroxy-3-phenanthroic acid and 3-hydroxy-2-phenanthroic acid [22].

Vapor Pressure: 5×10^{-9} mm at 20 °C [23]; 1.1×10^{-7} mm at 25 °C [7]; 2.2×10^{-8} mm at 20 °C [24].

FIRE HAZARDS

Flash Point: No data found.

Lower Explosive Limit (LEL): No data found.

Upper Explosive Limit (UEL): No data found.

HEALTH HAZARD DATA

Immediately Dangerous to Life or Health (IDLH): Potential human carcinogen [25].

Permissible Exposure Limits (PEL) in Air: No standards set, however, as a constituent in coal tar pitch volatiles, the following exposure limits have been established: 0.2 mg/m^3 (benzene-soluble fraction) [26]; 0.1 mg/m^3 10-hour TWA (cyclohexane-extractable fraction) [25]; 0.2 mg/m^3 TWA (benzene solubles) [27].

MANUFACTURING

Selected Manufacturers:

> Fluka Chemical Corp.
> 980 South Second St.
> Ronkonkoma, NY 11779
>
> Pfaltz & Bauer, Inc.
> 172 East Aurora St.
> Waterbury, CT 06708

Uses: Research chemical; organic synthesis. Not manufactured commercially but is derived from industrial and experimental coal gasification operations where maximum concentrations detected in gas, liquid, and coal tar streams were 28 mg/m^3, 4.1 mg/m^3, and 18 mg/m^3, respectively [28].

REFERENCES

1. *Catalog Handbook of Fine Chemicals* (Milwaukee, WI: Aldrich Chemical Co., 1988), 2212 p.
2. Sims, R.C., Doucette, W.C., McLean, J.E., Greeney, W.J., and R.R. Dupont. "Treatment Potential for 56 EPA Listed Hazardous Chemicals in Soil," National Technical Information Service, U.S. EPA Report-600/6-88-001 (1988), 105 p.
3. Lyman, W.J., Reehl, W.F., and D.H. Rosenblatt. *Handbook of Chemical Property Estimation Methods: Environmental Behavior of Organic Compounds* (New York: McGraw-Hill, Inc., 1982).
4. Southworth, G.R. "The Role Volatilization in Removing Polycyclic Aromatic Hydrocarbons from Aquatic Environments," *Bull. Environ. Contam. Toxicol.*, 21(4/5):507-514 (1979).
5. Franklin, J.L., Dillard, J.G., Rosenstock, H.M., Herron, J.T., Draxl K., and F.H. Field. "Ionization Potentials, Appearance Potentials and Heats of Formation of Gaseous Positive Ions," National Bureau of Standards Report NSRDS-NBS 26, U.S. Government Printing Office (1969), 289 p.

6. Yoshida, K., Shigeoka, T., and F. Yamauchi. "Non-Steady State Equilibrium Model for the Preliminary Prediction of the Fate of Chemicals in the Environment," *Ecotoxicol. Environ. Safety*, 7(2):179-190 (1983).

7. Radding, S.B., Mill, T., Gould, C.W., Lia, D.H., Johnson, H.L., Bomberger, D.S., and C.V. Fojo. "The Environmental Fate of Selected Polynuclear Aromatic Hydrocarbons," Office of Toxic Substances, U.S. EPA Report-560/5-75-009 (1976), 122 p.

8. Camilleri, Patrick, Watts, S.A., and J.A. Boraston. "A Surface Area Approach to Determination of Partition Coefficients," *J. Chem. Soc. Perkin Trans. II*, (September 1988), pp 1699-1707.

9. Weast, R.C., Ed. *CRC Handbook of Chemistry and Physics*, 67th ed. (Boca Raton, FL: CRC Press, Inc., 1986), 2406 p.

10. "Treatability Manual - Volume 1: Treatability Data," Office of Research and Development, U.S. EPA Report-600/8-80-042a (1980), 1035 p.

11. *Fluka Catalog 1988/89 - Chemika-Biochemika* (Ronkonkoma, NY: Fluka Chemical Corp., 1988), 1536 p.

12. "Chemical, Physical, and Biological Properties of Compounds Present at Hazardous Waste Sites," U.S. EPA Report-530/SW-89-010 (1985), 619 p.

13. Verschueren, K. *Handbook of Environmental Data on Organic Chemicals* (New York: Van Nostrand Reinhold Co., 1983), 1310 p.

14. Mackay, D., and W.Y. Shiu. "Aqueous Solubility of Polynuclear Aromatic Hydrocarbons," *J. Chem. Eng. Data*, 22(4):399-402 (1977).

15. May, W.E., Wasik, S.P., and D.H. Freeman. "Determination of the Solubility Behavior of Some Polycyclic Aromatic Hydrocarbons in Water," *Anal. Chem.*, 50(7):997-1000 (1978).

16. Davis, W.W., Krahl, M.E., and G.H.A. Clowes. "Solubility of Carcinogenic and Related Hydrocarbons in Water," *J. Am. Chem. Soc.*, 64(1):108-110 (1942).

17. Hollifield, H.C. "Rapid Nephelometric Estimate of Water Solubility of Highly Insoluble Organic Chemicals of Environmental Interest," *Bull. Environ. Contam. Toxicol.*, 23(4/5):579-586 (1979).

18. Smith, J.H., Mabey, W.R., Bohonos, N., Holt, B.R., Lee, S.S., Chou, T.-W., Bomberger, D.C., and T. Mill. "Environmental Pathways of Selected Chemicals in Freshwater Systems. Part II: Laboratory Studies," U.S. EPA Report-600/7-78-074 (1978), 406 p.

19. Hazardous Substances Data Bank. Benzo[*a*]pyrene, National Library of Medicine, Toxicology Information Program (1987).

20. Hinga, K.R., and M.E.Q. Pilson. "Persistence of Benz[*a*]anthracene Degradation Products in an Enclosed Marine Ecosystem," *Environ. Sci. Technol.*, 21(7):648-653 (1987).

21. Kobayashi, H., and B.E. Rittman. "Microbial Removal of Hazardous Organic Compounds," *Environ. Sci. Technol.*, 16(3):170A-183A (1982).

22. Mahaffey, W.R., Gibson, D.T., and C.E. Cerniglia. "Bacterial Oxidation of Chemical Carcinogens: Formation of Polycyclic Aromatic Acids from Benz[a]anthracene," *Appl. Environ. Microbiol.*, 54(10):2415-2423 (1988).

23. Pupp, C., Lao, R.C., Murray, J.J., and R.F. Pottie. "Equilibrium Vapor Concentrations of Some Polycyclic Hydrocarbons, Arsenic Trioxide (As_4O_6) and Selenium Dioxide and the Collection Efficiencies of these Air Pollutants," *Atmos. Environ.*, 8:915-925 (1974).

24. Mabey, W.R., Smith, J.H., Podoll, R.T., Johnson, H.L., Mill, T., Chou, T.-W., Gates, J., Partridge, I.W., Jaber, H., and D. Vandenberg. "Aquatic Fate Process Data for Organic Priority Pollutants - Final Report," Office of Regulations and Standards, U.S. EPA Report-440/4-81-014 (1982), 407 p.

25. "NIOSH Pocket Guide to Chemical Hazards," U.S. Department of Health and Human Services, U.S. Government Printing Office (1987), 241 p.

26. "General Industry Standards for Toxic and Hazardous Substances," U.S. Code of Federal Regulations 1910, Subpart Z, Section 1910.1000 (July 1982).

27. *Documentation of the Threshold Limit Values and Biological Exposure Indices for 1986-1987* (Cincinnati, OH: American Conference of Governmental Industrial Hygienists, 1986), 111 p.

28. Cleland, J.G. "Project Summary - Environmental Hazard Rankings of Pollutants Generated in Coal Gasification Processes," Office of Research and Development, U.S. EPA Report-600/S7-81-101 (1981), 19 p.

BENZO[b]FLUORANTHENE

Synonyms: **Benz[e]acephenanthrylene**; 3,4-Benz[e]acephenanthrylene; 2,3-Benzfluoranthene; 3,4-Benzfluoranthene; Benzo[b]fluoranthene; Benzo[e]fluoranthene; 2,3-Benzofluoranthene; 3,4-Benzofluoranthene; 3,4-Benzo[b]fluoranthene; B(b)F.

Structural Formula:

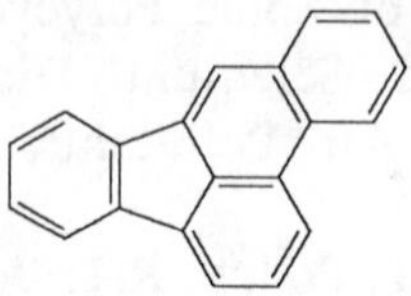

CHEMICAL DESIGNATIONS

CAS Registry Number: 205-99-2

DOT Designation: None assigned.

Empirical Formula: $C_{20}H_{12}$

Formula Weight: 252.32

RTECS Number: CU 1400000

PHYSICAL AND CHEMICAL PROPERTIES

Appearance: Solid.

Boiling Point: No data found.

Henry's Law Constant: 1.2×10^{-5} atm·m^3/mol at 20-25 °C (calculated).

Ionization Potential: No data found.

Log K$_{oc}$: 5.74 [1].

Log K$_{ow}$: 6.57 [2].

Melting Point: 168 °C [3]; 163-165 °C [4].

Solubility in Organics: Soluble in most solvents [5].

Solubility in Water: 0.014 mg/L [1]; 0.0012 mg/L at 25 °C [6].

Specific Density: No data found.

Transformation Products: No data found.

Vapor Pressure: 5×10^{-7} mm at 20 °C [1].

FIRE HAZARDS

Flash Point: No data found.

Lower Explosive Limit (LEL): No data found.

Upper Explosive Limit (UEL): No data found.

HEALTH HAZARD DATA

Immediately Dangerous to Life or Health (IDLH): Potential human carcinogen [7].

Permissible Exposure Limits (PEL) in Air: No individual standards have been set, however, as a constituent in coal tar pitch volatiles, the following exposure limits have been established: 0.2 mg/m^3 (benzene-soluble fraction) [8]; 0.1 mg/m^3 10-hour TWA (cyclohexane-extractable fraction) [7]; 0.2 mg/m^3 TWA (benzene solubles) [9].

MANUFACTURING

Use: Research chemical. Derived from industrial and experimental coal gasification operations where maximum concentrations detected in gas, liquid, and coal tar streams were 0.38 mg/m^3, 0.033 mg/m^3, and 3.2 mg/m^3, respectively [10].

REFERENCES

1. "Aquatic Fate Process Data for Organic Priority Pollutants," Office of Water Regulations and Standards, U.S. EPA Report-440/4-81-014

(1982), 407 p.

2. Walton, W.C. *Practical Aspects of Ground Water Modeling* (Worthington, OH: National Water Well Association, 1985), 587 p.

3. Weast, R.C., Ed. *CRC Handbook of Chemistry and Physics*, 67th ed. (Boca Raton, FL: CRC Press, Inc., 1986), 2406 p.

4. *Catalog Handbook of Fine Chemicals* (Milwaukee, WI: Aldrich Chemical Co., 1988), 2212 p.

5. "Chemical, Physical, and Biological Properties of Compounds Present at Hazardous Waste Sites," U.S. EPA Report-530/SW-89-010 (1985), 619 p.

6. "Treatability Manual - Volume 1: Treatability Data," Office of Research and Development, U.S. EPA Report-600/8-80-042a (1980), 1035 p.

7. "NIOSH Pocket Guide to Chemical Hazards," U.S. Department of Health and Human Services, U.S. Government Printing Office (1987), 241 p.

8. "General Industry Standards for Toxic and Hazardous Substances," U.S. Code of Federal Regulations 1910, Subpart Z, Section 1910.1000 (July 1982).

9. *Documentation of the Threshold Limit Values and Biological Exposure Indices for 1986-1987* (Cincinnati, OH: American Conference of Governmental Industrial Hygienists, 1986), 111 p.

10. Cleland, J.G. "Project Summary - Environmental Hazard Rankings of Pollutants Generated in Coal Gasification Processes," Office of Research and Development, U.S. EPA Report-600/S7-81-101 (1981), 19 p.

BENZO[k]FLUORANTHENE

Synonyms: 8,9-Benzfluoranthene; 8,9-Benzofluoranthene; 11,12-Benzofluoranthene; 11,12-Benzo[k]fluoranthene; B(k)F; 2,3,1',8'-Binaphthylene; Dibenzo[b,jk]fluorene.

Structural Formula:

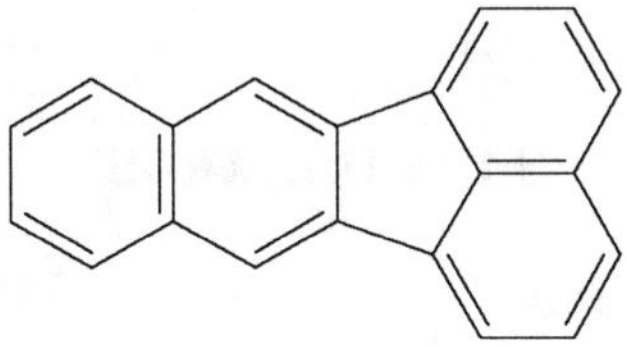

CHEMICAL DESIGNATIONS

CAS Registry Number: 207-08-9

DOT Designation: None assigned.

Empirical Formula: $C_{20}H_{12}$

Formula Weight: 252.32

RTECS Number: DF 6350000

PHYSICAL AND CHEMICAL PROPERTIES

Appearance: Solid.

Boiling Point: 480 °C [1].

Henry's Law Constant: 0.00104 atm·m^3/mol (calculated) [2].

Ionization Potential: No data found.

Log K$_{oc}$: 6.64 using method of Karickhoff and others [3].

Log K$_{ow}$: 6.85 [4].

Melting Point: 217 °C [1].

Solubility in Organics: Soluble in most solvents [5].

Solubility in Water: 0.00055 mg/L at 25 °C [6].

Specific Density: No data found.

Transformation Products: No data found.

Vapor Pressure: 9.59 x 10^{-11} mm at 25 °C [7].

FIRE HAZARDS

Flash Point: No data found.

Lower Explosive Limit (LEL): No data found.

Upper Explosive Limit (UEL): No data found.

HEALTH HAZARD DATA

Immediately Dangerous to Life or Health (IDLH): Potential human carcinogen [8].

Permissible Exposure Limits (PEL) in Air: No individual standards have been set, however, as a constituent in coal tar pitch volatiles, the following exposure limits have been established: 0.2 mg/m^3 (benzene-soluble fraction) [9]; 0.1 mg/m^3 10-hour TWA (cyclohexane-extractable fraction) [8]; 0.2 mg/m^3 TWA (benzene solubles) [10].

MANUFACTURING

Use: Research chemical. Derived from industrial and experimental coal gasification operations where maximum concentrations detected in liquid and coal tar streams were 0.017 mg/m^3 and 1.6 mg/m^3, respectively [11].

REFERENCES

1. Weast, R.C., Ed. *CRC Handbook of Chemistry and Physics*, 67th ed. (Boca Raton, FL: CRC Press, Inc., 1986), 2406 p.
2. "Treatability Manual - Volume 1: Treatability Data," Office of

Research and Development, U.S. EPA Report-600/8-80-042a (1980), 1035 p.

3. Karickhoff, S.W., Brown, D.S., and T.A. Scott. "Sorption of Hydrophobic Pollutants on Natural Sediments," *Water Res.*, 13:241-248 (1979).

4. Mills, W.B., Porcella, D.B., Ungs, M.J., Gherini, S.A., Summers, K.V., Mok, L., Rupp, G.L., and G.L. Bowie. "Water Quality Assessment: A Screening Procedure for Toxic and Conventional Pollutants in Surface and Groundwater-Part I," Office of Research and Development, U.S. EPA Report-600/6-85-002a (1985), 638 p.

5. "Chemical, Physical, and Biological Properties of Compounds Present at Hazardous Waste Sites," U.S. EPA Report-530/SW-89-010 (1985), 619 p.

6. Walton, W.C. *Practical Aspects of Ground Water Modeling* (Worthington, OH: National Water Well Association, 1985), 587 p.

7. Radding, S.B., Mill, T., Gould, C.W., Lia, D.H., Johnson, H.L., Bomberger, D.S., and C.V. Fojo. "The Environmental Fate of Selected Polynuclear Aromatic Hydrocarbons," Office of Toxic Substances, U.S. EPA Report-560/5-75-009 (1976), 122 p.

8. "NIOSH Pocket Guide to Chemical Hazards," U.S. Department of Health and Human Services, U.S. Government Printing Office (1987), 241 p.

9. "General Industry Standards for Toxic and Hazardous Substances," U.S. Code of Federal Regulations 1910, Subpart Z, Section 1910.1000 (July 1982).

10. *Documentation of the Threshold Limit Values and Biological Exposure Indices for 1986-1987* (Cincinnati, OH: American Conference of Governmental Industrial Hygienists, 1986), 111 p.

11. Cleland, J.G. "Project Summary - Environmental Hazard Rankings of Pollutants Generated in Coal Gasification Processes," Office of Research and Development, U.S. EPA Report-600/S7-81-101 (1981), 19 p.

BENZOIC ACID

Synonyms: Benzenecarboxylic acid; Benzeneformic acid; Benzene-methanoic acid; Benzoate; Carboxybenzene; Dracylic acid; NA 9,094; Phenylcarboxylic acid; Phenylformic acid; Retarder BA; Retardex; Salvo liquid; Salvo powder; Tenn-plas.

Structural Formula:

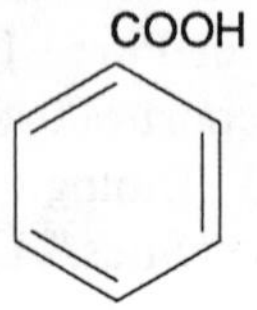

CHEMICAL DESIGNATIONS

CAS Registry Number: 65-85-0

DOT Designation: 9094

Empirical Formula: $C_7H_6O_2$

Formula Weight: 122.12

RTECS Number: DG 0875000

PHYSICAL AND CHEMICAL PROPERTIES

Appearance and Odor: Colorless to white needles, scales, or powder with a faint benzoin or benzaldehyde-like odor.

Boiling Point: 249.2 °C [1].

Dissociation Constant: 4.21 [2].

Henry's Law Constant: 7.02 x 10^{-8} atm·m^3/mol (calculated) [3].

Ionization Potential: 9.73 ± 0.09 eV [4].

Log K_{oc}: 1.48-2.70 with an average value of 2.26 [5].

Log K_{ow}: 1.87 [6]; 2.03 [7]; 1.81-1.88 [8].

52

Melting Point: 122.13 °C [1]; 122.375 [9].

Solubility in Organics: Soluble in acetone, ethanol, benzene, chloroform, ether [1], carbon tetrachloride, carbon disulfide, and turpentine [10].

Solubility in Water: 3,400 mg/L [11]; 2,900 mg/L at 20 °C [3]; 3,000 mg/L at 18 °C, 3,200 mg/L at 23.5 °C, 3,400 mg/L at 25 °C [12]; 345 parts of at 20 °C, 17 parts at 100 °C [13]; 1,700 mg/L at 0 °C, 2,100 mg/L at 10 °C, 2,900 mg/L at 20 °C, 3,400 mg/L at 25 °C, 4,200 mg/L at 30 °C, 6,000 mg/L at 40 °C, 8,500 mg/L at 50 °C, 12,000 mg/L at 60 °C, 17,700 mg/L at 70 °C, 27,500 mg/L at 80 °C, 45,500 mg/L at 90 °C, 68,000 mg/L at 95 °C [9]; 1,800 mg/L at 4 °C, 2,700 mg/L at 18 °C, 22,000 mg/L at 75 °C [14].

Specific Density: 1.2659 at 15/4 °C, 1.0749 at 130/4 °C [1]; 1.316 at 28/4 °C [15]; 1.316 at 28/4 °C, 1.0819 at 122.375/4 °C, 1.05218 at 155/4 °C, 1.02942 at 180/4 °C [9].

Transformation Products: Benzoic acid may degrade to catechol if it is the central metabolite whereas if protocatechuic acid (3,4-dihydroxybenzoic acid) is the central metabolite, the precursor is *m*-hydroxybenzoic acid [16]. Other compounds identified following degradation of benzoic acid to catechol include *cis,cis*-muconic acid, (+)-muconolactone, 3-oxoadipate enol lactone, and 3-oxoadipate [17].

Vapor Pressure: 1 mm at 96 °C, 10 mm at 132.1 °C, 40 mm at 162.6 °C, 100 mm at 186.2 °C, 400 mm at 227.0 °C, 760 mm at 249.2 °C [1]; 20 mm at 146.7 °C, 60 mm at 172.8 °C, 200 mm at 205.9 °C [9]; 0.0045 mm at 25 °C [18].

FIRE HAZARDS

Flash Point: 121 °C [19]; 121-131 °C [9].

Lower Explosive Limit (LEL): Unknown [20].

Upper Explosive Limit (UEL): Unknown [20].

HEALTH HAZARD DATA

Immediately Dangerous to Life or Health (IDLH): No data found.

Permissible Exposure Limits (PEL) in Air: No standards set.

MANUFACTURING

Selected Manufacturers:

> Monsanto Industrial Chemicals Co.
> 800 North Lindbergh Blvd.
> St. Louis, MO 63166

> Tenneco Chemical Inc.
> Tenneco Intermediates Division
> Piscataway, NJ 08854

Uses: Preparation of sodium and butyl benzoates, benzoyl chloride, phenol, caprolactum, and esters for perfume and flavor industry; plasticizers; manufacture of alkyl resins; preservative for food, fats, and fatty oils; seasoning; tobacco; dentifrices; standard in analytical chemistry; antifungal agent; synthetic resins and coatings; pharmaceutical and cosmetic preparations; plasticizer manufacturing (to modify resins such as polyvinyl chloride, polyvinyl acetate, phenol-formaldehyde).

REFERENCES

1. Weast, R.C., Ed. *CRC Handbook of Chemistry and Physics*, 67th ed. (Boca Raton, FL: CRC Press, Inc., 1986), 2406 p.
2. Dean, J.A., Ed., *Lange's Handbook of Chemistry*, 11th ed. (New York: McGraw-Hill, Inc., 1973), 1570 p.
3. "Treatability Manual - Volume 1: Treatability Data," Office of Research and Development, U.S. EPA Report-600/8-80-042a (1980), 1035 p.
4. Franklin, J.L., Dillard, J.G., Rosenstock, H.M., Herron, J.T., Draxl K., and F.H. Field. "Ionization Potentials, Appearance Potentials and Heats of Formation of Gaseous Positive Ions," National Bureau of Standards Report NSRDS-NBS 26, U.S. Government Printing Office (1969), 289 p.
5. Løkke, H. "Sorption of Selected Organic Pollutants in Danish Soils," *Ecotoxicol. Environ. Safety*, 8(5):395-409 (1984).
6. Leo, A., Hansch, C., and D. Elkins. "Partition Coefficients and Their Uses," *Chem. Rev.*, 71(6):525-616 (1971).

7. Lu, P.-Y., and R.L. Metcalf. "Environmental Fate and Biodegradability of Benzene Derivatives as Studied in a Model Ecosystem," *Environ. Health Perspect.*, 10:269-284 (1975).
8. Fujita, T., Iwasa, J., and C. Hansch. "A New Substituent Constant, π, Derived from Partition Coefficients," *J. Am. Chem. Soc.*, 86(23):5175-5180 (1964).
9. Standen, A., Ed. *Kirk-Othmer Encyclopedia of Chemical Technology, Volume 3*, 2nd ed. (New York: John Wiley and Sons, Inc., 1964), 927 p.
10. Yoshida, K., Shigeoka, T., and F. Yamauchi. "Non-Steady-State Equilibrium Model for the Preliminary Prediction of the Fate of Chemicals in the Environment," *Ecotoxicol. Environ. Safety*, 7(2):179-190 (1983).
11. Morrison, R.T., and R.N. Boyd. *Organic Chemistry* (Boston: Allyn and Bacon, Inc., 1971), 1258 p.
12. Stephen, H., and T. Stephen. *Solubilities of Inorganic and Organic Compounds - Part 1, Volume 1* (London: Pergamon Printing and Art Services, Ltd., 1963), 960 p.
13. Huntress, E.H., and S.P. Mulliken. *Identification of Pure Organic Compounds - Tables of Data on Selected Compounds of Order I* (New York: John Wiley and Sons, Inc., 1941), 691 p.
14. Hodgman, C.D., Weast, R.C., Shankland, R.S., and S.M. Selby. *Handbook of Chemistry and Physics* (Cleveland, OH: Chemical Rubber Publishing Co., 1961).
15. Weiss, G. *Hazardous Chemicals Data Book* (Park Ridge, NJ: Noyes Data Corp., 1986), 1069 p.
16. Chapman, P.J. "An Outline of Reaction Sequences Used for the Bacterial Degradation of Phenolic Compounds" in *Degradation of Synthetic Organic Molecules in the Biosphere: Natural, Pesticidal, and Various Other Man-Made Compounds* (Washington, DC: National Academy of Sciences, 1972), pp 17-55.
17. Verschueren, K. *Handbook of Environmental Data on Organic Chemicals* (New York: Van Nostrand Reinhold Co., 1983), 1310 p.
18. Howard, P.H. *Handbook of Environmental Fate and Exposure Data for Organic Chemicals* (Chelsea, MI: Lewis Publishers, Inc., 1989), 574 p.
19. *Catalog Handbook of Fine Chemicals* (Milwaukee, WI: Aldrich Chemical Co., 1988), 2212 p.
20. "NIOSH Pocket Guide to Chemical Hazards," U.S. Department of Health and Human Services, U.S. Government Printing Office (1987), 241 p.

BENZO[ghi]PERYLENE

Synonyms: 1,12-Benzoperylene; 1,12-Benzperylene; B(*ghi*)P.

Structural Formula:

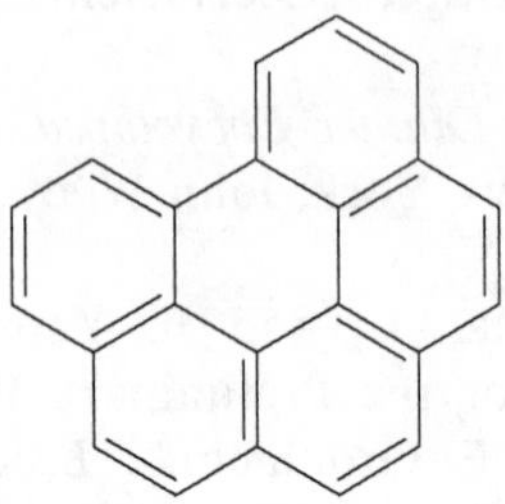

CHEMICAL DESIGNATIONS

CAS Registry Number: 191-24-2

DOT Designation: None assigned.

Empirical Formula: $C_{22}H_{12}$

Formula Weight: 276.34

RTECS Number: DI 6200500

PHYSICAL AND CHEMICAL PROPERTIES

Appearance: Solid.

Boiling Point: > 500 °C [1].

Henry's Law Constant: 1.4×10^{-7} atm·m^3/mol at 25 °C (calculated).

Ionization Potential: 7.24 eV [2].

Log K_{oc}: 6.89 using method of Karickhoff and others [3].

Log K_{ow}: 7.10 [4].

Melting Point: 222 °C [5]; 276.8 °C [6]; 278-280 °C [7].

Solubility in Organics: Soluble in most solvents [8].

Solubility in Water: 0.00026 mg/L at 25 °C [9].

Specific Density: No data found.

Transformation Products: No data found.

Vapor Pressure: 1.01 x 10^{-10} mm at 25 °C [10].

FIRE HAZARDS

Flash Point: No data found.

Lower Explosive Limit (LEL): No data found.

Upper Explosive Limit (UEL): No data found.

HEALTH HAZARD DATA

Immediately Dangerous to Life or Health (IDLH): Potential human carcinogen [11].

Permissible Exposure Limits (PEL) in Air: No individual standards have been set, however, as a constituent in coal tar pitch volatiles, the following exposure limits have been established: 0.2 mg/m^3 (benzene-soluble fraction) [12]; 0.1 mg/m^3 10-hour TWA (cyclohexane-extractable fraction) [11]; 0.2 mg/m^3 TWA (benzene solubles) [13].

MANUFACTURING

Selected Manufacturers:

Fluka Chemical Corp.
980 South Second St.
Ronkonkoma, NY 11779

Pfaltz & Bauer, Inc.
172 East Aurora St.
Waterbury, CT 06708

Use: Research chemical. Derived from industrial and experimental coal gasification operations where the maximum concentration detected in coal tar streams was 2.7 mg/m^3 [14].

REFERENCES

1. *Catalog Handbook of Fine Chemicals* (Milwaukee, WI: Aldrich Chemical Co., 1988), 2212 p.
2. Franklin, J.L., Dillard, J.G., Rosenstock, H.M., Herron, J.T., Draxl K., and F.H. Field. "Ionization Potentials, Appearance Potentials and Heats of Formation of Gaseous Positive Ions," National Bureau of Standards Report NSRDS-NBS 26, U.S. Government Printing Office (1969), 289 p.
3. Karickhoff, S.W., Brown, D.S., and T.A. Scott. "Sorption of Hydrophobic Pollutants on Natural Sediments," *Water Res.,* 13:241-248 (1979).
4. Mackay, D., Bobra, A., Shiu, W.Y., and S.H. Yalkowsky. "Relationships Between Aqueous Solubility and Octanol-Water Partition Coefficients," *Chemosphere*, 9:701-711 (1980).
5. Cleland, J.G., and G.L. Kingsbury. "Multimedia Environmental Goals for Environmental Assessment, Volume II. MEG charts and Background Information," Office of Research and Development, U.S. EPA Report-600/7-77-136b (1977), 454 p.
6. Lyman, W.J., Reehl, W.F., and D.H. Rosenblatt. *Handbook of Chemical Property Estimation Methods: Environmental Behavior of Organic Compounds* (New York: McGraw-Hill, Inc., 1982).
7. *Fluka Catalog 1988/89 - Chemika-Biochemika* (Ronkonkoma, NY: Fluka Chemical Corp., 1988), 1536 p.
8. "Chemical, Physical, and Biological Properties of Compounds Present at Hazardous Waste Sites," U.S. EPA Report-530/SW-89-010 (1985), 619 p.
9. Mackay, D., and W.Y. Shiu. "Aqueous Solubility of Polynuclear Aromatic Hydrocarbons," *J. Chem. Eng. Data*, 22(4):399-402 (1977).
10. Radding, S.B., Mill, T., Gould, C.W., Lia, D.H., Johnson, H.L., Bomberger, D.S., and C.V. Fojo. "The Environmental Fate of Selected Polynuclear Aromatic Hydrocarbons," Office of Toxic Substances, U.S. EPA Report-560/5-75-009 (1976), 122 p.
11. "NIOSH Pocket Guide to Chemical Hazards," U.S. Department of Health and Human Services, U.S. Government Printing Office (1987), 241 p.
12. "General Industry Standards for Toxic and Hazardous Substances," U.S. Code of Federal Regulations 1910, Subpart Z, Section 1910.1000 (July 1982).

13. *Documentation of the Threshold Limit Values and Biological Exposure Indices for 1986-1987* (Cincinnati, OH: American Conference of Governmental Industrial Hygienists, 1986), 111 p.

14. Cleland, J.G. "Project Summary - Environmental Hazard Rankings of Pollutants Generated in Coal Gasification Processes," Office of Research and Development, U.S. EPA Report-600/S7-81-101 (1981), 19 p.

BENZO[a]PYRENE

Synonyms: Benzo[*d,e,f*]chrysene; 1,2-Benzopyrene; 3,4-Benzopyrene; 6,7-Benzopyrene; Benz[*a*]pyrene; Benzo[alpha]pyrene; 1,2-Benzpyrene; 3,4-Benzpyrene; 3,4-Benz[*a*]pyrene; 3,4-Benzypyrene; BP; B(*a*)P; 3,4-BP; RCRA waste number U022.

Structural Formula:

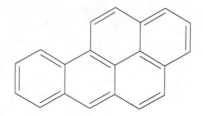

CHEMICAL DESIGNATIONS

CAS Registry Number: 50-32-8

DOT Designation: None assigned.

Empirical Formula: $C_{20}H_{12}$

Formula Weight: 252.32

RTECS Number: DJ 3675000

PHYSICAL AND CHEMICAL PROPERTIES

Appearance and Odor: Odorless, yellow crystals.

Boiling Point: 495 °C [1].

Henry's Law Constant: < 2.4 x 10^{-6} atm·m^3/mol [2].

Ionization Potential: No data found.

Log K_{oc}: 5.60-6.29 using method of Karickhoff and others [3].

Log K_{ow}: 5.99 [4]; 6.04 [5]; 6.50 [6]; 5.81 [7].

Melting Point: 179-179.3 °C [8]; 174-177 °C [9].

Solubility in Organics: Soluble in most solvents [10].

Solubility in Water: 0.003 mg/L, 0.005-0.010 mg/L in seawater at 22 °C [11]; 0.0038 mg/L at 25 °C [12]; 0.004 mg/L at 25 °C [13]; 1.2 ng/L at 22 °C [14].

Specific Density: 1.351 [15].

Transformation Products: Benzo[*a*]pyrene was biooxidized by *Beijerinckia* B836 to *cis*-9,10-dihydroxy-9,10-dihydrobenzo[*a*]pyrene. Under nonenzymatic conditions, this metabolite monodehydroxylated to form 9-hydroxybenzo[*a*]pyrene [11]. Under aerobic conditions, *Cunninghanella elegans* degraded benzo[*a*]pyrene to *trans*-7,8-dihydroxy-7,8-dihydrobenzo[*a*]pyrene [16]. Coated glass fibers exposed to air containing 0.1-0.2 ppm ozone yielded the metabolite benzo[*a*]pyrene-4,5-oxide. At 0.2 ppm ozone, conversion yields of 50% and 80% were observed after one and four hours, respectively [17].

Vapor Pressure: 5.49×10^{-9} mm at 25 °C [5]; 5.6×10^{-9} mm at 25 °C [18]; 5×10^{-9} mm at 25 °C [14]; 5.0×10^{-7} mm at 20 °C [19].

FIRE HAZARDS

Flash Point: No data found.

Lower Explosive Limit (LEL): No data found.

Upper Explosive Limit (UEL): No data found.

HEALTH HAZARD DATA

Immediately Dangerous to Life or Health (IDLH): Potential human carcinogen [20].

Permissible Exposure Limits (PEL) in Air: No individual standards have been set, however, as a constituent in coal tar pitch volatiles, the following exposure limits have been established: 0.2 mg/m^3 (benzene-soluble fraction) [21]; 0.1 mg/m^3 10-hour TWA (cyclohexane-extractable fraction) [20]; 0.2 mg/m^3 TWA (benzene solubles) [22].

MANUFACTURING

Selected Manufacturers:

> Fluka Chemical Corp.
> 980 South Second St.
> Ronkonkoma, NY 11779
>
> Pfaltz & Bauer, Inc.
> 172 East Aurora St.
> Waterbury, CT 06708

Use: Research chemical. Derived from industrial and experimental coal gasification operations where maximum concentrations detected in gas, liquid, and coal tar streams were 5.0 mg/m^3, 0.036 mg/m^3, and 3.5 mg/m^3, respectively [23].

REFERENCES

1. *Catalog Handbook of Fine Chemicals* (Milwaukee, WI: Aldrich Chemical Co., 1988), 2212 p.
2. Southworth, G.R. "The Role of Volatilization in Removing Polycyclic Aromatic Hydrocarbons from Aquatic Environments," *Bull. Environ. Contam. Toxicol.*, 21(4/5):507-514 (1979).
3. Karickhoff, S.W., Brown, D.S., and T.A. Scott. "Sorption of Hydrophobic Pollutants on Natural Sediments," *Water Res.*, 13:241-248 (1979).
4. Mallon, B.J., and F.L. Harrison. "Octanol-Water Partition Coefficient of Benzo(a)pyrene: Measurement, Calculation and Environmental Implications," *Bull. Environ. Contam. Toxicol.*, 32(3):316-323 (1984).
5. Radding, S.B., Mill, T., Gould, C.W., Lia, D.H., Johnson, H.L., Bomberger, D.S., and C.V. Fojo. "The Environmental Fate of Selected Polynuclear Aromatic Hydrocarbons," Office of Toxic Substances, U.S. EPA Report-560/5-75-009 (1976), 122 p.
6. Landrum, P.F., Nihart, S.R., Eadie, B.J., and W.S. Gardner. "Reverse-Phase Separation Method for Determining Pollutant Binding to Aldrich Humic Acid and Dissolved Organic Carbon of Natural Waters," *Environ. Sci. Technol.*, 19(3):187-192 (1984).
7. Zepp, R.G., and P.F. Scholtzhauer. "Photoreactivity of Selected Aromatic Hydrocarbons in Water," in *Polynuclear Aromatic Hydrocarbons, 3rd International Symposium on Chemistry, Biology,*

Carcinogenesis and Mutagenesis, Jones, P.W., and P. Leber, Eds. (Ann Arbor, MI: Ann Arbor Science Publishers, Inc., 1979), pp 141-158.

8. Weast, R.C., Ed. *CRC Handbook of Chemistry and Physics*, 67th ed. (Boca Raton, FL: CRC Press, Inc., 1986), 2406 p.

9. *Fluka Catalog 1988/89 - Chemika-Biochemika* (Ronkonkoma, NY: Fluka Chemical Corp., 1988), 1536 p.

10. "Chemical, Physical, and Biological Properties of Compounds Present at Hazardous Waste Sites," U.S. EPA Report-530/SW-89-010 (1985), 619 p.

11. Verschueren, K. *Handbook of Environmental Data on Organic Chemicals* (New York: Van Nostrand Reinhold Co., 1983), 1310 p.

12. Mackay, D., and W.Y. Shiu. "Aqueous Solubility of Polynuclear Aromatic Hydrocarbons," *J. Chem. Eng. Data*, 22(4):399-402 (1977).

13. Schwarz, F.P., and S.P. Wasik. "Fluorescence Measurements of Benzene, Naphthalene, Anthracene, Pyrene, Fluoranthene, and Benzo[e]pyrene in Water," *Anal. Chem.*, 48(3):524-528 (1976).

14. Smith, J.H., Mabey, W.R., Bohonos, N., Holt, B.R., Lee, S.S., Chou, T.-W., Bomberger, D.C., and T. Mill. "Environmental Pathways of Selected Chemicals in Freshwater Systems. Part II: Laboratory Studies," U.S. EPA Report-600/7-78-074 (1978), 406 p.

15. Kronberger, H., and J. Weiss. "Formation and Structure of Some Organic Molecular Compounds. III. The Dielectric Polarization of Some Solid Crystalline Molecular Compounds," *J. Chem. Soc. (London)*, (1944), pp 464-469.

16. Kobayashi, H., and B.E. Rittman. "Microbial Removal of Hazardous Organic Compounds," *Environ. Sci. Technol.*, 16(3):170A-183A (1982).

17. Pitts, J.N. Jr., Lokensgard, D.M., Ripley, P.S., van Cauwenberghe, K.A., van Vaeck, L., Schaffer, S.D., Thill, A.J., and W.L. Belser Jr. "'Atmospheric' Epoxidation of Benzo[a]pyrene by Ozone: Formation of the Metabolite Benzo[a]pyrene-4,5-oxide," *Science*, 210:1347-1349 (1980).

18. Murray, J.M., Pottie, R.F., and C. Pupp. "The Vapor Pressures and Enthalpies of Sublimation of Five Polycyclic Aromatic Hydrocarbons," *Can. J. Chem.* 52(4):557-563 (1974).

19. Sims, R.C., Doucette, W.C., McLean, J.E., Greeney, W.J., and R.R. Dupont. "Treatment Potential for 56 EPA Listed Hazardous Chemicals in Soil," National Technical Information Service, U.S. EPA Report-600/6-88-001 (1988), 105 p.

20. "NIOSH Pocket Guide to Chemical Hazards," U.S. Department of Health and Human Services, U.S. Government Printing Office (1987), 241 p.

21. "General Industry Standards for Toxic and Hazardous Substances," U.S. Code of Federal Regulations 1910, Subpart Z, Section

1910.1000 (July 1982).

22. *Documentation of the Threshold Limit Values and Biological Exposure Indices for 1986-1987* (Cincinnati, OH: American Conference of Governmental Industrial Hygienists, 1986), 111 p.

23. Cleland, J.G. "Project Summary - Environmental Hazard Rankings of Pollutants Generated in Coal Gasification Processes," Office of Research and Development, U.S. EPA Report-600/S7-81-101 (1981), 19 p.

BENZYL ALCOHOL

Synonyms: Benzal alcohol; Benzene carbinol; **Benzenemethanol**; Benzoyl alcohol; α-Hydroxy toluene; NCI-C06111; Phenol carbinol; Phenyl carbinol; Phenyl methanol; Phenyl methyl alcohol; α-Toluenol.

Structural Formula:

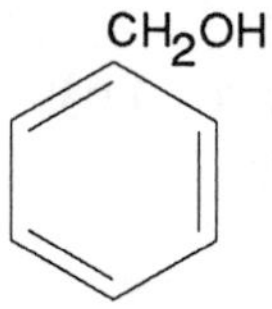

CH₂OH

CHEMICAL DESIGNATIONS

CAS Registry Number: 100-51-6

DOT Designation: None assigned.

Empirical Formula: C_7H_8O

Formula Weight: 108.14

RTECS Number: DN 3150000

PHYSICAL AND CHEMICAL PROPERTIES

Appearance and Odor: Colorless liquid with a faint, pleasant aromatic odor.

Boiling Point: 205.3 °C [1].

Henry's Law Constant: Insufficient vapor pressure data for calculation at 25 °C.

Ionization Potential: 9.14 ± 0.05 eV [2].

Log K_{oc}: 1.98 using method of Kenaga and Goring [3].

Log K_{ow}: 1.10 [4].

Melting Point: -15.3 °C [1]; -11 to -9 °C [5].

Solubility in Organics: Soluble in acetone, ethanol, benzene, and ether [1].

Solubility in Water: 40,000 mg/L at 17 °C, 35,000 mg/L at 20 °C [6]; 4 vol% at 20 °C [7]; 42,900 mg/L at 25 °C [8]; 3.66 wt% at 20 °C [9]; 25 parts at 17 °C [10]; 1 part in 30 parts water [11].

Specific Density: 1.04535 at 20/4 °C [12]; 1.050 at 15/15 °C [13]; 1.044 at 20/4 °C [5]; 1.0424 at 25/4 °C, 1.0383 at 30/4 °C, 1.0313 at 40/4 °C, 1.0232 at 50/4 °C, 1.0153 at 60/4 °C, 1.0075 at 70/4 °C [14].

Transformation Products: No data found.

Vapor Density: 4.42 g/L at 25 °C, 3.73 (air = 1).

Vapor Pressure: 1 mm at 58 °C, 10 mm at 92.6 °C, 40 mm at 119.8 °C, 100 mm at 141.9 °C, 400 mm at 183.0 °C, 760 mm at 204.7 °C [1].

FIRE HAZARDS

Flash Point: 93 °C [15].

Lower Explosive Limit (LEL): Unknown [16].

Upper Explosive Limit (UEL): Unknown [16].

HEALTH HAZARD DATA

Immediately Dangerous to Life or Health (IDLH): No data found.

Permissible Exposure Limits (PEL) in Air: No standards set.

MANUFACTURING

Selected Manufacturers:

> Stauffer Chemical Co.
> Specialty Chemicals Division
> Meadow Road
> Edison, NJ 08817

Velsicol Chemical Corp.
341 East Ohio St.
Chicago, IL 60161

Givaudan Corp.
125 Delaware Ave.
Clifton, NJ 07014

Uses: Perfumes and flavors; photographic developer for color movie films; dying nylon filament, textiles and sheet plastics; solvent for dyestuffs, cellulose, esters, casein, waxes, etc.; heat-sealing polyethylene films; bacteriostat, cosmetics, ointments, emulsions; ballpoint pen inks and stencil inks; surfactant.

REFERENCES

1. Weast, R.C., Ed. *CRC Handbook of Chemistry and Physics*, 67th ed. (Boca Raton, FL: CRC Press, Inc., 1986), 2406 p.
2. Franklin, J.L., Dillard, J.G., Rosenstock, H.M., Herron, J.T., Draxl K., and F.H. Field. "Ionization Potentials, Appearance Potentials and Heats of Formation of Gaseous Positive Ions," National Bureau of Standards Report NSRDS-NBS 26, U.S. Government Printing Office (1969), 289 p.
3. Kenaga, E.E., and C.A.I. Goring. "Relationship Between Water Solubility, Soil Sorption, Octanol-Water Partitioning and Concentration of Chemicals in Biota," in *Aquatic Toxicology, ASTM STP 707*, Eaton, J.G., Parrish, P.R., and A.C. Hendricks, Eds. (Philadelphia, PA: American Society for Testing and Materials, 1980), pp 78-115.
4. Fujita, T., Iwasa, J., and C. Hansch. "A New Substituent Constant, π, Derived from Partition Coefficients," *J. Am. Chem. Soc.,* 86(23):5175-5180 (1964).
5. *Fluka Catalog 1988/89 - Chemika-Biochemika* (Ronkonkoma, NY: Fluka Chemical Corp., 1988), 1536 p.
6. Verschueren, K. *Handbook of Environmental Data on Organic Chemicals* (New York: Van Nostrand Reinhold Co., 1983), 1310 p.
7. Meites, L., Ed. *Handbook of Analytical Chemistry*, 1st ed. (New York: McGraw-Hill, Inc., 1963), 1782 p.
8. Banerjee, S. "Solubility of Organic Mixtures in Water," *Environ. Sci. Technol.,* 18(8):587-591 (1984).
9. Stephen, H., and T. Stephen. *Solubilities of Inorganic and Organic Compounds - Part 1, Volume 1* (London: Pergamon Printing and Art

Services, Ltd., 1963), 960 p.

10. Huntress, E.H., and S.P. Mulliken. *Identification of Pure Organic Compounds - Tables of Data on Selected Compounds of Order I* (New York: John Wiley and Sons, Inc., 1941), 691 p.

11. Standen, A., Ed. *Kirk-Othmer Encyclopedia of Chemical Technology, Volume 3*, 2nd ed. (New York: John Wiley and Sons, Inc., 1964), 927 p.

12. Sax, N.I., Ed. *Dangerous Properties of Industrial Materials Report* (New York: Van Nostrand Reinhold Co., 1984), 4(6): 105 p.

13. Weiss, G. *Hazardous Chemicals Data Book* (Park Ridge, NJ: Noyes Data Corp., 1986), 1069 p.

14. Abraham, T., Bery, V., and A.P. Kudchadker. "Densities of Some Organic Substances," *J. Chem. Eng. Data*, 16(3):355-356 (1971).

15. *Fire Protection Guide on Hazardous Materials* (Quincy, MA: National Fire Protection Association, 1984), 443 p.

16. "NIOSH Pocket Guide to Chemical Hazards," U.S. Department of Health and Human Services, U.S. Government Printing Office (1987), 241 p.

BENZYL BUTYL PHTHALATE

Synonyms: BBP; 1,2-Benzenedicarboxylic acid, butyl phenylmethyl ester; Benzyl *n*-butyl phthalate; Butyl benzyl phthalate; *n*-Butyl benzyl phthalate; Butyl phenylmethyl 1,2-benzenedicarboxylate; Butyl phenylmethyl ester; NCI-C54375; Palatinol BB; Santicizer 160; Sicol 160; Unimoll BB.

Structural Formula:

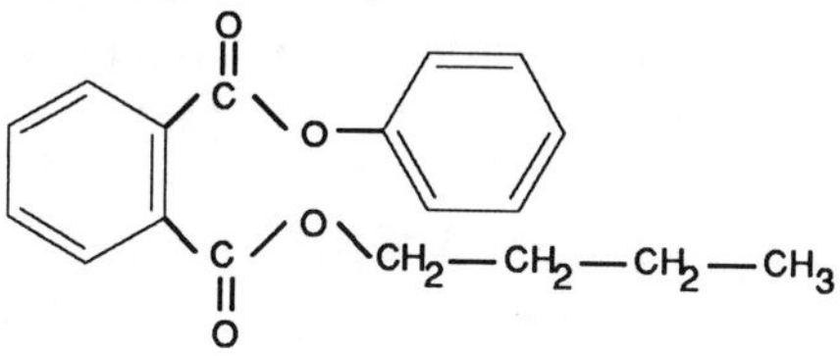

CHEMICAL DESIGNATIONS

CAS Registry Number: 85-68-7

DOT Designation: None assigned.

Empirical Formula: $C_{19}H_{20}O_4$

Formula Weight: 312.37

RTECS Number: TH 9990000

PHYSICAL AND CHEMICAL PROPERTIES

Appearance: Clear oily liquid.

Boiling Point: 370 °C [1]; 380 °C [2]; 377 °C [3].

Henry's Law Constant: 1.3 x 10^{-6} atm·m³/mol at 25 °C (calculated) [4].

Ionization Potential: No data found.

Log K_{oc}: 1.83-2.54 [5].

Log K_{ow}: 4.78 [1]; 4.05 [6]; 4.91 [7]; 4.77 [5].

Melting Point: -35 °C [3].

Solubility in Organics: No data found.

Solubility in Water: 2.9 mg/L [1]; 42.2 mg/L at 25 °C [8]; 2.82 mg/L at 20 °C [7]; 2.69 mg/L at 25 °C [9]; 0.710 mg/L at 24 °C [10]; 2.0 mg/L at 25 °C [11].

Specific Density: 1.12 at 20/4 °C [2]; 1.111 at 25/4 °C [12].

Transformation Products: In anaerobic sludge diluted to 10%, benzyl butyl phthalate biodegraded to monobutyl phthalate which subsequently degraded to phthalic acid [13].

Vapor Density: 12.76 g/L at 25 °C, 10.78 (air = 1).

Vapor Pressure: 8.6 x 10^{-6} mm at 20 °C, 1.9 mm at 200 °C [1]; 0.0011 Pa at 25 °C [9].

FIRE HAZARDS

Flash Point: 197 °C [2]; 110 °C [14].

Lower Explosive Limit (LEL): No data found.

Upper Explosive Limit (UEL): No data found.

HEALTH HAZARD DATA

Immediately Dangerous to Life or Health (IDLH): No data found.

Permissible Exposure Limits (PEL) in Air: No standards set.

MANUFACTURING

Selected Manufacturer:

> Pfaltz & Bauer, Inc.
> 172 East Aurora St.
> Waterbury, CT 06708

Uses: Plasticizer used in polyvinyl chloride (PVC) formulations;

additive in polyvinyl acetate emulsions, ethylene glycol, and ethyl cellulose.

REFERENCES

1. Verschueren, K. *Handbook of Environmental Data on Organic Chemicals* (New York: Van Nostrand Reinhold Co., 1983), 1310 p.
2. Weiss, G. *Hazardous Chemicals Data Book* (Park Ridge, NJ: Noyes Data Corp., 1986), 1069 p.
3. Fishbein, L., and P.W. Albro. "Chromatographic and Biological Aspects of the Phthalate Esters," *J. Chromatogr.*, 70(2):365-412 (1972).
4. Howard, P.H. *Handbook of Environmental Fate and Exposure Data for Organic Chemicals* (Chelsea, MI: Lewis Publishers, Inc., 1989), 574 p.
5. Gledhill, W.E., Kaley, R.G., Adams, W.J., Hicks, O., Michael, P.R., and V.W. Saeger. "An Environmental Safety Assessment of Butyl Benzyl Phthalate," *Environ. Sci. Technol.*, 14(3):301-305 (1980).
6. Veith, G.D., Macek, K.J., Petrocelli, S.R., and J. Carroll. "An Evaluation of Using Partition Coefficients and Water Solubility to Estimate Bioconcentration Factors for Organic Chemicals in Fish," in *Aquatic Toxicology, ASTM STP 707*, Eaton, J.G., Parrish, P.R., and A.C. Hendricks, Eds. (Philadelphia, PA: American Society for Testing and Materials, 1980), pp 116-129.
7. Leyder, F., and P. Boulanger. "Ultraviolet Absorption, Aqueous Solubility and Octanol-Water Partition for Several Phthalates," *Bull. Environ. Contam. Toxicol.*, 30(2):152-157 (1983).
8. Lyman, W.J., Reehl, W.F., and D.H. Rosenblatt. *Handbook of Chemical Property Estimation Methods: Environmental Behavior of Organic Compounds* (New York: McGraw-Hill, Inc., 1982).
9. Howard, P.H., Banerjee, S., and K.H. Robillard. "Measurement of Water Solubilities, Octanol/Water Partition Coefficients and Vapor Pressures of Commercial Phthalate Esters," *Environ. Toxicol. Chem.*, 4:653-661 (1985).
10. Hollifield, H.C. "Rapid Nephelometric Estimate of Water Solubility of Highly Insoluble Organic Chemicals of Environmental Interest," *Bull. Environ. Contam. Toxicol.*, 23(4/5):579-586 (1979).
11. Russell, D.J., and B. McDuffie. "Chemodynamic Properties of Phthalate Esters: Partitioning and Soil Migration," *Chemosphere*, 15(8):1003-1021 (1986).
12. Standen, A., Ed. *Kirk-Othmer Encyclopedia of Chemical Technology, Volume 15*, 2nd ed. (New York: John Wiley and Sons, Inc., 1968),

923 p.

13. Shelton, D.R., Boyd, S.A., and J.M. Tiedje. "Anaerobic Biodegradation of Phthalic Acid Esters in Sludge," *Environ. Sci. Technol.*, 18(2):93-97 (1984).
14. *Catalog Handbook of Fine Chemicals* (Milwaukee, WI: Aldrich Chemical Co., 1988), 2212 p.

α–BHC

Synonyms: Benzene hexachloride-α-isomer; α-Benzene hexachloride; ENT 9,232; α-HCH; α-Hexachloran; α-Hexachlorane; α-Hexachlorcyclohexane; α-Hexachlorocyclohexane; 1,2,3,4,5,6-Hexachloro-α-cyclohexane; **1α,2α,3β,4α,5β,6β-Hexachlorocyclohexane**; α-1,2,3,4,5,6-Hexachlorocyclohexane; α-Lindane; TBH.

Structural Formula:

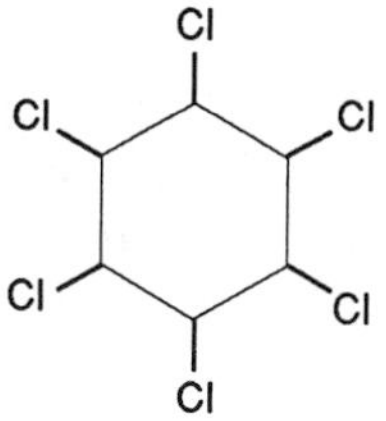

CHEMICAL DESIGNATIONS

CAS Registry Number: 319-84-6

DOT Designation: None assigned.

Empirical Formula: $C_6H_6Cl_6$

Formula Weight: 290.83

RTECS Number: GV 3500000

PHYSICAL AND CHEMICAL PROPERTIES

Appearance and Odor: Brownish-to-white crystalline solid with a phosgene-like odor (technical grade).

Boiling Point: 288 °C [1].

Henry's Law Constant: 5.3 x 10^{-6} atm·m^3/mol at 20 °C (calculated).

Ionization Potential: No data found.

Log K_{oc}: 3.279 [2].

Log K_{ow}: 3.81 [3]; 3.89 [4]; 3.46 [5]; 3.72 [6].

Melting Point: 159.1 °C [7].

Solubility in Organics: Soluble in ethanol, benzene, and chloroform [1].

Solubility in Water: 1.4 mg/L in salt water [8]; 1.21-2.03 mg/L at 28 °C [3]; 2.00 mg/L at 20 °C [9]; 1.63 mg/L at 20 °C [10]; 2.0 mg/L at 25 °C [11]; 10 ppm [12].

Specific Density: ≈1.87 [13].

Transformation Products: Under aerobic conditions, indigenous microbes in contaminated soil produced pentachlorocyclohexane. However under methanogenic conditions, α-BHC was converted to chlorobenzene, 3,5-dichlorophenol and the tentatively identified compound 2,4,5-trichlorophenol [14].

Vapor Pressure: 2.5 x 10^{-5} mm at 20 °C [15]; 2.15 x 10^{-5} mm at 20 °C [16].

FIRE HAZARDS

Flash Point: No data found.

Lower Explosive Limit (LEL): No data found.

Upper Explosive Limit (UEL): No data found.

HEALTH HAZARD DATA

Immediately Dangerous to Life or Health (IDLH): No data found.

Permissible Exposure Limits (PEL) in Air: No standards set.

MANUFACTURING

Selected Manufacturer:

Hindustan Insecticides Ltd.
Hans Bhawan, Wing-1
Bahadurshah Zafar Marg
New Delhi, 110 002 India

Use: Not produced commercially in the U.S. and its sale is prohibited by the U.S. EPA.

REFERENCES

1. Weast, R.C., Ed. *CRC Handbook of Chemistry and Physics*, 67th ed. (Boca Raton, FL: CRC Press, Inc., 1986), 2406 p.
2. Karickhoff, S.W. "Correspondence - On the Sorption of Neutral Organic Solutes in Soils," *J. Agric. Food Chem.*, 29(2):425-426 (1981).
3. Kurihara, N., Uchida, M., Fujita, T., and M. Nakajima. "Studies on BHC Isomers and Related Compounds. V. Some Physicochemical Properties of BHC isomers," *Pestic. Biochem. Physiol.*, 2(4):383-390 (1973).
4. Isnard, S., and S. Lambert. "Estimating Bioconcentration Factors from Octanol-Water Partition Coefficient and Aqueous Solubility," *Chemosphere*, 17(1):21-34 (1988).
5. Geyer, H.J., Scheunert, I., and F. Korte. "Correlation Between the Bioconcentration Potential of Organic Environmental Chemicals in Humans and Their *n*-Octanol/Water Partition Coefficients," *Chemosphere*, 16(1):239-252 (1987).
6. Schwarzenbach, R.P., Giger, W., Hoehn, E., and J.K. Schneider. "Behavior of Organic Compounds during Infiltration of River Water to Groundwater. Field Studies, " *Environ. Sci. Technol.*, 17(8):472-479 (1983).
7. Standen, A., Ed. *Kirk-Othmer Encyclopedia of Chemical Technology, Volume 4*, 2nd ed. (New York: John Wiley and Sons, Inc., 1964), 937 p.
8. Verschueren, K. *Handbook of Environmental Data on Organic Chemicals* (New York: Van Nostrand Reinhold Co., 1983), 1310 p.
9. Weil, L., Dure, G., and K.E. Quentin. "Solubility in Water of Insecticide Chlorinated Hydrocarbons and Polychlorinated Biphenyls in View of Water Pollution," *Z. Wasser Forsch.*, 7(6):169-175 (1974).
10. Brooks, G.T. *Chlorinated Insecticides, Volume I, Technology and Applications* (Cleveland, OH: CRC Press, 1974), 249 p.
11. Walton, W.C. *Practical Aspects of Ground Water Modeling* (Worthington, OH: National Water Well Association, 1985), 587 p.
12. Gunther, F.A., Westlake, W.E., and P.S. Jaglan. "Reported Solubilities of 738 Pesticide Chemicals in Water," *Res. Rev.*, 20:1-148 (1968).
13. Hawley, G.G. *The Condensed Chemical Dictionary* (New York: Van Nostrand Reinhold Co., 1981), 1135 p.
14. Bachmann, A., Wijnen, W.P., de Bruin, W., Huntjens, J.L.M.,

Roelofsen, W., and A.J.B. Zehnder. "Biodegradation of Alpha- and Beta-Hexachlorocyclohexane in a Soil Slurry under Different Redox Conditions," *Appl. Environ. Microbiol.*, 54(1):143-149 (1988).

15. Balson, E.W. "Studies in Vapour Pressure Measurement, Part III. - An Effusion Manometer Sensitive to 5 x 10^{-6} Millimetres of Mercury: Vapour Pressure of DDT and other Slightly Volatile Substances," *Trans. Faraday Soc.*, 43:54-60 (1947).

16. Sims, R.C., Doucette, W.C., McLean, J.E., Greeney, W.J., and R.R. Dupont. "Treatment Potential for 56 EPA Listed Hazardous Chemicals in Soil," National Technical Information Service, U.S. EPA Report-600/6-88-001 (1988), 105 p.

β-BHC

Synonyms: *trans*-α-Benzenehexachloride; β-Benzenehexachloride; Benzene-*cis*-hexachloride; ENT 9,233; β-HCH; β-Hexachlorobenzene; **1α,2β,3α,4β,5α,6β-Hexachlorocyclohexane**; β-Hexachlorocyclohexane; 1,2,3,4,5,6-Hexachloro-β-cyclohexane; 1,2,3,4,5,6-Hexachloro-*trans*-cyclohexane; β-1,2,3,4,5,6-Hexachlorocyclohexane; β-Isomer; β-Lindane; TBH.

Structural Formula:

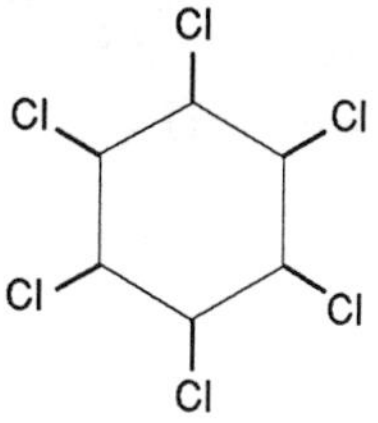

CHEMICAL DESIGNATIONS

CAS Registry Number: 319-85-7

DOT Designation: None assigned.

Empirical Formula: $C_6H_6Cl_6$

Formula Weight: 290.83

RTECS Number: GV 4375000

PHYSICAL AND CHEMICAL PROPERTIES

Appearance: Solid.

Boiling Point: 60 °C at 0.50 mm [1]; sublimes at 760 mm [2].

Henry's Law Constant: 2.3 x 10^{-7} atm·m^3/mol at 20 °C (calculated).

Ionization Potential: No data found.

Log K_{oc}: 3.462 [3]; 3.322 [4]; 3.553 [5].

Log K_{ow}: 3.80 [6]; 3.96 [7]; 4.50 [8].

Melting Point: 311.7 °C [9]; 309 °C [2].

Solubility in Organics: Soluble in ethanol, benzene, and chloroform [1].

Solubility in Water: 0.13-0.20 ppm at 28 °C [6]; 0.24 ppm at 25 °C [10]; 0.70 ppm at 25 °C [11]; 5 ppm at 20 °C [3].

Specific Density: 1.89 at 19/4 °C [1].

Transformation Products: No biodegradation of β-BHC was observed under denitrifying and sulfate-reducing conditions in a contaminated soil collected from the Netherlands [12].

Vapor Pressure: 2.8×10^{-7} mm at 20 °C [13]; 1.4×10^{-5} mm at 50 °C [14].

FIRE HAZARDS

Flash Point: No data found.

Lower Explosive Limit (LEL): No data found.

Upper Explosive Limit (UEL): No data found.

HEALTH HAZARD DATA

Immediately Dangerous to Life or Health (IDLH): No data found.

Permissible Exposure Limits (PEL) in Air: No standards set.

MANUFACTURING

Use: Insecticide.

REFERENCES

1. Weast, R.C., Ed. *CRC Handbook of Chemistry and Physics*, 67th ed. (Boca Raton, FL: CRC Press, Inc., 1986), 2406 p.
2. "Treatability Manual - Volume 1: Treatability Data," Office of

Research and Development, U.S. EPA Report-600/8-80-042a (1980), 1035 p.

3. Chiou, C.T., Peters, L.J., and V.H. Freed. "A Physical Concept of Soil-Water Equilibria for Nonionic Organic Compounds," *Science*, 206:831-832 (1979).

4. Karickhoff, S.W. "Correspondence - On the Sorption of Neutral Organic Solutes in Soils," *J. Agric. Food Chem.*, 29(2):425-426 (1981).

5. Reinbold, K.A., Hassett, J.J., Means, J.C., and W.L. Banwart. "Adsorption of Energy-Related Organic Pollutants: A Literature Review," Office of Research and Development, U.S. EPA Report-600/3-79-086 (1979), 180 p.

6. Kurihara, N., Uchida, M., Fujita, T., and M. Nakajima. "Studies on BHC Isomers and Related Compounds. V. Some Physicochemical Properties of BHC isomers," *Pestic. Biochem. Physiol.*, 2(4):383-390 (1973).

7. Isnard, S., and S. Lambert. "Estimating Bioconcentration Factors from Octanol-Water Partition Coefficient and Aqueous Solubility," *Chemosphere*, 17(1):21-34 (1988).

8. Geyer, H.J., Scheunert, I., and F. Korte. "Correlation Between the Bioconcentration Potential of Organic Environmental Chemicals in Humans and Their *n*-Octanol/Water Partition Coefficients," *Chemosphere*, 16(1):239-252 (1987).

9. Standen, A., Ed. *Kirk-Othmer Encyclopedia of Chemical Technology, Volume 4*, 2nd ed. (New York: John Wiley and Sons, Inc., 1964), 937 p.

10. Weil, L., Dure, G., and K.E. Quentin. "Solubility in Water of Insecticide Chlorinated Hydrocarbons and Polychlorinated Biphenyls in View of Water Pollution," *Z. Wasser Forsch.*, 7(6):169-175 (1974).

11. Brooks, G.T. *Chlorinated Insecticides, Volume I, Technology and Applications* (Cleveland, OH: CRC Press, 1974), 249 p.

12. Bachmann, A., Wijnen, W.P., de Bruin, W., Huntjens, J.L.M., Roelofsen, W., and A.J.B. Zehnder. "Biodegradation of Alpha- and Beta-Hexachlorocyclohexane in a Soil Slurry under Different Redox Conditions," *Appl. Environ. Microbiol.*, 54(1):143-149 (1988).

13. Balson, E.W. "Studies in Vapour Pressure Measurement, Part III. - An Effusion Manometer Sensitive to 5 x 10^{-6} Millimetres of Mercury: Vapour Pressure of DDT and other Slightly Volatile Substances," *Trans. Faraday Soc.*, 43:54-60 (1947).

14. Hazardous Substances Data Bank. *beta*-Hexachlorocyclohexane, National Library of Medicine, Toxicology Information Program (1989).

δ-BHC

Synonyms: δ-Benzenehexachloride; ENT 9,234; δ-HCH; δ-Hexachlorocyclohexane; δ-1,2,3,4,5,6-Hexachlorocyclohexane; δ-(aeeeee)-1,2,3,4,5,6-Hexachlorocyclohexane; **1α,2α,3α,4β,5β,6β-Hexachlorocyclohexane;** 1,2,3,4,5,6-Hexachloro-δ-cyclohexane; δ-Lindane; TBH.

Structural Formula:

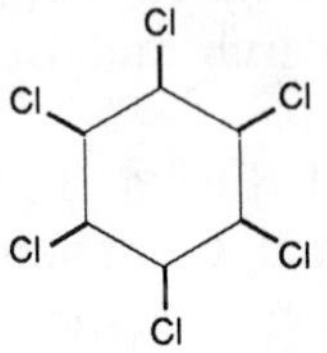

CHEMICAL DESIGNATIONS

CAS Registry Number: 319-86-8

DOT Designation: None assigned.

Empirical Formula: $C_6H_6Cl_6$

Formula Weight: 290.83

RTECS Number: GV 4550000

PHYSICAL AND CHEMICAL PROPERTIES

Appearance: Solid.

Boiling Point: 60 °C at 0.36 mm [1].

Henry's Law Constant: 2.5 x 10^{-7} atm·m³/mol at 20-25 °C (calculated).

Ionization Potential: No data found.

Log K_{oc}: 3.279 [2].

Log K_{ow}: 4.14 [3]; 2.80 [4].

Melting Point: 140.8 °C [5].

Solubility in Organics: Soluble in ethanol, benzene, and chloroform [1].

Solubility in Water: 8.64-15.7 ppm at 28 °C [3]; 31.4 ppm at 25 °C [6]; 21.3 ppm at 25 °C [7].

Specific Density: ≈1.87 [8].

Transformation Products: No data found.

Vapor Pressure: 1.7 x 10^{-5} mm at 20 °C [9].

FIRE HAZARDS

Flash Point: No data found.

Lower Explosive Limit (LEL): No data found.

Upper Explosive Limit (UEL): No data found.

HEALTH HAZARD DATA

Immediately Dangerous to Life or Health (IDLH): No data found.

Permissible Exposure Limits (PEL) in Air: No standards set.

MANUFACTURING

Use: Insecticide.

REFERENCES

1. Weast, R.C., Ed. *CRC Handbook of Chemistry and Physics*, 67th ed. (Boca Raton, FL: CRC Press, Inc., 1986), 2406 p.
2. Karickhoff, S.W. "Correspondence - On the Sorption of Neutral Organic Solutes in Soils," *J. Agric. Food Chem.*, 29(2):425-426 (1981).
3. Kurihara, N., Uchida, M., Fujita, T., and M. Nakajima. "Studies on BHC Isomers and Related Compounds. V. Some Physicochemical Properties of BHC isomers," *Pestic. Biochem. Physiol.*, 2(4):383-390 (1973).
4. Geyer, H.J., Scheunert, I., and F. Korte. "Correlation Between the

Bioconcentration Potential of Organic Environmental Chemicals in Humans and Their *n*-Octanol/Water Partition Coefficients," *Chemosphere*, 16(1):239-252 (1987).

5. Standen, A., Ed. *Kirk-Othmer Encyclopedia of Chemical Technology, Volume 4*, 2nd ed. (New York: John Wiley and Sons, Inc., 1964), 937 p.

6. Weil, L., Dure, G., and K.E. Quentin. "Solubility in Water of Insecticide Chlorinated Hydrocarbons and Polychlorinated Biphenyls in View of Water Pollution," *Z. Wasser Forsch.*, 7(6):169-175 (1974).

7. Brooks, G.T. *Chlorinated Insecticides, Volume I, Technology and Applications* (Cleveland, OH: CRC Press, 1974), 249 p.

8. Hawley, G.G. *The Condensed Chemical Dictionary* (New York: Van Nostrand Reinhold Co., 1981), 1135 p.

9. Balson, E.W. "Studies in Vapour Pressure Measurement, Part III. - An Effusion Manometer Sensitive to 5×10^{-6} Millimetres of Mercury: Vapour Pressure of DDT and other Slightly Volatile Substances," *Trans. Faraday Soc.*, 43:54-60 (1947).

BIS(2-CHLOROETHOXY)METHANE

Synonyms: BCEXM; Bis(2-chloroethyl)formal; Bis(β-chloroethyl)formal; Dichlorodiethyl formal; Dichlorodiethyl methylal; Dichloroethyl formal; Di-2-chloroethyl formal; **1,1'-[Methylenebis(oxy)]-bis(2-chloroethane);** 1,1'-[Methylenebis(oxy)]bis(2-chloroformaldehyde); Bis(β-chloroethyl)acetal ethane; Formaldehyde bis(β-chloroethylacetal); RCRA waste number U024.

Structural Formula:

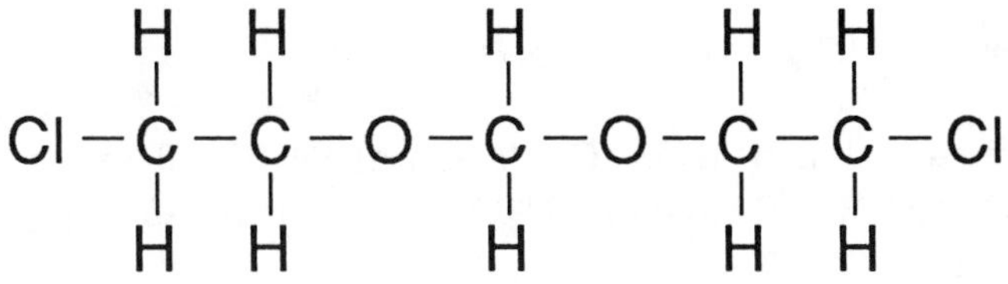

$$\text{Cl} - \underset{\underset{\text{H}}{|}}{\overset{\overset{\text{H}}{|}}{\text{C}}} - \underset{\underset{\text{H}}{|}}{\overset{\overset{\text{H}}{|}}{\text{C}}} - \text{O} - \underset{\underset{\text{H}}{|}}{\overset{\overset{\text{H}}{|}}{\text{C}}} - \text{O} - \underset{\underset{\text{H}}{|}}{\overset{\overset{\text{H}}{|}}{\text{C}}} - \underset{\underset{\text{H}}{|}}{\overset{\overset{\text{H}}{|}}{\text{C}}} - \text{Cl}$$

CHEMICAL DESIGNATIONS

CAS Registry Number: 111-91-1

DOT Designation: 1916

Empirical Formula: $C_5H_{10}Cl_2O_2$

Formula Weight: 173.04

RTECS Number: PA 3675000

PHYSICAL AND CHEMICAL PROPERTIES

Appearance: Colorless liquid.

Boiling Point: 218.1 °C [1].

Henry's Law Constant: 3.78 x 10^{-7} atm·m^3/mol (calculated) [2].

Ionization Potential: No data found.

Log K_{oc}: 2.06 using method of Kenaga and Goring [3].

Log K_{ow}: 1.26 (calculated) [4].

83

Melting Point: -32.8 °C [5].

Solubility in Organics: No data found.

Solubility in Water: Calculated as 81,000 mg/L at 25 °C using method of Moriguchi [6].

Specific Density: 1.2339 at 20/20 °C [5].

Transformation Products: No data found.

Vapor Density: 7.07 g/L at 25 °C, 5.97 (air = 1).

Vapor Pressure: 1 mm at 53 °C, 10 mm at 94 °C, 40 mm at 125.5 °C, 100 mm at 149.6 °C, 400 mm at 192.0 °C, 760 mm at 215 °C [7].

FIRE HAZARDS

Flash Point: 110 °C (open cup) [5].

Lower Explosive Limit (LEL): Unknown [8].

Upper Explosive Limit (UEL): Unknown [8].

HEALTH HAZARD DATA

Immediately Dangerous to Life or Health (IDLH): No data found.

Permissible Exposure Limits (PEL) in Air: No standards set.

MANUFACTURING

Selected Manufacturer:

> Pfaltz & Bauer, Inc.
> 172 East Aurora St.
> Waterbury, CT 06708

Uses: Manufacturing of insecticides, polymers; degreasing solvent; intermediate for polysulfide rubber.

REFERENCES

1. Webb, R.F., Duke, A.J., and L.S.A. Smith. "Acetals and Oligoacetals. Part I. Preparation and Properties of Reactive Oligoformals," *J. Chem. Soc. (London)*, pp 4307-4319 (1962).
2. "Treatability Manual - Volume 1: Treatability Data," Office of Research and Development, U.S. EPA Report-600/8-80-042a (1980), 1035 p.
3. Kenaga, E.E., and C.A.I. Goring. "Relationship Between Water Solubility, Soil Sorption, Octanol-Water Partitioning and Concentration of Chemicals in Biota," in *Aquatic Toxicology, ASTM STP 707*, Eaton, J.G., Parrish, P.R., and A.C. Hendricks, Eds. (Philadelphia, PA: American Society for Testing and Materials, 1980), pp 78-115.
4. Leo, A., Hansch, C., and D. Elkins. "Partition Coefficients and Their Uses," *Chem. Rev.*, 71(6):525-616 (1971).
5. Hawley, G.G. *The Condensed Chemical Dictionary* (New York: Van Nostrand Reinhold Co., 1981), 1135 p.
6. Moriguchi, I. "Quantitative Structure-Activity Studies on Parameters Related to Hydrophobicity," *Chem. Pharm. Bull.*, 23:247-257 (1975).
7. Weast, R.C., Ed. *CRC Handbook of Chemistry and Physics*, 67th ed. (Boca Raton, FL: CRC Press, Inc., 1986), 2406 p.
8. "NIOSH Pocket Guide to Chemical Hazards," U.S. Department of Health and Human Services, U.S. Government Printing Office (1987), 241 p.

BIS(2-CHLOROETHYL)ETHER

Synonyms: Bis-(β-chloroethyl)ether; Chlorex; 1-Chloro-2-(β-chloroethoxy)ethane; Chloroethyl ether; 2-Chloroethyl ether; (β-Chloroethyl)ether; DCEE; Dichlorodiethyl ether; 2,2'-Dichlorodiethyl ether; β,β'-Dichlorodiethyl ether; Dichloroether; Dichloroethyl ether; α,α'-Dichloroethyl ether; Di(β-chloroethyl)ether; Di(2-chloroethyl)ether; sym-Dichloroethyl ether; 2,2'-Dichloroethyl ether; Dichloroethyl oxide; ENT 4,504; **1,1'-Oxybis(2-chloroethane)**; RCRA waste number U025; UN 1916.

Structural Formula:

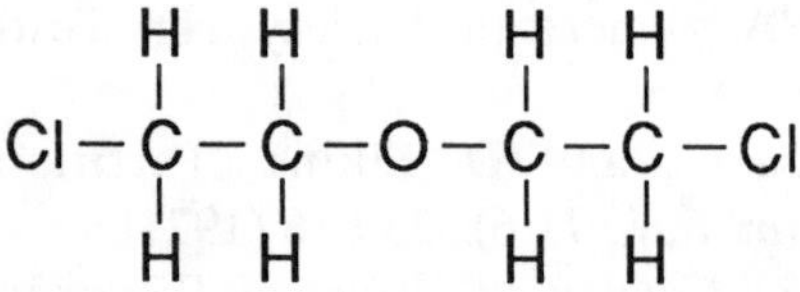

$$\text{Cl}-\underset{\underset{\displaystyle H}{|}}{\overset{\overset{\displaystyle H}{|}}{C}}-\underset{\underset{\displaystyle H}{|}}{\overset{\overset{\displaystyle H}{|}}{C}}-O-\underset{\underset{\displaystyle H}{|}}{\overset{\overset{\displaystyle H}{|}}{C}}-\underset{\underset{\displaystyle H}{|}}{\overset{\overset{\displaystyle H}{|}}{C}}-\text{Cl}$$

CHEMICAL DESIGNATIONS

CAS Registry Number: 111-44-4

DOT Designation: 1916

Empirical Formula: $C_4H_8Cl_2O$

Formula Weight: 143.01

RTECS Number: KN 0875000

PHYSICAL AND CHEMICAL PROPERTIES

Appearance and Odor: Colorless liquid with a strong fruity odor.

Boiling Point: 178.5 °C [1]; 176-178 °C [2].

Henry's Law Constant: 1.3 x 10^{-5} atm·m^3/mol [3].

Ionization Potential: 9.85 eV [4].

Log K$_{oc}$: 1.15 [3].

Log K$_{ow}$: 1.12 [5]; 1.58 [6].

Melting Point: -24.5 °C [7]; -47 °C [8]; -52.2 °C [9].

Solubility in Organics: Soluble in acetone, ethanol, benzene, and ether [7].

Solubility in Water: 10,200 mg/L at 20 °C [3]; 10,700 mg/L at 20 °C [1]; 10,200 mg/L at 25 °C [6]; 11,000 mg/L at 20 °C [10].

Specific Density: 1.2199 at 20/4 °C [7].

Transformation Products: No data found.

Vapor Density: 5.84 g/L at 25 °C, 4.94 (air = 1).

Vapor Pressure: 0.71 mm at 20 °C, 1.4 mm at 25 °C [11]; 1 mm at 23.5 °C, 10 mm at 62.0 °C, 40 mm at 91.5 °C, 100 mm at 114.5 °C, 400 mm at 155.4 °C, 760 mm at 178.5 °C [7]; 1.55 mm at 25 °C [12].

FIRE HAZARDS

Flash Point: 55 °C [13].

Lower Explosive Limit (LEL): No data found.

Upper Explosive Limit (UEL): No data found.

HEALTH HAZARD DATA

Immediately Dangerous to Life or Health (IDLH): No data found.

Permissible Exposure Limits (PEL) in Air: No standards set.

MANUFACTURING

Selected Manufacturer:

> Aldrich Chemical Co.
> 940 West Saint Paul Ave.
> Milwaukee, WI 53233

Uses: Scouring and cleaning textiles; fumigants; processing fats, waxes, greases, cellulose esters; preparation of insecticides, butadiene, pharmaceuticals; solvent in paints, varnishes, and lacquers; selective solvent for production of high-grade lubricating oils; fulling, wetting and penetrating compounds; finish removers; spotting and dry cleaning; soil fumigant; organic synthesis.

REFERENCES

1. Dean, J.A., Ed., *Lange's Handbook of Chemistry*, 11th ed. (New York: McGraw-Hill, Inc., 1973), 1570 p.
2. Standen, A., Ed. *Kirk-Othmer Encyclopedia of Chemical Technology, Volume 5*, 2nd ed. (New York: John Wiley and Sons, Inc., 1965), 884 p.
3. Schwille, F. *Dense Chlorinated Solvents* (Chelsea, MI: Lewis Publishers, Inc., 1988), 146 p.
4. Franklin, J.L., Dillard, J.G., Rosenstock, H.M., Herron, J.T., Draxl K., and F.H. Field. "Ionization Potentials, Appearance Potentials and Heats of Formation of Gaseous Positive Ions," National Bureau of Standards Report NSRDS-NBS 26, U.S. Government Printing Office (1969), 289 p.
5. Veith, G.D., Macek, K.J., Petrocelli, S.R., and J. Carroll. "An Evaluation of Using Partition Coefficients and Water Solubility to Estimate Bioconcentration Factors for Organic Chemicals in Fish," in *Aquatic Toxicology, ASTM STP 707*, Eaton, J.G., Parrish, P.R., and A.C. Hendricks, Eds. (Philadelphia, PA: American Society for Testing and Materials, 1980), pp 116-129.
6. "Treatability Manual - Volume 1: Treatability Data," Office of Research and Development, U.S. EPA Report-600/8-80-042a (1980), 1035 p.
7. Weast, R.C., Ed. *CRC Handbook of Chemistry and Physics*, 67th ed. (Boca Raton, FL: CRC Press, Inc., 1986), 2406 p.
8. *Catalog Handbook of Fine Chemicals* (Milwaukee, WI: Aldrich Chemical Co., 1988), 2212 p.
9. "NIOSH Pocket Guide to Chemical Hazards," U.S. Department of Health and Human Services, U.S. Government Printing Office (1987), 241 p.
10. Gunther, F.A., Westlake, W.E., and P.S. Jaglan. "Reported Solubilities of 738 Pesticide Chemicals in Water," *Res. Rev.*, 20:1-148 (1968).
11. Verschueren, K. *Handbook of Environmental Data on Organic Chemicals* (New York: Van Nostrand Reinhold Co., 1983), 1310 p.
12. Howard, P.H. *Handbook of Environmental Fate and Exposure Data*

for Organic Chemicals (Chelsea, MI: Lewis Publishers, Inc., 1989), 574 p.

13. *Fire Protection Guide on Hazardous Materials* (Quincy, MA: National Fire Protection Association, 1984), 443 p.

BIS(2-CHLOROISOPROPYL)ETHER

Synonyms: BCIE; BCMEE; Bis-(β-chloroisopropyl)ether; Bis(2-chloro-1-methylethyl)ether; 1-Chloro-2-(β-chloroisopropoxy)propane; 2-Chloroisopropyl ether; β-Chloroisopropylether; (2-Chloro-1-methylethyl)ether; Dichlorodiisopropyl ether; Dichloroisopropyl ether; 2,2'-Dichloroisopropyl ether; NCI-C50044; **2,2'-Oxybis(1-chloropropane)**; RCRA waste number U027; UN 2490.

Structural Formula:

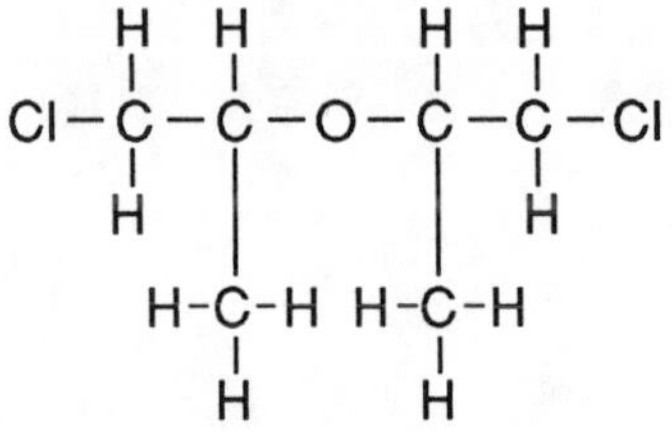

$$Cl-\overset{\overset{\displaystyle H}{|}}{\underset{\underset{\displaystyle H}{|}}{C}}-\overset{\overset{\displaystyle H}{|}}{\underset{\underset{\displaystyle H-C-H}{|}}{\underset{\underset{\displaystyle H}{|}}{C}}}-O-\overset{\overset{\displaystyle H}{|}}{\underset{\underset{\displaystyle H-C-H}{|}}{\underset{\underset{\displaystyle H}{|}}{C}}}-\overset{\overset{\displaystyle H}{|}}{\underset{\underset{\displaystyle H}{|}}{C}}-Cl$$

CHEMICAL DESIGNATIONS

CAS Registry Number: 108-60-1

DOT Designation: None assigned.

Empirical Formula: $C_6H_{12}Cl_2O$

Formula Weight: 171.07

RTECS Number: KN 1750000

PHYSICAL AND CHEMICAL PROPERTIES

Appearance: Colorless liquid.

Boiling Point: 187 °C [1]; 189 °C [2].

Henry's Law Constant: 1.1 x 10^{-4} atm·m^3/mol [3].

Ionization Potential: No data found.

Log K_{oc}: 1.79 [4].

Log K_{ow}: 2.58 [5].

Melting Point: 96.8-101.8 °C [6].

Solubility in Organics: Soluble in acetone, ethanol, benzene, and ether [1].

Solubility in Water: 1,700 mg/L at 20 °C [7]; 0.17 wt% at 20 °C [8].

Specific Density: 1.103 at 20/4 °C [1]; 1.1127 at 25/4 °C [8].

Transformation Products: No data found.

Vapor Density: 6.99 g/L at 25 °C, 5.91 (air = 1).

Vapor Pressure: 0.85 mm at 20 °C, 44.5 mm at 100 °C [2]; 1 mm at 29.6 °C, 10 mm at 68.2 °C, 40 mm at 97.3 °C, 100 mm at 119.7 °C, 400 mm at 159.8 °C, 760 mm at 182.7 °C [1].

FIRE HAZARDS

Flash Point: 85 °C [9].

Lower Explosive Limit (LEL): Unknown [10].

Upper Explosive Limit (UEL): Unknown [10].

HEALTH HAZARD DATA

Immediately Dangerous to Life or Health (IDLH): No data found.

Permissible Exposure Limits (PEL) in Air: No standards set.

MANUFACTURING

Uses: Chemical intermediate in the manufacturing of dyes, resins, and pharmaceuticals; solvent and extractant for fats, waxes, and greases; textile manufacturing; agent in paint and varnish removers, spotting and cleaning agents; a combatant in liver fluke infections; preparation of glycol esters in fungicidal preparations and as an insecticidal wood preservative; apparently used as a nematocide in Japan but is not registered in the U.S. for use as a pesticide.

REFERENCES

1. Weast, R.C., Ed. *CRC Handbook of Chemistry and Physics*, 67th ed. (Boca Raton, FL: CRC Press, Inc., 1986), 2406 p.
2. Verschueren, K. *Handbook of Environmental Data on Organic Chemicals* (New York: Van Nostrand Reinhold Co., 1983), 1310 p.
3. Pankow, J.F., and M.E. Rosen. "Determination of Volatile Compounds in Water by Purging Directly to a Capillary Column with Whole Column Cryotrapping," *Environ. Sci. Technol.*, 22(4):398-405 (1988).
4. Schwille, F. *Dense Chlorinated Solvents* (Chelsea, MI: Lewis Publishers, Inc., 1988), 146 p.
5. "Treatability Manual - Volume 1: Treatability Data," Office of Research and Development, U.S. EPA Report-600/8-80-042a (1980), 1035 p.
6. Clayton, G.D., and F.E. Clayton. Eds. *Patty's Industrial Hygiene and Toxicology*, 3rd ed. (New York, John Wiley and Sons, Inc., 1981), 2878 p.
7. Dean, J.A., Ed., *Lange's Handbook of Chemistry*, 11th ed. (New York: McGraw-Hill, Inc., 1973), 1570 p.
8. Standen, A., Ed. *Kirk-Othmer Encyclopedia of Chemical Technology, Volume 5*, 2nd ed. (New York: John Wiley and Sons, Inc., 1965), 884 p.
9. Hawley, G.G. *The Condensed Chemical Dictionary* (New York: Van Nostrand Reinhold Co., 1981), 1135 p.
10. "NIOSH Pocket Guide to Chemical Hazards," U.S. Department of Health and Human Services, U.S. Government Printing Office (1987), 241 p.

BIS(2-ETHYLHEXYL)PHTHALATE

Synonyms: **1,2-Benzenedicarboxylic acid, bis(2-ethylhexyl)ester;** Bioflex 81; Bioflex DOP; Bis(2-ethylhexyl)-1,2-benzenedicarboxylate; Compound 889; DAF 68; DEHP; Di(2-ethylhexyl)orthophthalate; Di(2-ethylhexyl)-phthalate; Dioctyl phthalate; Di-*sec*-octyl phthalate; DOP; Ergoplast FDO; Ethylhexyl phthalate; 2-Ethylhexyl phthalate; Eviplast 80; Eviplast 81; Fleximel; Flexol DOP; Flexol plasticizer DOP; Good-rite GP 264; Hatcol DOP; Hercoflex 260; Kodaflex DOP; Mollan 0; NCI-C52733; Nuoplaz DOP; Octoil; Octyl phthalate; Palantinol AH; Phthalic acid, bis(2-ethylhexyl) ester; Phthalic acid, dioctyl ester; Pittsburgh PX-138; Platinol AH; Platinol DOP; RC plasticizer DOP; RCRA waste number U028; Reomol D 79P; Reomol DOP; Sicol 150; Staflex DOP; Truflex DOP; Vestinol 80; Witicizer 312.

Structural Formula:

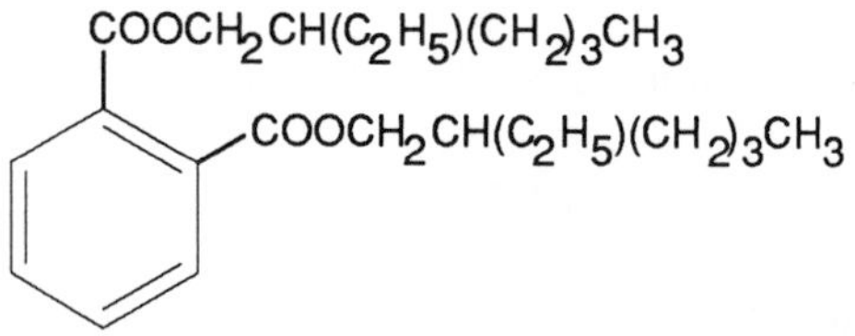

$$COOCH_2CH(C_2H_5)(CH_2)_3CH_3$$
$$COOCH_2CH(C_2H_5)(CH_2)_3CH_3$$

CHEMICAL DESIGNATIONS

CAS Registry Number: 117-81-7

DOT Designation: None assigned.

Empirical Formula: $C_{24}H_{38}O_4$

Formula Weight: 390.57

RTECS Number: TI 0350000

PHYSICAL AND CHEMICAL PROPERTIES

Appearance and Odor: Colorless, oily liquid with a very faint odor.

Boiling Point: 385 °C [1]; 386.9 °C at 5 mm [2]; 230 °C at 5 mm [3].

Henry's Law Constant: 1.1 x 10^{-5} atm·m^3/mol at 25 °C (calculated) [3].

Ionization Potential: No data found.

Log K$_{oc}$: 5.0 [4].

Log K$_{ow}$: 4.20 [5]; 5.11 [6].

Melting Point: -55 °C [1]; -50 °C [7]; -46 °C [8].

Solubility in Organics: Miscible with mineral oil and hexane [9].

Solubility in Water: 0.4 mg/L at 25 °C [1]; 0.005 wt% at 20 °C [10]; 0.041 mg/L at 20 °C [11]; 0.34 mg/L at 25 °C, 0.30 mg/L in well water at 25 °C, 0.16 mg/L in natural seawater at 25 °C [12]; 0.285 mg/L at 24 °C [13]; 0.047 mg/L at 25 °C [14]; 0.01 wt% at 20 °C [2].

Specific Density: 0.9861 at 20/20 °C [15]; 0.985 at 20/4 °C [16].

Transformation Products: No data found.

Vapor Density: 15.96 g/L at 25 °C, 13.48 (air = 1).

Vapor Pressure: 1.2 mm at 200 °C [1]; 2 x 10^{-7} mm at 20 °C [17]; 5 x 10^{-8} mm at 68 °C, 5 x 10^{-6} mm at 120 °C [18]; 8.6 (± 6.6) x 10^{-4} Pa at 25 °C [12]; 1 x 10^{-7} mm at 20 °C [19]; 6.2 x 10^{-8} mm at 25 °C [20].

FIRE HAZARDS

Flash Point: 207 °C [21]; 196 °C (open cup) [19].

Lower Explosive Limit (LEL): 0.3% [22].

Upper Explosive Limit (UEL): Unknown [10].

HEALTH HAZARD DATA

Immediately Dangerous to Life or Health (IDLH): Potential human carcinogen [10].

Permissible Exposure Limits (PEL) in Air: 5 mg/m^3 [23]; reduce exposure to lowest feasible limit [10]; 5 mg/m^3 TWA, 10 mg/m^3 STEL [24].

MANUFACTURING

Selected Manufacturers:

Allied Chemical Corp.
Plastics Division
Morristown, NJ 07960

Monsanto Industrial Chemicals Co.
800 North Lindbergh Blvd.
St. Louis, MO 63166

W.R. Grace and Co.
Hatco Chemical Division
Fords, NJ 08863

Uses: Plasticizer; in vacuum pumps.

REFERENCES

1. Verschueren, K. *Handbook of Environmental Data on Organic Chemicals* (New York: Van Nostrand Reinhold Co., 1983), 1310 p.
2. Fishbein, L., and P.W. Albro. "Chromatographic and Biological Aspects of the Phthalate Esters," *J. Chromatogr.*, 70(2):365-412 (1972).
3. Howard, P.H. *Handbook of Environmental Fate and Exposure Data for Organic Chemicals* (Chelsea, MI: Lewis Publishers, Inc., 1989), 574 p.
4. Neely, W.B., and G.E. Blau, Eds. *Environmental Exposure from Chemicals. Volume 1* (Boca Raton, FL: CRC Press, Inc. 1985), 245 p.
5. Mackay, D. "Correlation of Bioconcentration Factors," *Environ. Sci. Technol.*, 16(5):274-278 (1982).
6. Geyer, H., Politzki, G., and D. Freitag. "Prediction of Ecotoxicological Behaviour of Chemicals: Relationship Between *n*-Octanol/Water Partition Coefficient and Bioaccumulation of Organic Chemicals by Alga *Chlorella*," *Chemosphere*, 13(2):269-284 (1984).
7. "Treatability Manual - Volume 1: Treatability Data," Office of Research and Development, U.S. EPA Report-600/8-80-042a (1980), 1035 p.
8. Standen, A., Ed. *Kirk-Othmer Encyclopedia of Chemical Technology, Volume 15*, 2nd ed. (New York: John Wiley and Sons, Inc., 1968),

923 p.

9. "Chemical, Physical, and Biological Properties of Compounds Present at Hazardous Waste Sites," U.S. EPA Report-530/SW-89-010 (1985), 619 p.

10. "NIOSH Pocket Guide to Chemical Hazards," U.S. Department of Health and Human Services, U.S. Government Printing Office (1987), 241 p.

11. Leyder, F., and P. Boulanger. "Ultraviolet Absorption, Aqueous Solubility and Octanol-Water Partition for Several Phthalates," *Bull. Environ. Contam. Toxicol.*, 30(2):152-157 (1983).

12. Howard, P.H., Banerjee, S., and K.H. Robillard. "Measurement of Water Solubilities, Octanol/Water Partition Coefficients and Vapor Pressures of Commercial Phthalate Esters," *Environ. Toxicol. Chem.*, 4:653-661 (1985).

13. Hollifield, H.C. "Rapid Nephelometric Estimate of Water Solubility of Highly Insoluble Organic Chemicals of Environmental Interest," *Bull. Environ. Contam. Toxicol.*, 23(4/5):579-586 (1979).

14. Klöpffer, W., Kaufman, G., Rippen, G., and H.-P. Poremski. "A Laboratory Method for Testing the Volatility from Aqueous Solution: First Results and Comparison with Theory," *Ecotoxicol. Environ. Safety*, 6(6):545-559 (1982).

15. Hawley, G.G. *The Condensed Chemical Dictionary* (New York: Van Nostrand Reinhold Co., 1981), 1135 p.

16. *Fluka Catalog 1988/89 - Chemika-Biochemika* (Ronkonkoma, NY: Fluka Chemical Corp., 1988), 1536 p.

17. Hirzy, J.W., Adams, W.J., Gledhill, W.E., and J.P. Mieure. "Phthalate Esters: The Environmental Issues," unpublished seminar document, Monsanto Industrial Chemicals Co. (1978), 54 p.

18. Gross, F.C., and J.A. Colony. "The Ubiquitous Nature and Objectionable Characteristics of Phthalate Esters in Aerospace Industry," *Environ. Health Perspect.*, (January 1973), pp 37-48.

19. Broadhurst, M.G. "Use and Replaceability of Polychlorinated Biphenyls," *Environ. Health Perspect.*, (October 1972), pp 81-102.

20. Giam, C.S., Atlas, E., Chan, H.S., and G.S. Neff. "Phthalate Esters, PCB and DDT Residues in the Gulf of Mexico Atmosphere," *Atmos. Environ.*, 14:65-69 (1980).

21. *Catalog Handbook of Fine Chemicals* (Milwaukee, WI: Aldrich Chemical Co., 1988), 2212 p.

22. *Fire Protection Guide on Hazardous Materials* (Quincy, MA: National Fire Protection Association, 1984), 443 p.

23. "General Industry Standards for Toxic and Hazardous Substances," U.S. Code of Federal Regulations 1910, Subpart Z, Section 1910.1000 (July 1982).

24. *Documentation of the Threshold Limit Values and Biological Exposure Indices for 1986-1987* (Cincinnati, OH: American Conference of Governmental Industrial Hygienists, 1986), 111 p.

BROMODICHLOROMETHANE

Synonyms: BDCM; Dichlorobromomethane; NCI-C55243.

Structural Formula:

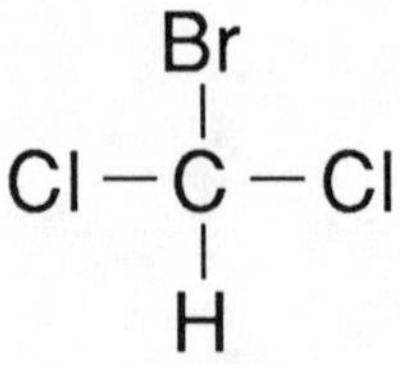

$$Cl - \underset{\underset{H}{|}}{\overset{\overset{Br}{|}}{C}} - Cl$$

CHEMICAL DESIGNATIONS

CAS Registry Number: 75-27-4

DOT Designation: None assigned.

Empirical Formula: $CHBrCl_2$

Formula Weight: 163.83

RTECS Number: PA 5310000

PHYSICAL AND CHEMICAL PROPERTIES

Appearance: Colorless liquid.

Boiling Point: 90.1 °C [1]; 87-90 °C [2].

Henry's Law Constant: 0.0024 atm·m³/mol [3]; 2.12×10^{-4} atm·m³/mol [4].

Ionization Potential: 10.88 ± 0.05 eV [5].

Log K_{oc}: 1.79 [3].

Log K_{ow}: 1.88 [6].

Melting Point: -57.1 °C [7]; -55 °C [8].

Solubility in Organics: Soluble in acetone, ethanol, benzene, chloroform, and ether [7].

Solubility in Water: 4,500 mg/L at 0 °C [3].

Specific Density: 1.980 at 20/4 °C [7]; 1.986 at 20/4 °C [2]; 1.9945 at 20/4 °C [9].

Transformation Products: No data found.

Vapor Density: 6.70 g/L at 25 °C, 5.66 (air = 1).

Vapor Pressure: 50 mm at 20 °C [6].

FIRE HAZARDS

Flash Point: No data found.

Lower Explosive Limit (LEL): No data found.

Upper Explosive Limit (UEL): No data found.

HEALTH HAZARD DATA

Immediately Dangerous to Life or Health (IDLH): No data found.

Permissible Exposure Limits (PEL) in Air: No standards set.

MANUFACTURING

Selected Manufacturers:

Aldrich Chemical Co.
940 West Saint Paul Ave.
Milwaukee, WI 53233

Fluka Chemical Corp.
980 South Second St.
Ronkonkoma, NY 11779

Pfaltz & Bauer, Inc.
172 East Aurora St.
Waterbury, CT 06708

Uses: Component of fire extinguisher fluids; solvent for waxes, fats, and resins; degreaser; flame retardant; heavy liquid for mineral and salt separations; chemical intermediate; laboratory use.

REFERENCES

1. Dean, J.A., Ed., *Lange's Handbook of Chemistry*, 11th ed. (New York: McGraw-Hill, Inc., 1973), 1570 p.
2. *Fluka Catalog 1988/89 - Chemika-Biochemika* (Ronkonkoma, NY: Fluka Chemical Corp., 1988), 1536 p.
3. Schwille, F. *Dense Chlorinated Solvents* (Chelsea, MI: Lewis Publishers, Inc., 1988), 146 p.
4. Warner, H.P., Cohen, J.M., and J.C. Ireland. "Determination of Henry's Law Constants of Selected Priority Pollutants," Office of Science and Development, U.S. EPA Report-600/D-87/229 (1987), 14 p.
5. Franklin, J.L., Dillard, J.G., Rosenstock, H.M., Herron, J.T., Draxl K., and F.H. Field. "Ionization Potentials, Appearance Potentials and Heats of Formation of Gaseous Positive Ions," National Bureau of Standards Report NSRDS-NBS 26, U.S. Government Printing Office (1969), 289 p.
6. Mills, W.B., Porcella, D.B., Ungs, M.J., Gherini, S.A., Summers, K.V., Mok, L., Rupp, G.L., and G.L. Bowie. "Water Quality Assessment: A Screening Procedure for Toxic and Conventional Pollutants in Surface and Groundwater-Part I," Office of Research and Development, U.S. EPA Report-600/6-85-002a (1985), 638 p.
7. Weast, R.C., Ed. *CRC Handbook of Chemistry and Physics*, 67th ed. (Boca Raton, FL: CRC Press, Inc., 1986), 2406 p.
8. *Catalog Handbook of Fine Chemicals* (Milwaukee, WI: Aldrich Chemical Co., 1988), 2212 p.
9. Standen, A., Ed. *Kirk-Othmer Encyclopedia of Chemical Technology, Volume 3*, 2nd ed. (New York: John Wiley and Sons, Inc., 1964), 927 p.

BROMOFORM

Synonyms: Methenyl tribromide; Methyl tribromide; NCI-C55130; RCRA waste number U225; **Tribromomethane**; UN 2515.

Structural Formula:

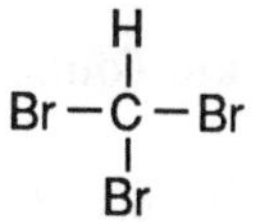

$$Br - \overset{\displaystyle H}{\underset{\displaystyle Br}{C}} - Br$$

CHEMICAL DESIGNATIONS

CAS Registry Number: 75-25-2

DOT Designation: 2515

Empirical Formula: $CHBr_3$

Formula Weight: 252.73

RTECS Number: PB 5600000

PHYSICAL AND CHEMICAL PROPERTIES

Appearance and Odor: Colorless liquid with an odor similar to chloroform.

Boiling Point: 149.5 °C [1]; 149.21 °C [2].

Henry's Law Constant: 5.6 x 10^{-4} atm·m^3/mol [3]; 5.32 x 10^{-4} atm·m^3/mol [4].

Ionization Potential: 10.51 eV [5].

Log K_{oc}: 2.45 [6]; 2.06 [7].

Log K_{ow}: 2.30 [8]; 2.38 [9].

Melting Point: 8.3 °C [1].

Solubility in Organics: Soluble in ethanol, benzene, chloroform, ether, and ligroin [1].

Solubility in Water: 3,010 mg/L at 15 °C, 3,190 mg/L at 30 °C [10]; 3,010 mg/L 20 °C [7]; 3,130 mg/L at 25 °C [9]; 3,050 mg/L 20 °C [11]; 0.318 wt% at 30 °C [12]; 3,100 ppm at 25 °C [13].

Specific Density: 2.8899 at 20/4 °C [1]; 2.89165 at 20/4 °C [2].

Transformation Products: No data found.

Vapor Density: 10.33 g/L at 25 °C, 8.72 (air = 1).

Vapor Pressure: 10 mm at 34 °C, 40 mm at 63.6 °C, 100 mm at 85.9 °C, 400 mm at 127.9 °C, 760 mm at 150.5 °C [1]; 4 mm at 20 °C [11]; 5.6 mm at 25 °C [14].

FIRE HAZARDS

Flash Point: Non-combustible [5].

Lower Explosive Limit (LEL): Not applicable [5].

Upper Explosive Limit (UEL): Not applicable [5].

HEALTH HAZARD DATA

Immediately Dangerous to Life or Health (IDLH): No data found.

Permissible Exposure Limits (PEL) in Air: 0.5 ppm ($\approx$5 mg/m^3) STEL [15].

MANUFACTURING

Selected Manufacturers:

Aldrich Chemical Co.
940 West Saint Paul Ave.
Milwaukee, WI 53233

Fluka Chemical Corp.
980 South Second St.
Ronkonkoma, NY 11779

Pfaltz & Bauer, Inc.
172 East Aurora St.
Waterbury, CT 06708

Uses: Solvent for waxes, greases, and oils; separating solids with lower densities; component of fire-resistant chemicals; geological assaying; medicine (sedative); intermediate in organic synthesis.

REFERENCES

1. Weast, R.C., Ed. *CRC Handbook of Chemistry and Physics*, 67th ed. (Boca Raton, FL: CRC Press, Inc., 1986), 2406 p.
2. Kudchadker, A.P., Kudchadker, S.A., Shukla, R.P., and P.R. Patnaik. "Vapor Pressures and Boiling Points of Selected Halomethanes," *J. Phys. Chem. Ref. Data*, 8(2):499-517 (1979).
3. Pankow, J.F., and M.E. Rosen. "Determination of Volatile Compounds in Water by Purging Directly to a Capillary Column with Whole Column Cryotrapping," *Environ. Sci. Technol.*, 22(4):398-405 (1988).
4. Warner, H.P., Cohen, J.M., and J.C. Ireland. "Determination of Henry's Law Constants of Selected Priority Pollutants," Office of Science and Development, U.S. EPA Report-600/D-87/229 (1987), 14 p.
5. "NIOSH Pocket Guide to Chemical Hazards," U.S. Department of Health and Human Services, U.S. Government Printing Office (1987), 241 p.
6. Abdul, S.A., Gibson, T.L., and D.N. Rai. "Statistical Correlations for Predicting the Partition Coefficient for Nonpolar Organic Contaminants Between Aquifer Organic Carbon and Water," *Haz. Waste Haz. Mater.*, 4(3):211-222 (1987).
7. Schwille, F. *Dense Chlorinated Solvents* (Chelsea, MI: Lewis Publishers, Inc., 1988), 146 p.
8. Mills, W.B., Porcella, D.B., Ungs, M.J., Gherini, S.A., Summers, K.V., Mok, L., Rupp, G.L., and G.L. Bowie. "Water Quality Assessment: A Screening Procedure for Toxic and Conventional Pollutants in Surface and Groundwater-Part I," Office of Research and Development, U.S. EPA Report-600/6-85-002a (1985), 638 p.
9. Valsaraj, K.T. "On the Physio-Chemical Aspects of Partitioning of Non-Polar Hydrophobic Organics at the Air-Water Interface," *Chemosphere*, 17(5):875-887 (1988).
10. Gross, P.M., and J.H. Saylor. "The Solubilities of Certain Slightly Soluble Organic Compounds in Water," *J. Am. Chem. Soc.*,

53(5):1744-1751 (1931).

11. Munz, C., and P.V. Roberts. "Air-Water Phase Equilibria of Volatile Organic Solutes," *J. Am. Water Works Assoc.*, 79(5):62-69 (1987).

12. Riddick, J.A., Bunger, W.B., and T.K. Sakano. *Organic Solvents - Physical Properties and Methods of Purification. Volume II* (New York: John Wiley and Sons, Inc., 1986), 1325 p.

13. Amoore, J.E., and E. Hautala. "Odor as an Aide to Chemical Safety: Odor Thresholds Compared with Threshold Limit Values and Volatilities for 214 Industrial Chemicals in Air and Water Dilution," *J. Appl. Toxicol.*, 3(6):272-290 (1983).

14. Verschueren, K. *Handbook of Environmental Data on Organic Chemicals* (New York: Van Nostrand Reinhold Co., 1983), 1310 p.

15. *Documentation of the Threshold Limit Values and Biological Exposure Indices for 1986-1987* (Cincinnati, OH: American Conference of Governmental Industrial Hygienists, 1986), 111 p.

4–BROMOPHENYL PHENYL ETHER

Synonyms: 4-Bromodiphenyl ether; *p*-Bromodiphenyl ether; **1-Bromo-4-phenoxybenzene**; 1-Bromo-*p*-phenoxybenzene; 4-Bromophenyl ether; *p*-Bromophenyl ether; *p*-Bromophenyl phenyl ether; Phenyl-4-bromophenyl ether; Phenyl-*p*-bromophenyl ether.

Structural Formula:

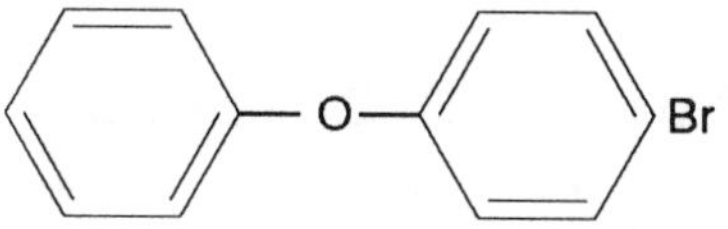

CHEMICAL DESIGNATIONS

CAS Registry Number: 101-55-3

DOT Designation: None assigned.

Empirical Formula: $C_{12}H_9BrO$

Formula Weight: 249.20

RTECS Number: Not assigned.

PHYSICAL AND CHEMICAL PROPERTIES

Appearance: Liquid.

Boiling Point: 310.1 °C [1]; 305 °C [2].

Henry's Law Constant: 1.0×10^{-4} atm·m^3/mol [3].

Ionization Potential: No data found.

Log K_{oc}: 4.94 using method of Karickhoff and others [4].

Log K_{ow}: 5.15 [5].

Melting Point: 18.7 °C [1].

Solubility in Organics: Soluble in ether [1].

Solubility in Water: No data found.

Specific Density: 1.4208 at 20/4 °C [1].

Transformation Products: No data found.

Vapor Density: 10.19 g/L at 25 °C, 8.60 (air = 1).

Vapor Pressure: 0.0015 mm at 20 °C (calculated) [6].

FIRE HAZARDS

Flash Point: > 110 °C [2].

Lower Explosive Limit (LEL): No data found.

Upper Explosive Limit (UEL): No data found.

HEALTH HAZARD DATA

Immediately Dangerous to Life or Health (IDLH): No data found.

Permissible Exposure Limits (PEL) in Air: No standards set.

MANUFACTURING

Selected Manufacturers:

Aldrich Chemical Co.
940 West Saint Paul Ave.
Milwaukee, WI 53233

Pfaltz & Bauer, Inc.
172 East Aurora St.
Waterbury, CT 06708

Use: Research chemical.

REFERENCES

1. Weast, R.C., Ed. *CRC Handbook of Chemistry and Physics*, 67th ed. (Boca Raton, FL: CRC Press, Inc., 1986), 2406 p.
2. *Catalog Handbook of Fine Chemicals* (Milwaukee, WI: Aldrich Chemical Co., 1988), 2212 p.
3. Pankow, J.F., and M.E. Rosen. "Determination of Volatile Compounds in Water by Purging Directly to a Capillary Column with Whole Column Cryotrapping," *Environ. Sci. Technol.*, 22(4):398-405 (1988).
4. Karickhoff, S.W., Brown, D.S., and T.A. Scott. "Sorption of Hydrophobic Pollutants on Natural Sediments," *Water Res.*, 13:241-248 (1979).
5. Walton, W.C. *Practical Aspects of Ground Water Modeling* (Worthington, OH: National Water Well Association, 1985), 587 p.
6. Dreisbach, R.R. *Pressure-Volume-Temperature Relationships of Organic Compounds* (Sandusky, OH: Handbook Publishers, 1952), 349 p.

2-BUTANONE

Synonyms: Butanone; Ethyl methyl ketone; Meetco; MEK; Methyl acetone; Methyl ethyl ketone; RCRA waste number U159; UN 1193; UN 1232.

Structural Formula:

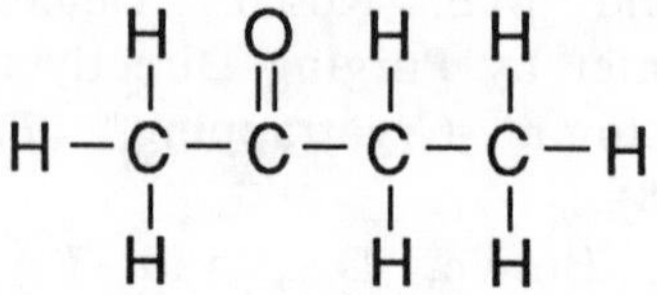

$$H-\overset{\overset{\displaystyle H}{|}}{\underset{\underset{\displaystyle H}{|}}{C}}-\overset{\overset{\displaystyle O}{\parallel}}{C}-\overset{\overset{\displaystyle H}{|}}{\underset{\underset{\displaystyle H}{|}}{C}}-\overset{\overset{\displaystyle H}{|}}{\underset{\underset{\displaystyle H}{|}}{C}}-H$$

CHEMICAL DESIGNATIONS

CAS Registry Number: 78-93-3

DOT Designation: 1193

Empirical Formula: C_4H_8O

Formula Weight: 72.11

RTECS Number: EL 6475000

PHYSICAL AND CHEMICAL PROPERTIES

Appearance and Odor: Colorless liquid with a sweet mint-like odor.

Boiling Point: 79.6 °C [1].

Henry's Law Constant: 4.66 x 10^{-5} atm·m^3/mol at 25 °C [2].

Ionization Potential: 9.48 eV [3].

Log K_{oc}: 0.09 using method of Rao and Davidson [4].

Log K_{ow}: 0.26 and 0.29 [5].

Melting Point: -86.9 °C [6]; -85.9 °C [7].

Solubility in Organics: Miscible with acetone, ethanol, benzene, and ether [8].

Solubility in Water: 353 g/L at 10 °C, 190 g/L at 90 °C [9]; 27 wt% at 20 °C [3]; 28 vol% at 20 °C [10]; 24.00 wt% at 20 °C [11]; 26.7 wt% at 20 °C [12]; 27.33 wt% at 20 °C, 25.57 wt% at 25 °C, 24.07 wt% at 30 °C [13].

Specific Density: 0.8054 at 20/4 °C [1]; 0.804 at 20/4 °C [14]; 0.8061 at 20/4 °C, 0.8047 at 20/20 °C [15].

Transformation Products: No data found.

Vapor Density: 2.94 g/L at 25 °C, 2.49 (air = 1).

Vapor Pressure: 77.5 mm at 20 °C [9]; 1 mm at -48.3 °C, 10 mm at -17.7 °C, 40 mm at 6.0 °C, 100 mm at 25.0 °C, 400 mm at 60.0 °C, 760 mm at 79.6 °C [1]; 70 mm at 20 °C [3]; 71.2 mm at 20 °C [16].

FIRE HAZARDS

Flash Point: -9 °C [17].

Lower Explosive Limit (LEL): 2% [3].

Upper Explosive Limit (UEL): 10% [3].

HEALTH HAZARD DATA

Immediately Dangerous to Life or Health (IDLH): 3,000 ppm [3].

Permissible Exposure Limits (PEL) in Air: 200 ppm ($\approx$590 mg/m^3) [18]; 200 ppm 10-hour TWA, 300 ppm ($\approx$885 mg/m^3) 15-minute ceiling [3]; 200 ppm TLV, 300 ppm STEL [19].

MANUFACTURING

Selected Manufacturers:

Celanese Chemical Co.
245 Park Ave.
New York, NY 10017

Exxon Chemical Co.
P.O. Box 3272
Houston, TX 77001

Shell Chemical Co.
Industrial Chemicals Division
Houston, TX 77001

Uses: Solvent in nitrocellulose coatings, vinyl films, and "Glyptal" resins; paint removers; cements and adhesives; organic synthesis; manufacturing of smokeless powders, ketones and amines; cleaning fluids; printing; catalyst carrier; acrylic coatings.

REFERENCES

1. Weast, R.C., Ed. *CRC Handbook of Chemistry and Physics*, 67th ed. (Boca Raton, FL: CRC Press, Inc., 1986), 2406 p.
2. Hine, J., and P.K. Mookerjee. "The Intrinsic Hydrophilic Character of Organic Compounds. Correlations in Terms of Structural Contributions," *J. Org. Chem.*, 40(3):292-298 (1975).
3. "NIOSH Pocket Guide to Chemical Hazards," U.S. Department of Health and Human Services, U.S. Government Printing Office (1987), 241 p.
4. Rao, P.S.C., and J.M. Davidson. "Estimation of Pesticide Retention and Transformation Parameters Required in Nonpoint Source Pollution Models," in *Environmental Impact of Nonpoint Source Pollution*, Overcash, M.R., and J.M. Davidson, Eds. (Ann Arbor, MI: Ann Arbor Science Publishers, Inc., 1980), pp 23-67.
5. Leo, A., Hansch, C., and D. Elkins. "Partition Coefficients and Their Uses," *Chem. Rev.*, 71(6):525-616 (1971).
6. Dean, J.A., Ed., *Lange's Handbook of Chemistry*, 11th ed. (New York: McGraw-Hill, Inc., 1973), 1570 p.
7. Sax, N.I. *Dangerous Properties of Industrial Materials* (New York: Van Nostrand Reinhold Co., 1984), 3124 p.
8. "Chemical, Physical, and Biological Properties of Compounds Present at Hazardous Waste Sites," U.S. EPA Report-530/SW-89-010 (1985), 619 p.
9. Verschueren, K. *Handbook of Environmental Data on Organic Chemicals* (New York: Van Nostrand Reinhold Co., 1983), 1310 p.
10. Meites, L., Ed. *Handbook of Analytical Chemistry*, 1st ed. (New York: McGraw-Hill, Inc., 1963), 1782 p.
11. Riddick, J.A., Bunger, W.B., and T.K. Sakano. *Organic Solvents -*

Physical Properties and Methods of Purification. Volume II (New York: John Wiley and Sons, Inc., 1986), 1325 p.

12. Stephen, H., and T. Stephen. *Solubilities of Inorganic and Organic Compounds - Part 1, Volume 1* (London: Pergamon Printing and Art Services, Ltd., 1963), 960 p.

13. Ginnings, P.M., Plonk, D., and E. Carter. "Aqueous Solubilities of Some Aliphatic Ketones," *J. Am. Chem. Soc.*, 62(8):1923-1924 (1940).

14. *Fluka Catalog 1988/89 - Chemika-Biochemika* (Ronkonkoma, NY: Fluka Chemical Corp., 1988), 1536 p.

15. Standen, A., Ed. *Kirk-Othmer Encyclopedia of Chemical Technology, Volume 4*, 2nd ed. (New York: John Wiley and Sons, Inc., 1964), 937 p.

16. Standen, A., Ed. *Kirk-Othmer Encyclopedia of Chemical Technology, Volume 12*, 2nd ed. (New York: John Wiley and Sons, Inc., 1967), 905 p.

17. *Fire Protection Guide on Hazardous Materials* (Quincy, MA: National Fire Protection Association, 1984), 443 p.

18. "General Industry Standards for Toxic and Hazardous Substances," U.S. Code of Federal Regulations 1910, Subpart Z, Section 1910.1000 (July 1982).

19. *Documentation of the Threshold Limit Values and Biological Exposure Indices for 1986-1987* (Cincinnati, OH: American Conference of Governmental Industrial Hygienists, 1986), 111 p.

CARBON DISULFIDE

Synonyms: Carbon bisulfide; Carbon bisulphide; Carbon disulphide; Carbon sulfide; Carbon sulphide; Dithiocarbonic anhydride; NCI-C04591; RCRA waste number P022; Sulphocarbonic anhydride; UN 1131; Weeviltox.

Structural Formula:

$$S=C=S$$

CHEMICAL DESIGNATIONS

CAS Registry Number: 75-15-0

DOT Designation: 1131

Empirical Formula: CS_2

Formula Weight: 76.13

RTECS Number: FF 6650000

PHYSICAL AND CHEMICAL PROPERTIES

Appearance and Odor: Clear, water-white to pale yellow liquid; ethereal odor when pure; technical and reagent grades have foul odors.

Boiling Point: 46.2 °C [1].

Henry's Law Constant: 0.0133 atm·m^3/mol (calculated) [2].

Ionization Potential: 10.06 eV [3].

Log K_{oc}: 2.38-2.55 using method of Kenaga and Goring [4].

Log K_{ow}: 1.84, 2.16 (calculated) [5].

Melting Point: -111.5 °C [1].

Solubility in Organics: Soluble in ethanol, chloroform, and ether [1].

Solubility in Water: 2,300 mg/L at 22 °C [6]; 2,000 mg/L at 0 °C [7]; 0.2 wt% at 20 °C [3]; 2,200 mg/L at 22 °C [8]; 2,940 mg/L at 20 °C [2]; 2,000 mg/L at 20 °C [9]; 0.1185 wt% at 25 °C [10]; 0.210 wt% at 20 °C [11]; 1,700 ppm at 25 °C [12].

Specific Density: 1.2632 at 20/4 °C [1]; 1.260 at 20/4 °C [13]; 1.2931 at 0/4 °C, 1.2632 at 15/4 °C, 1.2559 at 25/4 °C, 1.2500 at 30/4 °C, 1.2250 at 46.25/4 °C [14].

Transformation Products: No data found.

Vapor Density: 3.11 g/L at 25 °C, 2.63 (air = 1).

Vapor Pressure: 3.5 °C at -60 °C, 14.0 mm at -40 °C, 46.5 mm at -20 °C, 78.8 mm at -10 °C, 127.3 mm at 0 °C, 198.1 mm at 10 °C, 297.5 mm at 20 °C, 616.7 mm at 40 °C, 1 atm at 46.25 °C, 1.54 atm at 60 °C, 2.69 atm at 80 °C, 4.42 atm at 100 °C, 12.4 atm at 150 °C, 28.3 atm at 200 °C, 56.5 atm at 250 °C, 75 atm at 273 °C [14]; 430 mm at 30 °C [6]; 1 mm at -73.8 °C, 10 mm at -44.7 °C, 40 mm at -22.5 °C, 100 mm at -5.1 °C, 400 mm at 28.0 °C, 760 mm at 46.5 °C [1]; 300 mm at 20 °C [8]; 360 mm at 25 °C [2].

FIRE HAZARDS

Flash Point: -30 °C [3].

Lower Explosive Limit (LEL): 1.3% [3].

Upper Explosive Limit (UEL): 50% [3].

HEALTH HAZARD DATA

Immediately Dangerous to Life or Health (IDLH): 500 ppm [3].

Permissible Exposure Limits (PEL) in Air: 20 ppm, 30 ppm ceiling, 100 ppm peak for 30 minutes [15]; 1 ppm 10-hour TWA, 10 ppm ($\approx$30 mg/m^3) 15-minute ceiling [3]; 10 ppm TLV [8].

MANUFACTURING

Selected Manufacturers:

Aldrich Chemical Co.
940 West Saint Paul Ave.
Milwaukee, WI 53233

Pennwalt Corp.
Chemicals Group
Organic Chemicals Division
Greens Bayou, TX 77015

Uses: Manufacturing of viscose rayon, cellophane, and flotation agents; soil disinfectants; electronic vacuum tubes; herbicides; grain fumigants; solvent for fats, resins, phosphorus, sulfur, bromine, iodine, and rubber.

REFERENCES

1. Weast, R.C., Ed. *CRC Handbook of Chemistry and Physics*, 67th ed. (Boca Raton, FL: CRC Press, Inc., 1986), 2406 p.
2. "Treatability Manual - Volume 1: Treatability Data," Office of Research and Development, U.S. EPA Report-600/8-80-042a (1980), 1035 p.
3. "NIOSH Pocket Guide to Chemical Hazards," U.S. Department of Health and Human Services, U.S. Government Printing Office (1987), 241 p.
4. Kenaga, E.E., and C.A.I. Goring. "Relationship Between Water Solubility, Soil Sorption, Octanol-Water Partitioning and Concentration of Chemicals in Biota," in *Aquatic Toxicology, ASTM STP 707*, Eaton, J.G., Parrish, P.R., and A.C. Hendricks, Eds. (Philadelphia, PA: American Society for Testing and Materials, 1980), pp 78-115.
5. Leo, A., Hansch, C., and D. Elkins. "Partition Coefficients and Their Uses," *Chem. Rev.*, 71(6):525-616 (1971).
6. Verschueren, K. *Handbook of Environmental Data on Organic Chemicals* (New York: Van Nostrand Reinhold Co., 1983), 1310 p.
7. Dean, J.A., Ed., *Lange's Handbook of Chemistry*, 11th ed. (New York: McGraw-Hill, Inc., 1973), 1570 p.
8. *Documentation of the Threshold Limit Values and Biological Exposure Indices for 1986-1987* (Cincinnati, OH: American Conference of Governmental Industrial Hygienists, 1986), 111 p.

9. *Environmental Health Criteria 10: Carbon disulfide* (Geneva: World Health Organization, 1979), 100 p.

10. Stephen, H., and T. Stephen. *Solubilities of Inorganic and Organic Compounds - Part 1, Volume 1* (London: Pergamon Printing and Art Services, Ltd., 1963), 960 p.

11. Riddick, J.A., Bunger, W.B., and T.K. Sakano. *Organic Solvents - Physical Properties and Methods of Purification. Volume II* (New York: John Wiley and Sons, Inc., 1986), 1325 p.

12. Amoore, J.E., and E. Hautala. "Odor as an Aide to Chemical Safety: Odor Thresholds Compared with Threshold Limit Values and Volatilities for 214 Industrial Chemicals in Air and Water Dilution," *J. Appl. Toxicol.*, 3(6):272-290 (1983).

13. *Fluka Catalog 1988/89 - Chemika-Biochemika* (Ronkonkoma, NY: Fluka Chemical Corp., 1988), 1536 p.

14. Standen, A., Ed. *Kirk-Othmer Encyclopedia of Chemical Technology, Volume 4*, 2nd ed. (New York: John Wiley and Sons, Inc., 1964), 937 p.

15. "General Industry Standards for Toxic and Hazardous Substances," U.S. Code of Federal Regulations 1910, Subpart Z, Section 1910.1000 (July 1982).

CARBON TETRACHLORIDE

Synonyms: Benzinoform; Carbona; Carbon chloride; Carbon tet; ENT 4,705; Fasciolin; Flukoids; Freon 10; Halon 104; Methane tetrachloride; Necatorina; Necatorine; Perchloromethane; R 10; RCRA waste number U211; Tetrachloormetaan; Tetrachlorocarbon; **Tetrachloromethane**; Tetrafinol; Tetraform; Tetrasol; UN 1846; Univerm; Vermoestricid.

Structural Formula:

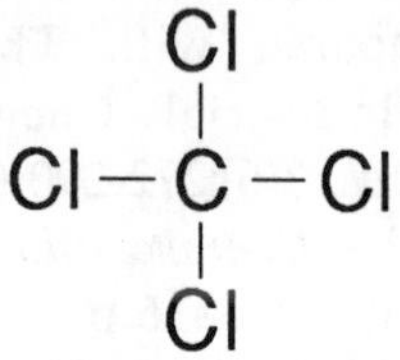

$$Cl - \underset{\underset{Cl}{|}}{\overset{\overset{Cl}{|}}{C}} - Cl$$

CHEMICAL DESIGNATIONS

CAS Registry Number: 56-23-5

DOT Designation: 1846

Empirical Formula: CCl_4

Formula Weight: 153.82

RTECS Number: FG 4900000

PHYSICAL AND CHEMICAL PROPERTIES

Appearance and Odor: Clear, colorless, heavy, watery-liquid with a strong sweetish, distinctive odor resembling ether.

Boiling Point: 76.5 °C [1]; 78-79 °C [2]; 76.72 °C [3].

Henry's Law Constant: 0.0302 atm·m³/mol [4]; 0.024 atm·m³/mol at 20 °C [5]; 0.023 atm·m³/mol [6]; 0.102 atm·m³/mol at 37 °C [7].

Ionization Potential: 11.47 eV [8].

Log K_{oc}: 2.35 [9]; 2.64 [10]; 2.62 [11].

Log K_{ow}: 2.83 [12]; 2.73 [13].

Melting Point: -22.96 °C [14].

Solubility in Organics: Miscible with ethanol, benzene, chloroform, ether, carbon disulfide [15], petroleum ether, solvent naphtha, and volatile oils [16].

Solubility in Water: 800 mg/L at 20 °C, 1,160 mg/L at 25 °C [17]; 785 mg/L at 20 °C [18]; 970 mg/L at 0 °C [14]; 0.05 vol% at 20 °C [19]; 800 mg/L at 25 °C [20]; 757 mg/L at 25 °C [13]; 770 mg/L at 15 °C, 810 mg/L at 30 °C [21]; 803 mg/L at 20 °C [22]; 0.097 wt% at 0 °C, 0.083 wt% at 10 °C, 0.080 wt% at 20 °C, 0.085 wt% at 30 °C [23]; 0.077 wt% at 25 °C [24]; 770 mg/L at 25 °C [25]; 927 mg/L at 25 °C [26].

Specific Density: 1.5940 at 20/4 °C [1]; 1.592 at 20/4 °C [2]; 1.63195 at 0/4 °C, 1.59472 at 20/4 °C, 1.58445 at 24/4 °C, 1.58828 at 25/25 °C, 1.48020 at 76/4 °C [3]; 1.5844 at 25/4 °C [27].

Transformation Products: Under laboratory conditions, carbon tetrachloride hydrolyzed partially to chloroform and carbon dioxide. Complete hydrolysis yielded carbon dioxide and hydrochloric acid. Chloroform was also formed by microbial degradation of carbon tetrachloride using denitrifying bacteria [28]. An anaerobic species of *Clostridium* biodegraded carbon tetrachloride by reductive dechlorination yielding trichloromethane (chloroform), dichloromethane, and unidentified products [29].

Vapor Density: 6.29 g/L at 25 °C, 5.31 (air = 1).

Vapor Pressure: 56 mm at 10 °C, 90 mm at 20 °C, 113 mm at 25 °C, 137 mm at 30 °C [17]; 760 mm at 76.7 °C, 1,520 mm at 102 °C, 3,800 mm at 141.7 °C, 7,600 mm at 178 °C, 15,200 mm at 222 °C, 30,400 mm at 276 °C [1]; 100 mm at 23 °C [30]; 0.925 mm at -50.1 °C, 9.92 mm at -20 °C, 18.81 mm at -10 °C, 33.08 mm at 0 °C, 55.65 mm at 10 °C, 89.55 mm at 20 °C, 139.6 mm at 30 °C, 210.9 mm at 40 °C, 309.0 mm at 50 °C, 439.0 mm at 60 °C, 613.8 mm at 70 °C, 760 mm at 76.72 °C, 4,555 mm at 150 °C, 10,936 mm at 200 °C [3]; 115 mm at 25 °C [31].

FIRE HAZARDS

Flash Point: Non-combustible [32].

Lower Explosive Limit (LEL): Not applicable [32].

Upper Explosive Limit (UEL): Not applicable [32].

HEALTH HAZARD DATA

Immediately Dangerous to Life or Health (IDLH): 300 ppm [33].

Permissible Exposure Limits (PEL) in Air: 10 ppm, 25 ppm ceiling, 200 ppm 5-minute/4-hour peak [34]; 2 ppm 1-hour ceiling [32]; 5 ppm ($\approx$30 mg/m^3) TLV [35].

MANUFACTURING

Selected Manufacturers:

Dow Chemical Co.
Midland, MI 48640

FMC Corp.
633 Third Ave.
New York, NY 10017

Stauffer Chemical Corp.
Industrial Chemical Division
Le Mayne, AL 36505

Uses: Preparation of refrigerants, aerosols and propellants; metal degreasing; agricultural fumigant; chlorinating organic compounds; production of semiconductors; solvent for fats, oils, rubber, etc; dry cleaning operations; industrial extractant; spot remover; fire extinguisher manufacturing; preparation of chlorofluoromethanes; veterinary medicine; organic synthesis.

REFERENCES

1. Weast, R.C., Ed. *CRC Handbook of Chemistry and Physics*, 67th ed. (Boca Raton, FL: CRC Press, Inc., 1986), 2406 p.
2. *Fluka Catalog 1988/89 - Chemika-Biochemika* (Ronkonkoma, NY: Fluka Chemical Corp., 1988), 1536 p.
3. Standen, A., Ed. *Kirk-Othmer Encyclopedia of Chemical Technology, Volume 4*, 2nd ed. (New York: John Wiley and Sons, Inc., 1964), 937

p.
4. Warner, H.P., Cohen, J.M., and J.C. Ireland. "Determination of Henry's Law Constants of Selected Priority Pollutants," Office of Science and Development, U.S. EPA Report-600/D-87/229 (1987), 14 p.
5. Roberts, P.V., and P.G. Dändliker. "Mass Transfer of Volatile Organic Contaminants from Aqueous Solution to the Atmosphere During Surface Aeration," *Environ. Sci. Technol.*, 17(8):484-489 (1983).
6. Jury, W.A., Spencer, W.F., and W.J. Farmer. "Behavior Assessment Model for Trace Organics in Soil: III. Application of Screening Model," *J. Environ. Qual.*, 13(4):573-579 (1984).
7. Sato, A., and T. Nakajima. "A Structure-Activity Relationship of Some Chlorinated Hydrocarbons," *Arch. Environ. Health*, 34(2):69-75 (1979).
8. Franklin, J.L., Dillard, J.G., Rosenstock, H.M., Herron, J.T., Draxl K., and F.H. Field. "Ionization Potentials, Appearance Potentials and Heats of Formation of Gaseous Positive Ions," National Bureau of Standards Report NSRDS-NBS 26, U.S. Government Printing Office (1969), 289 p.
9. Abdul, S.A., Gibson, T.L., and D.N. Rai. "Statistical Correlations for Predicting the Partition Coefficient for Nonpolar Organic Contaminants Between Aquifer Organic Carbon and Water," *Haz. Waste Haz. Mater.*, 4(3):211-222 (1987).
10. Schwille, F. *Dense Chlorinated Solvents* (Chelsea, MI: Lewis Publishers, Inc., 1988), 146 p.
11. Chin, Y.-P., Peven, C.S., and W.J. Weber. "Estimating Soil/Sediment Partition Coefficients for Organic Compounds by High Performance Reverse Phase Liquid Chromatography," *Water Res.*, 22(7):873-881 (1988).
12. Chou, J.T., and P.C. Jurs. "Computer Assisted Computation of Partition Coefficients from Molecular Structures using Fragment Constants," *J. Chem. Info. Comp. Sci.*, 19:172-178 (1979).
13. Banerjee, S., Yalkowsky, S.H., and S.C. Valvani. "Water Solubility and Octanol/Water Partition Coefficients of Organics. Limitations of the Solubility-Partition Coefficient Correlation," *Environ. Sci. Technol.*, 14(10):1227-1229 (1980).
14. Dean, J.A., Ed., *Lange's Handbook of Chemistry*, 11th ed. (New York: McGraw-Hill, Inc., 1973), 1570 p.
15. "Chemical, Physical, and Biological Properties of Compounds Present at Hazardous Waste Sites," U.S. EPA Report-530/SW-89-010 (1985), 619 p.
16. Yoshida, K., Shigeoka, T., and F. Yamauchi. "Non-Steady-State

Equilibrium Model for the Preliminary Prediction of the Fate of Chemicals in the Environment," *Ecotoxicol. Environ. Safety*, 7(2):179-190 (1983).

17. Verschueren, K. *Handbook of Environmental Data on Organic Chemicals* (New York: Van Nostrand Reinhold Co., 1983), 1310 p.

18. Pearson, C.R., and G. McConnell. "Chlorinated C_1 and C_2 Hydrocarbons in the Marine Environment," in *Proc. R. Soc. London*, B189(1096):305-322 (1975).

19. Meites, L., Ed. *Handbook of Analytical Chemistry*, 1st ed. (New York: McGraw-Hill, Inc., 1963), 1782 p.

20. Valsaraj, K.T. "On the Physio-Chemical Aspects of Partitioning of Non-Polar Hydrophobic Organics at the Air-Water Interface," *Chemosphere*, 17(5):875-887 (1988).

21. Gross, P.M., and J.H. Saylor. "The Solubilities of Certain Slightly Soluble Organic Compounds in Water," *J. Am. Chem. Soc.*, 53(5):1744-1751 (1931).

22. Munz, C., and P.V. Roberts. "Air-Water Phase Equilibria of Volatile Organic Solutes," *J. Am. Water Works Assoc.*, 79(5):62-69 (1987).

23. Stephen, H., and T. Stephen. *Solubilities of Inorganic and Organic Compounds - Part 1, Volume 1* (London: Pergamon Printing and Art Services, Ltd., 1963), 960 p.

24. Riddick, J.A., Bunger, W.B., and T.K. Sakano. *Organic Solvents - Physical Properties and Methods of Purification. Volume II* (New York: John Wiley and Sons, Inc., 1986), 1325 p.

25. Gross, P. "The Determination of the Solubility of Slightly Soluble Liquids in Water and the Solubilities of the Dichloro- Ethanes and -Propanes," *J. Am. Chem. Soc.*, 51(8):2362-2366 (1929).

26. Kamlet, M.J., Doherty, R.M., Abraham, M.H., Carr, P.W., Doherty, R.F., and R.W. Taft. "Linear Solvation Energy Relationships. 41. Important Differences Between Aqueous Solubility Relationships for Aliphatic and Aromatic Solutes," *J. Phys. Chem.*, 91(7):1996-2004 (1987).

27. Kirchnerová, J., and G.C.B. Cave. "The Solubility of Water in Low-Dielectric Solvents," *Can. J. Chem.*, 54(24):3909-3916 (1976).

28. Smith, L.R., and J. Dragun. "Degradation of Volatile Chlorinated Aliphatic Priority Pollutants in Groundwater," *Environ. Int.*, 19(4):291-298 (1984).

29. Gälli, R., and P.L. McCarty. "Biotransformation of 1,1,1-Trichloroethane, Trichloromethane, and Tetrachloromethane by a *Clostridium* sp.," *Appl. Environ. Microbiol.*, 55(4):837-844 (1989).

30. Sax, N.I. *Dangerous Properties of Industrial Materials* (New York: Van Nostrand Reinhold Co., 1984), 3124 p.

31. Rogers, R.D., and J.C. McFarlane. "Sorption of Carbon

Tetrachloride, Ethylene Dibromide, and Trichloroethylene on Soil and Clay," *Environ. Monitor. Assess.*, 1(2):155-162 (1981).

32. "NIOSH Pocket Guide to Chemical Hazards," U.S. Department of Health and Human Services, U.S. Government Printing Office (1987), 241 p.

33. Weiss, G. *Hazardous Chemicals Data Book* (Park Ridge, NJ: Noyes Data Corp., 1986), 1069 p.

34. "General Industry Standards for Toxic and Hazardous Substances," U.S. Code of Federal Regulations 1910, Subpart Z, Section 1910.1000 (July 1982).

35. *Documentation of the Threshold Limit Values and Biological Exposure Indices for 1986-1987* (Cincinnati, OH: American Conference of Governmental Industrial Hygienists, 1986), 111 p.

CHLORDANE

Synonyms: 1,068; Aspon-chlordane; Belt; CD-68; Chlordan; γ-Chlordan; Chloridan; Chlorindan; Chlor kil; Chlorodane; Chlortox; Corodane; Cortilan-neu; Dichlorochlordene; Dowklor; ENT 9,932; ENT 25,552-X; HCS 3,260; Kypchlor; M 140; M 410; NA 2,762; NCI-C00099; Niran; Octachlor; 1,2,4,5,6,7,8,8-Octachlor-2,3,3a,4,7,7a-hexahydro-4,7-methanoindane; Octachlorodihydrodicyclopentadiene; 1,2,4,5,6,7,8,8-Octachloro-2,3,3a,4,7,7a-hexahydro-4,7-methanoindene; **1,2,4,5,6,7,8,8-Octachloro-2,3,3a,4,7,7a-hexahydro-4,7-methano-1*H*-indene**; 1,2,4,5,6,7,8,8-Octachloro-3a,4,7,7a-hexahydro-4,7-methyleneindane; Octachloro-4,7-methanohydroindane; Octachloro-4,7-methano-tetrahydroindane; 1,2,4,5,6,7,8,8-Octachloro-4,7-methano-3a,4,7,7a-tetrahydroindane; 1,2,4,5,6,7,8,8-Octachloro-3a,4,7,7a-tetrahydro-4,7-methanoindan; 1,2,4,5,6,7,8,8-Octachloro-3a,4,7,7a-tetrahydro-4,7-methanoindane; Octaklor; Octaterr; Orthoklor; RCRA waste number U036; SD 5,532; Shell SD-5532; Synklor; Tat chlor 4; Topichlor 20; Topiclor; Topiclor 20; Toxichlor; Velsicol 1,068.

Structural Formula:

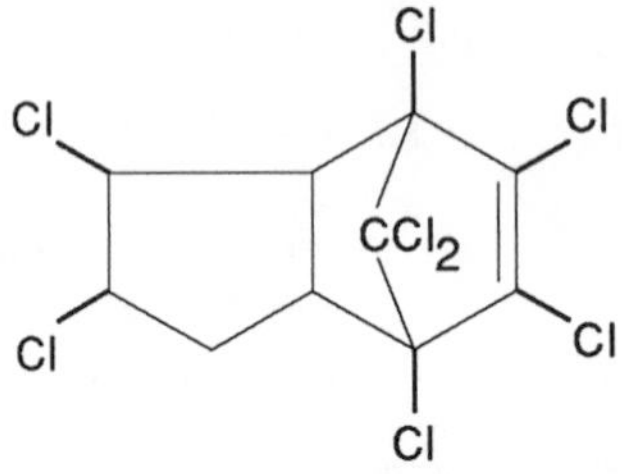

Chlordane is a mixture of *cis*-chlordane, *trans*-chlordane, and other complex chlorinated hydrocarbons including heptachlor and nonachlor.

CHEMICAL DESIGNATIONS

CAS Registry Number: 57-74-9

DOT Designation: 2762

Empirical Formula: $C_{10}H_6Cl_8$

Formula Weight: 409.78

RTECS Number: PB 9800000

PHYSICAL AND CHEMICAL PROPERTIES

Appearance and Odor: Colorless to amber to yellowish-brown viscous liquid with an aromatic, slight pungent odor similar to chlorine.

Boiling Point: 175 °C at 2 mm [1]; decomposes under atmospheric pressure [2].

Henry's Law Constant: 4.8×10^{-5} atm·m^3/mol [3].

Ionization Potential: No data found.

Log K_{oc}: 5.15, 5.57 [4].

Log K_{ow}: 6.00 [5].

Melting Point: < 25 °C [6].

Solubility in Organics: Technical grades are miscible with aliphatic and aromatic solvents [7].

Solubility in Water: 9 µg/L (technical grade), *cis:trans* (75:25) chlordane: 0.056 ppm [8]; 0.056 ppm at 25 °C [9]; 1.85 ppm at 25 °C [10]; 0.009 ppm at 25 °C [11].

Specific Density: 1.59-1.63 at 20/4 °C [12]; 1.57-1.63 at 15/15 °C [13]; ≈1.65 at 16/16 °C [14].

Transformation Products: No data found.

Vapor Density: 16.75 g/L at 25 °C, 14.15 (air = 1).

Vapor Pressure: 1×10^{-5} mm at 20 °C [2]; 1×10^{-6} mm at 20 °C [15]; 1×10^{-5} mm at 25 °C [16].

FIRE HAZARDS

Flash Point: The solid is non-combustible, however, in solution the open cup and closed cup flash points, respectively are 107.2 °C and 55.6 °C [17]; > 81 °C (technical grade) [7].

Lower Explosive Limit (LEL): 0.7% (kerosene solution) [17].

Upper Explosive Limit (UEL): 5% (kerosene solution) [17].

HEALTH HAZARD DATA

Immediately Dangerous to Life or Health (IDLH): 500 mg/m^3 [2].

Permissible Exposure Limits (PEL) in Air: 0.5 mg/m^3 [18]; 0.5 mg/m^3 TLV, 2 mg/m^3 STEL [19].

MANUFACTURING

Selected Manufacturers:

Velsicol Chemical Corp.
341 East Ohio St.
Chicago, IL 60611

S.B. Penick and Co.
260 Madison Ave.
New York, NY 10016

Chempar Chemical Co.
260 Madison Ave.
New York, NY 10016

Uses: Insecticide and fumigant.

REFERENCES

1. Roark, R.C. "A Digest of Information on Chlordane," Bureau of Entomology and Plant Quarantine, U.S. Dept. of Agr. Report E-817 (1951), 132 p.
2. "NIOSH Pocket Guide to Chemical Hazards," U.S. Department of Health and Human Services, U.S. Government Printing Office (1987), 241 p.
3. Warner, H.P., Cohen, J.M., and J.C. Ireland. "Determination of Henry's Law Constants of Selected Priority Pollutants," Office of Science and Development, U.S. EPA Report-600/D-87/229 (1987), 14 p.
4. Chin, Y.-P., Peven, C.S., and W.J. Weber. "Estimating Soil/Sediment

Partition Coefficients for Organic Compounds by High Performance Reverse Phase Liquid Chromatography," *Water Res.*, 22(7):873-881 (1988).

5. Travis, C.C., and A.D. Arms. "Bioconcentration of Organics in Beef, Milk and Vegetation," *Environ. Sci. Technol.*, 22(3):271-274 (1988).

6. Windholz, M., Budavari, S., Blumetti, R.F., and E.S. Otterbein, Eds. *The Merck Index*, 10th ed., (Rahway, NJ: Merck and Co., 1983), 1463 p.

7. "Chemical, Physical, and Biological Properties of Compounds Present at Hazardous Waste Sites," U.S. EPA Report-530/SW-89-010 (1985), 619 p.

8. Verschueren, K. *Handbook of Environmental Data on Organic Chemicals* (New York: Van Nostrand Reinhold Co., 1983), 1310 p.

9. Sanborn, J.R., Metcalf, R.L., Bruce, W.N., and P.-Y. Lu. "The Fate of Chlordane and Toxaphene in a Terrestrial-Aquatic Model Ecosystem," *Environ. Entomol.*, 5(3):533-538 (1976).

10. Weil, L., Dure, G., and K.E. Quentin. "Solubility in Water of Insecticide Chlorinated Hydrocarbons and Polychlorinated Biphenyls in View of Water Pollution," *Z. Wasser Forsch.*, 7(6):169-175 (1974).

11. *Drinking Water and Health* (Washington, DC: National Academy of Sciences, 1977), 939 p.

12. Melnikov, N.N. *Chemistry of Pesticides* (New York: Springer-Verlag, Inc., 1971), 480 p.

13. Sax, N.I. *Dangerous Properties of Industrial Materials* (New York: Van Nostrand Reinhold Co., 1984), 3124 p.

14. Standen, A., Ed. *Kirk-Othmer Encyclopedia of Chemical Technology, Volume 4*, 2nd ed. (New York: John Wiley and Sons, Inc., 1964), 937 p.

15. *Documentation of the Threshold Limit Values and Biological Exposure Indices* (Cincinnati, OH: American Conference of Governmental Industrial Hygienists, 1986), 744 p.

16. Sunshine, I., Ed. *Handbook of Analytical Toxicology* (Cleveland, OH: The Chemical Rubber Co., 1969), 1081 p.

17. Weiss, G. *Hazardous Chemicals Data Book* (Park Ridge, NJ: Noyes Data Corp., 1986), 1069 p.

18. "General Industry Standards for Toxic and Hazardous Substances," U.S. Code of Federal Regulations 1910, Subpart Z, Section 1910.1000 (July 1982).

19. *Documentation of the Threshold Limit Values and Biological Exposure Indices for 1986-1987* (Cincinnati, OH: American Conference of Governmental Industrial Hygienists, 1986), 111 p.

cis-CHLORDANE

Synonyms: α-Chlordane; β-Chlordane; **1,2,4,5,6,7,8,8-Octachloro-2,3,3a,4,7,7a-hexahydro-4,7-methano-1*H*-indene;** α-1,2,4,5,6,7,8,8-Octachloro-3a,4,7,7a-tetrahydro-4,7-methanoindan.

Structural Formula:

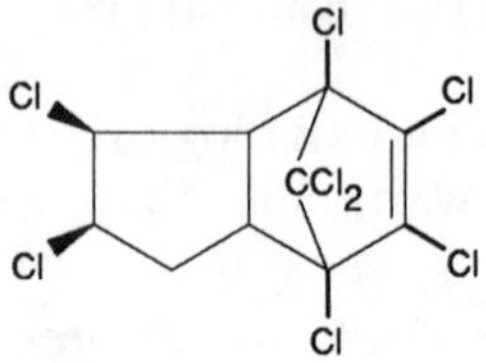

CHEMICAL DESIGNATIONS

CAS Registry Number: 5103-74-2

DOT Designation: None assigned.

Empirical Formula: $C_{10}H_6Cl_8$

Formula Weight: 409.78

RTECS Number: PC 01750000

PHYSICAL AND CHEMICAL PROPERTIES

Appearance: Solid.

Boiling Point: Technical grade containing both *cis-* and *trans-* isomers boils at 175 °C [1].

Henry's Law Constant: Insufficient vapor pressure data for calculation at 25 °C.

Ionization Potential: No data found.

Log K_{oc}: 6.00 [2]; 5.40 [3]; 5.57 [4].

Log K_{ow}: 5.93 using method of Kenaga and Goring [5].

Melting Point: 107.0-108.8 °C [6].

Solubility in Organics: Technical grades are miscible with aliphatic and aromatic solvents [7].

Solubility in Water: 0.051 mg/L at 20-25 °C [8].

Specific Density: No data found.

Transformation Products: No data found.

Vapor Pressure: No data found.

FIRE HAZARDS

Flash Point: No data found.

Lower Explosive Limit (LEL): No data found.

Upper Explosive Limit (UEL): No data found.

HEALTH HAZARD DATA

Immediately Dangerous to Life or Health (IDLH): No data found.

Permissible Exposure Limits (PEL) in Air: No standards set.

MANUFACTURING

Use: Insecticide.

REFERENCES

1. Sims, R.C., Doucette, W.C., McLean, J.E., Greeney, W.J., and R.R. Dupont. "Treatment Potential for 56 EPA Listed Hazardous Chemicals in Soil," National Technical Information Service, U.S. EPA Report-600/6-88-001 (1988), 105 p.
2. Oliver, B.G., and A.J. Niimi. "Bioconcentration Factors of Some Halogenated Organics for Rainbow Trout: Limitations in Their Use for Prediction of Environmental Residues," *Environ. Sci. Technol.*, 19(9):842-849 (1985).

3. Oliver, B.G., and M.N. Charlton. "Chlorinated Organic Contaminants on Settling Particulates in the Niagara River Vicinity of Lake Ontario," *Environ. Sci. Technol.*, 18(12):903-908 1984).

4. Chin, Y.-P., Peven, C.S., and W.J. Weber. "Estimating Soil/Sediment Partition Coefficients for Organic Compounds by High Performance Reverse Phase Liquid Chromatography," *Water Res.*, 22(7):873-881 (1988).

5. Kenaga, E.E., and C.A.I. Goring. "Relationship Between Water Solubility, Soil Sorption, Octanol-Water Partitioning and Concentration of Chemicals in Biota," in *Aquatic Toxicology, ASTM STP 707*, Eaton, J.G., Parrish, P.R., and A.C. Hendricks, Eds. (Philadelphia, PA: American Society for Testing and Materials, 1980), pp 78-115.

6. Callahan, M.A., Slimak, M.W., Gable, N.W., May, I.P., Fowler, C.F., Freed, J.R., Jennings, P., Durfee, R.L., Whitmore, F.C., Maestri, B., Mabey, W.R., Holt, B.R., and C. Gould. "Water-Related Environmental Fate of 129 Priority Pollutants Volumes I and II," National Technical Information Service, U.S. EPA Report-440/4-79-029 (1979), 1160 p.

7. "Chemical, Physical, and Biological Properties of Compounds Present at Hazardous Waste Sites," U.S. EPA Report-530/SW-89-010 (1985), 619 p.

8. Geyer, H., Politzki, G., and D. Freitag. "Prediction of Ecotoxicological Behaviour of Chemicals: Relationship Between *n*-Octanol/Water Partition Coefficient and Bioaccumulation of Organic Chemicals by Alga *Chlorella*," *Chemosphere*, 13(2):269-284 (1984).

trans-CHLORDANE

Synonyms: α-Chlordan; *cis*-Chlordan; α-Chlordane; α(*cis*)-Chlordane; γ-Chlordane; 1,2,4,5,6,7,8,8-Octachloro-3a,4,7,7a-tetrahydro-4,7-methanoindan.

Structural Formula:

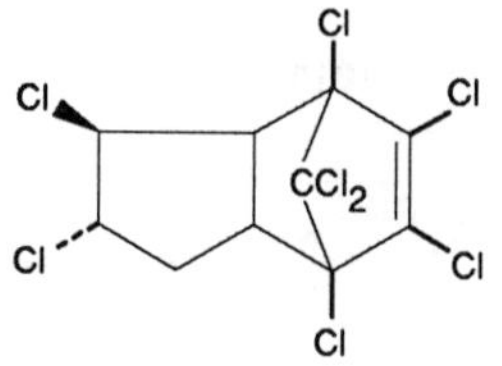

CHEMICAL DESIGNATIONS

CAS Registry Number: 5103-71-9

DOT Designation: None assigned.

Empirical Formula: $C_{10}H_6Cl_8$

Formula Weight: 409.78

RTECS Number: PB 9705000

PHYSICAL AND CHEMICAL PROPERTIES

Appearance: Solid.

Boiling Point: Technical grade containing both *cis-* and *trans-* isomers boils at 175 °C [1].

Henry's Law Constant: Insufficient vapor pressure and solubility data for calculation at 25 °C.

Ionization Potential: No data found.

Log K_{oc}: 6.00 [2]; 5.48 [3].

Log K_{ow}: 8.69, 9.65 using method of Kenaga and Goring [4].

Melting Point: 103.0-105.0 °C [5].

Solubility in Organics: Technical grades are miscible in aliphatic and aromatic solvents [6].

Solubility in Water: No data found.

Specific Density: No data found.

Transformation Products: No data found.

Vapor Pressure: No data found.

FIRE HAZARDS

Flash Point: No data found.

Lower Explosive Limit (LEL): No data found.

Upper Explosive Limit (UEL): No data found.

HEALTH HAZARD DATA

Immediately Dangerous to Life or Health (IDLH): No data found.

Permissible Exposure Limits (PEL) in Air: No standards set.

MANUFACTURING

Use: Insecticide.

REFERENCES

1. Sims, R.C., Doucette, W.C., McLean, J.E., Greeney, W.J., and R.R. Dupont. "Treatment Potential for 56 EPA Listed Hazardous Chemicals in Soil," National Technical Information Service, U.S. EPA Report-600/6-88-001 (1988), 105 p.
2. Oliver, B.G., and A.J. Niimi. "Bioconcentration Factors of Some Halogenated Organics for Rainbow Trout: Limitations in Their Use for Prediction of Environmental Residues," *Environ. Sci. Technol.*, 19(9):842-849 (1985).

3. Oliver, B.G., and M.N. Charlton. "Chlorinated Organic Contaminants on Settling Particulates in the Niagara River Vicinity of Lake Ontario," *Environ. Sci. Technol.*, 18(12):903-908 (1984).

4. Kenaga, E.E., and C.A.I. Goring. "Relationship Between Water Solubility, Soil Sorption, Octanol-Water Partitioning and Concentration of Chemicals in Biota," in *Aquatic Toxicology, ASTM STP 707*, Eaton, J.G., Parrish, P.R., and A.C. Hendricks, Eds. (Philadelphia, PA: American Society for Testing and Materials, 1980), pp 78-115.

5. Callahan, M.A., Slimak, M.W., Gable, N.W., May, I.P., Fowler, C.F., Freed, J.R., Jennings, P., Durfee, R.L., Whitmore, F.C., Maestri, B., Mabey, W.R., Holt, B.R., and C. Gould. "Water-Related Environmental Fate of 129 Priority Pollutants Volumes I and II," National Technical Information Service, U.S. EPA Report-440/4-79-029 (1979), 1160 p.

6. "Chemical, Physical, and Biological Properties of Compounds Present at Hazardous Waste Sites," U.S. EPA Report-530/SW-89-010 (1985), 619 p.

4–CHLOROANILINE

Synonyms: 1-Amino-4-chlorobenzene; 1-Amino-*p*-chlorobenzene; 4-Aminochlorobenzene; *p*-Aminochlorobenzene; 4-Chloraniline; *p*-Chloraniline; *p*-Chloroaniline; **4-Chlorobenzamine**; *p*-Chlorobenzamine; 4-Chlorophenylamine; *p*-Chlorophenylamine; NCI-C02039; RCRA waste number P024; UN 2018; UN 2019.

Structural Formula:

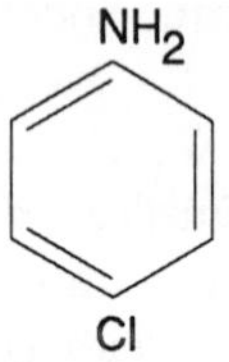

CHEMICAL DESIGNATIONS

CAS Registry Number: 106-47-8

DOT Designation: 2018 (solid); 2019 (liquid).

Empirical Formula: C_6H_6ClN

Formula Weight: 127.57

RTECS Number: BX 0700000

PHYSICAL AND CHEMICAL PROPERTIES

Appearance and Odor: Yellowish-white solid with a mild, sweetish odor.

Boiling Point: 232 °C [1].

Dissociation Constants: 4.15 [1]; 3.98 [2].

Henry's Law Constant: 1.07 x 10^{-5} atm·m^3/mol at 25 °C (calculated) [2].

Ionization Potential: No data found.

Log K_{oc}: 2.42 [3]; 1.98, 2.05, 3.10, 3.13, 3.18 [4]; 2.75 [5].

Log K_{ow}: 1.83 [6]; 2.78 [7].

Melting Point: 72.5 °C [1].

Solubility in Organics: Soluble in ethanol and ether [1].

Solubility in Water: 3.9 g/L at 20-25 °C [8].

Specific Density: 1.429 at 19/4 °C [1].

Transformation Products: Under artificial sunlight, river water containing 2-5 ppm 4-chloroaniline photodegraded to 4-aminophenol and unidentified polymers [9].

Vapor Pressure: 0.015 mm at 20 °C, 0.05 mm at 30 °C [10]; 1 mm at 59.3 °C, 10 mm at 102.1 °C, 40 mm at 135.0 °C, 100 mm at 159.9 °C, 760 mm at 230.5 °C [1]; 0.025 mm at 25 °C [2].

FIRE HAZARDS

Flash Point: > 1,205 °C [11].

Lower Explosive Limit (LEL): Not pertinent [11].

Upper Explosive Limit (UEL): Not pertinent [11].

HEALTH HAZARD DATA

Immediately Dangerous to Life or Health (IDLH): No data found.

Permissible Exposure Limits (PEL) in Air: No standards set.

MANUFACTURING

Selected Manufacturers:

Aldrich Chemical Co.
940 West Saint Paul Ave.
Milwaukee, WI 53233

Monsanto Industrial Chemicals Co.
800 North Lindbergh Blvd.
St. Louis, MO 63166

Uses: Dye intermediate; pharmaceuticals; agricultural chemicals.

REFERENCES

1. Weast, R.C., Ed. *CRC Handbook of Chemistry and Physics*, 67th ed. (Boca Raton, FL: CRC Press, Inc., 1986), 2406 p.
2. Howard, P.H. *Handbook of Environmental Fate and Exposure Data for Organic Chemicals* (Chelsea, MI: Lewis Publishers, Inc., 1989), 574 p.
3. Hodson, J., and N.A. Williams. "The Estimation of the Adsorption Coefficient (K_{oc}) for Soils by High Performance Liquid Chromatography," *Chemosphere*, 19(1):67-77 (1988).
4. Rippen, G., Ilgenstein, M., and W. Klöpffer. "Screening of the Adsorption Behavior of New Chemicals: Natural Soils and Model Adsorbents," *Ecotoxicol. Environ. Safety*, 6(3):236-245 (1982).
5. Rao, P.S.C., and J.M. Davidson. "Estimation of Pesticide Retention and Transformation Parameters Required in Nonpoint Source Pollution Models," in *Environmental Impact of Nonpoint Source Pollution*, Overcash, M.R., and J.M. Davidson, Eds. (Ann Arbor, MI: Ann Arbor Science Publishers, Inc., 1980), pp 23-67.
6. Leo, A., Hansch, C., and D. Elkins. "Partition Coefficients and Their Uses," *Chem. Rev.*, 71(6):525-616 (1971).
7. Geyer, H., Politzki, G., and D. Freitag. "Prediction of Ecotoxicological Behaviour of Chemicals: Relationship Between *n*-Octanol/Water Partition Coefficient and Bioaccumulation of Organic Chemicals by Alga *Chlorella*," *Chemosphere*, 13(2):269-284 (1984).
8. Kilzer, L., Scheunert, I., Geyer, H., Klein, W., and F. Korte. "Laboratory Screening of the Volatilization Rates of Organic Chemicals from Water and Soil," *Chemosphere*, 8(10):751-761 (1979).
9. Mansour, M., Feicht, E. and P. Méallier. "Improvement of the Photostability of Selected Substances in Aqueous Medium," *Toxicol. Environ. Chem.*, 20-21:139-147 (1989).
10. Verschueren, K. *Handbook of Environmental Data on Organic Chemicals* (New York: Van Nostrand Reinhold Co., 1983), 1310 p.
11. Weiss, G. *Hazardous Chemicals Data Book* (Park Ridge, NJ: Noyes Data Corp., 1986), 1069 p.

CHLOROBENZENE

Synonyms: Benzene chloride; Chlorbenzene; Chlorbenzol; Chlorobenzol; MCB; Monochlorbenzene; Monochlorobenzene; NCI-C54886; Phenyl chloride; RCRA waste number U037; UN 1134.

Structural Formula:

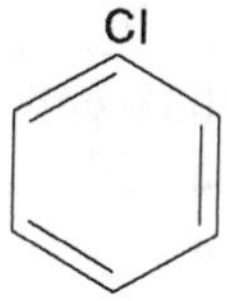

CHEMICAL DESIGNATIONS

CAS Registry Number: 108-90-7

DOT Designation: 1134

Empirical Formula: C_6H_5Cl

Formula Weight: 112.56

RTECS Number: CZ 0175000

PHYSICAL AND CHEMICAL PROPERTIES

Appearance and Odor: Clear, colorless, watery-liquid with a sweet almond odor.

Boiling Point: 132 °C [1]; 131.5 °C [2].

Henry's Law Constant: 0.00393 atm·m³/mol [3]; 0.0036 atm·m³/mol [4]; 0.0037 atm·m³/mol [5]; 0.00445 atm·m³/mol at 25 °C [6]; 0.00621 atm·m³/mol at 37 °C [7].

Ionization Potential: 9.07 eV [8]; 9.14 eV [9].

Log K_{oc}: 1.68 [10]; 2.52 [11].

Log K_{ow}: 2.84 [12]; 2.98 [13]; 2.71 [14]; 2.83 [9].

Melting Point: -45.6 °C [1].

Solubility in Organics: Soluble in ethanol, benzene, chloroform, ether, and carbon tetrachloride [15].

Solubility in Water: 500 mg/L at 20 °C, 488 mg/L at 30 °C [16]; 502 mg/L at 25 °C [17]; 295 mg/L at 25 °C [13]; 0.04 vol% at 20 °C [18]; 471.7 mg/L at 25 °C [19]; 488 mg/L at 25 °C [20]; 500 mg/L at 25 °C [21]; 446 mg/L at 30 °C [22]; 0.036 vol% at 25 °C, 0.0488 wt% at 30 °C [23]; 448 ppm [24]; 448 mg/L at 30 °C [25]; 503 mg/L at 25 °C [26]; 471.7 mg/L at 25 °C [27].

Specific Density: 1.1058 at 20/4 °C [1]; 1.113 at 15/15 °C [28]; 1.1293 at 0/4 °C, 1.1167 at 10/4 °C, 1.1058 at 16.5/4 °C [2]; 1.1009 at 25/4 °C [29].

Transformation Products: Under artificial sunlight, river water containing 2-5 ppm chlorobenzene photodegraded to phenol and chlorophenol [30].

Vapor Density: 4.60 g/L at 25 °C, 3.88 (air = 1).

Vapor Pressure: 3 mm at 0 °C, 9 mm at 20 °C, 26 mm at 40 °C, 65 mm at 60 °C, 144 mm at 80 °C, 292 mm at 100 °C, 543 mm at 120 °C [2]; 11.8 mm at 25 °C, 15 mm at 30 °C [16]; 1 mm at -13.0 °C, 10 mm at 22.2 °C, 40 mm at 49.7 °C, 100 mm at 70.7 °C, 400 mm at 110.0 °C, 760 mm at 132.2 °C [1].

FIRE HAZARDS

Flash Point: 28 °C [8].

Lower Explosive Limit (LEL): 1.3% [8].

Upper Explosive Limit (UEL): 7.1% [8].

HEALTH HAZARD DATA

Immediately Dangerous to Life or Health (IDLH): 2,400 ppm [8].

Permissible Exposure Limits (PEL) in Air: 75 ppm ($\approx$350 mg/m^3) [31]; 75 ppm TLV [32].

MANUFACTURING

Selected Manufacturers:

Dow Chemical Co.
Midland, MI 48640

Monsanto Industrial Chemicals Co.
800 North Lindbergh Blvd.
St. Louis, MO 63166

Uses: Preparation of phenol, chloronitrobenzene, and aniline; solvent carrier for methylene diisocyanate; solvent; insecticide, pesticide, and dyestuffs intermediate; heat transfer agent.

REFERENCES

1. Weast, R.C., Ed. *CRC Handbook of Chemistry and Physics*, 67th ed. (Boca Raton, FL: CRC Press, Inc., 1986), 2406 p.
2. Standen, A., Ed. *Kirk-Othmer Encyclopedia of Chemical Technology, Volume 4*, 2nd ed. (New York: John Wiley and Sons, Inc., 1964), 937 p.
3. Warner, H.P., Cohen, J.M., and J.C. Ireland. "Determination of Henry's Law Constants of Selected Priority Pollutants," Office of Science and Development, U.S. EPA Report-600/D-87/229 (1987), 14 p.
4. Pankow, J.F., and M.E. Rosen. "Determination of Volatile Compounds in Water by Purging Directly to a Capillary Column with Whole Column Cryotrapping," *Environ. Sci. Technol.*, 22(4):398-405 (1988).
5. Valsaraj, K.T. "On the Physio-Chemical Aspects of Partitioning of Non-Polar Hydrophobic Organics at the Air-Water Interface," *Chemosphere*, 17(5):875-887 (1988).
6. Hine, J., and P.K. Mookerjee. "The Intrinsic Hydrophilic Character of Organic Compounds. Correlations in Terms of Structural Contributions," *J. Org. Chem.*, 40(3):292-298 (1975).
7. Sato, A., and T. Nakajima. "A Structure-Activity Relationship of Some Chlorinated Hydrocarbons," *Arch. Environ. Health*, 34(2):69-75 (1979).
8. "NIOSH Pocket Guide to Chemical Hazards," U.S. Department of Health and Human Services, U.S. Government Printing Office (1987), 241 p.

9.   Yoshida, K., Shigeoka, T., and F. Yamauchi. "Non-Steady State Equilibrium Model for the Preliminary Prediction of the Fate of Chemicals in the Environment," *Ecotoxicol. Environ. Safety*, 7(2):179-190 (1983).

10. Abdul, S.A., Gibson, T.L., and D.N. Rai. "Statistical Correlations for Predicting the Partition Coefficient for Nonpolar Organic Contaminants Between Aquifer Organic Carbon and Water," *Haz. Waste Haz. Mater.*, 4(3):211-222 (1987).

11. Stephen, H., and T. Stephen. *Solubilities of Inorganic and Organic Compounds - Part 1, Volume 1* (London: Pergamon Printing and Art Services, Ltd., 1963), 960 p.

12. Fujita, T., Iwasa, J., and C. Hansch. "A New Substituent Constant, π, Derived from Partition Coefficients," *J. Am. Chem. Soc.*, 86(23):5175-5180 (1964).

13. Tewari, Y.B., Miller, M.M., Wasik, S.P., and D.E. Martire. "Aqueous Solubility and Octanol/Water Partition Coefficient of Organic Compounds at 25.0 °C," *J. Chem. Eng. Data*, 27(4):451-454 (1982).

14. Schwarzenbach, R.P., and J. Westall. "Transport of Nonpolar Organic Compounds from Surface Water to Groundwater. Laboratory Sorption Studies," *Environ. Sci. Technol.*, 15(11):1360-1367 (1981).

15. "Chemical, Physical, and Biological Properties of Compounds Present at Hazardous Waste Sites," U.S. EPA Report-530/SW-89-010 (1985), 619 p.

16. Verschueren, K. *Handbook of Environmental Data on Organic Chemicals* (New York: Van Nostrand Reinhold Co., 1983), 1310 p.

17. Banerjee, S. "Solubility of Organic Mixtures in Water," *Environ. Sci. Technol.*, 18(8):587-591 (1984).

18. Meites, L., Ed. *Handbook of Analytical Chemistry*, 1st ed. (New York: McGraw-Hill, Inc., 1963), 1782 p.

19. Aquan-Yuen, M., Mackay, D., and W.Y. Shiu. "Solubility of Hexane, Phenanthrene, Chlorobenzene, and *p*-Dichlorobenzene in Aqueous Electrolyte Solutions," *J. Chem. Eng. Data*, 24(1):30-34 (1979).

20. "Treatability Manual - Volume 1: Treatability Data," Office of Research and Development, U.S. EPA Report-600/8-80-042a (1980), 1035 p.

21. Andrews, L.J., and R.M. Keefer. "Cation Complexes of Compounds Containing Carbon-Carbon Double Bonds. VI. The Argentation of Substituted Benzenes," *J. Am. Chem. Soc.*, 72(7):3113-3116 (1950).

22. Gross, P.M., and J.H. Saylor. "The Solubilities of Certain Slightly Soluble Organic Compounds in Water," *J. Am. Chem. Soc.*, 53(5):1744-1751 (1931).

23. Stephen, H., and T. Stephen. *Solubilities of Inorganic and Organic*

Compounds - Part 1, Volume 1 (London: Pergamon Printing and Art Services, Ltd., 1963), 960 p.

24. Kenaga, E.E., and C.A.I. Goring. "Relationship Between Water Solubility, Soil Sorption, Octanol-Water Partitioning and Concentration of Chemicals in Biota," in *Aquatic Toxicology, ASTM STP 707*, Eaton, J.G., Parrish, P.R., and A.C. Hendricks, Eds. (Philadelphia, PA: American Society for Testing and Materials, 1980), pp 78-115.

25. Freed, V.H., Chiou, C.T., and R. Haque. "Chemodynamics: Transport and Behavior of Chemicals in the Environment - A Problem in Environmental Health," *Environ. Health Perspect.*, 20:55-70 (1977).

26. Yalkowsky, S.H., Orr, R.J., and S.C. Valvani. "Solubility and Partitioning. 3. The Solubility of Halobenzenes in Water," *Indust. Eng. Chem. Fund.*, 18(4):351-353 (1979).

27. Howard, P.H. *Handbook of Environmental Fate and Exposure Data for Organic Chemicals* (Chelsea, MI: Lewis Publishers, Inc., 1989), 574 p.

28. Sax, N.I. *Dangerous Properties of Industrial Materials* (New York: Van Nostrand Reinhold Co., 1984), 3124 p.

29. Kirchnerová, J., and G.C.B. Cave. "The Solubility of Water in Low-Dielectric Solvents," *Can. J. Chem.*, 54(24):3909-3916 (1976).

30. Mansour, M., Feicht, E. and P. Méallier. "Improvement of the Photostability of Selected Substances in Aqueous Medium," *Toxicol. Environ. Chem.*, 20-21:139-147 (1989).

31. "General Industry Standards for Toxic and Hazardous Substances," U.S. Code of Federal Regulations 1910, Subpart Z, Section 1910.1000 (July 1982).

32. *Documentation of the Threshold Limit Values and Biological Exposure Indices for 1986-1987* (Cincinnati, OH: American Conference of Governmental Industrial Hygienists, 1986), 111 p.

p–CHLORO–m–CRESOL

Synonyms: Aptal; Baktol; Baktolan; Candaseptic; *p*-Chlor-*m*-cresol; Chlorocresol; 4-Chlorocresol; *p*-Chlorocresol; 4-Chloro-*m*-cresol; 6-Chloro-*m*-cresol; 4-Chloro-1-hydroxy-3-methylbenzene; 2-Chlorohydroxytoluene; 2-Chloro-5-hydroxytoluene; 4-Chloro-3-hydroxytoluene; 6-Chloro-3-hydroxytoluene; *p*-Chlorometacresol; **4-Chloro-3-methylphenol**; *p*-Chloro-3-methylphenol; 3-Methyl-4-chlorophenol; Ottafact; Parmetol; Parol; PCMC; Peritonan; Preventol CMK; Raschit; Raschit K; Rasenanicon; RCRA waste number U039.

Structural Formula:

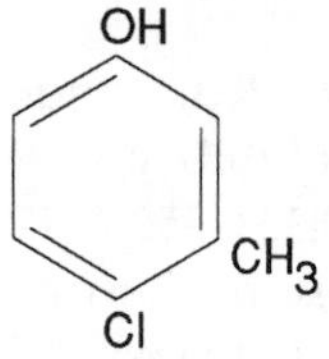

CHEMICAL DESIGNATIONS

CAS Registry Number: 59-50-7

DOT Designation: None assigned.

Empirical Formula: C_7H_7ClO

Formula Weight: 142.59

RTECS Number: GO 7100000

PHYSICAL AND CHEMICAL PROPERTIES

Appearance and Odor: Colorless, white, or pinkish crystals with a slight phenolic odor. On exposure to air it slowly becomes light brown.

Boiling Point: 235 °C [1].

Dissociation Constant: 9.549 [2].

Henry's Law Constant: 1.78×10^{-6} atm·m^3/mol (calculated).

Ionization Potential: No data found.

Log K$_{oc}$: 2.89 using method of Karickhoff and others [3].

Log K$_{ow}$: 3.10 [4]; 2.95 [5].

Melting Point: 66-68 °C [1]; 63-65 °C [6].

Solubility in Organics: Soluble in ethanol and ether [1].

Solubility in Water: 3,850 mg/L at 25 °C [7]; 1 g/.260L at 20 °C [8]; 3,850 mg/L at 20 °C [9].

Specific Density: No data found.

Transformation Products: No data found.

Vapor Pressure: No data was found, however, a value of 0.05 mm at 20 °C was assigned by analogy [10].

FIRE HAZARDS

Flash Point: No data found.

Lower Explosive Limit (LEL): No data found.

Upper Explosive Limit (UEL): No data found.

HEALTH HAZARD DATA

Immediately Dangerous to Life or Health (IDLH): No data found.

Permissible Exposure Limits (PEL) in Air: No standards set.

MANUFACTURING

Selected Manufacturers:

Aldrich Chemical Co.
940 West Saint Paul Ave.
Milwaukee, WI 53233

Fluka Chemical Corp.
980 South Second St.
Ronkonkoma, NY 11779

Uses: External germicide; preservative for gums, glues, paints, inks, textile, and leather products; topical antiseptic (veterinarian).

REFERENCES

1. Weast, R.C., Ed. *CRC Handbook of Chemistry and Physics*, 67th ed. (Boca Raton, FL: CRC Press, Inc., 1986), 2406 p.
2. Dean, J.A., Ed., *Lange's Handbook of Chemistry*, 11th ed. (New York: McGraw-Hill, Inc., 1973), 1570 p.
3. Karickhoff, S.W., Brown, D.S., and T.A. Scott. "Sorption of Hydrophobic Pollutants on Natural Sediments," *Water Res.*, 13:241-248 (1979).
4. Leo, A., Hansch, C., and D. Elkins. "Partition Coefficients and Their Uses," *Chem. Rev.*, 71(6):525-616 (1971).
5. Walton, W.C. *Practical Aspects of Ground Water Modeling* (Worthington, OH: National Water Well Association, 1985), 587 p.
6. *Fluka Catalog 1988/89 - Chemika-Biochemika* (Ronkonkoma, NY: Fluka Chemical Corp., 1988), 1536 p.
7. Verschueren, K. *Handbook of Environmental Data on Organic Chemicals* (New York: Van Nostrand Reinhold Co., 1983), 1310 p.
8. Windholz, M., Budavari, S., Blumetti, R.F., and E.S. Otterbein, Eds. *The Merck Index*, 10th ed., (Rahway, NJ: Merck and Co., 1983), 1463 p.
9. "Treatability Manual - Volume 1: Treatability Data," Office of Research and Development, U.S. EPA Report-600/8-80-042a (1980), 1035 p.
10. Mabey, W.R., Smith, J.H., Podoll, R.T., Johnson, H.L., Mill, T., Chou, T.-W., Gates, J., Partridge, I.W., Jaber, H., and D. Vandenberg. "Aquatic Fate Process Data for Organic Priority Pollutants - Final Report," Office of Regulations and Standards, U.S. EPA Report-440/4-81-014 (1982), 407 p.

CHLOROETHANE

Synonyms: Aethylis; Aethylis chloridum; Anodynon; Chelen; Chlorethyl; Chloridum; Chloryl; Chloryl anesthetic; Ether chloratus; Ether hydrochloric; Ether muriatic; Ethyl chloride; Hydrochloric ether; Kelene; Monochlorethane; Monochloroethane; Muriatic ether; Narcotile; NCI-C06224; UN 1037.

Structural Formula:

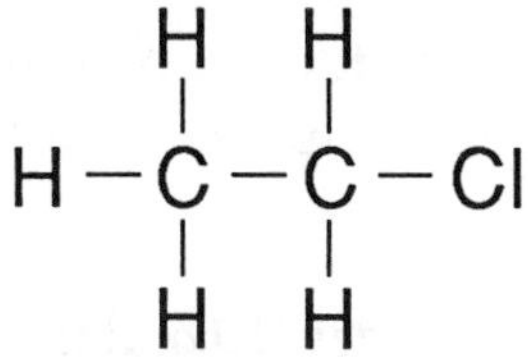

$$H - \underset{\underset{H}{|}}{\overset{\overset{H}{|}}{C}} - \underset{\underset{H}{|}}{\overset{\overset{H}{|}}{C}} - Cl$$

CHEMICAL DESIGNATIONS

CAS Registry Number: 75-00-3

DOT Designation: 1037

Empirical Formula: C_2H_5Cl

Formula Weight: 64.52

RTECS Number: KH 75250000

PHYSICAL AND CHEMICAL PROPERTIES

Appearance and Odor: Colorless liquid with an ethereal odor.

Boiling Point: 12.3 °C [1].

Henry's Law Constant: 0.0111 atm·m^3/mol [2]; 0.0146 atm·m^3/mol [3]; 0.0093 atm·m^3/mol [4]; 0.0085 atm·m^3/mol at 25 °C [5].

Ionization Potential: 10.97 eV [6].

Log K$_{oc}$: 0.51 using method of Chiou and others [7].

Log K$_{ow}$: 1.43 [8].

Melting Point: -136.4 °C [1]; -138.3 °C [9].

Solubility in Organics: Soluble in ethanol and ether [10].

Solubility in Water: 3,330 mg/L at 0 °C, 5,740 mg/L at 20 °C [11]; 6,000 mg/L at 18 °C [12]; 5,700 mg/L at 17.5 °C [13]; 4,470 mg/L at 0 °C [9]; 0.57 wt% at 17.5 °C [14]; 0.455 mass% at 0 °C [15]; 4,700 ppm at 25 °C [16].

Specific Density: 0.8978 at 20/4 °C [1]; 0.92138 at 0/0 °C, 0.92390 at 0/4 °C, 0.92295 at 2/2 °C, 0.91708 at 6/6 °C, 0.9028 at 15/4 °C, 0.8970 at 20/4 °C [9].

Transformation Product: Under laboratory conditions, chloroethane hydrolyzed to ethanol [17]; in the atmosphere, formyl chloride is the initial photooxidation product [10].

Vapor Density: 2.76 kg/m^3 at 20 °C [15].

Vapor Pressure: 114 mm at -30 °C, 190 mm at -20 °C, 304 mm at -10 °C, 464 mm at 0 °C, 692 mm at 10 °C, 760 mm at 12.2 °C, 1,011 mm at 20 °C, 1,938 mm at 40 °C, 3,420 mm at 60 °C, 5,632 mm at 80 °C, 8,740 mm at 100 °C [9]; 1,444 mm at 30 °C [11]; 1,520 mm at 32.5 °C, 3,800 mm at 64.0 °C, 7,600 mm at 92.6 °C, 15,200 mm at 127.3 °C, 30,400 mm at 167.0 °C [1]; 1,064 mm at 20 °C [6].

FIRE HAZARDS

Flash Point: -50 °C [18]; -43 °C (open cup) [19].

Lower Explosive Limit (LEL): 3.16% [9].

Upper Explosive Limit (UEL): 14% [9].

HEALTH HAZARD DATA

Immediately Dangerous to Life or Health (IDLH): 20,000 ppm [6].

Permissible Exposure Limits (PEL) in Air: 1,000 ppm ($\approx$2,600 mg/m^3) [20]; 1,000 ppm TLV [21].

MANUFACTURING

Selected Manufacturers:

Aldrich Chemical Co.
940 West Saint Paul Ave.
Milwaukee, WI 53233

Fluka Chemical Corp.
980 South Second St.
Ronkonkoma, NY 11779

Uses: Intermediate for tetraethyl lead and ethylcellulose; anesthetic; organic synthesis; alkylating agent; refrigeration; analytical reagent; solvent for phosphorus, sulfur, fats, oils, resins, and waxes; insecticides.

REFERENCES

1. Weast, R.C., Ed. *CRC Handbook of Chemistry and Physics*, 67th ed. (Boca Raton, FL: CRC Press, Inc., 1986), 2406 p.
2. Gossett, J.M. "Measurement of Henry's Law Constants for C_1 and C_2 Chlorinated Hydrocarbons," *Environ. Sci. Technol.*, 21(2):202-208 (1987).
3. "Treatability Manual - Volume 1: Treatability Data," Office of Research and Development, U.S. EPA Report-600/8-80-042a (1980), 1035 p.
4. Pankow, J.F., and M.E. Rosen. "Determination of Volatile Compounds in Water by Purging Directly to a Capillary Column with Whole Column Cryotrapping," *Environ. Sci. Technol.*, 22(4):398-405 (1988).
5. Hine, J., and P.K. Mookerjee. "The Intrinsic Hydrophilic Character of Organic Compounds. Correlations in Terms of Structural Contributions," *J. Org. Chem.*, 40(3):292-298 (1975).
6. "NIOSH Pocket Guide to Chemical Hazards," U.S. Department of Health and Human Services, U.S. Government Printing Office (1987), 241 p.
7. Chiou, C.T., Peters, L.J., and V.H. Freed. "A Physical Concept of Soil-Water Equilibria for Nonionic Organic Compounds," *Science*, 206:831-832 (1979).
8. Valvani, S.C., Yalkowsky, S.H., and T.J. Roseman. "Solubility and Partitioning IV: Aqueous Solubility and Octanol-Water Partition Coefficients of Liquid Nonelectrolytes," *J. Pharm. Sci.*, 70(5):502-506

(1981).

9. Standen, A., Ed. *Kirk-Othmer Encyclopedia of Chemical Technology, Volume 4,* 2nd ed. (New York: John Wiley and Sons, Inc., 1964), 937 p.

10. "Chemical, Physical, and Biological Properties of Compounds Present at Hazardous Waste Sites," U.S. EPA Report-530/SW-89-010 (1985), 619 p.

11. Verschueren, K. *Handbook of Environmental Data on Organic Chemicals* (New York: Van Nostrand Reinhold Co., 1983), 1310 p.

12. Gordon, A.J., and R.A. Ford. *The Chemist's Companion* (New York: John Wiley and Sons, Inc., 1972), 551 p.

13. Meites, L., Ed. *Handbook of Analytical Chemistry,* 1st ed. (New York: McGraw-Hill, Inc., 1963), 1782 p.

14. Stephen, H., and T. Stephen. *Solubilities of Inorganic and Organic Compounds - Part 1, Volume 1* (London: Pergamon Printing and Art Services, Ltd., 1963), 960 p.

15. Konietzko, H. "Chlorinated Ethanes: Sources, Distribution, Environmental Impact, and Health Effects," in *Hazard Assessment of Chemicals, Volume 3,* J. Saxena, Ed. (New York: Academic Press, Inc., 1984), pp 401-448.

16. Amoore, J.E., and E. Hautala. "Odor as an Aide to Chemical Safety: Odor Thresholds Compared with Threshold Limit Values and Volatilities for 214 Industrial Chemicals in Air and Water Dilution," *J. Appl. Toxicol.,* 3(6):272-290 (1983).

17. Smith, L.R., and J. Dragun. "Degradation of Volatile Chlorinated Aliphatic Priority Pollutants in Groundwater," *Environ. Int.,* 19(4):291-298 (1984).

18. "NIOSH Pocket Guide to Chemical Hazards," U.S. Department of Health and Human Services, U.S. Government Printing Office (1987), 241 p.

19. Windholz, M., Budavari, S., Blumetti, R.F., and E.S. Otterbein, Eds. *The Merck Index,* 10th ed., (Rahway, NJ: Merck and Co., 1983), 1463 p.

20. "General Industry Standards for Toxic and Hazardous Substances," U.S. Code of Federal Regulations 1910, Subpart Z, Section 1910.1000 (July 1982).

21. *Documentation of the Threshold Limit Values and Biological Exposure Indices for 1986-1987* (Cincinnati, OH: American Conference of Governmental Industrial Hygienists, 1986), 111 p.

2-CHLOROETHYL VINYL ETHER

Synonyms: 2-Chlorethyl vinyl ether; **(2-Chloroethoxy)ethene**; RCRA waste number U042; Vinyl 2-chloroethyl ether; Vinyl β-chloroethyl ether.

Structural Formula:

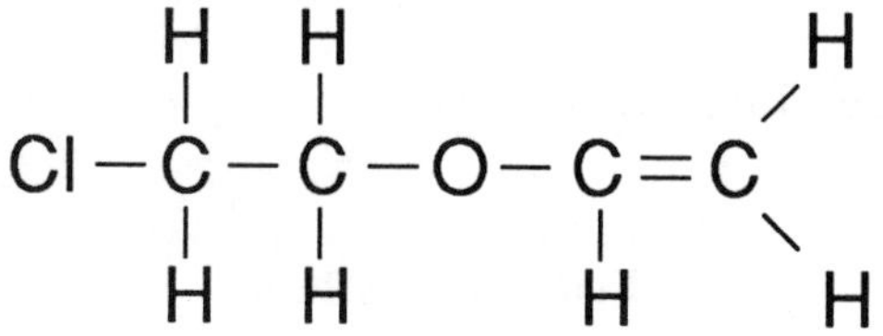

$$Cl - \underset{\underset{H}{|}}{\overset{\overset{H}{|}}{C}} - \underset{\underset{H}{|}}{\overset{\overset{H}{|}}{C}} - O - \underset{\underset{H}{|}}{\overset{}{C}} = C \overset{\diagup H}{\diagdown H}$$

CHEMICAL DESIGNATIONS

CAS Registry Number: 110-75-8

DOT Designation: None assigned.

Empirical Formula: C_4H_7ClO

Formula Weight: 106.55

RTECS Number: KN 6300000

PHYSICAL AND CHEMICAL PROPERTIES

Appearance: Colorless liquid.

Boiling Point: 108 °C [1]; 109 °C [2].

Henry's Law Constant: 2.5 x 10^{-4} atm·m^3/mol [3].

Ionization Potential: No data found.

Log K_{oc}: 0.82 [4].

Log K_{ow}: 1.28 (calculated) [5].

Melting Point: -70.3 °C [6].

Solubility in Organics: Soluble in ethanol and ether [1].

Solubility in Water: 15,000 mg/L at 20 °C [4]; 0.6% at 20 °C [2].

Specific Density: 1.0475 at 20/4 °C [1]; 1.0493 at 20/4 °C [2].

Transformation Products: No data found.

Vapor Density: 4.36 g/L at 25 °C, 3.68 (air = 1).

Vapor Pressure: 26.75 mm at 20 °C [7].

FIRE HAZARDS

Flash Point: 16 °C [8].

Lower Explosive Limit (LEL): No data found.

Upper Explosive Limit (UEL): No data found.

HEALTH HAZARD DATA

Immediately Dangerous to Life or Health (IDLH): No data found.

Permissible Exposure Limits (PEL) in Air: No standards set.

MANUFACTURING

Selected Manufacturer:

> Aldrich Chemical Co.
> 940 West Saint Paul Ave.
> Milwaukee, WI 53233

Uses: Anesthetics, sedatives, and cellulose ethers; copolymer of 95% ethyl acrylate with 5% 2-chloroethyl vinyl ether is used to produce an acrylic elastomer.

REFERENCES

1. Weast, R.C., Ed. *CRC Handbook of Chemistry and Physics*, 67th ed.

(Boca Raton, FL: CRC Press, Inc., 1986), 2406 p.

2. Standen, A., Ed. *Kirk-Othmer Encyclopedia of Chemical Technology, Volume 21*, 2nd ed. (New York: John Wiley and Sons, Inc., 1970), 707 p.

3. Pankow, J.F., and M.E. Rosen. "Determination of Volatile Compounds in Water by Purging Directly to a Capillary Column with Whole Column Cryotrapping," *Environ. Sci. Technol.*, 22(4):398-405 (1988).

4. Schwille, F. *Dense Chlorinated Solvents* (Chelsea, MI: Lewis Publishers, Inc., 1988), 146 p.

5. Leo, A., Hansch, C., and D. Elkins. "Partition Coefficients and Their Uses," *Chem. Rev.*, 71(6):525-616 (1971).

6. Sax, N.I. *Dangerous Properties of Industrial Materials* (New York: Van Nostrand Reinhold Co., 1984), 3124 p.

7. "Treatability Manual - Volume 1: Treatability Data," Office of Research and Development, U.S. EPA Report-600/8-80-042a (1980), 1035 p.

8. *Catalog Handbook of Fine Chemicals* (Milwaukee, WI: Aldrich Chemical Co., 1988), 2212 p.

CHLOROFORM

Synonyms: Formyl trichloride; Freon 20; Methane trichloride; Methenyl chloride; Methenyl trichloride; Methyl trichloride; NCI-C02686; R 20; R 20 (refrigerant); RCRA waste number U044; TCM; Trichloroform; **Trichloromethane**; UN 1888.

Structural Formula:

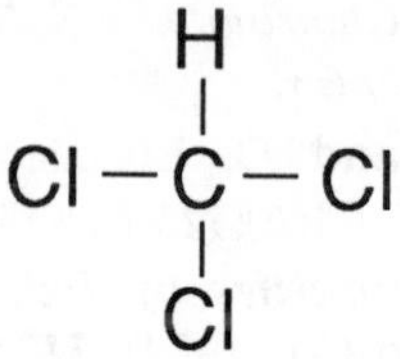

$$Cl-\underset{\underset{\textstyle Cl}{\displaystyle |}}{\overset{\overset{\textstyle H}{\displaystyle |}}{C}}-Cl$$

CHEMICAL DESIGNATIONS

CAS Registry Number: 67-66-3

DOT Designation: 1888

Empirical Formula: $CHCl_3$

Formula Weight: 119.38

RTECS Number: FS 9100000

PHYSICAL AND CHEMICAL PROPERTIES

Appearance and Odor: Clear, water-white, volatile liquid with a strong, sweet, ethereal odor.

Boiling Point: 61.7 °C [1].

Henry's Law Constant: 0.0029 atm·m^3/mol [2]; 0.00339 atm·m^3/mol [3]; 0.00323 atm·m^3/mol [4]; 0.0053 atm·m^3/mol at 20 °C [5]; 0.0032 at 25 °C [6]; 0.00727 atm·m^3/mol at 37 °C [7].

Ionization Potential: 11.42 eV [8].

Log K$_{oc}$: 1.64 [9].

Log K$_{ow}$: 1.97 [10]; 1.90 [11]; 1.95 [12].

Melting Point: -63.5 °C [1].

Solubility in Organics: Soluble in acetone; miscible with ethanol, ether, benzene, and ligroin [13].

Solubility in Water: 8,000 mg/L at 20 °C, 9,300 mg/L at 25 °C [14]; 8,200 mg/L at 20 °C [15]; 7,950 mg/L at 25 °C [16]; 8.22 mL/L at 20 °C [17]; 8,220 mg/L at 20 °C [18]; 9,600 mg/L at 25 °C [19]; 7,222 mg/L at 25 °C [20]; 7,840 mg/L at 25 °C [6]; 8,520 mg/L at 15 °C, 7,710 mg/L at 30 °C [21]; 0.84 wt% at 15 °C, 0.8 wt% at 20 °C, 0.76 wt% at 30 °C [22]; 0.815 wt% at 20 °C [23]; 10,620 mg/L at 0 °C, 8,950 mg/L at 10 °C, 8,220 mg/L at 20 °C, 7,760 mg/L at 30 °C [24]; 7,100 ppm at 25 °C [25]; 1% at 15 °C [26].

Specific Density: 1.4832 at 20/4 °C [1]; 1.49845 15/4 °C [27]; 1.52637 at 0/4 °C, 1.4890 at 20/4 °C, 1.48069 at 25/4 °C, 1.4081 at 60.9/4 °C [24].

Transformation Products: An anaerobic species of *Clostridium* biodegraded chloroform (a metabolite of carbon tetrachloride) by reductive dechlorination yielding methylene chloride and unidentified products [28].

Vapor Density: 4.88 g/L at 25 °C, 4.12 (air = 1).

Vapor Pressure: 160 mm at 20 °C, 245 mm at 30 °C [14]; 760 mm at 61.3 °C, 1,520 mm at 83.9 °C, 3,800 mm at 120.0 °C, 7,600 mm at 152.3 °C, 15,200 mm at 191.8 °C, 30,400 mm at 237.5 °C [1]; 150.5 mm at 20 °C [29]; 198 mm at 25 °C [3]; 100 mm at 10.4 °C [27]; 0.81 mm at -60 °C, 2.06 mm at -50 °C, 4.20 mm at -40 °C, 10.00 mm at -30 °C, 19.60 mm at -20 °C, 34.75 mm at -10 °C, 61.0 mm at 0 °C, 100.5 mm at 10 °C, 159.6 mm at 20 °C, 246.0 mm at 30 °C, 366.4 mm at 40 °C, 526.0 mm at 50 °C [24].

FIRE HAZARDS

Flash Point: Non-combustible [8].

Lower Explosive Limit (LEL): Not applicable [8].

Upper Explosive Limit (UEL): Not applicable [8].

HEALTH HAZARD DATA

Immediately Dangerous to Life or Health (IDLH): 1,000 ppm (carcinogen) [30].

Permissible Exposure Limits (PEL) in Air: 50 ppm ceiling ($\approx$240 mg/m^3) [31]; 2 ppm 1-hour ceiling [8]; 10 ppm ($\approx$50 mg/m^3) TLV [17].

MANUFACTURING

Selected Manufacturers:

> Allied Chemical Corp.
> Specialty Chemical Division
> Moundville, WV 26041

> Dow Chemical Co.
> Midland, MI 48640

> Stauffer Chemical Co.
> Industrial Chemicals Division
> Louisville, KY 40200

Uses: Fluorocarbon refrigerants; fluorocarbon plastics; solvent for natural products; analytical chemistry; soil fumigant; insecticides; preparation of chlorodifluoromethane; cleaning electronic circuit boards; anesthetics.

REFERENCES

1. Weast, R.C., Ed. *CRC Handbook of Chemistry and Physics*, 67th ed. (Boca Raton, FL: CRC Press, Inc., 1986), 2406 p.
2. Jury, W.A., Spencer, W.F., and W.J. Farmer. "Behavior Assessment Model for Trace Organics in Soil: III. Application of Screening Model," *J. Environ. Qual.*, 13(4):573-579 (1984).
3. Warner, H.P., Cohen, J.M., and J.C. Ireland. "Determination of Henry's Law Constants of Selected Priority Pollutants," Office of Science and Development, U.S. EPA Report-600/D-87/229 (1987), 14 p.
4. Valsaraj, K.T. "On the Physio-Chemical Aspects of Partitioning of Non-Polar Hydrophobic Organics at the Air-Water Interface,"

Chemosphere, 17(5):875-887 (1988).

5. Roberts, P.V., and P.G. Dändliker. "Mass Transfer of Volatile Organic Contaminants from Aqueous Solution to the Atmosphere During Surface Aeration," *Environ. Sci. Technol.*, 17(8):484-489 (1983).

6. Dilling, W.L. "Interphase Transfer Processes. II. Evaporation Rates of Chloro Methanes, Ethanes, Ethylenes, Propanes, and Propylenes from Dilute Aqueous Solutions. Comparisons with Theoretical Predictions," *Environ. Sci. Technol.*, 11(4):405-409 (1977).

7. Sato, A., and T. Nakajima. "A Structure-Activity Relationship of Some Chlorinated Hydrocarbons," *Arch. Environ. Health*, 34(2):69-75 (1979).

8. "NIOSH Pocket Guide to Chemical Hazards," U.S. Department of Health and Human Services, U.S. Government Printing Office (1987), 241 p.

9. Schwille, F. *Dense Chlorinated Solvents* (Chelsea, MI: Lewis Publishers, Inc., 1988), 146 p.

10. Hansch, C., and S.M. Anderson. "The Effect of Intramolecular Hydrophobic Bonding on Partition Coefficients," *J. Org. Chem.*, 32(8):2583-2586 (1967).

11. Veith, G.D., Macek, K.J., Petrocelli, S.R., and J. Carroll. "An Evaluation of Using Partition Coefficients and Water Solubility to Estimate Bioconcentration Factors for Organic Chemicals in Fish," in *Aquatic Toxicology, ASTM STP 707*, Eaton, J.G., Parrish, P.R., and A.C. Hendricks, Eds. (Philadelphia, PA: American Society for Testing and Materials, 1980), pp 116-129.

12. Mackay, D. "Correlation of Bioconcentration Factors," *Environ. Sci. Technol.*, 16(5):274-278 (1982).

13. "Chemical, Physical, and Biological Properties of Compounds Present at Hazardous Waste Sites," U.S. EPA Report-530/SW-89-010 (1985), 619 p.

14. Verschueren, K. *Handbook of Environmental Data on Organic Chemicals* (New York: Van Nostrand Reinhold Co., 1983), 1310 p.

15. Pearson, C.R., and G. McConnell. "Chlorinated C_1 and C_2 Hydrocarbons in the Marine Environment," in *Proc. R. Soc. London*, B189(1096):305-322 (1975).

16. Kenaga, E.E. *Environmental Dynamics of Pesticides* (New York: Plenum Press, 1975), 243 p.

17. *Documentation of the Threshold Limit Values and Biological Exposure Indices for 1986-1987* (Cincinnati, OH: American Conference of Governmental Industrial Hygienists, 1986), 111 p.

18. Meites, L., Ed. *Handbook of Analytical Chemistry*, 1st ed. (New York: McGraw-Hill, Inc., 1963), 1782 p.

19. Walton, W.C. *Practical Aspects of Ground Water Modeling* (Worthington, OH: National Water Well Association, 1985), 587 p.

20. Banerjee, S., Yalkowsky, S.H., and S.C. Valvani. "Water Solubility and Octanol/Water Partition Coefficients of Organics. Limitations of the Solubility-Partition Coefficient Correlation," *Environ. Sci. Technol.*, 14(10):1227-1229 (1980).

21. Gross, P.M., and J.H. Saylor. "The Solubilities of Certain Slightly Soluble Organic Compounds in Water," *J. Am. Chem. Soc.*, 53(5):1744-1751 (1931).

22. Stephen, H., and T. Stephen. *Solubilities of Inorganic and Organic Compounds - Part 1, Volume 1* (London: Pergamon Printing and Art Services, Ltd., 1963), 960 p.

23. Riddick, J.A., Bunger, W.B., and T.K. Sakano. *Organic Solvents - Physical Properties and Methods of Purification. Volume II* (New York: John Wiley and Sons, Inc., 1986), 1325 p.

24. Standen, A., Ed. *Kirk-Othmer Encyclopedia of Chemical Technology, Volume 4*, 2nd ed. (New York: John Wiley and Sons, Inc., 1964), 937 p.

25. Amoore, J.E., and E. Hautala. "Odor as an Aide to Chemical Safety: Odor Thresholds Compared with Threshold Limit Values and Volatilities for 214 Industrial Chemicals in Air and Water Dilution," *J. Appl. Toxicol.*, 3(6):272-290 (1983).

26. Gunther, F.A., Westlake, W.E., and P.S. Jaglan. "Reported Solubilities of 738 Pesticide Chemicals in Water," *Res. Rev.*, 20:1-148 (1968).

27. Sax, N.I. *Dangerous Properties of Industrial Materials* (New York: Van Nostrand Reinhold Co., 1984), 3124 p.

28. Gälli, R., and P.L. McCarty. "Biotransformation of 1,1,1-Trichloroethane, Trichloromethane, and Tetrachloromethane by a *Clostridium* sp.," *Appl. Environ. Microbiol.*, 55(4):837-844 (1989).

29. McConnell, G., Ferguson, D.M., and C.R. Pearson. "Chlorinated Hydrocarbons and the Environment," *Endeavour*, 34(121):13-18 (1975).

30. Weiss, G. *Hazardous Chemicals Data Book* (Park Ridge, NJ: Noyes Data Corp., 1986), 1069 p.

31. "General Industry Standards for Toxic and Hazardous Substances," U.S. Code of Federal Regulations 1910, Subpart Z, Section 1910.1000 (July 1982).

2-CHLORONAPHTHALENE

Synonyms: β-Chloronaphthalene; RCRA waste number U047.

Structural Formula:

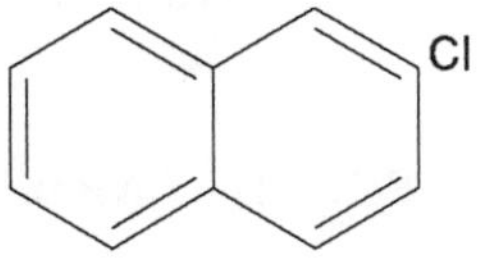

CHEMICAL DESIGNATIONS

CAS Registry Number: 91-58-7

DOT Designation: None assigned.

Empirical Formula: $C_{10}H_7Cl$

Formula Weight: 162.62

RTECS Number: QJ 2275000

PHYSICAL AND CHEMICAL PROPERTIES

Appearance: Monoclinic plates or leaflets.

Boiling Point: 256 °C [1].

Henry's Law Constant: 6.12×10^{-4} atm·m^3/mol (calculated) [2].

Ionization Potential: No data found.

Log K_{oc}: 3.93 using method of Karickhoff and others [3].

Log K_{ow}: 4.07 [4].

Melting Point: 61 °C [1].

Solubility in Organics: Soluble in ethanol, benzene, and ether [1].

Solubility in Water: 6.74 mg/L at 25 °C (calculated) [2].

Specific Density: 1.2656 at 16/4 °C [5].

Transformation Products: No data found.

Vapor Pressure: 0.017 mm at 25 °C (calculated) [2].

FIRE HAZARDS

Flash Point: Non-flammable [6].

Lower Explosive Limit (LEL): No data found.

Upper Explosive Limit (UEL): No data found.

HEALTH HAZARD DATA

Immediately Dangerous to Life or Health (IDLH): No data found.

Permissible Exposure Limits (PEL) in Air: No standards set.

MANUFACTURING

Selected Manufacturers:

> The Foxboro Co.
> 80 Republic Drive
> New Have, CT 06473

> Pfaltz & Bauer, Inc.
> 375 Fairfield Ave.
> Stamford, CT 06902

> Koppers Co., Inc.
> P.O. Box 219
> Bridgeville, PA 15015

Uses: Chlorinated naphthalenes were formerly used in the production of electric condensers, insulating electric condensers, electric cables, and wires; additive for high pressure lubricants.

REFERENCES

1. Weast, R.C., Ed. *CRC Handbook of Chemistry and Physics*, 67th ed. (Boca Raton, FL: CRC Press, Inc., 1986), 2406 p.
2. "Treatability Manual - Volume 1: Treatability Data," Office of Research and Development, U.S. EPA Report-600/8-80-042a (1980), 1035 p.
3. Karickhoff, S.W., Brown, D.S., and T.A. Scott. "Sorption of Hydrophobic Pollutants on Natural Sediments," *Water Res.*, 13:241-248 (1979).
4. Isnard, S., and S. Lambert. "Estimating Bioconcentration Factors from Octanol-Water Partition Coefficient and Aqueous Solubility," *Chemosphere*, 17(1):21-34 (1988).
5. Sax, N.I., Ed. *Dangerous Properties of Industrial Materials Report* (New York: Van Nostrand Reinhold Co., 1984), 4(6): 105 p.
6. Sittig, M. *Handbook of Toxic and Hazardous Chemicals and Carcinogens* (Park Ridge, NJ: Noyes Publications, 1985), 950 p.

2-CHLOROPHENOL

Synonyms: 1-Chloro-2-hydroxybenzene; 1-Chloro-*o*-hydroxybenzene; 2-Chlorohydroxybenzene; *o*-**Chlorophenol**; 1-Hydroxy-2-chlorobenzene; 1-Hydroxy-*o*-chlorobenzene; 2-Hydroxychlorobenzene; RCRA waste number U048; UN 2020; UN 2021.

Structural Formula:

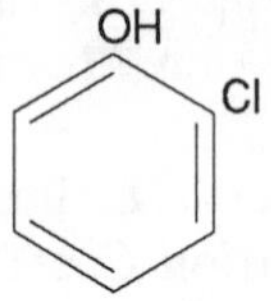

CHEMICAL DESIGNATIONS

CAS Registry Number: 95-57-8

DOT Designation: 2020 (liquid); 2021 (solid).

Empirical Formula: C_6H_5ClO

Formula Weight: 128.56

RTECS Number: SK 2625000

PHYSICAL AND CHEMICAL PROPERTIES

Appearance and Odor: Pale amber liquid with a slight phenolic odor.

Boiling Point: 174.9 °C [1].

Dissociation Constant: 8.48 [2].

Henry's Law Constant: 8.28×10^{-6} atm·m^3/mol (calculated) [3]; 5.6×10^{-7} atm·m^3/mol at 25 °C (estimated) [4].

Ionization Potential: 9.28 eV [5].

Log K_{oc}: 2.56 using method of Kenaga and Goring [6].

Log K_{ow}: 2.16 [7]; 2.19 [8]; 2.15 [9].

158

Melting Point: 9.0 °C [1].

Solubility in Organics: Soluble in ethanol, benzene, and ether [1].

Solubility in Water: 28,500 mg/L at 20 °C [10]; 28,000 mg/L at 25 °C [11]; 11,350 mg/L at 25 °C [12]; 0.2 M at 25 °C [13].

Specific Density: 1.2634 at 20/4 °C [1]; 1.256 at 25/25 °C [14]; 1.257 at 25/4 °C [15]; 1.262 at 20/4 °C [16].

Transformation Products: Monochlorophenols exposed to sunlight (ultraviolet light radiation) produced catechol and other hydroxybenzenes [17].

Vapor Density: 5.25 g/L at 25 °C, 4.44 (air = 1).

Vapor Pressure: 1 mm at 12.1 °C, 10 mm at 51.2 °C, 40 mm at 82.0 °C, 100 mm at 106.0 °C, 400 mm at 149.8 °C, 760 mm at 174.5 °C [1]; 1.42 mm at 25 °C [4].

FIRE HAZARDS

Flash Point: 64 °C [18].

Lower Explosive Limit (LEL): Unknown [19].

Upper Explosive Limit (UEL): Unknown [19].

HEALTH HAZARD DATA

Immediately Dangerous to Life or Health (IDLH): No data found.

Permissible Exposure Limits (PEL) in Air: No standards set.

MANUFACTURING

Selected Manufacturers:

Dow Chemical Co.
Midland, MI 48640

Aldrich Chemical Co.
940 West Saint Paul Ave.
Milwaukee, WI 53233

Fluka Chemical Corp.
980 South Second St.
Ronkonkoma, NY 11779

Uses: Component of disinfectant formulations; chemical intermediate for phenolic resins; solvent for polyester fibers, antiseptic (veterinarian); organic synthesis.

REFERENCES

1. Weast, R.C., Ed. *CRC Handbook of Chemistry and Physics*, 67th ed. (Boca Raton, FL: CRC Press, Inc., 1986), 2406 p.
2. Dean, J.A., Ed., *Lange's Handbook of Chemistry*, 11th ed. (New York: McGraw-Hill, Inc., 1973), 1570 p.
3. "Treatability Manual - Volume 1: Treatability Data," Office of Research and Development, U.S. EPA Report-600/8-80-042a (1980), 1035 p.
4. Howard, P.H. *Handbook of Environmental Fate and Exposure Data for Organic Chemicals* (Chelsea, MI: Lewis Publishers, Inc., 1989), 574 p.
5. Franklin, J.L., Dillard, J.G., Rosenstock, H.M., Herron, J.T., Draxl K., and F.H. Field. "Ionization Potentials, Appearance Potentials and Heats of Formation of Gaseous Positive Ions," National Bureau of Standards Report NSRDS-NBS 26, U.S. Government Printing Office (1969), 289 p.
6. Kenaga, E.E., and C.A.I. Goring. "Relationship Between Water Solubility, Soil Sorption, Octanol-Water Partitioning and Concentration of Chemicals in Biota," in *Aquatic Toxicology, ASTM STP 707*, Eaton, J.G., Parrish, P.R., and A.C. Hendricks, Eds. (Philadelphia, PA: American Society for Testing and Materials, 1980), pp 78-115.
7. Veith, G.D., Macek, K.J., Petrocelli, S.R., and J. Carroll. "An Evaluation of Using Partition Coefficients and Water Solubility to Estimate Bioconcentration Factors for Organic Chemicals in Fish," in *Aquatic Toxicology, ASTM STP 707*, Eaton, J.G., Parrish, P.R., and A.C. Hendricks, Eds. (Philadelphia, PA: American Society for Testing and Materials, 1980), pp 116-129.
8. Leo, A., Hansch, C., and D. Elkins. "Partition Coefficients and

Their Uses," *Chem. Rev.*, 71(6):525-616 (1971).

9. Fujita, T., Iwasa, J., and C. Hansch. "A New Substituent Constant, π, Derived from Partition Coefficients," *J. Am. Chem. Soc.*, 86(23):5175-5180 (1964).

10. Verschueren, K. *Handbook of Environmental Data on Organic Chemicals* (New York: Van Nostrand Reinhold Co., 1983), 1310 p.

11. Morrison, R.T., and R.N. Boyd. *Organic Chemistry* (Boston: Allyn and Bacon, Inc., 1971), 1258 p.

12. Banerjee, S., Yalkowsky, S.H., and S.C. Valvani. "Water Solubility and Octanol/Water Partition Coefficients of Organics. Limitations of the Solubility-Partition Coefficient Correlation," *Environ. Sci. Technol.*, 14(10):1227-1229 (1980).

13. Caturla, F., Martin-Martinez, J.M., Molina-Sabio, M., Rodriguez-Reinoso, F., and R. Torregrosa. "Adsorption of Substituted Phenols on Activated Carbon," *J. Colloid Interface Sci.*, 124(2):528-534 (1988).

14. Sax, N.I. *Dangerous Properties of Industrial Materials* (New York: Van Nostrand Reinhold Co., 1984), 3124 p.

15. Krijgsheld, K.R., and A. van der Gen. "Assessment of the Impact of the Emission of Certain Organochlorine Compounds on the Aquatic Environment - Part I: Monochlorophenols and 2,4-Dichlorophenol," *Chemosphere*, 15(7):825-860 (1986).

16. *Fluka Catalog 1988/89 - Chemika-Biochemika* (Ronkonkoma, NY: Fluka Chemical Corp., 1988), 1536 p.

17. Hwang, H.-M., Hodson, R.E., and R.F. Lee. "Degradation of Phenol and Chlorophenols by Sunlight and Microbes in Estuarine Water," *Environ. Sci. Technol.*, 20(10):1002-1007 (1986).

18. *Fire Protection Guide on Hazardous Materials* (Quincy, MA: National Fire Protection Association, 1984), 443 p.

19. "NIOSH Pocket Guide to Chemical Hazards," U.S. Department of Health and Human Services, U.S. Government Printing Office (1987), 241 p.

4-CHLOROPHENYL PHENYL ETHER

Synonyms: 4-Chlorodiphenyl ether; *p*-Chlorodiphenyl ether; **1-Chloro-4-phenoxybenzene**; 1-Chloro-*p*-phenoxybenzene; 4-Chlorophenyl ether; *p*-Chlorophenyl ether; *p*-Chlorophenyl phenyl ether; Monochlorodiphenyl oxide.

Structural Formula:

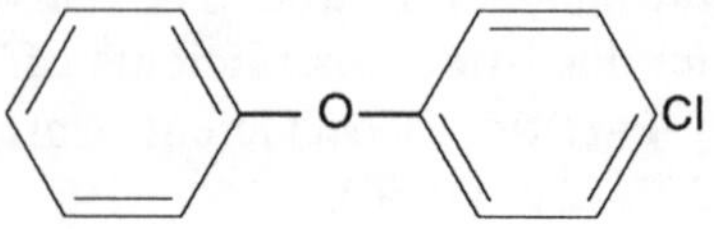

CHEMICAL DESIGNATIONS

CAS Registry Number: 7005-72-3

DOT Designation: None assigned.

Empirical Formula: $C_{12}H_9ClO$

Formula Weight: 204.66

RTECS Number: None assigned.

PHYSICAL AND CHEMICAL PROPERTIES

Appearance: Liquid.

Boiling Point: 284-285 °C [1].

Henry's Law Constant: 2.2×10^{-4} atm·m^3/mol [2].

Ionization Potential: No data found.

Log K_{oc}: 3.60 using method of Kenaga and Goring [3].

Log K_{ow}: 4.08 [4].

Melting Point: -8 °C [5].

162

Solubility in Organics: No data found.

Solubility in Water: 3.3 mg/L at 25 °C [4].

Specific Density: 1.2026 at 15/4 °C [1].

Transformation Products: No data found.

Vapor Density: 8.36 g/L at 25 °C, 7.06 (air = 1).

Vapor Pressure: 0.0027 mm at 25 °C (calculated) [4].

FIRE HAZARDS

Flash Point: No data found.

Lower Explosive Limit (LEL): No data found.

Upper Explosive Limit (UEL): No data found.

HEALTH HAZARD DATA

Immediately Dangerous to Life or Health (IDLH): No data found.

Permissible Exposure Limits (PEL) in Air: No standards set.

MANUFACTURING

Use: Research chemical.

REFERENCES

1. Weast, R.C., Ed. *CRC Handbook of Chemistry and Physics*, 67th ed. (Boca Raton, FL: CRC Press, Inc., 1986), 2406 p.
2. Pankow, J.F., and M.E. Rosen. "Determination of Volatile Compounds in Water by Purging Directly to a Capillary Column with Whole Column Cryotrapping," *Environ. Sci. Technol.*, 22(4):398-405 (1988).
3. Kenaga, E.E., and C.A.I. Goring. "Relationship Between Water

Solubility, Soil Sorption, Octanol-Water Partitioning and Concentration of Chemicals in Biota," in *Aquatic Toxicology, ASTM STP 707*, Eaton, J.G., Parrish, P.R., and A.C. Hendricks, Eds. (Philadelphia, PA: American Society for Testing and Materials, 1980), pp 78-115.

4.   Branson, D.R. "Predicting the Fate of Chemicals in the Aquatic Environment from Laboratory Data," in *Estimating the Hazard of Chemical Substances to Aquatic Life*, Cairns, J.Jr., Dickson, K.L., and A.W. Maki, Eds. (Philadelphia, PA: American Society for Testing and Materials, 1978), pp 55-70.

5.   "Treatability Manual - Volume 1: Treatability Data," Office of Research and Development, U.S. EPA Report-600/8-80-042a (1980), 1035 p.

CHRYSENE

Synonyms: Benz[*a*]phenanthrene; Benzo[*a*]phenanthrene; Benzo[alpha]-phenanthrene; 1,2-Benzophenanthrene; 1,2-Benzphenanthrene; 1,2-Dibenzonaphthalene; 1,2,5,6-Dibenzonaphthalene; RCRA waste number U050.

Structural Formula:

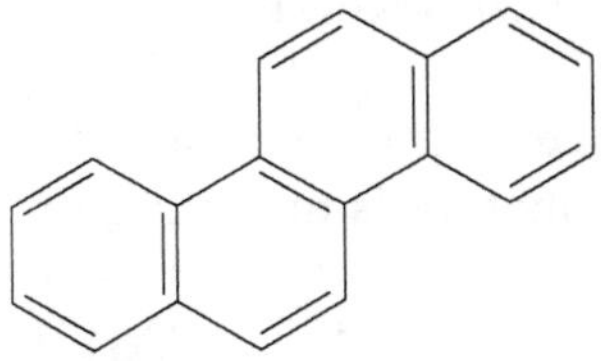

CHEMICAL DESIGNATIONS

CAS Registry Number: 218-01-9

DOT Designation: None assigned.

Empirical Formula: $C_{18}H_{12}$

Formula Weight: 228.30

RTECS Number: GC 0700000

PHYSICAL AND CHEMICAL PROPERTIES

Appearance: Orthorhombic plates exhibiting strong fluorescence under ultraviolet light.

Boiling Point: 448 °C [1].

Henry's Law Constant: 7.26 x 10^{-20} atm·m^3/mol (calculated).

Ionization Potential: 7.85 ± 0.15 eV [2].

Log K$_{oc}$: 5.39 using method of Karickhoff and others [3].

Log K$_{ow}$: 5.60 [4]; 5.61 [5]; 5.91 [6].

Melting Point: 255-256 °C [1].

Solubility in Organics: Soluble in ether, ethanol, and acetic acid [7].

Solubility in Water: 0.0015 mg/L at 15 °C, 0.006 mg/L at 25 °C; 0.001-0.05 ppm in seawater at 22 °C, 0.017 mg/L in seawater at 24 °C (practical grade) [8]; 0.0020 mg/L at 25 °C [9]; 0.0018 mg/L at 25 °C, 0.0022 mg/L at 29 °C [10]; 0.0015 mg/L [11]; 0.017 mg/L at 24 °C [12].

Specific Density: 1.274 at 20/4 °C [1].

Transformation Products: Based on structurally related compounds, chrysene may undergo photolysis to yield quinones [7].

Vapor Pressure: 6.3×10^{-9} mm at 25 °C [13]; 6.3×10^{-7} mm at 20 °C [14].

FIRE HAZARDS

Flash Point: No data found.

Lower Explosive Limit (LEL): No data found.

Upper Explosive Limit (UEL): No data found.

HEALTH HAZARD DATA

Immediately Dangerous to Life or Health (IDLH): Potential human carcinogen [15].

Permissible Exposure Limits (PEL) in Air: No individual standards have been set, however, as a constituent in coal tar pitch volatiles, the following exposure limits have been established: 0.2 mg/m^3 (benzene-soluble fraction) [16]; 0.1 mg/m^3 10-hour TWA (cyclohexane-extractable fraction) [15]; 0.2 mg/m^3 TWA (benzene solubles) [17].

MANUFACTURING

Selected Manufacturers:

Aldrich Chemical Co.
940 West Saint Paul Ave.
Milwaukee, WI 53233

Fluka Chemical Corp.
980 South Second St.
Ronkonkoma, NY 11779

Use: Organic synthesis. Derived from industrial and experimental coal gasification operations where maximum concentrations detected in gas, liquid, and coal tar streams were 7.3 mg/m^3, 0.16 mg/m^3, and 8.6 mg/m^3, respectively [18].

REFERENCES

1. Weast, R.C., Ed. *CRC Handbook of Chemistry and Physics*, 67th ed. (Boca Raton, FL: CRC Press, Inc., 1986), 2406 p.
2. Franklin, J.L., Dillard, J.G., Rosenstock, H.M., Herron, J.T., Draxl K., and F.H. Field. "Ionization Potentials, Appearance Potentials and Heats of Formation of Gaseous Positive Ions," National Bureau of Standards Report NSRDS-NBS 26, U.S. Government Printing Office (1969), 289 p.
3. Karickhoff, S.W., Brown, D.S., and T.A. Scott. "Sorption of Hydrophobic Pollutants on Natural Sediments," *Water Res.*, 13:241-248 (1979).
4. Mills, W.B., Porcella, D.B., Ungs, M.J., Gherini, S.A., Summers, K.V., Mok, L., Rupp, G.L., and G.L. Bowie. "Water Quality Assessment: A Screening Procedure for Toxic and Conventional Pollutants in Surface and Groundwater-Part I," Office of Research and Development, U.S. EPA Report-600/6-85-002a (1985), 638 p.
5. Walton, W.C. *Practical Aspects of Ground Water Modeling* (Worthington, OH: National Water Well Association, 1985), 587 p.
6. Yoshida, K., Shigeoka, T., and F. Yamauchi. "Non-Steady State Equilibrium Model for the Preliminary Prediction of the Fate of Chemicals in the Environment," *Ecotoxicol. Environ. Safety*, 7(2):179-190 (1983).
7. "Chemical, Physical, and Biological Properties of Compounds Present at Hazardous Waste Sites," U.S. EPA Report-530/SW-89-010 (1985), 619 p.
8. Verschueren, K. *Handbook of Environmental Data on Organic Chemicals* (New York: Van Nostrand Reinhold Co., 1983), 1310 p.
9. Mackay, D., and W.Y. Shiu. "Aqueous Solubility of Polynuclear Aromatic Hydrocarbons," *J. Chem. Eng. Data*, 22(4):399-402 (1977).
10. May, W.E., Wasik, S.P., and D.H. Freeman. "Determination of the Solubility Behavior of Some Polycyclic Aromatic Hydrocarbons in Water," *Anal. Chem.*, 50(7):997-1000 (1978).

11. Davis, W.W., Krahl, M.E., and G.H.A. Clowes. "Solubility of Carcinogenic and Related Hydrocarbons in Water," *J. Am. Chem. Soc.*, 64(1):108-110 (1942).

12. Hollifield, H.C. "Rapid Nephelometric Estimate of Water Solubility of Highly Insoluble Organic Chemicals of Environmental Interest," *Bull. Environ. Contam. Toxicol.*, 23(4/5):579-586 (1979).

13. Mabey, W.R., Smith, J.H., Podoll, R.T., Johnson, H.L., Mill, T., Chou, T.-W., Gates, J., Partridge, I.W., Jaber, H., and D. Vandenberg. "Aquatic Fate Process Data for Organic Priority Pollutants - Final Report," Office of Regulations and Standards, U.S. EPA Report-440/4-81-014 (1982), 407 p.

14. Sims, R.C., Doucette, W.C., McLean, J.E., Greeney, W.J., and R.R. Dupont. "Treatment Potential for 56 EPA Listed Hazardous Chemicals in Soil," National Technical Information Service, U.S. EPA Report-600/6-88-001 (1988), 105 p.

15. "NIOSH Pocket Guide to Chemical Hazards," U.S. Department of Health and Human Services, U.S. Government Printing Office (1987), 241 p.

16. "General Industry Standards for Toxic and Hazardous Substances," U.S. Code of Federal Regulations 1910, Subpart Z, Section 1910.1000 (July 1982).

17. *Documentation of the Threshold Limit Values and Biological Exposure Indices for 1986-1987* (Cincinnati, OH: American Conference of Governmental Industrial Hygienists, 1986), 111 p.

18. Cleland, J.G. "Project Summary - Environmental Hazard Rankings of Pollutants Generated in Coal Gasification Processes," Office of Research and Development, U.S. EPA Report-600/S7-81-101 (1981), 19 p.

p,p'-DDD

Synonyms: 1,1-Bis(4-chlorophenyl)-2,2-dichloroethane; 1,1-Bis(*p*-chlorophenyl)-2,2-dichloroethane; 2,2-Bis(4-chlorophenyl)-1,1-dichloroethane; 2,2-Bis(*p*-chlorophenyl)-1,1-dichloroethane; DDD; 4,4'-DDD; 1,1-Dichloro-2,2-bis(*p*-chlorophenyl)ethane; 1,1-Dichloro-2,2-bis(parachlorophenyl)ethane; 1,1-Dichloro-2,2-di(4-chlorophenyl)-ethane; 1,1-Dichloro-2,2-di(*p*-chlorophenyl)ethane; Dichlorodiphenyl-dichloroethane; 4,4'-Dichlorodiphenyldichloroethane; *p,p*'-Dichlorodi-phenyldichloroethane; **1,1'-(2,2-Dichloroethylidene)bis-[4-chloro-benzene]**; Dilene; ENT 4,225; ME-1,700; NA 2,761; NCI-C00475; RCRA waste number U060; Rhothane; Rhothane D-3; Rothane; TDE; 4,4'-TDE; *p,p*'-TDE; Tetrachlorodiphenylethane.

Structural Formula:

CHEMICAL DESIGNATIONS

CAS Registry Number: 72-54-8

DOT Designation: 2761

Empirical Formula: $C_{14}H_{10}Cl_4$

Formula Weight: 320.05

RTECS Number: KI 0700000

PHYSICAL AND CHEMICAL PROPERTIES

Appearance: Crystalline white solid.

Boiling Point: 193 °C [1].

Henry's Law Constant: 2.16 x 10^{-5} atm·m³/mol (calculated) [2].

Ionization Potential: No data found.

169

Log K$_{oc}$: 4.64 using method of Kenaga and Goring [3].

Log K$_{ow}$: 5.99 [4]; 6.02 [3]; 5.061 [5].

Melting Point: 112 °C [6]; 110 °C [7]; 107-109 °C [8].

Solubility in Organics: No data found.

Solubility in Water: 0.160 mg/L at 24 °C [6]; 20 ppb at 25 °C [9]; 100 ppm at 25 °C [1]; 0.02-0.09 mg/L at 25 °C [2]; 0.005 ppm [3]; 50 ppb at 15 °C, 90 ppb at 25 °C, 150 ppb at 35 °C, 240 ppb at 45 °C (particle sizes ≤ 5μ) [10].

Specific Density: 1.476 at 20/4 °C [11].

Transformation Products: It was reported that *p,p'*-DDD, a major biodegradation product of *p,p'*-DDT, was degraded by *Aerobacter aerogenes* under aerobic conditions to yield 1-chloro-2,2-bis(*p*-chlorophenyl)ethylene, 1-chloro-2,2-bis(*p*-chlorophenyl)ethane, and 1,1-bis(*p*-chlorophenyl)ethylene. Under anaerobic conditions, however, four additional compounds were identified. These are bis(*p*-chlorophenyl)-acetic acid, *p,p'*-dichlorodiphenylmethane, *p,p'*-dichlorobenzhydrol, and *p,p'*-dichlorobenzophenone [12]. Under reducing conditions, indigenous microbes in Lake Michigan sediments were able to degrade DDD to 2,2-bis(*p*-chlorophenyl)ethane and 2,2-bis(*p*-chlorophenyl)ethanol [13].

Vapor Density: 17.2 ng/L at 30 °C [14].

Vapor Pressure: 1.02 x 10^{-6} mm at 30 °C [14].

FIRE HAZARDS

Flash Point: Not pertinent [11].

Lower Explosive Limit (LEL): Not pertinent [11].

Upper Explosive Limit (UEL): Not pertinent [11].

HEALTH HAZARD DATA

Immediately Dangerous to Life or Health (IDLH): No data found.

Permissible Exposure Limits (PEL) in Air: No standards set.

MANUFACTURING

Selected Manufacturer:

> Aldrich Chemical Co.
> 940 West Saint Paul Ave.
> Milwaukee, WI 53233

Uses: Dusts, emulsions, and wettable powders for contact control of leaf rollers and other insects on vegetables and tobacco.

REFERENCES

1. Sax, N.I., Ed. *Dangerous Properties of Industrial Materials Report* (New York: Van Nostrand Reinhold Co., 1985), 5(3): 88 p.
2. "Treatability Manual - Volume 1: Treatability Data," Office of Research and Development, U.S. EPA Report-600/8-80-042a (1980), 1035 p.
3. Kenaga, E.E., and C.A.I. Goring. "Relationship Between Water Solubility, Soil Sorption, Octanol-Water Partitioning and Concentration of Chemicals in Biota," in *Aquatic Toxicology, ASTM STP 707*, Eaton, J.G., Parrish, P.R., and A.C. Hendricks, Eds. (Philadelphia, PA: American Society for Testing and Materials, 1980), pp 78-115.
4. Callahan, M.A., Slimak, M.W., Gable, N.W., May, I.P., Fowler, C.F., Freed, J.R., Jennings, P., Durfee, R.L., Whitmore, F.C., Maestri, B., Mabey, W.R., Holt, B.R., and C. Gould. "Water-Related Environmental Fate of 129 Priority Pollutants Volumes I and II," National Technical Information Service, U.S. EPA Report-440/4-79-029 (1979), 1160 p.
5. Rao, P.S.C., and J.M. Davidson. "Estimation of Pesticide Retention and Transformation Parameters Required in Nonpoint Source Pollution Models," in *Environmental Impact of Nonpoint Source Pollution*, Overcash, M.R., and J.M. Davidson, Eds. (Ann Arbor, MI: Ann Arbor Science Publishers, Inc., 1980), pp 23-67.
6. Verschueren, K. *Handbook of Environmental Data on Organic Chemicals* (New York: Van Nostrand Reinhold Co., 1983), 1310 p.
7. Sax, N.I. *Dangerous Properties of Industrial Materials* (New York: Van Nostrand Reinhold Co., 1984), 3124 p.

8. *Catalog Handbook of Fine Chemicals* (Milwaukee, WI: Aldrich Chemical Co., 1988), 2212 p.

9. Weil, L., Dure, G., and K.E. Quentin. "Solubility in Water of Insecticide Chlorinated Hydrocarbons and Polychlorinated Biphenyls in View of Water Pollution," *Z. Wasser Forsch.*, 7(6):169-175 (1974).

10. Biggar, J.W., and I.R. Riggs. "Apparent Solubility of Organochlorine Insecticides in Water at Various Temperatures," *Hilgardia*, 42(10):383-391 (1974).

11. Weiss, G. *Hazardous Chemicals Data Book* (Park Ridge, NJ: Noyes Data Corp., 1986), 1069 p.

12. Fries, G.F. "Degradation of Chlorinated Hydrocarbons under Anaerobic Conditions," in *Fate of Organic Pesticides in the Aquatic Environment*, Advances in Chemistry Series, R.F. Gould, Ed. (Washington, D.C.: American Chemical Society, 1972), pp 256-270.

13. Leland, H.V., Bruce, W.N., and N.F. Shimp. "Chlorinated Hydrocarbon Insecticides in Sediments in Southern Lake Michigan," *Environ. Sci. Technol.*, 7(9):833-838 (1973).

14. Spencer, W.F., and M.M. Cliath. "Volatility of DDT and Related Compounds," *J. Agric. Food Chem.*, 20(3):645-649 (1972).

p,p'-DDE

Synonyms: 2,2-Bis(4-chlorophenyl)-1,1-dichloroethene; 2,2-Bis(*p*-chlorophenyl)-1,1-dichloroethene; DDE; 4,4'-DDE; DDT dehydrochloride; 1,1-Dichloro-2,2-bis(*p*-chlorophenyl)ethylene; Dichlorodiphenyldichloroethylene; *p,p'*-Dichlorodiphenyl-dichloroethylene; **1,1'-(Dichloroethenylidene)bis(4-chlorobenzene)**; NCI-C00555.

Structural Formula:

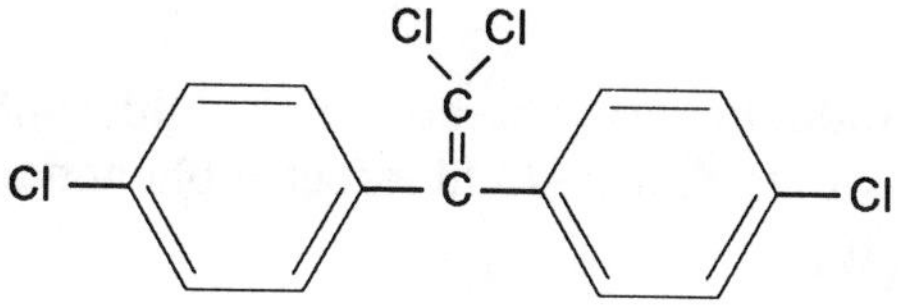

CHEMICAL DESIGNATIONS

CAS Registry Number: 72-55-9

DOT Designation: 2761

Empirical Formula: $C_{14}H_8Cl_4$

Formula Weight: 319.03

RTECS Number: KV 9450000

PHYSICAL AND CHEMICAL PROPERTIES

Appearance: Solid.

Boiling Point: No data found.

Henry's Law Constant: 2.34 x 10^{-5} atm·m^3/mol (calculated) [1].

Ionization Potential: No data found.

Log K_{oc}: 6.00 [2]; 5.386 [3].

Log K_{ow}: 5.83 [4]; 5.69 [5]; 5.766 [6].

Melting Point: 88-90 °C [7]; 112 °C [8].

Solubility in Organics: Soluble in fats and most solvents [9].

Solubility in Water: 0.040 mg/L at 20 °C, 0.065 mg/L at 24 °C, 0.0013 mg/L at 25 °C [10]; 14 ppb at 20 °C [11]; 40 ppb at 20 °C [12]; 0.12 mg/L at 25 °C [13]; 0.010 ppm [6]; 55 ppb at 15 °C, 120 ppb at 25 °C, 235 ppb at 35 °C, 450 ppb at 45 °C (particle sizes $\leq$ 5μ) [14].

Specific Density: No data found.

Transformation Product: May degrade to bis(chlorophenyl)acetic acid in water [15], or oxidize to *p,p'*-dichlorobenzophenone using ultraviolet light as a catalyst [16].

Vapor Density: 109 ng/L at 30 °C [17].

Vapor Pressure: 6.49 x 10^{-6} mm at 30 °C [17].

FIRE HAZARDS

Flash Point: No data found.

Lower Explosive Limit (LEL): No data found.

Upper Explosive Limit (UEL): No data found.

HEALTH HAZARD DATA

Immediately Dangerous to Life or Health (IDLH): No data found.

Permissible Exposure Limits (PEL) in Air: No standards set.

MANUFACTURING

Selected Manufacturer:

>Aldrich Chemical Co.
>940 West Saint Paul Ave.
>Milwaukee, WI 53233

Uses: Military product; chemical research.

REFERENCES

1. "Treatability Manual - Volume 1: Treatability Data," Office of Research and Development, U.S. EPA Report-600/8-80-042a (1980), 1035 p.
2. Oliver, B.G., and M.N. Charlton. "Chlorinated Organic Contaminants on Settling Particulates in the Niagara River Vicinity of Lake Ontario," *Environ. Sci. Technol.*, 18(12):903-908 (1984).
3. Reinbold, K.A., Hassett, J.J., Means, J.C., and W.L. Banwart. "Adsorption of Energy-Related Organic Pollutants: A Literature Review," Office of Research and Development, U.S. EPA Report-600/3-79-086 (1979), 180 p.
4. Travis, C.C., and A.D. Arms. "Bioconcentration of Organics in Beef, Milk and Vegetation," *Environ. Sci. Technol.*, 22(3):271-274 (1988).
5. Freed, V.H., Chiou, C.T., and R. Haque. "Chemodynamics: Transport and Behavior of Chemicals in the Environment - A Problem in Environmental Health," *Environ. Health Perspect.*, 20:55-70 (1977).
6. Kenaga, E.E., and C.A.I. Goring. "Relationship Between Water Solubility, Soil Sorption, Octanol-Water Partitioning and Concentration of Chemicals in Biota," in *Aquatic Toxicology, ASTM STP 707*, Eaton, J.G., Parrish, P.R., and A.C. Hendricks, Eds. (Philadelphia, PA: American Society for Testing and Materials, 1980), pp 78-115.
7. Leffingwell, J.T. "The Photolysis of DDT in Water," PhD Thesis, University of California, Davis, CA (1975).
8. Melnikov, N.N. *Chemistry of Pesticides* (New York: Springer-Verlag, Inc., 1971), 480 p.
9. *IARC Monographs on the Evaluation of Carcinogenic Risk of Chemicals to Man. Some Organochlorine Pesticides, Volume 5* (Lyon, France: International Agency for Research on Cancer, 1974), 241 p.
10. Metcalf, R.L., Kapoor, I.P., Lu, P.-Y., Schuth, C.K., and P. Sherman. "Model Ecosystem Studies of the Environmental Fate of Six Organochlorine Pesticides," *Environ. Health Perspect.*, (June 1973), pp 35-44.
11. Weil, L., Dure, G., and K.E. Quentin. "Solubility in Water of Insecticide Chlorinated Hydrocarbons and Polychlorinated Biphenyls in View of Water Pollution," *Z. Wasser Forsch.*, 7(6):169-175 (1974).

12. Chiou, C.T., Freed, V.H., Schmedding, D.W., and R.L. Kohnert. "Partition Coefficients and Bioaccumulation of Selected Organic Chemicals," *Environ. Sci. Technol.*, 11(5):475-478 (1977).
13. Walton, W.C. *Practical Aspects of Ground Water Modeling* (Worthington, OH: National Water Well Association, 1985), 587 p.
14. Biggar, J.W., and I.R. Riggs. "Apparent Solubility of Organochlorine Insecticides in Water at Various Temperatures," *Hilgardia*, 42(10):383-391 (1974).
15. Verschueren, K. *Handbook of Environmental Data on Organic Chemicals* (New York: Van Nostrand Reinhold Co., 1983), 1310 p.
16. Hazardous Substances Data Bank. *p,p'*-DDE, National Library of Medicine, Toxicology Information Program (1989).
17. Spencer, W.F., and M.M. Cliath. "Volatility of DDT and Related Compounds," *J. Agric. Food Chem.*, 20(3):645-649 (1972).

p,p'-DDT

Synonyms: Agritan; Anofex; Arkotine; Azotox; 2,2-Bis(4-chlorophenyl)-1,1,1-trichloroethane; 2,2-Bis(*p*-chlorophenyl)-1,1,1-trichloroethane; α,α-Bis(*p*-chlorophenyl)-β,β,β-trichloroethane; 1,1-Bis(*p*-chlorophenyl)-2,2,2-trichloroethane; Bosan Supra; Bovidermol; Chlorophenothan; Chlorophenothane; Chlorophenotoxum; Citox; Clofenotane; DDT; 4,4'-DDT; Dedelo; Deoval; Detox; Detoxan; Dibovan; Dichlorodiphenyltrichloroethane; *p,p*'-Dichlorodiphenyltrichloroethane; 4,4'-Dichlorodiphenyltrichloroethane; Dicophane; Didigam; Didimac; Diphenyltrichloroethane; Dodat; Dykol; ENT 1,506; Estonate; Genitox; Gesafid; Gesapon; Gesarex; Gesarol; Guesapon; Gyron; Havero-extra; Ivoran; Ixodex; Kopsol; Mutoxin; NCI-C00464; Neocid; Parachlorocidum; PEB1; Pentachlorin; Pentech; PPzeidan; RCRA waste number U061; Rukseam; Santobane; Trichlorobis(4-chlorophenyl)-ethane; Trichlorobis(*p*-chlorophenyl)ethane; 1,1,1-Trichloro-2,2-bis(*p*-chlorophenyl)ethane; 1,1,1-Trichloro-2,2-di(4-chlorophenyl)ethane; 1,1,1-Trichloro-2,2-di(*p*-chlorophenyl)ethane; **1,1'-(2,2,2-Trichloroethylidene)bis[4-chlorobenzene]**; Zeidane; Zerdane.

Structural Formula:

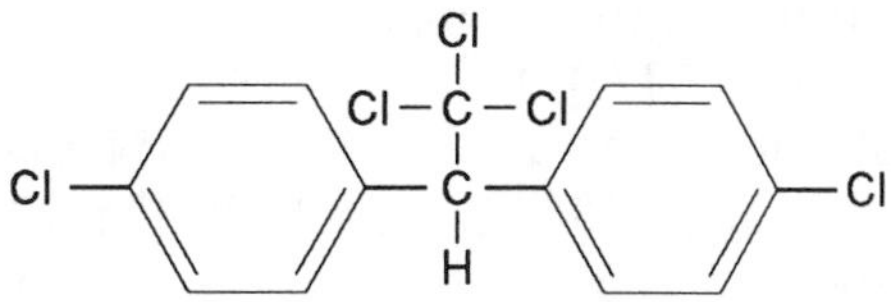

CHEMICAL DESIGNATIONS

CAS Registry Number: 50-29-3

DOT Designation: None assigned.

Empirical Formula: $C_{14}H_9Cl_5$

Formula Weight: 354.49

RTECS Number: KJ 3325000

PHYSICAL AND CHEMICAL PROPERTIES

Appearance and Odor: Colorless crystals or white powder, odorless to slightly fragrant powder.

Boiling Point: 260 °C [1]; 185 °C [2].

Henry's Law Constant: 3.8 x 10^{-5} atm·m^3/mol [3]; 4.89 x 10^{-5} atm·m^3/mol [4]; 1.03 x 10^{-4} atm·m^3/mol [5]; 5.2 x 10^{-5} atm·m^3/mol [6].

Ionization Potential: No data found.

Log K_{oc}: 5.38 [4]; 5.146 [7]; 5.14 [8]; 5.18 [9]; 6.26 [10]; 5.39 [11].

Log K_{ow}: 6.36 [12]; 6.19 [13]; 5.76 [14]; 5.98 [15]; 5.38 [16]; 6.16, 6.17, 6.22, 6.44 [17]; 6.28 [18]; 5.982 [19]; 4.89 [20]; 4.96 [21].

Melting Point: 108.5 °C [22]; 108-109 °C [1].

Solubility in Organics: Soluble in acetone, benzene, cyclohexane, morpholine, pyridine, dioxane [23], ether, and carbon tetrachloride [24].

Solubility in Water: 0.0031-0.0034 mg/L at 25 °C [25]; 5.5 ppb at 25 °C [26]; 0.0017 mg/L at 25 °C [27]; 1 x 10^{-5} wt% at 20 °C [28]; 0.0012 mg/L at 25 °C [29]; 0.007 mg/L at 20 °C [30]; 0.0054 mg/L at 24 °C [31]; 0.004 mg/L at 24 °C [32]; 0.004 ppm at 25 °C [7]; 5.9 ppb at 2 °C, 37.4 ppb at 25 °C, 45 ppb at 37.5 °C [33]; 0.01-0.1 ppm at 22 °C [34]; 0.002 ppm [35]; 17 ppb at 15 °C, 25 ppb at 25 °C, 37 ppb at 35 °C, 45 ppb at 45 °C (particle sizes ≤ 5µ) [36]; 0.26 ppm at 25 °C [37]; at room temperature, solubilities of 40 ppb (maximum particle size 5µ) and 16 ppb (maximum particle size 0.05µ) were reported [38]; 0.0077 mg/L at 20 °C [39].

Specific Density: 1.56 at 15/4 °C [40].

Transformation Products: In soils under anaerobic conditions, *p,p'*-DDT is rapidly converted to *p,p'*-DDD and very slowly to *p,p'*-DDE under aerobic conditions [41]. Other reported degradation products under aerobic and anaerobic conditions by various microbes include 1,1'-bis(*p*-chlorophenyl)-2-chloroethane, 1,1'-bis(*p*-chlorophenyl)-2-hydroxyethane, and *p*-chlorophenyl acetic acid [42]. It was also reported that *p,p'*-DDE formed by hydrolyzing *p,p'*-DDT [20]. The white rot fungus *Phanerochaete chrysosporium* degraded *p,p'*-DDT yielding the following metabolites: 1,1-dichloro-2,2-bis(4-chlorophenyl)ethane (DDD), 2,2,2-trichloro-1,1-bis(4-chloro-phenyl)ethanol (dicofol), 2,2-dichloro-1,1-bis(4-chlorophenyl)ethanol, and 4,4'-dichlorobenzophenone. Mineralization of *p,p*'-DDT by the white rot fungi *Pleurotus ostreatus*, *Phellinus weirri*, and *Polyporus versicolor* was also demonstrated [43]. Fries [44] reported that *Aerobacter aerogenes* degraded *p,p*'-DDT under aerobic conditions to yield *p,p*'-DDD, *p,p*'-DDE, 1-chloro-2,2-bis(*p*-chlorophenyl)ethylene, 1-chloro-2,2-bis(*p*-chlorophenyl)ethane, and 1,1-bis(*p*-chlorophenyl)ethylene. Under anaerobic conditions the same organism produced four additional compounds. These are bis(*p*-chlorophenyl)acetic acid, *p,p*'-dichlorodiphenylmethane, *p,p*'-dichlorobenzhydrol, and *p,p*'-dichlorobenzophenone. Other degradation products of *p,p*'-DDT under aerobic and anaerobic conditions in soils using various cultures not previously mentioned included 1,1-bis(*p*-chlorophenyl)-2,2,2-trichloroethanol (Kelthane), and *p*-chlorobenzoic acid.

Vapor Density: 13.6 ng/L at 30 °C [45].

Vapor Pressure: 1.9×10^{-7} mm at 20 °C [25]; 1×10^{-7} mm at 25 °C [46]; 1.5×10^{-7} mm at 20 °C [47]; 1.9×10^{-7} mm at 25 °C [48]; 7.26×10^{-7} mm at 30 °C [45]; 1.7×10^{-10} mm at 20 °C, 6.2×10^{-10} mm at 50 °C, 8.9×10^{-10} mm at 100 °C [49].

FIRE HAZARDS

Flash Point: 72.2-77.2 °C [40].

Lower Explosive Limit (LEL): Unknown [28].

Upper Explosive Limit (UEL): Unknown [28].

HEALTH HAZARD DATA

Immediately Dangerous to Life or Health (IDLH): Potential human carcinogen [28].

Permissible Exposure Limits (PEL) in Air: 1 mg/m^3 [50]; lowest detectable limit (0.5 mg/m^3 TWA by NIOSH validated method) [28]; 1 mg/m^3 TWA [51].

MANUFACTURING

Selected Manufacturers:

Aldrich Chemical Co.
940 West Saint Paul Ave.
Milwaukee, WI 53233

Lebanon Chemical Corp.
Lebanon, PA 17042

Montrose Chemical Corp. of California
500 South Virgil Ave.
Los Angeles, CA 90005

Uses: Use as an insecticide is now prohibited; chemical research; nonsystemic stomach and contact insecticide.

REFERENCES

1. Weast, R.C., Ed. *CRC Handbook of Chemistry and Physics*, 67th ed. (Boca Raton, FL: CRC Press, Inc., 1986), 2406 p.
2. "Treatability Manual - Volume 1: Treatability Data," Office of Research and Development, U.S. EPA Report-600/8-80-042a (1980), 1035 p.
3. Eisenreich, S.J., Looney, B.B., and J.D. Thornton. "Airborne Organic Contaminants in the Great Lakes Ecosystem," *Environ. Sci. Technol.*, 15(1):30-38 (1981).
4. Jury, W.A., Spencer, W.F., and W.J. Farmer. "Behavior Assessment Model for Trace Organics in Soil: III. Application of Screening Model," *J. Environ. Qual.*, 13(4):573-579 (1984).
5. Jury, W.A., Spencer, W.F., and W.J. Farmer. "Behavior Assessment Model for Trace Organics in Soil: IV. Review of Experimental Evidence," *J. Environ. Qual.*, 13(4):580-586 (1984).
6. Jury, W.A., Spencer, W.F., and W.J. Farmer. "Use of Models for Assessing Relative Volatility, Mobility, and Persistence of Pesticides and other Trace Organics in Soil Systems," in *Hazard Assessment of Chemicals, Volume 2*, J. Saxena, Ed. (New York: Academic Press, Inc., 1983), pp 1-43.
7. Chiou, C.T., Peters, L.J., and V.H. Freed. "A Physical Concept of Soil-Water Equilibria for Nonionic Organic Compounds," *Science*, 206:831-832 (1979).

8. Schwarzenbach, R.P., and J. Westall. "Transport of Nonpolar Organic Compounds from Surface Water to Groundwater. Laboratory Sorption Studies," *Environ. Sci. Technol.*, 15(11):1360-1367 (1981).

9. McCall, P.J., Swann, R.L., Laskowski, D.A., Unger, S.M., Vrona, S.A., and H.J. Dishburger. "Estimation of Chemical Mobility in Soil from Liquid Chromatographic Retention Times," *Bull. Environ. Contam. Toxicol.*, 24(2):190-195 (1980).

10. Pierce, R.H., Olney, C.E., and G.T. Felbeck. "*p,p'*-DDT Adsorption to Suspended Particulate Matter in Sea Water," *Geochim. Cosmochim. Acta*, 38(7):1061-1073 (1974).

11. Rao, P.S.C., and J.M. Davidson. "Estimation of Pesticide Retention and Transformation Parameters Required in Nonpoint Source Pollution Models," in *Environmental Impact of Nonpoint Source Pollution*, Overcash, M.R., and J.M. Davidson, Eds. (Ann Arbor, MI: Ann Arbor Science Publishers, Inc., 1980), pp 23-67.

12. Chiou, C.T., Schmedding D.W., and M. Manes. "Partitioning of Organic Compounds in Octanol-Water Systems," *Environ. Sci. Technol.*, 16(1):4-10 (1982).

13. Freed, V.H., Chiou, C.T., and R. Haque. "Chemodynamics: Transport and Behavior of Chemicals in the Environment - A Problem in Environmental Health," *Environ. Health Perspect.*, 20:55-70 (1977).

14. Travis, C.C., and A.D. Arms. "Bioconcentration of Organics in Beef, Milk and Vegetation," *Environ. Sci. Technol.*, 22(3):271-274 (1988).

15. Mackay, D., and S. Paterson. "Calculating Fugacity," *Environ. Sci. Technol.*, 15(9):1006-1014 (1981).

16. Kenaga, E.E. "Predicted Bioconcentration Factors and Soil Sorption Coefficients of Pesticides and other Chemicals," *Ecotoxicol. Environ. Safety*, 4(1):26-38 (1980).

17. Brooke, D.N., Dobbs, A.J., and N. Williams. "Octanol:Water Partition Coefficients (P): Measurement, Estimation, and Interpretation, Particularly for Chemicals with P > 10^5," *Ecotoxicol. Environ. Safety*, 11(3):251-260 (1986).

18. Geyer, H., Politzki, G., and D. Freitag. "Prediction of Ecotoxicological Behaviour of Chemicals: Relationship Between *n*-Octanol/Water Partition Coefficient and Bioaccumulation of Organic Chemicals by Alga *Chlorella*," *Chemosphere*, 13(2):269-284 (1984).

19. Kenaga, E.E., and C.A.I. Goring. "Relationship Between Water Solubility, Soil Sorption, Octanol-Water Partitioning and Concentration of Chemicals in Biota," in *Aquatic Toxicology, ASTM STP 707*, Eaton, J.G., Parrish, P.R., and A.C. Hendricks, Eds.

(Philadelphia, PA: American Society for Testing and Materials, 1980), pp 78-115.

20. Wolfe, N.L., Zepp, R.G., Paris, D.F., Baughman, G.L., and R.C. Hollis. "Methoxychlor and DDT Degradation in Water: Rates and Products," *Environ. Sci. Technol.*, 11(12):1077-1081 (1977).

21. *Special Occupational Hazard Review for DDT* (Rockville, MD: National Institute for Occupational Safety and Health, 1978), 205 p.

22. Caswell, R.L., DeBold, K.J., and L.S. Gilbert, Eds. *Pesticide Handbook*, 29th ed. (College Park, MD: The Entomological Society of America, 1981), 286 p.

23. "Chemical, Physical, and Biological Properties of Compounds Present at Hazardous Waste Sites," U.S. EPA Report-530/SW-89-010 (1985), 619 p.

24. *Toxic and Hazardous Industrial Chemicals Safety Manual for Handling and Disposal with Toxicity and Hazard Data* (Tokyo, Japan: International Technical Information Institute, 1986), 700 p.

25. Verschueren, K. *Handbook of Environmental Data on Organic Chemicals* (New York: Van Nostrand Reinhold Co., 1983), 1310 p.

26. Weil, L., Dure, G., and K.E. Quentin. "Solubility in Water of Insecticide Chlorinated Hydrocarbons and Polychlorinated Biphenyls in View of Water Pollution," *Z. Wasser Forsch.*, 7(6):169-175 (1974).

27. Lyman, W.J., Reehl, W.F., and D.H. Rosenblatt. *Handbook of Chemical Property Estimation Methods: Environmental Behavior of Organic Compounds* (New York: McGraw-Hill, Inc., 1982).

28. "NIOSH Pocket Guide to Chemical Hazards," U.S. Department of Health and Human Services, U.S. Government Printing Office (1987), 241 p.

29. Bowman, M.C., Acree, Fred Jr., and M.K. Corbett. "Solubility of Carbon-14 DDT in Water," *J. Agric. Food Chem.*, 8(5):406-408 (1960).

30. Nisbet, I.C., and A.F. Sarofim. "Rates and Routes of Transport of PCBs in the Environment," *Environ. Health Perspect.*, (April 1972), pp 21-38.

31. Chiou, C.T., Malcolm, R.L., Brinton, T.I., and D.E. Kile. "Water Solubility Enhancement of Some Organic Pollutants and Pesticides by Dissolved Humic and Fulvic Acids," *Environ. Sci. Technol.*, 20(5):502-508 (1986).

32. Hollifield, H.C. "Rapid Nephelometric Estimate of Water Solubility of Highly Insoluble Organic Chemicals of Environmental Interest," *Bull. Environ. Contam. Toxicol.*, 23(4/5):579-586 (1979).

33. Babers, F.H. "The Solubility of DDT in Water Determined Radiometrically," *J. Am. Chem. Soc.*, 77(17):4666 (1955).

34. Roeder, K.D., and E.A. Weiant. "The Site of Action of DDT in the Cockroach," *Science*, 103(2671):304-305 (1946).

35. Kapoor, I.P., Metcalf, R.L., Hirwe, A.S., Coats, J.R., and M.S. Khalsa. "Structure Activity Correlations of Biodegradability of DDT Analogs," *J. Agric. Food Chem.*, 21(2):310-315 (1973).

36. Biggar, J.W., and I.R. Riggs. "Apparent Solubility of Organochlorine Insecticides in Water at Various Temperatures," *Hilgardia*, 42(10):383-391 (1974).

37. *Drinking Water and Health* (Washington, DC: National Academy of Sciences, 1977), 939 p.

38. Robeck, G.G., Dostal, K.A., Cohen, J.M., and J.F. Kreissl. "Effectiveness of Water Treatment Processes in Pesticide Removal," *J. Am. Water Works Assoc.*, 57(2):181-200 (1965).

39. Friesen, K.J., Sarna, L.P., and G.R.B. Webster. "Aqueous Solubility of Polychlorinated Dibenzo-*p*-dioxins Determined by High Pressure Liquid Chromatography," *Chemosphere*, 14(9):1267-1274 (1985).

40. Weiss, G. *Hazardous Chemicals Data Book* (Park Ridge, NJ: Noyes Data Corp., 1986), 1069 p.

41. Kearney, P.C., and D.D. Kaufman. *Herbicides: Chemistry, Degradation and Mode of Action* (New York: Marcel Dekker, Inc., 1976), 1036 p.

42. Kobayashi, H., and B.E. Rittman. "Microbial Removal of Hazardous Organic Compounds," *Environ. Sci. Technol.*, 16(3):170A-183A (1982).

43. Bumpus, J.A., and S.D. Aust. "Biodegradation of DDT [1,1,1-Trichloro-2,2-bis(4-chlorophenyl)ethane] by the White Rot Fungus *Phanerochaete chrysosporium*," *Appl. Environ. Microbiol.*, 53(9):2001-2008 (1987).

44. Fries, G.F. "Degradation of Chlorinated Hydrocarbons under Anaerobic Conditions," in *Fate of Organic Pesticides in the Aquatic Environment, Advances in Chemistry Series*, R.F. Gould, Ed. (Washington, D.C.: American Chemical Society, 1972), pp 256-270.

45. Spencer, W.F., and M.M. Cliath. "Volatility of DDT and Related Compounds," *J. Agric. Food Chem.*, 20(3):645-649 (1972).

46. Mackay, D., and A.W. Wolkoff. "Rate of Evaporation of Low-Solubility Contaminants from Water Bodies to Atmosphere," *Environ. Sci. Technol.*, 7(7):611-614 (1973).

47. Balson, E.W. "Studies in Vapour Pressure Measurement, Part III. - An Effusion Manometer Sensitive to 5 x 10⁻⁶ Millimetres of Mercury: Vapour Pressure of DDT and other Slightly Volatile Substances," *Trans. Faraday Soc.*, 43:54-60 (1947).

48. Martin, H., Ed. *Pesticide Manual*, 3rd ed. (Worcester, England: British Crop Protection Council, 1972).

49. Webster, G.R.B., Friesen, K.J., Sarna, L.P., and D.C.G. Muir.

"Environmental Fate Modelling of Chlorodioxins: Determination of Physical Constants," *Chemosphere*, 14(6/7):609-622.

50. "General Industry Standards for Toxic and Hazardous Substances," U.S. Code of Federal Regulations 1910, Subpart Z Section 1910.1000 (July 1982).

51. *Documentation of the Threshold Limit Values and Biological Exposure Indices for 1986-1987* (Cincinnati, OH: American Conference of Governmental Industrial Hygienists, 1986), 111 p.

DIBENZ[a,h]ANTHRACENE

Synonyms: 1,2:5,6-Benzanthracene; DBA; 1,2,5,6-DBA; DB[a,h]A; 1,2:5,6-Dibenzanthracene; 1,2:5,6-Dibenz[a]anthracene; 1,2:5,6-Dibenzoanthracene; Dibenzo[a,h]anthracene; RCRA waste number U063.

Structural Formula:

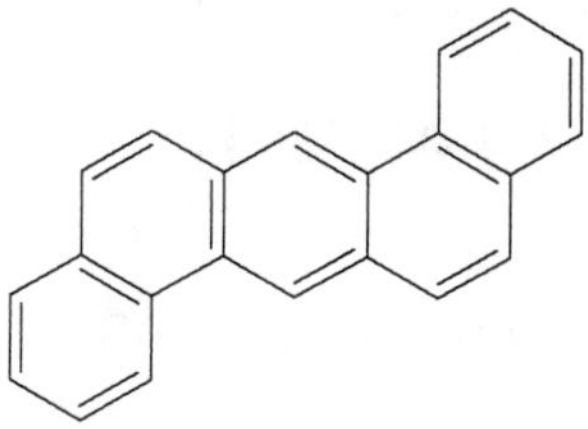

CHEMICAL DESIGNATIONS

CAS Registry Number: 53-70-3

DOT Designation: None assigned.

Empirical Formula: $C_{22}H_{14}$

Formula Weight: 278.36

RTECS Number: HN 2625000

PHYSICAL AND CHEMICAL PROPERTIES

Appearance: Monoclinic or orthorhombic crystals or leaflets.

Boiling Point: 524 °C [1].

Henry's Law Constant: 7.33 x 10^{-9} atm·m^3/mol at 20-25 °C (calculated).

Ionization Potential: 7.28 ± 0.29 eV [2].

Log K$_{oc}$: 6.22 [3].

Log K$_{ow}$: 6.36 [4]; 6.50 [3]; 5.97 [5].

Melting Point: 269-270 °C [6]; 262-265 °C [7].

Solubility in Organics: Soluble in most solvents [8].

Solubility in Water: 0.0005 mg/L at 25 °C [9]; 0.005 mg/L [10]; 0.00249 mg/L at 25 °C [11].

Specific Density: 1.282 [12].

Transformation Products: No data found.

Vapor Pressure: $\approx 10^{-10}$ mm at 20 °C [13].

FIRE HAZARDS

Flash Point: No data found.

Lower Explosive Limit (LEL): No data found.

Upper Explosive Limit (UEL): No data found.

HEALTH HAZARD DATA

Immediately Dangerous to Life or Health (IDLH): Potential human carcinogen [14].

Permissible Exposure Limits (PEL) in Air: No individual standards have been set, however, as a constituent in coal tar pitch volatiles, the following exposure limits have been established: 0.2 mg/m^3 (benzene-soluble fraction) [15]; 0.1 mg/m^3 10-hour TWA (cyclohexane-extractable fraction) [14]; 0.2 mg/m^3 TWA (benzene solubles) [16].

MANUFACTURING

Selected Manufacturer:

Aldrich Chemical Co.
940 West Saint Paul Ave.
Milwaukee, WI 53233

Use: Research chemical. Though not produced commercially in the U.S., dibenz[a,h]anthracene is derived from industrial and experimental coal gasification operations where maximum concentrations detected in gas and coal tar streams were 0.0061 mg/m^3 and 3.4 mg/m^3, respectively [17].

REFERENCES

1. Verschueren, K. *Handbook of Environmental Data on Organic Chemicals* (New York: Van Nostrand Reinhold Co., 1983), 1310 p.
2. Franklin, J.L., Dillard, J.G., Rosenstock, H.M., Herron, J.T., Draxl K., and F.H. Field. "Ionization Potentials, Appearance Potentials and Heats of Formation of Gaseous Positive Ions," National Bureau of Standards Report NSRDS-NBS 26, U.S. Government Printing Office (1969), 289 p.
3. Abdul, S.A., Gibson, T.L., and D.N. Rai. "Statistical Correlations for Predicting the Partition Coefficient for Nonpolar Organic Contaminants Between Aquifer Organic Carbon and Water," *Haz. Waste Haz. Mater.*, 4(3):211-222 (1987).
4. Chiou, C.T., Schmedding D.W., and M. Manes. "Partitioning of Organic Compounds in Octanol-Water Systems," *Environ. Sci. Technol.*, 16(1):4-10 (1982).
5. Sims, R.C., Doucette, W.C., McLean, J.E., Grenney, W.J., and R.R. Dupont. "Treatment Potential for 56 EPA Listed Hazardous Chemicals in Soil," U.S. EPA Report-600/6-88-001 (1988), 105 p.
6. Weast, R.C., Ed. *CRC Handbook of Chemistry and Physics*, 67th ed. (Boca Raton, FL: CRC Press, Inc., 1986), 2406 p.
7. *Fluka Catalog 1988/89 - Chemika-Biochemika* (Ronkonkoma, NY: Fluka Chemical Corp., 1988), 1536 p.
8. "Chemical, Physical, and Biological Properties of Compounds Present at Hazardous Waste Sites," U.S. EPA Report-530/SW-89-010 (1985), 619 p.
9. Walton, W.C. *Practical Aspects of Ground Water Modeling* (Worthington, OH: National Water Well Association, 1985), 587 p.
10. Davis, W.W., Krahl, M.E., and G.H.A. Clowes. "Solubility of Carcinogenic and Related Hydrocarbons in Water," *J. Am. Chem. Soc.*, 64(1):108-110 (1942).
11. Means, J.C., Wood, S.G., Hassett, J.J., and W.L. Banwart. "Sorption of Polynuclear Aromatic Hydrocarbons by Sediments and Soils," *Environ. Sci. Technol.*, 14(2):1524-1528 (1980).
12. *IARC Monographs on the Evaluation of Carcinogenic Risk of Chemicals to Man. Certain Polycyclic Aromatic Hydrocarbons and Heterocyclic Compounds, Volume 3* (Lyon, France: International Agency for Research on Cancer, 1973), 271 p.

13. Callahan, M.A., Slimak, M.W., Gable, N.W., May, I.P., Fowler, C.F., Freed, J.R., Jennings, P., Durfee, R.L., Whitmore, F.C., Maestri, B., Mabey, W.R., Holt, B.R., and C. Gould. "Water-Related Environmental Fate of 129 Priority Pollutants Volumes I and II," National Technical Information Service, U.S. EPA Report-440/4-79-029 (1979), 1160 p.

14. "NIOSH Pocket Guide to Chemical Hazards," U.S. Department of Health and Human Services, U.S. Government Printing Office (1987), 241 p.

15. "General Industry Standards for Toxic and Hazardous Substances," U.S. Code of Federal Regulations 1910, Subpart Z Section 1910.1000 (July 1982).

16. *Documentation of the Threshold Limit Values and Biological Exposure Indices* (Cincinnati, OH: American Conference of Governmental Industrial Hygienists, 1986), 744 p.

17. Cleland, J.G. "Project Summary - Environmental Hazard Rankings of Pollutants Generated in Coal Gasification Processes," Office of Research and Development, U.S. EPA Report-600/S7-81-101 (1981), 19 p.

DIBENZOFURAN

Synonyms: Biphenylene oxide; Diphenylene oxide.

Structural Formula:

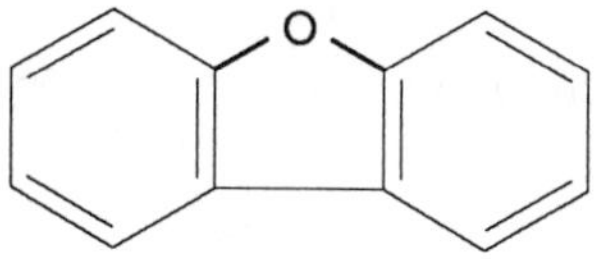

CHEMICAL DESIGNATIONS

CAS Registry Number: 132-64-9

DOT Designation: None assigned.

Empirical Formula: $C_{12}H_8O$

Formula Weight: 168.20

RTECS Number: Not assigned.

PHYSICAL AND CHEMICAL PROPERTIES

Appearance: Solid.

Boiling Point: 287 °C [1].

Henry's Law Constant: Insufficient vapor pressure data for calculation at 25 °C.

Ionization Potential: 8.59 eV [2].

Log K_{oc}: 3.91-4.10 using method of Karickhoff and others [3].

Log K_{ow}: 4.17 [4]; 4.12 [5]; 4.31 [6].

Melting Point: 86-87 °C [1]; 82 °C [4].

Solubility in Organics: Soluble in acetic acid, acetone, ethanol, and ether [1].

Solubility in Water: 10 mg/L at 25 °C [4].

Specific Density: 1.0886 at 99/4 °C [1].

Transformation Products: No data found.

Vapor Pressure: No data found.

FIRE HAZARDS

Flash Point: No data found.

Lower Explosive Limit (LEL): No data found.

Upper Explosive Limit (UEL): No data found.

HEALTH HAZARD DATA

Immediately Dangerous to Life or Health (IDLH): No data found.

Permissible Exposure Limits (PEL) in Air: No standards set.

MANUFACTURING

Selected Manufacturers:

Aldrich Chemical Co.
940 West Saint Paul Ave.
Milwaukee, WI 53233

Fluka Chemical Corp.
980 South Second St.
Ronkonkoma, NY 11779

Pfaltz & Bauer, Inc.
172 East Aurora St.
Waterbury, CT 06708

Use: Research chemical. Derived from industrial and experimental coal gasification operations where the maximum concentration detected in gas tar streams was 12 mg/m^3 [7].

REFERENCES

1. Weast, R.C., Ed. *CRC Handbook of Chemistry and Physics*, 67th ed. (Boca Raton, FL: CRC Press, Inc., 1986), 2406 p.
2. Franklin, J.L., Dillard, J.G., Rosenstock, H.M., Herron, J.T., Draxl K., and F.H. Field. "Ionization Potentials, Appearance Potentials and Heats of Formation of Gaseous Positive Ions," National Bureau of Standards Report NSRDS-NBS 26, U.S. Government Printing Office (1969), 289 p.
3. Karickhoff, S.W., Brown, D.S., and T.A. Scott. "Sorption of Hydrophobic Pollutants on Natural Sediments," *Water Res.*, 13:241-248 (1979).
4. Banerjee, S., Yalkowsky, S.H., and S.C. Valvani. "Water Solubility and Octanol/Water Partition Coefficients of Organics. Limitations of the Solubility-Partition Coefficient Correlation," *Environ. Sci. Technol.*, 14(10):1227-1229 (1980).
5. Leo, A., Hansch, C., and D. Elkins. "Partition Coefficients and Their Uses," *Chem. Rev.*, 71(6):525-616 (1971).
6. Doucette, W.J., and A.W. Andren. "Estimation of Octanol/Water Partition Coefficients: Evaluation of Six Methods for Highly Hydrophobic Aromatic Hydrocarbons," *Chemosphere*, 17(2):345-359 (1988).
7. Cleland, J.G. "Project Summary - Environmental Hazard Rankings of Pollutants Generated in Coal Gasification Processes," Office of Research and Development, U.S. EPA Report-600/S7-81-101 (1981), 19 p.

DIBROMOCHLOROMETHANE

Synonyms: Chlorodibromomethane; CDBM; NCI-C55254.

Structural Formula:

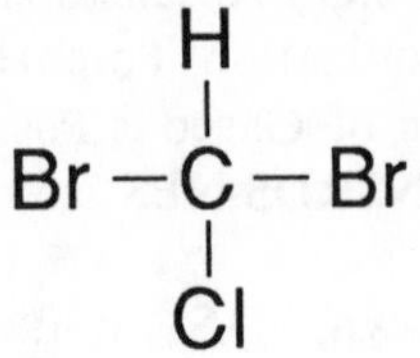

$$Br - \overset{\displaystyle H}{\underset{\displaystyle Cl}{C}} - Br$$

CHEMICAL DESIGNATIONS

CAS Registry Number: 124-48-1

DOT Designation: None assigned.

Empirical Formula: $CHBr_2Cl$

Formula Weight: 208.28

RTECS Number: PA 6360000

PHYSICAL AND CHEMICAL PROPERTIES

Appearance: Clear, colorless, heavy liquid.

Boiling Point: 118-122 °C [1]; 116 °C [2]; 117-120 °C [3].

Henry's Law Constant: 9.9 x 10^{-4} atm·m^3/mol [4]; 0.00783 atm·m^3/mol [5].

Ionization Potential: 10.59 eV [6].

Log K_{oc}: 1.92 [4].

Log K_{ow}: 2.08 [7].

Melting Point: -23° to -21 °C [8].

Solubility in Organics: Miscible with oils, dichloropropane, and isopropanol [9].

Solubility in Water: 4,000 mg/L at 20 °C [4].

Specific Density: 2.451 at 20/4 °C [10]; 2.440 at 25/25 °C [1].

Transformation Products: No data found.

Vapor Density: 8.51 g/L at 25 °C, 7.19 (air = 1).

Vapor Pressure: 76 mm at 20 °C [4]; 15 mm at 10.5 °C [11].

FIRE HAZARDS

Flash Point: Non-combustible [12].

Lower Explosive Limit (LEL): No data found.

Upper Explosive Limit (UEL): No data found.

HEALTH HAZARD DATA

Immediately Dangerous to Life or Health (IDLH): No data found.

Permissible Exposure Limits (PEL) in Air: No standards set.

MANUFACTURING

Selected Manufacturer:

> Pfaltz & Bauer, Inc.
> 172 East Aurora St.
> Waterbury, CT 06708

Uses: Manufacturing of fire extinguishing agents, propellants, refrigerants, and pesticides; organic synthesis.

REFERENCES

1. Sax, N.I. *Dangerous Properties of Industrial Materials* (New York: Van Nostrand Reinhold Co., 1984), 3124 p.

2.   Hawley, G.G. *The Condensed Chemical Dictionary* (New York: Van Nostrand Reinhold Co., 1981), 1135 p.

3.   *Fluka Catalog 1988/89 - Chemika-Biochemika* (Ronkonkoma, NY: Fluka Chemical Corp., 1988), 1536 p.

4.   Schwille, F. *Dense Chlorinated Solvents* (Chelsea, MI: Lewis Publishers, Inc., 1988), 146 p.

5.   Warner, H.P., Cohen, J.M., and J.C. Ireland. "Determination of Henry's Law Constants of Selected Priority Pollutants," Office of Science and Development, U.S. EPA Report-600/D-87/229 (1987), 14 p.

6.   *Instruction Manual - Model ISP1 101: Intrinsically Safe Portable Photoionization Analyzer* (Newton, MA: HNU Systems, Inc., 1986), 86 p.

7.   Mills, W.B., Porcella, D.B., Ungs, M.J., Gherini, S.A., Summers, K.V., Mok, L., Rupp, G.L., and G.L. Bowie. "Water Quality Assessment: A Screening Procedure for Toxic and Conventional Pollutants in Surface and Groundwater-Part I," Office of Research and Development, U.S. EPA Report-600/6-85-002a (1985), 638 p.

8.   Dean, J.A., Ed., *Lange's Handbook of Chemistry*, 11th ed. (New York: McGraw-Hill, Inc., 1973), 1570 p.

9.   "Chemical, Physical, and Biological Properties of Compounds Present at Hazardous Waste Sites," U.S. EPA Report-530/SW-89-010 (1985), 619 p.

10. Weast, R.C., Ed. *CRC Handbook of Chemistry and Physics*, 67th ed. (Boca Raton, FL: CRC Press, Inc., 1986), 2406 p.

11. "Treatability Manual - Volume 1: Treatability Data," Office of Research and Development, U.S. EPA Report-600/8-80-042a (1980), 1035 p.

12. *Catalog Handbook of Fine Chemicals* (Milwaukee, WI: Aldrich Chemical Co., 1988), 2212 p.

DI–n–BUTYL PHTHALATE

Synonyms: 1,2-Benzenedicarboxylate; **1,2-Benzenedicarboxylic acid, dibutyl ester;** o-Benzenedicarboxylic acid, dibutyl ester; Benzene-o-dicarboxylic acid di-n-butyl ester; Butyl phthalate; n-Butyl phthalate; Celluflex DPB; DBP; Dibutyl-1,2-benzenedicarboxylate; Dibutyl phthalate; Elaol; Hexaplas M/B; Palatinol C; Phthalic acid dibutyl ester; Polycizer DBP; PX 104; RCRA waste number U069; Staflex DBP; Witicizer 300.

Structural Formula:

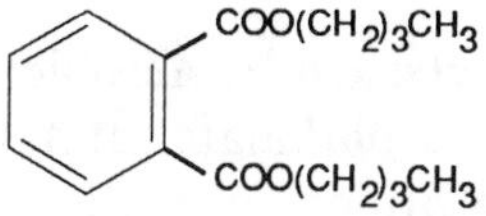

CHEMICAL DESIGNATIONS

CAS Registry Number: 84-74-2

DOT Designation: 9095

Empirical Formula: $C_{16}H_{22}O_4$

Formula Weight: 278.35

RTECS Number: TI 0875000

PHYSICAL AND CHEMICAL PROPERTIES

Appearance and Odor: Colorless oily liquid with a mild aromatic odor.

Boiling Point: 340 °C [1]; 335 °C [2].

Henry's Law Constant: 6.3×10^{-5} atm·m^3/mol [3].

Ionization Potential: No data found.

Log K_{oc}: 3.14 [4].

Log K_{ow}: 4.31 [5]; 4.57 [6]; 4.79 [7].

Melting Point: -35 °C [8]; -40 °C [9]; ≈71 °C [10].

Solubility in Organics: Soluble in ethanol, benzene, and ether [1].

Solubility in Water: 400 mg/L at 25 °C, 4,500 mg/L at 25 °C, 28 mg/L at 26 °C [8]; 0.45 wt% at 20 °C [11]; 13 ppm at 25 °C [12]; 10.1 mg/L at 20 °C [13]; 11.2 mg/L at 25 °C [7]; 100 mg/L at 22 °C [14]; 1.300 ppm at 20-25 °C in Narragansett Bay water containing 1.8 mg/L dissolved organic carbon [15]; 0.00183 wt% at 23.5 °C [16]; 9.2 mg/L at 25 °C [4].

Specific Density: 1.0457 at 20/20 °C [1]; 1.047 at 20/20 °C [17]; 1.046 at 20/4 °C [18]; 1.042 at 25/4 °C [9].

Transformation Products: Under aerobic conditions using a freshwater hydrosoil, mono-*n*-butyl phthalate, and phthalic acid were produced. Under anaerobic conditions, phthalic acid was not present [8]. In anaerobic sludge, di-*n*-butyl phthalate degraded as follows: monobutylphthalate to phthalic acid to protocatechuic acid followed by ring cleavage and mineralization [19].

Vapor Density: 11.38 g/L at 25 °C, 9.61 (air = 1).

Vapor Pressure: < 0.01 mm at 20 °C [11]; 0.1 mm at 115 °C, 2.0 mm at 150 °C [8]; 1 mm at 148.2 °C, 10 mm at 198.2 °C, 40 mm at 235.8 °C, 100 mm at 263.7 °C, 400 mm at 313.5 °C, 760 mm at 340 °C [1]; 0.00001 mm at 25 °C [20]; 0.0097 (± 3.3) Pa at 25 °C [7]; 1.4×10^{-5} mm at 25 °C [21].

FIRE HAZARDS

Flash Point: 157 °C [22]; 157.2 °C, 179.4 °C (open cup) [2]; 159 °C (open cup) [10].

Lower Explosive Limit (LEL): 0.5% at 235 °C [22].

Upper Explosive Limit (UEL): 2.5% (calculated) [2].

HEALTH HAZARD DATA

Immediately Dangerous to Life or Health (IDLH): 9,300 mg/m^3 [11].

Permissible Exposure Limits (PEL) in Air: 5 mg/m^3 [23]; 5 mg/m^3 TWA [24].

MANUFACTURING

Selected Manufacturer:

Aldrich Chemical Co.
940 West Saint Paul Ave.
Milwaukee, WI 53233

Uses: Plasticizer; insect repellant; organic synthesis.

REFERENCES

1. Weast, Robert C., Ed. *CRC Handbook of Chemistry and Physics*, 67th ed. (Boca Raton, FL: CRC Press, Inc., 1986), 2406 p.
2. Weiss, G. *Hazardous Chemicals Data Book* (Park Ridge, NJ: Noyes Data Corp., 1986), 1069 p.
3. Petrasek, A.C., Kugelman, I.J., Austern, B.M., Pressley, T.A., Winslow, L.A., and R.H. Wise. "Fate of Toxic Organic Compounds in Wastewater Treatment Plants," *J. Water Poll. Control Fed.*, 55(10):1286-1296 (1983).
4. Russell, D.J., and B. McDuffie. "Chemodynamic Properties of Phthalate Esters: Partitioning and Soil Migration," *Chemosphere*, 15(8):1003-1021 (1986).
5. Doucette, W.J., and A.W. Andren. "Estimation of Octanol/Water Partition Coefficients: Evaluation of Six Methods for Highly Hydrophobic Aromatic Hydrocarbons," *Chemosphere*, 17(2):345-359 (1988).
6. Leyder, F., and P. Boulanger. "Ultraviolet Absorption, Aqueous Solubility and Octanol-Water Partition for Several Phthalates," *Bull. Environ. Contam. Toxicol.*, 30(2):152-157 (1983).
7. Howard, P.H., Banerjee, S., and K.H. Robillard. "Measurement of Water Solubilities, Octanol/Water Partition Coefficients and Vapor Pressures of Commercial Phthalate Esters," *Environ. Toxicol. Chem.*, 4:653-661 (1985).
8. Verschueren, K. *Handbook of Environmental Data on Organic Chemicals* (New York: Van Nostrand Reinhold Co., 1983), 1310 p.
9. Standen, A., Ed. *Kirk-Othmer Encyclopedia of Chemical Technology, Volume 15*, 2nd ed. (New York: John Wiley and Sons, Inc., 1968), 923 p.
10. Broadhurst, M.G. "Use and Replaceability of Polychlorinated Biphenyls," *Environ. Health Perspect.*, (October 1972), pp 81-102.
11. "NIOSH Pocket Guide to Chemical Hazards," U.S. Department of

Health and Human Services, U.S. Government Printing Office (1987), 241 p.

12. Fukano, I., and Y. Obata. "Solubility of Phthalates in Water," [Chemical abstract 120601u, 86(17):486 (1977)]: *Purasuchikkusu*, 27(7):48-49 (1976).

13. Leyder, F., and P. Boulanger. "Ultraviolet Absorption, Aqueous Solubility and Octanol-Water Partition for Several Phthalates," *Bull. Environ. Contam. Toxicol.*, 30(2):152-157 (1983).

14. Nyssen, G.A., Miller, E.T., Glass, T.F., Quinn II, C.R., Underwood, J., and D.J. Wilson. "Solubilities of Hydrophobic Compounds in Aqueous-Organic Solvent Mixtures," *Environ. Monit. Assess.*, 9(1):1-11 (1987).

15. Boehm, P.D., and J.G. Quinn. "Solubilization of Hydrocarbons by the Dissolved Organic Matter in Sea Water," *Geochim. Cosmochim. Acta*, 37(11):2459-2477 (1973).

16. Schwarz, F.P. "Measurement of the Solubilities of Slightly Soluble Organic Liquids in Water by Elution Chromatography," *Anal. Chem.*, 52(1):10-15 (1980).

17. Huntress, E.H., and S.P. Mulliken. *Identification of Pure Organic Compounds - Tables of Data on Selected Compounds of Order I* (New York: John Wiley and Sons, Inc., 1941), 691 p.

18. *Fluka Catalog 1988/89 - Chemika-Biochemika* (Ronkonkoma, NY: Fluka Chemical Corp., 1988), 1536 p.

19. Shelton, D.R., Boyd, S.A., and J.M. Tiedje. "Anaerobic Biodegradation of Phthalic Acid Esters in Sludge," *Environ. Sci. Technol.*, 18(2):93-97 (1984).

20. Mabey, W.R., Smith, J.H., Podoll, R.T., Johnson, H.L., Mill, T., Chou, T.-W., Gates, J., Partridge, I.W., Jaber, H., and D. Vandenberg. "Aquatic Fate Process Data for Organic Priority Pollutants - Final Report," Office of Regulations and Standards, U.S. EPA Report-440/4-81-014 (1982), 407 p.

21. Giam, C.S., Atlas, E., Chan, H.S., and G.S. Neff. "Phthalate Esters, PCB and DDT Residues in the Gulf of Mexico Atmosphere," *Atmos. Environ.*, 14:65-69 (1980).

22. *Fire Protection Guide on Hazardous Materials* (Quincy, MA: National Fire Protection Association, 1984), 443 p.

23. "General Industry Standards for Toxic and Hazardous Substances," U.S. Code of Federal Regulations 1910, Subpart Z Section 1910.1000 (July 1982).

24. *Documentation of the Threshold Limit Values and Biological Exposure Indices for 1986-1987* (Cincinnati, OH: American Conference of Governmental Industrial Hygienists, 1986), 111 p.

1,2-DICHLOROBENZENE

Synonyms: Chloroben; Chloroden; Cloroben; DCB; 1,2-DCB; o-DCB; 1,2-Dichlorbenzene; o-Dichlorbenzene; 1,2-Dichlorbenzol; o-Dichlorbenzol; o-Dichlorobenzene; 1,2-Dichlorobenzol; o-Dichlorobenzol; Dilantin DB; Dilatin DB; Dizene; Dowtherm E; NCI-C54944; ODB; ODCB; Orthodichlorobenzene; Orthodichlorobenzol; RCRA waste number U070; Special termite fluid; Termitkil; UN 1591.

Structural Formula:

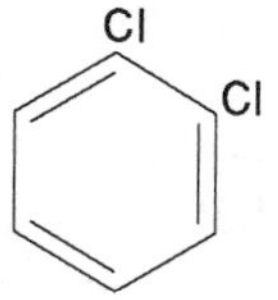

CHEMICAL DESIGNATIONS

CAS Registry Number: 95-50-1

DOT Designation: 1591

Empirical Formula: $C_6H_4Cl_2$

Formula Weight: 147.00

RTECS Number: CZ 4500000

PHYSICAL AND CHEMICAL PROPERTIES

Appearance and Odor: Clear, colorless liquid with a pleasant aromatic odor.

Boiling Point: 180.5 °C [1]; 180-183 °C [2]; 179.5 °C [3].

Henry's Law Constant: 0.0019 atm·m³/mol [4]; 0.0012 atm·m³/mol at 20 °C [5]; 0.0024 atm·m³/mol at 25 °C [6]; 0.00283 atm·m³/mol at 37 °C [7].

Ionization Potential: 9.06 eV [8].

Log K_{oc}: 2.27 [9]; 3.23 [10]; 2.255 [11].

Log K_{ow}: 3.38 [12]; 3.40 [13]; 3.55 [14].

Melting Point: -17.5 °C [2]; -16° to -14 °C [15].

Solubility in Organics: Soluble in ethanol, acetone, ether, benzene, carbon tetrachloride, and ligroin [16].

Solubility in Water: 100 mg/L at 20 °C [17]; 137 mg/L at 25 °C [18]; 0.015 wt% at 20 °C [8]; 156 mg/L at 25 °C [13]; 0.0309 vol% at 25 °C [19]; 0.0156 wt% at 25 °C [20]; 148 ppm at 20 °C [11]; 92.7 mg/L at 25 °C [21]; 0.000628 M at 25 °C [22]; 145 mg/L at 25 °C [23].

Specific Density: 1.3048 at 20/4 °C [1]; 1.307 at 20/20 °C [2]; 1.3100 at 15/4 °C, 1.3064 at 20/4 °C [3]; 1.30024 at 25/4 °C [24].

Transformation Products: *Pseudomonas* sp. isolated from sewage samples produced 3,4-dichloro-*cis*-1,2-dihydroxycyclohexa-3,5-diene. Subsequent degradation of this metabolite yielded 3,4-dichlorocatechol which underwent ring cleavage to form 2,3-dichloro-*cis,cis*-muconate followed by hydrolysis to form 5-chloromaleylacetic acid [25].

Vapor Density: 6.01 g/L at 25 °C, 5.07 (air = 1).

Vapor Pressure: 1.5 mm at 25 °C, 1.9 mm at 30 °C [17]; 1 mm at 20 °C, 10 mm at 59.1 °C, 40 mm at 89.4 °C, 100 mm at 112.9 °C, 400 mm at 155.8 °C, 760 mm at 179 °C [1]; 62 mm at 100 °C [23].

FIRE HAZARDS

Flash Point: 66 °C [8].

Lower Explosive Limit (LEL): 2.2% [8].

Upper Explosive Limit (UEL): 9.2% [8].

HEALTH HAZARD DATA

Immediately Dangerous to Life or Health (IDLH): 1,700 ppm [8].

Permissible Exposure Limits (PEL) in Air: 50 ppm ceiling ($\approx$300 mg/m^3) [26,27].

MANUFACTURING

Selected Manufacturers:

Monsanto Industrial Chemicals Co.
800 North Lindbergh Blvd.
St. Louis, MO 63166

Pfaltz & Bauer, Inc.
172 East Aurora St.
Waterbury, CT 06708

Standard Chlorine Chemical Co., Inc.
1015-25 Belleville Turnpike
Kearny, NJ 07032

Uses: Preparation of 3,4-dichloroaniline; solvent for a wide variety of organic compounds and for oxides of nonferrous metals; solvent carrier in products of toluene diisocyanate; intermediate for dyes; fumigant and insecticide; degreasing hides and wool; metal polishes; industrial air control; disinfectant; heat transfer medium.

REFERENCES

1. Weast, R.C., Ed. *CRC Handbook of Chemistry and Physics*, 67th ed. (Boca Raton, FL: CRC Press, Inc., 1986), 2406 p.
2. Sax, N.I. *Dangerous Properties of Industrial Materials* (New York: Van Nostrand Reinhold Co., 1984), 3124 p.
3. Standen, A., Ed. *Kirk-Othmer Encyclopedia of Chemical Technology, Volume 4*, 2nd ed. (New York: John Wiley and Sons, Inc., 1964), 937 p.
4. Pankow, J.F., and M.E. Rosen. "Determination of Volatile Compounds in Water by Purging Directly to a Capillary Column with Whole Column Cryotrapping," *Environ. Sci. Technol.*, 22(4):398-405 (1988).
5. Howard, P.H. *Handbook of Environmental Fate and Exposure Data for Organic Chemicals* (Chelsea, MI: Lewis Publishers, Inc., 1989), 574 p.
6. Hine, J., and P.K. Mookerjee. "The Intrinsic Hydrophilic Character of Organic Compounds. Correlations in Terms of Structural Contributions," *J. Org. Chem.*, 40(3):292-298 (1975).
7. Sato, A., and T. Nakajima. "A Structure-Activity Relationship of

Some Chlorinated Hydrocarbons," *Arch. Environ. Health*, 34(2):69-75 (1979).

8. "NIOSH Pocket Guide to Chemical Hazards," U.S. Department of Health and Human Services, U.S. Government Printing Office (1987), 241 p.

9. Abdul, S.A., Gibson, T.L., and D.N. Rai. "Statistical Correlations for Predicting the Partition Coefficient for Nonpolar Organic Contaminants Between Aquifer Organic Carbon and Water," *Haz. Waste Haz. Mater.*, 4(3):211-222 (1987).

10. Schwille, F. *Dense Chlorinated Solvents* (Chelsea, MI: Lewis Publishers, Inc., 1988), 146 p.

11. Chiou, C.T., Peters, L.J., and V.H. Freed. "A Physical Concept of Soil-Water Equilibria for Nonionic Organic Compounds," *Science*, 206:831-832 (1979).

12. Leo, A., Hansch, C., and D. Elkins. "Partition Coefficients and Their Uses," *Chem. Rev.*, 71(6):525-616 (1971).

13. Banerjee, S., Yalkowsky, S.H., and S.C. Valvani. "Water Solubility and Octanol/Water Partition Coefficients of Organics. Limitations of the Solubility-Partition Coefficient Correlation," *Environ. Sci. Technol.*, 14(10):1227-1229 (1980).

14. Könemann, H., Zelle, R., and F. Busser. "Determination of Log P_{oct} Values of Chloro-Substituted Benzenes, Toluenes, and Anilines by High-Performance Liquid Chromatography on ODS-Silica," *J. Chromatogr.*, 178:559-565 (1979).

15. *Fluka Catalog 1988/89 - Chemika-Biochemika* (Ronkonkoma, NY: Fluka Chemical Corp., 1988), 1536 p.

16. "Chemical, Physical, and Biological Properties of Compounds Present at Hazardous Waste Sites," U.S. EPA Report-530/SW-89-010 (1985), 619 p.

17. Verschueren, K. *Handbook of Environmental Data on Organic Chemicals* (New York: Van Nostrand Reinhold Co., 1983), 1310 p.

18. Banerjee, S. "Solubility of Organic Mixtures in Water," *Environ. Sci. Technol.*, 18(8):587-591 (1984).

19. Stephen, H., and T. Stephen. *Solubilities of Inorganic and Organic Compounds - Part 1, Volume 1* (London: Pergamon Printing and Art Services, Ltd., 1963), 960 p.

20. Riddick, J.A., Bunger, W.B., and T.K. Sakano. *Organic Solvents - Physical Properties and Methods of Purification. Volume II* (New York: John Wiley and Sons, Inc., 1986), 1325 p.

21. Yalkowsky, S.H., Orr, R.J., and S.C. Valvani. "Solubility and Partitioning. 3. The Solubility of Halobenzenes in Water," *Indust. Eng. Chem. Fund.*, 18(4):351-353 (1979).

22. Miller, M.M., Ghodbane, S., Wasik, S.P., Tewari, Y.B., and D.E.

Martire. "Aqueous Solubilities, Octanol/Water Partition Coefficients, and Entropies of Melting of Chlorinated Benzenes and Biphenyls," *J. Chem. Eng. Data*, 29(2):184-190 (1984).

23. Bailey, G.W., and J.L. White. "Herbicides: A Compilation of Their Physical, Chemical, and Biological Properties," *Res. Rev.*, 10:97-122 (1965).

24. Kirchnerová, J., and G.C.B. Cave. "The Solubility of Water in Low-Dielectric Solvents," *Can. J. Chem.*, 54(24):3909-3916 (1976).

25. Haigler, B.E., Nishino, S.F., and J.C. Spain. "Degradation of 1,2-Dichlorobenzene by a *Pseudomonas* sp.," *Appl. Environ. Microbiol.*, 54(2):294-301 (1988).

26. "General Industry Standards for Toxic and Hazardous Substances," U.S. Code of Federal Regulations 1910, Subpart Z Section 1910.1000 (July 1982).

27. *Documentation of the Threshold Limit Values and Biological Exposure Indices for 1986-1987* (Cincinnati, OH: American Conference of Governmental Industrial Hygienists, 1986), 111 p.

1,3-DICHLOROBENZENE

Synonyms: 1,3-DCB; *m*-DCB; 1,3-Dichlorbenzene; *m*-Dichlorbenzene; 1,3-Dichlorbenzol; *m*-Dichlorbenzol; *m*-Dichlorobenzene; 1,3-Dichlorobenzol; *m*-Dichlorobenzol; RCRA waste number U071; UN 1591.

Structural Formula:

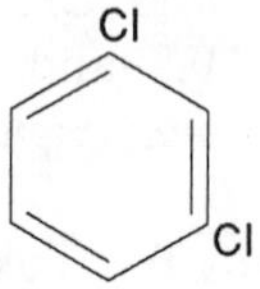

CHEMICAL DESIGNATIONS

CAS Registry Number: 541-73-1

DOT Designation: None assigned.

Empirical Formula: $C_6H_4Cl_2$

Formula Weight: 147.00

RTECS Number: CZ 4499000

PHYSICAL AND CHEMICAL PROPERTIES

Appearance: Colorless liquid.

Boiling Point: 173 °C [1]; 172 °C [2].

Henry's Law Constant: 0.0036 atm·m^3/mol [3]; 0.00263 atm·m^3/mol [4]; 0.0047 atm·m^3/mol at 25 °C [5]; 0.00463 atm·m^3/mol at 37 °C [6].

Ionization Potential: 9.12 eV [7].

Log K_{oc}: 2.23 [8]; 3.23 [9].

Log K_{ow}: 3.38 [10]; 3.43 [11]; 3.48 [12]; 3.44 [13]; 3.53 [14]; 3.72 at 13 °C, 3.55 at 19 °C, 3.48 at 28 °C, 3.42 at 33 °C [15]; 3.60 [16].

Melting Point: -24.7 °C [1]; -25 to -22 °C [17]; -26.4 °C [2].

Solubility in Organics: Soluble in ethanol, acetone, ether, benzene, carbon tetrachloride, and ligroin [18].

Solubility in Water: 69 mg/L at 22 °C, 123 mg/L at 25 °C [19]; 143 mg/L at 25 °C [13]; 133 mg/L at 25 °C [20]; 0.0111 wt% at 20 °C [21]; 0.000847 M at 25 °C [12]; 0.01465 wt% at 23.5 °C [22].

Specific Density: 1.2884 at 20/4 °C [1]; 1.2881 at 20/4 °C, 1.2799 at 25/4 °C [2].

Transformation Products: No data found.

Vapor Density: 6.01 g/L at 25 °C, 5.07 (air = 1).

Vapor Pressure: 1 mm at 12.1 °C, 10 mm at 52.0 °C, 40 mm at 82.0 °C, 100 mm at 105.0 °C, 400 mm at 149.0 °C, 760 mm at 173.0 °C [1]; 2.3 mm at 25 °C [9]; 1.9 mm at 25 °C [4].

FIRE HAZARDS

Flash Point: 63 °C [23].

Lower Explosive Limit (LEL): 2.02% (estimated) [24].

Upper Explosive Limit (UEL): 9.2% (estimated) [24].

HEALTH HAZARD DATA

Immediately Dangerous to Life or Health (IDLH): No data found.

Permissible Exposure Limits (PEL) in Air: No standards set.

MANUFACTURING

Selected Manufacturers:

Aldrich Chemical Co.
940 West Saint Paul Ave.
Milwaukee, WI 53233

Fluka Chemical Corp.
980 South Second St.
Ronkonkoma, NY 11779

Uses: Fumigant and insecticide; organic synthesis.

REFERENCES

1. Weast, R.C., Ed. *CRC Handbook of Chemistry and Physics*, 67th ed. (Boca Raton, FL: CRC Press, Inc., 1986), 2406 p.
2. Standen, A., Ed. *Kirk-Othmer Encyclopedia of Chemical Technology, Volume 4*, 2nd ed. (New York: John Wiley and Sons, Inc., 1964), 937 p.
3. Pankow, J.F., and M.E. Rosen. "Determination of Volatile Compounds in Water by Purging Directly to a Capillary Column with Whole Column Cryotrapping," *Environ. Sci. Technol.*, 22(4):398-405 (1988).
4. Warner, H.P., Cohen, J.M., and J.C. Ireland. "Determination of Henry's Law Constants of Selected Priority Pollutants," Office of Science and Development, U.S. EPA Report-600/D-87/229 (1987), 14 p.
5. Hine, J., and P.K. Mookerjee. "The Intrinsic Hydrophilic Character of Organic Compounds. Correlations in Terms of Structural Contributions," *J. Org. Chem.*, 40(3):292-298 (1975).
6. Sato, A., and T. Nakajima. "A Structure-Activity Relationship of Some Chlorinated Hydrocarbons," *Arch. Environ. Health*, 34(2):69-75 (1979).
7. Franklin, J.L., Dillard, J.G., Rosenstock, H.M., Herron, J.T., Draxl K., and F.H. Field. "Ionization Potentials, Appearance Potentials and Heats of Formation of Gaseous Positive Ions," National Bureau of Standards Report NSRDS-NBS 26, U.S. Government Printing Office (1969), 289 p.
8. Abdul, S.A., Gibson, T.L., and D.N. Rai. "Statistical Correlations for Predicting the Partition Coefficient for Nonpolar Organic Contaminants Between Aquifer Organic Carbon and Water," *Haz. Waste Haz. Mater.*, 4(3):211-222 (1987).
9. Schwille, F. *Dense Chlorinated Solvents* (Chelsea, MI: Lewis Publishers, Inc., 1988), 146 p.
10. Leo, A., Hansch, C., and D. Elkins. "Partition Coefficients and Their Uses," *Chem. Rev.*, 71(6):525-616 (1971).
11. Miller, M.M., Wasik, S.P., Huang, G.-L., Shiu, W.-Y., and D. Mackay. "Relationships Between Octanol-Water Partition Coefficient and

Aqueous Solubility," *Environ. Sci. Technol.*, 19(6):522-529 (1985).

12. Miller, M.M., Ghodbane, S., Wasik, S.P., Tewari, Y.B., and D.E. Martire. "Aqueous Solubilities, Octanol/Water Partition Coefficients, and Entropies of Melting of Chlorinated Benzenes and Biphenyls," *J. Chem. Eng. Data*, 29(2):184-190 (1984).

13. Banerjee, S. "Solubility of Organic Mixtures in Water," *Environ. Sci. Technol.*, 18(8):587-591 (1984).

14. Watarai, H., Tanaka, M., and N. Suzuki. "Determination of Partition Coefficients of Halobenzenes in Heptane/Water and 1-Octanol/Water Systems and Comparison with the Scaled Particle Calculation," *Anal. Chem.*, 54(4):702-705 (1982).

15. Opperhuizen, A., Serné, P., and J.M.D. Van der Steen. "Thermodynamics of Fish/Water Octan-1-ol/Water Partitioning of Some Chlorinated Benzenes," *Environ. Sci. Technol.*, 22(3):286-298 (1988).

16. Könemann, H., Zelle, R., and F. Busser. "Determination of Log P_{oct} Values of Chloro-Substituted Benzenes, Toluenes, and Anilines by High-Performance Liquid Chromatography on ODS-Silica," *J. Chromatogr.*, 178:559-565 (1979).

17. *Catalog Handbook of Fine Chemicals* (Milwaukee, WI: Aldrich Chemical Co., 1988), 2212 p.

18. "Chemical, Physical, and Biological Properties of Compounds Present at Hazardous Waste Sites," U.S. EPA Report-530/SW-89-010 (1985), 619 p.

19. Verschueren, K. *Handbook of Environmental Data on Organic Chemicals* (New York: Van Nostrand Reinhold Co., 1983), 1310 p.

20. Banerjee, S., Yalkowsky, S.H., and S.C. Valvani. "Water Solubility and Octanol/Water Partition Coefficients of Organics. Limitations of the Solubility-Partition Coefficient Correlation," *Environ. Sci. Technol.*, 14(10):1227-1229 (1980).

21. Riddick, J.A., Bunger, W.B., and T.K. Sakano. *Organic Solvents - Physical Properties and Methods of Purification. Volume II* (New York: John Wiley and Sons, Inc., 1986), 1325 p.

22. Schwarz, F.P. "Measurement of the Solubilities of Slightly Soluble Organic Liquids in Water by Elution Chromatography," *Anal. Chem.*, 52(1):10-15 (1980).

23. *Catalog Handbook of Fine Chemicals* (Milwaukee, WI: Aldrich Chemical Co., 1988), 2212 p.

24. Weiss, G. *Hazardous Chemicals Data Book* (Park Ridge, NJ: Noyes Data Corp., 1986), 1069 p.

1,4–DICHLOROBENZENE

Synonyms: 4-Chlorophenyl chloride; *p*-Chlorophenyl chloride; 1,4-DCB; 4-DCB; *p*-DCB; Di-chloricide; 4-Dichlorobenzene; *p*-Dichlorobenzene; 4-Dichlorobenzol; *p*-Dichlorobenzol; Evola; NCI-C54955; Paracide; Para crystals; Paradi; Paradichlorobenzene; Paradichlorobenzol; Paradow; Paramoth; Paranuggetts; Parazene; Parodi; PDB; PDCB; Persia-Perazol; RCRA waste number U072; Santochlor; UN 1592.

Structural Formula:

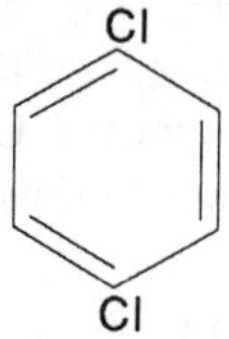

CHEMICAL DESIGNATIONS

CAS Registry Number: 106-46-7

DOT Designation: 1592

Empirical Formula: $C_6H_4Cl_2$

Formula Weight: 147.00

RTECS Number: CZ 4550000

PHYSICAL AND CHEMICAL PROPERTIES

Appearance and Odor: White crystals with a penetrating mothball-like odor.

Boiling Point: 174.4 °C [1]; 173.4 °C [2].

Henry's Law Constant: 0.0031 atm·m³/mol [3]; 0.00272 atm·m³/mol [4]; 0.00445 atm·m³/mol at 25 °C [5].

Ionization Potential: 8.94 eV [6]; 9.07 eV [7].

Log K_{oc}: 2.20 [8].

Log K_{ow}: 3.39 [9]; 3.37 [10]; 3.38 [11]; 3.53 [12]; 3.62 [13].

Melting Point: 53.1 °C [14].

Solubility in Organics: Soluble in ethanol, acetone, ether, benzene, carbon tetrachloride, and ligroin [15].

Solubility in Water: 49 mg/L at 22 °C, 79 mg/L at 25 °C [16]; 65.3 mg/L at 25 °C [17]; 0.008% at 25 °C [6]; 74 mg/L at 25 °C [10]; 77 mg/L at 30 °C [18]; 76 mg/L at 25 °C [19]; 87.15 mg/L at 25 °C [20]; 80 mg/L at 25 °C [21]; 90.6 mg/L at 25 °C [22]; 0.00021 M at 25 °C [23].

Specific Density: 1.2475 at 20/4 °C [14]; 1.4581 at 20.5/4 °C [2]; 1.238 at 55/4 °C [24].

Transformation Products: Under artificial sunlight, river water containing 2-5 ppm of 1,4-dichlorobenzene photodegraded to chlorophenol and phenol [25].

Vapor Pressure: 0.6 mm at 20 °C, 1.8 mm at 30 °C [16]; 10 mm at 54.8 °C, 40 mm at 84.8 °C, 100 mm at 108.4 °C, 400 mm at 150.2 °C, 760 mm at 173.9 °C [14]; 0.4 mm at 25 °C, 6.6 mm at 50 °C, 11.5 mm at 60 °C, 67 mm at 100 °C [24].

FIRE HAZARDS

Flash Point: 65.6 °C [6].

Lower Explosive Limit (LEL): 2.5% [6].

Upper Explosive Limit (UEL): No data found.

HEALTH HAZARD DATA

Immediately Dangerous to Life or Health (IDLH): 1,000 ppm [6].

Permissible Exposure Limits (PEL) in Air: 75 ppm ($\approx$450 mg/m^3) [26]; 75 ppm TLV, 110 ppm ($\approx$675 mg/m^3) STEL [27].

MANUFACTURING

Selected Manufacturers:

Aldrich Chemical Co.
940 West Saint Paul Ave.
Milwaukee, WI 53233

Fluka Chemical Corp.
980 South Second St.
Ronkonkoma, NY 11779

PPG Industries, Inc.
Chemical Division
1 PPG Plaza
Pittsburgh, PA 15272

Uses: Moth repellent; general insecticide, fumigant, and germicide; space odorant; manufacture of 2,5-dichloroaniline and dyes; pharmacy; agriculture (fumigating soil); disinfectant and chemical intermediate.

REFERENCES

1. Dean, J.A., Ed., *Lange's Handbook of Chemistry*, 11th ed. (New York: McGraw-Hill, Inc., 1973), 1570 p.
2. Sax, N.I. *Dangerous Properties of Industrial Materials* (New York: Van Nostrand Reinhold Co., 1984), 3124 p.
3. Pankow, J.F., and M.E. Rosen. "Determination of Volatile Compounds in Water by Purging Directly to a Capillary Column with Whole Column Cryotrapping," *Environ. Sci. Technol.*, 22(4):398-405 (1988).
4. Warner, H.P., Cohen, J.M., and J.C. Ireland. "Determination of Henry's Law Constants of Selected Priority Pollutants," Office of Science and Development, U.S. EPA Report-600/D-87/229 (1987), 14 p.
5. Hine, J., and P.K. Mookerjee. "The Intrinsic Hydrophilic Character of Organic Compounds. Correlations in Terms of Structural Contributions," *J. Org. Chem.*, 40(3):292-298 (1975).
6. "NIOSH Pocket Guide to Chemical Hazards," U.S. Department of Health and Human Services, U.S. Government Printing Office (1987), 241 p.
7. Yoshida, K., Shigeoka, T., and F. Yamauchi. "Non-Steady State

Equilibrium Model for the Preliminary Prediction of the Fate of Chemicals in the Environment," *Ecotoxicol. Environ. Safety*, 7(2):179-190 (1983).

8. Abdul, S.A., Gibson, T.L., and D.N. Rai. "Statistical Correlations for Predicting the Partition Coefficient for Nonpolar Organic Contaminants Between Aquifer Organic Carbon and Water," *Haz. Waste Haz. Mater.*, 4(3):211-222 (1987).

9. Leo, A., Hansch, C., and D. Elkins. "Partition Coefficients and Their Uses," *Chem. Rev.*, 71(6):525-616 (1971).

10. Banerjee, S., Yalkowsky, S.H., and S.C. Valvani. "Water Solubility and Octanol/Water Partition Coefficients of Organics. Limitations of the Solubility-Partition Coefficient Correlation," *Environ. Sci. Technol.*, 14(10):1227-1229 (1980).

11. Chiou, C.T., Freed, V.H., Schmedding, D.W., and R.L. Kohnert. "Partition Coefficients and Bioaccumulation of Selected Organic Chemicals," *Environ. Sci. Technol.*, 11(5):475-478 (1977).

12. Mackay, D. "Correlation of Bioconcentration Factors," *Environ. Sci. Technol.*, 16(5):274-278 (1982).

13. Könemann, H., Zelle, R., and F. Busser. "Determination of Log P_{oct} Values of Chloro-Substituted Benzenes, Toluenes, and Anilines by High-Performance Liquid Chromatography on ODS-Silica," *J. Chromatogr.*, 178:559-565 (1979).

14. Weast, R.C., Ed. *CRC Handbook of Chemistry and Physics*, 67th ed. (Boca Raton, FL: CRC Press, Inc., 1986), 2406 p.

15. "Chemical, Physical, and Biological Properties of Compounds Present at Hazardous Waste Sites," U.S. EPA Report-530/SW-89-010 (1985), 619 p.

16. Verschueren, K. *Handbook of Environmental Data on Organic Chemicals* (New York: Van Nostrand Reinhold Co., 1983), 1310 p.

17. Banerjee, S. "Solubility of Organic Mixtures in Water," *Environ. Sci. Technol.*, 18(8):587-591 (1984).

18. Gross, P.M., and J.H. Saylor. "The Solubilities of Certain Slightly Soluble Organic Compounds in Water," *J. Am. Chem. Soc.*, 53(5):1744-1751 (1931).

19. Andrews, L.J., and R.M. Keefer. "Cation Complexes of Compounds Containing Carbon-Carbon Double Bonds. VI. The Argentation of Substituted Benzenes," *J. Am. Chem. Soc.*, 72(7):3113-3116 (1950).

20. Riddick, J.A., Bunger, W.B., and T.K. Sakano. *Organic Solvents - Physical Properties and Methods of Purification. Volume II* (New York: John Wiley and Sons, Inc., 1986), 1325 p.

21. Gunther, F.A., Westlake, W.E., and P.S. Jaglan. "Reported Solubilities of 738 Pesticide Chemicals in Water," *Res. Rev.*, 20:1-148 (1968).

22. Yalkowsky, S.H., Orr, R.J., and S.C. Valvani. "Solubility and

Partitioning. 3. The Solubility of Halobenzenes in Water," *Indust. Eng. Chem. Fund.*, 18(4):351-353 (1979).

23. Miller, M.M., Ghodbane, S., Wasik, S.P., Tewari, Y.B., and D.E. Martire. "Aqueous Solubilities, Octanol/Water Partition Coefficients, and Entropies of Melting of Chlorinated Benzenes and Biphenyls," *J. Chem. Eng. Data*, 29(2):184-190 (1984).

24. Standen, A., Ed. *Kirk-Othmer Encyclopedia of Chemical Technology, Volume 4*, 2nd ed. (New York: John Wiley and Sons, Inc., 1964), 937 p.

25. Mansour, M., Feicht, E. and P. Méallier. "Improvement of the Photostability of Selected Substances in Aqueous Medium," *Toxicol. Environ. Chem.*, 20-21:139-147 (1989).

26. "General Industry Standards for Toxic and Hazardous Substances," U.S. Code of Federal Regulations 1910, Subpart Z Section 1910.1000 (July 1982).

27. *Documentation of the Threshold Limit Values and Biological Exposure Indices for 1986-1987* (Cincinnati, OH: American Conference of Governmental Industrial Hygienists, 1986), 111 p.

3,3'-DICHLOROBENZIDINE

Synonyms: C.I. 23,060; Curithane C126; DCB; 4,4'-Diamino-3,3'-dichlorobiphenyl; 4,4'-Diamino-3,3'-dichlorodiphenyl; Dichloro-benzidine; Dichlorobenzidine base; *m,m'*-Dichlorobenzidine; **3,3'-Dichloro-1,1'-(biphenyl)-4,4'-diamine**; 3,3'-Dichlorobiphenyl-4,4'-diamine; 3,3'-Dichloro-4,4'-biphenyldiamine; 3,3'-Dichloro-4,4'-diaminobiphenyl; 3,3'-Dichloro-4,4'-diamino-(1,1-biphenyl); RCRA waste number U073.

Structural Formula:

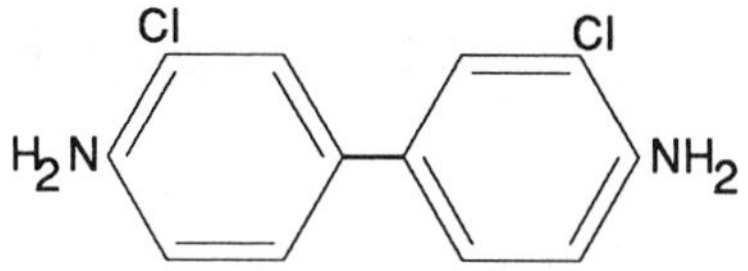

Note: Normally found as the dihydrochloride.

CHEMICAL DESIGNATIONS

CAS Registry Number: 91-94-1

DOT Designation: None assigned.

Empirical Formula: $C_{12}H_{10}Cl_2N_2$

Formula Weight: 253.13

RTECS Number: DD 0525000

PHYSICAL AND CHEMICAL PROPERTIES

Appearance and Odor: Colorless to grayish-purple crystals with a mild odor.

Boiling Point: No data found.

Dissociation Constant: No data found.

Henry's Law Constant: 4.5×10^{-8} atm·m^3/mol at 25 °C (estimated) [1].

Ionization Potential: No data found.

Log K$_{oc}$: 3.30 using method of Karickhoff and others [2].

Log K$_{ow}$: 3.51 [3].

Melting Point: 132-133 °C [4].

Solubility in Organics: Soluble in ethanol, benzene, and glacial acetic acid [4].

Solubility in Water: 3.11 ppm at 25 °C [3]; 4.0 mg/L at 22 °C as the dihydrochloride [5].

Specific Density: No data found.

Transformation Products: An aqueous solution subjected to ultraviolet radiation caused a rapid degradation to monochlorobenzidine, benzidine, and several unidentified chromaphores [6].

Vapor Pressure: No data was found however a value of 1×10^{-5} mm at 22 °C was assigned by analogy [7]; 4.2×10^{-7} mm at 25 °C (estimated) [1].

FIRE HAZARDS

Flash Point: No data found.

Lower Explosive Limit (LEL): No data found.

Upper Explosive Limit (UEL): No data found.

HEALTH HAZARD DATA

Immediately Dangerous to Life or Health (IDLH): Potential human carcinogen [8].

Permissible Exposure Limits (PEL) in Air: No standards set.

MANUFACTURING

Selected Manufacturers:

CTC Organics
P.O. Box 6933
Atlanta, GA 30315

Pfaltz & Bauer, Inc.
172 East Aurora St.
Waterbury, CT 06708

The Upjohn Co.
Fine Chemicals Division
North Haven, CT 06473

Uses: Intermediate for azo dyes and pigments; curing agent for isocyanate-terminated polymers and resins; rubber and plastic compounding ingredient; formerly used as chemical intermediate for direct red 61 dye.

REFERENCES

1. Howard, P.H. *Handbook of Environmental Fate and Exposure Data for Organic Chemicals* (Chelsea, MI: Lewis Publishers, Inc., 1989), 574 p.
2. Karickhoff, S.W., Brown, D.S., and T.A. Scott. "Sorption of Hydrophobic Pollutants on Natural Sediments," *Water Res.*, 13:241-248 (1979).
3. Banerjee, S., Yalkowsky, S.H., and S.C. Valvani. "Water Solubility and Octanol/Water Partition Coefficients of Organics. Limitations of the Solubility-Partition Coefficient Correlation," *Environ. Sci. Technol.*, 14(10):1227-1229 (1980).
4. Windholz, M., Budavari, S., Blumetti, R.F., and E.S. Otterbein, Eds. *The Merck Index*, 10th ed., (Rahway, NJ: Merck and Co., 1983), 1463 p.
5. "Treatability Manual - Volume 1: Treatability Data," Office of Research and Development, U.S. EPA Report-600/8-80-042a (1980), 1035 p.
6. Banerjee, S., Sikka, H.C., Gray, R., and C.M. Kelly. "Photodegradation of 3,3'-Dichlorobenzidine," *Environ. Sci. Technol.*, 12(13):1425-1427 (1978).

7. Mabey, W.R., Smith, J.H., Podoll, R.T., Johnson, H.L., Mill, T., Chou, T.-W., Gates, J., Partridge, I.W., Jaber, H., and D. Vandenberg. "Aquatic Fate Process Data for Organic Priority Pollutants - Final Report," Office of Regulations and Standards, U.S. EPA Report-440/4-81-014 (1982), 407 p.

8. "NIOSH Pocket Guide to Chemical Hazards," U.S. Department of Health and Human Services, U.S. Government Printing Office (1987), 241 p.

DICHLORODIFLUOROMETHANE

Synonyms: Algofrene type 2; Arcton 6; Difluorodichloromethane; Electro-CF 12; Eskimon 12; F 12; FC 12; Fluorocarbon 12; Freon 12; Freon F-12; Frigen 12; Genetron 12; Halon; Halon 122; Isceon 122; Isotron 2; Kaiser chemicals 12; Ledon 12; Propellant 12; R 12; RCRA waste number U075; Refrigerant 12; Ucon 12; Ucon 12/halocarbon 12; UN 1028.

Structural Formula:

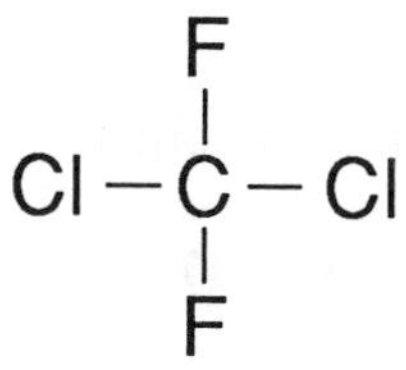

$$Cl-\underset{\underset{\displaystyle F}{|}}{\overset{\overset{\displaystyle F}{|}}{C}}-Cl$$

CHEMICAL DESIGNATIONS

CAS Registry Number: 75-71-8

DOT Designation: 1028

Empirical Formula: CCl_2F_2

Formula Weight: 120.91

RTECS Number: PA 8200000

PHYSICAL AND CHEMICAL PROPERTIES

Appearance and Odor: Colorless liquid or gas with an ethereal odor.

Boiling Point: -29.8 °C [1].

Henry's Law Constant: 3.0 atm·m^3/mol [2]; 0.425 atm·m^3/mol at 25 °C [3].

Ionization Potential: 11.97 eV [4].

Log K_{oc}: 2.56 using method of Kenaga and Goring [5].

Log K_{ow}: 2.16 [6].

Melting Point: -158 °C [1].

Solubility in Organics: Soluble in acetic acid, acetone, chloroform, ether [1], and ethanol [7].

Solubility in Water: 280 mg/L at 25 °C [8]; 0.008 wt% at 20 °C [4]; 301 mg/L at 25 °C [9]; 57 ml/L at 26 °C [10].

Specific Density: 1.75 at -115/4 °C, 1.1834 at 57/4 °C [1]; 1.35 at 15/4 °C [11].

Transformation Products: No data found.

Vapor Density: 4.94 g/L at 25 °C, 4.17 (air = 1).

Vapor Pressure: 4,250 mm at 20 °C, 5,776 mm at 30 °C [12]; 760 mm at -29.8 °C, 1,520 mm at -12.2 °C, 3,800 at 16.1 °C, 7,600 mm at 42.4 °C, 15,200 mm at 74.0 °C [1]; 4,306 mm at 20 °C [13]; 3,800 mm at 16.1 °C [14]; 4,870 mm at 25 °C [15]; 4,306 mm at 20 °C [16]; 4,887 mm at 25 °C [10].

FIRE HAZARDS

Flash Point: Not flammable [11].

Lower Explosive Limit (LEL): Not flammable [11].

Upper Explosive Limit (UEL): Not flammable [11].

HEALTH HAZARD DATA

Immediately Dangerous to Life or Health (IDLH): 50,000 ppm [4].

Permissible Exposure Limits (PEL) in Air: 1,000 ppm ($\approx$4,950 mg/m^3) [17]; 1,000 ppm TLV [18].

MANUFACTURING

Selected Manufacturers:

Allied Chemical Corp.
Baton Rouge, LA 70821

E.I. duPont de Nemours and Co.
Freon Products Division
Wilmington, DE 19898

Pennwalt Corp.
Chemical Division
Calvert City, KY 42029

Uses: Refrigerant; aerosol propellant; plastics; blowing agent; low temperature solvent; chilling cocktail glasses; freezing of foods by direct contact; leak-detecting agent.

REFERENCES

1. Weast, R.C., Ed. *CRC Handbook of Chemistry and Physics*, 67th ed. (Boca Raton, FL: CRC Press, Inc., 1986), 2406 p.
2. Pankow, J.F., and M.E. Rosen. "Determination of Volatile Compounds in Water by Purging Directly to a Capillary Column with Whole Column Cryotrapping," *Environ. Sci. Technol.*, 22(4):398-405 (1988).
3. Hine, J., and P.K. Mookerjee. "The Intrinsic Hydrophilic Character of Organic Compounds. Correlations in Terms of Structural Contributions," *J. Org. Chem.*, 40(3):292-298 (1975).
4. "NIOSH Pocket Guide to Chemical Hazards," U.S. Department of Health and Human Services, U.S. Government Printing Office (1987), 241 p.
5. Kenaga, E.E., and C.A.I. Goring. "Relationship Between Water Solubility, Soil Sorption, Octanol-Water Partitioning and Concentration of Chemicals in Biota," in *Aquatic Toxicology, ASTM STP 707*, Eaton, J.G., Parrish, P.R., and A.C. Hendricks, Eds. (Philadelphia, PA: American Society for Testing and Materials, 1980), pp 78-115.
6. Hansch, C., Vittoria, A., Silipo, C., and P.Y.C. Jow. "Partition Coefficients and the Structure-Activity Relationship of the Anesthetic Gases," *J. Med. Chem.*, 18(6):546-548 (1975).
7. *Toxic and Hazardous Industrial Chemicals Safety Manual for Handling and Disposal with Toxicity and Hazard Data* (Tokyo, Japan: International Technical Information Institute, 1986), 700 p.
8. Pearson, C.R., and G. McConnell. "Chlorinated C_1 and C_2 Hydrocarbons in the Marine Environment," in *Proc. R. Soc. London*, B189(1096):305-322 (1975).
9. Munz, C., and P.V. Roberts. "Air-Water Phase Equilibria of Volatile

Organic Solutes," *J. Am. Water Works Assoc.*, 79(5):62-69 (1987).

10. *Drinking Water and Health* (Washington, DC: National Academy of Sciences, 1980), 415 p.

11. Weiss, G. *Hazardous Chemicals Data Book* (Park Ridge, NJ: Noyes Data Corp., 1986), 1069 p.

12. Verschueren, K. *Handbook of Environmental Data on Organic Chemicals* (New York: Van Nostrand Reinhold Co., 1983), 1310 p.

13. "Treatability Manual - Volume 1: Treatability Data," Office of Research and Development, U.S. EPA Report-600/8-80-042a (1980), 1035 p.

14. Sax, N.I. *Dangerous Properties of Industrial Materials* (New York: Van Nostrand Reinhold Co., 1984), 3124 p.

15. Jordan, T.E. *Vapor Pressures of Organic Compounds* (New York: Interscience Publishers, Inc., 1954), 266 p.

16. McConnell, G., Ferguson, D.M., and C.R. Pearson. "Chlorinated Hydrocarbons and the Environment," *Endeavour*, 34(121):13-18 (1975).

17. "General Industry Standards for Toxic and Hazardous Substances," U.S. Code of Federal Regulations 1910, Subpart Z Section 1910.1000 (July 1982).

18. *Documentation of the Threshold Limit Values and Biological Exposure Indices for 1986-1987* (Cincinnati, OH: American Conference of Governmental Industrial Hygienists, 1986), 111 p.

1,1-DICHLOROETHANE

Synonyms: Chlorinated hydrochloric ether; 1,1-Dichlorethane; *asym*-Dichloroethane; Ethylidene chloride; Ethylidene dichloride; 1,1-Ethylidene dichloride; NCI-C04535; RCRA waste number U076; UN 2362.

Structural Formula:

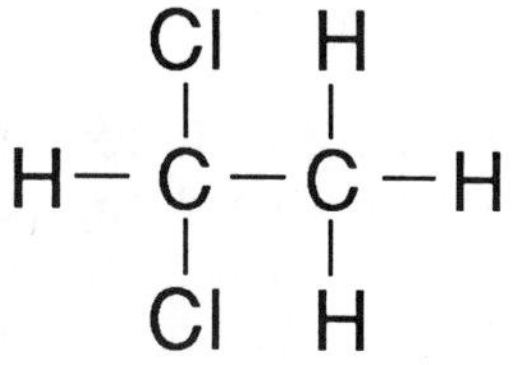

$$
\begin{array}{ccc}
Cl & & H \\
| & & | \\
H-C & - & C-H \\
| & & | \\
Cl & & H
\end{array}
$$

CHEMICAL DESIGNATIONS

CAS Registry Number: 75-34-3

DOT Designation: 2362

Empirical Formula: $C_2H_4Cl_2$

Formula Weight: 98.96

RTECS Number: KI 0175000

PHYSICAL AND CHEMICAL PROPERTIES

Appearance and Odor: Colorless liquid with a chloroform-like odor.

Boiling Point: 57.3 °C [1]; 55-58 °C [2].

Henry's Law Constant: 0.0043 atm·m^3/mol [3]; 0.00545 atm·m^3/mol [4]; 0.00587 atm·m^3/mol at 25 °C [5]; 0.00943 atm·m^3/mol at 37 °C [6].

Ionization Potential: No data found.

Log K$_{oc}$: 1.48 [7].

Log K$_{ow}$: 1.78 [8]; 1.79 [9].

Melting Point: -97.4 °C [10].

Solubility in Organics: Miscible with ethanol [11].

Solubility in Water: 5,500 mg/L at 20 °C [12]; 7,000 mg/L at 0 °C [10]; 5.03 wt% at 20 °C [13]; 5,060 mg/L at 25 °C [14].

Specific Density: 1.1757 at 20/4 °C [1]; 1.174 at 20/4 °C [15]; 1.173 at 20/4 °C [2]; 1.20685 at 0/4 °C, 1.1830 at 15/4 °C, 1.17600 at 20/4 °C, 1.60010 at 30/4 °C [16].

Transformation Product: Under anoxic conditions, indigenous microbes in uncontaminated sediments produced vinyl chloride [17].

Vapor Density: 4.04 g/L at 25 °C, 3.42 (air = 1).

Vapor Pressure: 70.1 mm at 0 °C, 115.3 mm at 10 °C, 182.1 mm at 20 °C, 277.2 mm at 30 °C [16]; 760 mm at 57.3 °C, 1,520 mm at 80.2 °C, 3,800 at 117.3 °C, 7,600 mm at 150.3 °C, 15,200 mm at 192.7 °C, 30,400 mm at 243.0 °C [1]; 234 mm at 25 °C, 270 mm at 30 °C [12].

FIRE HAZARDS

Flash Point: -6 °C [18].

Lower Explosive Limit (LEL): 5.6% [18].

Upper Explosive Limit (UEL): 16% [19].

HEALTH HAZARD DATA

Immediately Dangerous to Life or Health (IDLH): 4,000 ppm [19].

Permissible Exposure Limits (PEL) in Air: 100 ppm ($\approx$400 mg/m^3) [20]; 200 ppm ($\approx$810 mg/m^3) TWA, 250 ppm ($\approx$1,010 mg/m^3) STEL [21].

MANUFACTURING

Selected Manufacturers:

Fluka Chemical Corp.
980 South Second St.
Ronkonkoma, NY 11779

PPG Industries, Inc.
Chemicals Group
Industrial Chemical Division
Lake Charles, LA 70601

Vulcan Materials Co.
Vulcan Chemicals Division
Geismar, LA 70734

Uses: Extraction solvent; insecticide and fumigant; preparation of vinyl chloride; paint, varnish, and finish removers; degreasing and drying metal parts; ore flotation; solvent for plastics, oils, and fats; chemical intermediate for 1,1,1-trichloroethane; in rubber cementing, fabric spreading, and fire extinguishers; formerly used as an anesthetic; organic synthesis.

REFERENCES

1. Weast, R.C., Ed. *CRC Handbook of Chemistry and Physics*, 67th ed. (Boca Raton, FL: CRC Press, Inc., 1986), 2406 p.
2. *Fluka Catalog 1988/89 - Chemika-Biochemika* (Ronkonkoma, NY: Fluka Chemical Corp., 1988), 1536 p.
3. Pankow, J.F., and M.E. Rosen. "Determination of Volatile Compounds in Water by Purging Directly to a Capillary Column with Whole Column Cryotrapping," *Environ. Sci. Technol.*, 22(4):398-405 (1988).
4. Warner, H.P., Cohen, J.M., and J.C. Ireland. "Determination of Henry's Law Constants of Selected Priority Pollutants," Office of Science and Development, U.S. EPA Report-600/D-87/229 (1987), 14 p.
5. Hine, J., and P.K. Mookerjee. "The Intrinsic Hydrophilic Character of Organic Compounds. Correlations in Terms of Structural Contributions," *J. Org. Chem.*, 40(3):292-298 (1975).
6. Sato, A., and T. Nakajima. "A Structure-Activity Relationship of Some Chlorinated Hydrocarbons," *Arch. Environ. Health*, 34(2):69-75 (1979).
7. Schwille, F. *Dense Chlorinated Solvents* (Chelsea, MI: Lewis Publishers, Inc., 1988), 146 p.
8. Mills, W.B., Porcella, D.B., Ungs, M.J., Gherini, S.A., Summers, K.V., Mok, L., Rupp, G.L., and G.L. Bowie. "Water Quality Assessment: A Screening Procedure for Toxic and Conventional Pollutants in Surface and Groundwater-Part I," Office of Research and

Development, U.S. EPA Report-600/6-85-002a (1985), 638 p.

9. Hansch, C., and A. Leo. *Substituent Constants for Correlation Analysis in Chemistry and Biology* (New York: John Wiley and Sons, Inc., 1979), 339 p.

10. Dean, J.A., Ed., *Lange's Handbook of Chemistry*, 11th ed. (New York: McGraw-Hill, Inc., 1973), 1570 p.

11. "Chemical, Physical, and Biological Properties of Compounds Present at Hazardous Waste Sites," U.S. EPA Report-530/SW-89-010 (1985), 619 p.

12. Verschueren, K. *Handbook of Environmental Data on Organic Chemicals* (New York: Van Nostrand Reinhold Co., 1983), 1310 p.

13. Riddick, J.A., Bunger, W.B., and T.K. Sakano. *Organic Solvents - Physical Properties and Methods of Purification. Volume II* (New York: John Wiley and Sons, Inc., 1986), 1325 p.

14. Gross, P. "The Determination of the Solubility of Slightly Soluble Liquids in Water and the Solubilities of the Dichloro- Ethanes and -Propanes," *J. Am. Chem. Soc.*, 51(8):2362-2366 (1929).

15. Sax, N.I. *Dangerous Properties of Industrial Materials* (New York: Van Nostrand Reinhold Co., 1984), 3124 p.

16. Standen, A., Ed. *Kirk-Othmer Encyclopedia of Chemical Technology, Volume 4*, 2nd ed. (New York: John Wiley and Sons, Inc., 1964), 937 p.

17. Barrio-Lage, G., Parsons, F.Z., Nassar, R.S., and P.A. Lorenzo. "Sequential Dehalogenation of Chlorinated Ethenes," *Environ. Sci. Technol.*, 20(1):96-99 (1986).

18. *Fire Protection Guide on Hazardous Materials* (Quincy, MA: National Fire Protection Association, 1984), 443 p.

19. "NIOSH Pocket Guide to Chemical Hazards," U.S. Department of Health and Human Services, U.S. Government Printing Office (1987), 241 p.

20. "General Industry Standards for Toxic and Hazardous Substances," U.S. Code of Federal Regulations 1910, Subpart Z, Section 1910.1000 (July 1982).

21. *Documentation of the Threshold Limit Values and Biological Exposure Indices for 1986-1987* (Cincinnati, OH: American Conference of Governmental Industrial Hygienists, 1986), 111 p.

1,2-DICHLOROETHANE

Synonyms: 1,2-Bichloroethane; Borer sol; Brocide; 1,2-DCA; 1,2-DCE; Destruxol borer-sol; Dichloremulsion; 1,2-Dichlorethane; Dichlormulsion; α,β-Dichloroethane; *sym*-Dichloroethane; Dichloroethylene; Dutch liquid; Dutch oil; EDC; ENT 1,656; Ethane dichloride; Ethene dichloride; Ethylene chloride; Ethylene dichloride; 1,2-Ethylene dichloride; Freon 150; Glycol dichloride; NCI-C00511; RCRA waste number U077; UN 1184.

Structural Formula:

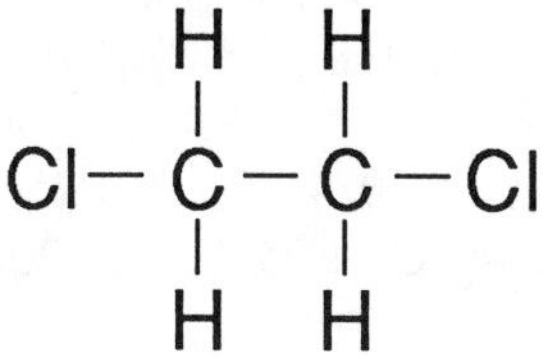

$$CI-\overset{\overset{\displaystyle H}{|}}{\underset{\underset{\displaystyle H}{|}}{C}}-\overset{\overset{\displaystyle H}{|}}{\underset{\underset{\displaystyle H}{|}}{C}}-CI$$

CHEMICAL DESIGNATIONS

CAS Registry Number: 107-06-2

DOT Designation: 2362

Empirical Formula: $C_2H_4Cl_2$

Formula Weight: 98.96

RTECS Number: KI 0525000

PHYSICAL AND CHEMICAL PROPERTIES

Appearance and Odor: Colorless liquid with a pleasant odor.

Boiling Point: 83.5 °C [1].

Henry's Law Constant: 9.1 x 10^{-4} atm·m^3/mol [2]; 0.0011 atm·m^3/mol [3]; 9.8 x 10^{-4} atm·m^3/mol at 25 °C [4]; 0.00131 atm·m^3/mol at 25 °C [5]; 0.00225 atm·m^3/mol at 37 °C [6].

Ionization Potential: 9.64 eV [7]; 11.04 eV [8].

Log K$_{oc}$: 1.15 [2]; 1.279 [9].

225

Log K$_{ow}$: 1.48 [10]; 1.45 [11].

Melting Point: -35.3 °C [1].

Solubility in Organics: Miscible with ethanol, chloroform, and ether [12].

Solubility in Water: 9,200 mg/L at 0 °C, 8,690 mg/L at 20 °C [13]; in 120 parts water [14]; 8,300 mg/L at 25 °C [3]; 7,986 mg/L at 25 °C [11]; 8,720 mg/L at 15 °C, 9,000 mg/L at 30 °C [15]; 8,800 mg/L at 20 °C [16]; 0.81 wt% at 20 °C [17]; 8,450 ppm at 20 °C [9]; 0.873 mass% at 0 °C [10]; 8,650 mg/L at 25 °C [18]; 0.87% [19].

Specific Density: 1.2351 at 20/4 °C [1]; 1.253 at 20/4 °C [20]; 1.28164 at 0/4 °C, 1.26000 at 15/4 °C, 1.25280 at 20/4 °C, 1.24530 at 25/4 °C, 1.23831 at 30/4 °C [21].

Transformation Products: No data found.

Vapor Density: 4.04 g/L at 25 °C, 3.42 (air = 1).

Vapor Pressure: 25 mm at 0 °C, 40 mm at 10 °C, 64 mm at 20 °C, 100 mm at 30 °C 160 mm at 40 °C, 240 mm at 50 °C, 350 mm at 60 °C, 500 mm at 70 °C, 700 mm at 80 °C, 760 mm at 83.5 °C [21]; 87 mm at 25 °C [14].

FIRE HAZARDS

Flash Point: 13 °C (open cup) [7]; 17 °C, 21 °C (open cup) [21].

Lower Explosive Limit (LEL): 6.2% [7].

Upper Explosive Limit (UEL): 16% [7].

HEALTH HAZARD DATA

Immediately Dangerous to Life or Health (IDLH): Potential human carcinogen [7].

Permissible Exposure Limits (PEL) in Air: 50 ppm, 100 ppm ceiling, 200 ppm 5-minute/3-hour peak [22]; 1 ppm 10-hour TWA, 2 ppm 15-minute ceiling [7]; 10 ppm ($\approx$40 mg/m^3) TWA [14].

MANUFACTURING

Selected Manufacturers:

Aldrich Chemical Co.
940 West Saint Paul Ave.
Milwaukee, WI 53233

Fluka Chemical Corp.
980 South Second St.
Ronkonkoma, NY 11779

Pfaltz & Bauer, Inc.
172 East Aurora St.
Waterbury, CT 06708

Uses: Vinyl chloride solvent; lead scavenger in antiknock unleaded gasoline; paint, varnish, and finish remover; metal degreasers; soaps and scouring compounds; wetting and penetrating agents; organic synthesis; ore flotation; tobacco flavoring; soil and foodstuff fumigant.

REFERENCES

1. Weast, R.C., Ed. *CRC Handbook of Chemistry and Physics*, 67th ed. (Boca Raton, FL: CRC Press, Inc., 1986), 2406 p.
2. Schwille, F. *Dense Chlorinated Solvents* (Chelsea, MI: Lewis Publishers, Inc., 1988), 146 p.
3. Warner, H.P., Cohen, J.M., and J.C. Ireland. "Determination of Henry's Law Constants of Selected Priority Pollutants," Office of Science and Development, U.S. EPA Report-600/D-87/229 (1987), 14 p.
4. Dilling, W.L. "Interphase Transfer Processes. II. Evaporation Rates of Chloro Methanes, Ethanes, Ethylenes, Propanes, and Propylenes from Dilute Aqueous Solutions. Comparisons with Theoretical Predictions," *Environ. Sci. Technol.*, 11(4):405-409 (1977).
5. Hine, J., and P.K. Mookerjee. "The Intrinsic Hydrophilic Character of Organic Compounds. Correlations in Terms of Structural Contributions," *J. Org. Chem.*, 40(3):292-298 (1975).
6. Sato, A., and T. Nakajima. "A Structure-Activity Relationship of Some Chlorinated Hydrocarbons," *Arch. Environ. Health*, 34(2):69-75 (1979).
7. "NIOSH Pocket Guide to Chemical Hazards," U.S. Department of

Health and Human Services, U.S. Government Printing Office (1987), 241 p.

8. Hazardous Substances Data Bank. 1,2-Dichloroethane, National Library of Medicine, Toxicology Information Program (1989).

9. Chiou, C.T., Peters, L.J., and V.H. Freed. "A Physical Concept of Soil-Water Equilibria for Nonionic Organic Compounds," *Science*, 206:831-832 (1979).

10. Konietzko, H. "Chlorinated Ethanes: Sources, Distribution, Environmental Impact, and Health Effects," in *Hazard Assessment of Chemicals, Volume 3*, J. Saxena, Ed. (New York: Academic Press, Inc., 1984), pp 401-448.

11. Banerjee, S., Yalkowsky, S.H., and S.C. Valvani. "Water Solubility and Octanol/Water Partition Coefficients of Organics. Limitations of the Solubility-Partition Coefficient Correlation," *Environ. Sci. Technol.*, 14(10):1227-1229 (1980).

12. "Chemical, Physical, and Biological Properties of Compounds Present at Hazardous Waste Sites," U.S. EPA Report-530/SW-89-010 (1985), 619 p.

13. Verschueren, K. *Handbook of Environmental Data on Organic Chemicals* (New York: Van Nostrand Reinhold Co., 1983), 1310 p.

14. *Documentation of the Threshold Limit Values and Biological Exposure Indices for 1986-1987* (Cincinnati, OH: American Conference of Governmental Industrial Hygienists, 1986), 111 p.

15. Gross, P. "The Determination of the Solubility of Slightly Soluble Liquids in Water and the Solubilities of the Dichloro- Ethanes and -Propanes," *J. Am. Chem. Soc.*, 51(8):2362-2366 (1929).

16. McConnell, G., Ferguson, D.M., and C.R. Pearson. "Chlorinated Hydrocarbons and the Environment," *Endeavour*, 34(121):13-18 (1975).

17. Riddick, J.A., Bunger, W.B., and T.K. Sakano. *Organic Solvents - Physical Properties and Methods of Purification. Volume II* (New York: John Wiley and Sons, Inc., 1986), 1325 p.

18. Gross, P.M., and J.H. Saylor. "The Solubilities of Certain Slightly Soluble Organic Compounds in Water," *J. Am. Chem. Soc.*, 53(5):1744-1751 (1931).

19. Gunther, F.A., Westlake, W.E., and P.S. Jaglan. "Reported Solubilities of 738 Pesticide Chemicals in Water," *Res. Rev.*, 20:1-148 (1968).

20. *Fluka Catalog 1988/89 - Chemika-Biochemika* (Ronkonkoma, NY: Fluka Chemical Corp., 1988), 1536 p.

21. Standen, A., Ed. *Kirk-Othmer Encyclopedia of Chemical Technology, Volume 4*, 2nd ed. (New York: John Wiley and Sons, Inc., 1964), 937 p.

22. "General Industry Standards for Toxic and Hazardous Substances,"

U.S. Code of Federal Regulations 1910, Subpart Z, Section 1910.1000 (July 1982).

1,1-DICHLOROETHYLENE

Synonyms: 1,1-DCE; **1,1-Dichloroethene;** *asym*-Dichloroethylene; NCI-C54262; RCRA waste number U078; Sconatex; VC; VDC; Vinylidene chloride; Vinylidene chloride (II); Vinylidene dichloride; Vinylidine chloride.

Structural Formula:

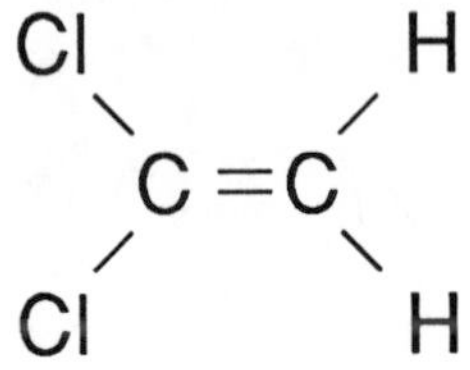

$$
\begin{array}{ccc}
Cl & & H \\
\diagdown & & \diagup \\
& C = C & \\
\diagup & & \diagdown \\
Cl & & H
\end{array}
$$

CHEMICAL DESIGNATIONS

CAS Registry Number: 75-35-4

DOT Designation: 1303

Empirical Formula: $C_2H_2Cl_2$

Formula Weight: 96.94

RTECS Number: KV 9275000

PHYSICAL AND CHEMICAL PROPERTIES

Appearance and Odor: Colorless liquid with an mild, sweet odor resembling that of chloroform.

Boiling Point: 37 °C [1]; 32 °C [2].

Henry's Law Constant: 0.021 atm·m^3/mol [3]; 0.015 atm·m^3/mol [4]; 0.19 atm·m^3/mol [5].

Ionization Potential: 9.81 ± 0.35 eV [6].

Log K_{oc}: 1.81 [3].

Log K_{ow}: 2.13 [7]; 1.48 [8].

Melting Point: -122.1 °C [1].

Solubility in Organics: Slightly soluble in ethanol, ether, acetone, benzene, and chloroform [9].

Solubility in Water: 400 mg/L at 20 °C [10]; 273 mg/L at 25 °C [11]; 5,000 mg/L at 25 °C [12]; 0.021 wt% at 25 °C [13]; 0.24 wt% at 15 °C, 0.255 wt% at 17 °C, 0.25 wt% at 20 °C, 0.225 wt% at 25 °C, 0.24 wt% at 28.5 °C, 0.255 wt% at 29.5 °C, 0.22 wt% at 38.5 °C, 0.21 wt% at 45 °C, 0.23 wt% at 51 °C, 0.24 wt% at 60 °C, 0.225 wt% at 65 °C, 0.295 wt% at 71 °C, 0.25 wt% at 74.5 °C, 0.295 wt% at 81 °C, 0.37 wt% at 85.5 °C, 0.35 wt% at 90.5 [14]; 6,400 ppm at 25 °C [15].

Specific Density: 1.218 at 20/4 °C [1]; 1.213 at 20/4 °C [2].

Transformation Product: In a methanogenic aquifer material, 1,1-dichloroethylene biodegraded to vinyl chloride [16]. Under anoxic conditions, indigenous microbes in uncontaminated sediments degraded 1,1-dichloroethylene to vinyl chloride [17]. Photooxidation of 1,1-dichloroethylene in the presence of nitrogen dioxide and air yielded phosgene, chloroacetyl chloride, formic acid, hydrochloric acid, carbon monoxide, formaldehyde, and ozone [18].

Vapor Density: 3.96 g/L at 25 °C, 3.35 (air = 1).

Vapor Pressure: 591 mm at 25 °C, 720 mm at 30 °C [19]; 760 mm at 87.3 °C, 1,520 mm at 108.1 °C, 3,800 at 147.8 °C, 7,600 mm at 183.5 °C, 15,200 mm at 226.5 °C, 30,400 mm at 272.0 °C [1]; 80 mm at -20 °C, 135 mm at -10 °C, 215 mm at 0 °C, 495 mm at 20 °C, 760 mm at 31.8 °C [2].

FIRE HAZARDS

Flash Point: -15 °C [20].

Lower Explosive Limit (LEL): 6.5% [21].

Upper Explosive Limit (UEL): 15.5% [21].

HEALTH HAZARD DATA

Immediately Dangerous to Life or Health (IDLH): No data found.

Permissible Exposure Limits (PEL) in Air: 5 ppm ($\approx$20 mg/m^3) TWA, 20 ppm ($\approx$80 mg/m^3) STEL [22].

MANUFACTURING

Selected Manufacturers:

Aldrich Chemical Co.
940 West Saint Paul Ave.
Milwaukee, WI 53233

Dow Chemical Co.
Midland, MI 48640

PPG Industries, Inc.
Industrial Chemicals Division
1 PPG Place
Pittsburgh, PA 15272

Uses: Synthetic fibers and adhesives; chemical intermediate in vinylidene fluoride synthesis; comonomer for food packaging, coating resins, and modacrylic fibers.

REFERENCES

1. Weast, R.C., Ed. *CRC Handbook of Chemistry and Physics*, 67th ed. (Boca Raton, FL: CRC Press, Inc., 1986), 2406 p.
2. Standen, A., Ed. *Kirk-Othmer Encyclopedia of Chemical Technology, Volume 4*, 2nd ed. (New York: John Wiley and Sons, Inc., 1964), 937 p.
3. Schwille, F. *Dense Chlorinated Solvents* (Chelsea, MI: Lewis Publishers, Inc., 1988), 146 p.
4. Warner, H.P., Cohen, J.M, and J.C. Ireland. "Determination of Henry's Law Constants of Selected Priority Pollutants," Office of Science and Development, U.S. EPA Report-600/D-87/229 (1987), 14 p.
5. Pankow, J.F., and M.E. Rosen. "Determination of Volatile Compounds in Water by Purging Directly to a Capillary Column with Whole Column Cryotrapping," *Environ. Sci. Technol.*, 22(4):398-405 (1988).

6. Franklin, J.L., Dillard, J.G., Rosenstock, H.M., Herron, J.T., Draxl K., and F.H. Field. "Ionization Potentials, Appearance Potentials and Heats of Formation of Gaseous Positive Ions," National Bureau of Standards Report NSRDS-NBS 26, U.S. Government Printing Office (1969), 289 p.

7. Mabey, W.R., Smith, J.H., Podoll, R.T., Johnson, H.L., Mill, T., Chou, T.-W., Gates, J., Partridge, I.W., Jaber, H., and D. Vandenberg. "Aquatic Fate Process Data for Organic Priority Pollutants - Final Report," Office of Regulations and Standards, U.S. EPA Report-440/4-81-014 (1982), 407 p.

8. Hazardous Substances Data Bank. 1,1-Dichloroethylene, National Library of Medicine, Toxicology Information Program (1989).

9. "Chemical, Physical, and Biological Properties of Compounds Present at Hazardous Waste Sites," U.S. EPA Report-530/SW-89-010 (1985), 619 p.

10. Pearson, C.R., and G. McConnell. "Chlorinated C_1 and C_2 Hydrocarbons in the Marine Environment," in *Proc. R. Soc. London*, B189(1096):305-322 (1975).

11. Lyman, W.J., Reehl, W.F., and D.H. Rosenblatt. *Handbook of Chemical Property Estimation Methods: Environmental Behavior of Organic Compounds* (New York: McGraw-Hill, Inc., 1982).

12. Walton, W.C. *Practical Aspects of Ground Water Modeling* (Worthington, OH: National Water Well Association, 1985), 587 p.

13. Riddick, J.A., Bunger, W.B., and T.K. Sakano. *Organic Solvents - Physical Properties and Methods of Purification. Volume II* (New York: John Wiley and Sons, Inc., 1986), 1325 p.

14. DeLassus, P.T., and D.D. Schmidt. "Solubilities of Vinyl Chloride and Vinylidene Chloride in Water," *J. Chem. Eng. Data*, 26(3):274-276 (1981).

15. Amoore, J.E., and E. Hautala. "Odor as an Aide to Chemical Safety: Odor Thresholds Compared with Threshold Limit Values and Volatilities for 214 Industrial Chemicals in Air and Water Dilution," *J. Appl. Toxicol.*, 3(6):272-290 (1983).

16. Wilson, B.H., Smith, G.B., and J.F. Rees. "Biotransformations of Selected Alkylbenzenes and Halogenated Aliphatic Hydrocarbons in Methanogenic Aquifer Material: A Microcosm Study," *Environ. Sci. Technol.*, 20(10):997-1002 (1986).

17. Barrio-Lage, G., Parsons, F.Z., Nassar, R.S., and P.A. Lorenzo. "Sequential Dehalogenation of Chlorinated Ethenes," *Environ. Sci. Technol.*, 20(1):96-99 (1986).

18. Gay, B.W., Jr., Hanst, P.L., Bufalini, J.J., and R.C. Noonan. "Atmospheric Oxidation of Chlorinated Ethylenes," *Environ. Sci. Technol.*, 10(1):58-67 (1976).

19. Verschueren, K. *Handbook of Environmental Data on Organic Chemicals* (New York: Van Nostrand Reinhold Co., 1983), 1310 p.
20. "NIOSH Pocket Guide to Chemical Hazards," U.S. Department of Health and Human Services, U.S. Government Printing Office (1987), 241 p.
21. *Fire Protection Guide on Hazardous Materials* (Quincy, MA: National Fire Protection Association, 1984), 443 p.
22. *Documentation of the Threshold Limit Values and Biological Exposure Indices for 1986-1987* (Cincinnati, OH: American Conference of Governmental Industrial Hygienists, 1986), 111 p.

trans-1,2-DICHLOROETHYLENE

Synonyms: Acetylene dichloride; *trans*-Acetylene dichloride; 1,2-Dichloroethene; (*E*)-1,2-Dichloroethene; *trans*-Dichloroethylene; 1,2-*trans*-Dichloroethene; 1,2-*trans*-Dichloroethylene; *sym*-Dichloroethylene; Dioform.

Structural Formula:

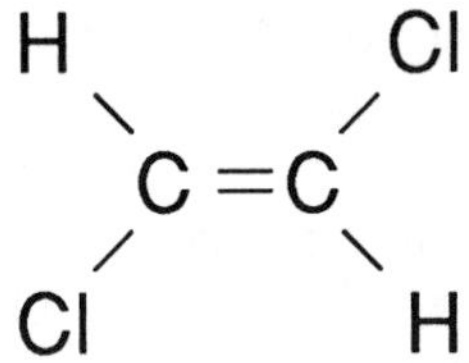

$$\begin{array}{ccc} H & & Cl \\ \diagdown & & \diagup \\ & C = C & \\ \diagup & & \diagdown \\ Cl & & H \end{array}$$

CHEMICAL DESIGNATIONS

CAS Registry Number: 156-60-5

DOT Designation: 1150 for a mixture containing *cis*- and *trans*- isomers.

Empirical Formula: $C_2H_2Cl_2$

Formula Weight: 96.94

RTECS Number: KV 9400000

PHYSICAL AND CHEMICAL PROPERTIES

Appearance and Odor: Colorless liquid with a sweet pleasant odor.

Boiling Point: 47.5 °C [1].

Henry's Law Constant: 0.384 atm·m^3/mol [2]; 0.0072 atm·m^3/mol [3]; 0.00532 atm·m^3/mol [4]; 0.00674 atm·m^3/mol at 25 °C [5]; 0.0121 atm·m^3/mol at 37 °C [6].

Ionization Potential: 9.64 eV [7].

Log K_{oc}: 1.77 [8].

Log K_{ow}: 2.09 [9].

Melting Point: -50 °C [1].

Solubility in Organics: Miscible with acetone, ethanol, ether, and very soluble in benzene and chloroform [10].

Solubility in Water: 600 mg/L at 20 °C [11]; 6,300 mg/L at 25 °C [12]; 0.63 wt% at 25 °C [13]; 6,260 mg/L at 25 °C [14].

Specific Density: 1.2565 at 20/4 °C [1]; 1.27 at 25/4 °C (mixture of *cis*- and *trans*- isomers) [15]; 1.2631 at 10/4 °C [16].

Transformation Product: In a methanogenic aquifer material, *trans*-1,2-dichloroethylene biodegraded to vinyl chloride [17]. Under anoxic conditions *trans*-1,2-dichloroethylene, when subjected to indigenous microbes in uncontaminated sediments, degraded to vinyl chloride [18].

Vapor Density: 3.96 g/L at 25 °C, 3.35 (air = 1).

Vapor Pressure: 200 mm at 14 °C [11]; 760 mm at 47.8 °C, 1,520 mm at 69.8 °C, 3,800 at 104.0 °C, 7,600 mm at 135.7 °C, 15,200 mm at 174.0 °C, 30,400 mm at 220.0 °C [1]; 40 mm at -20 °C, 64 mm at -10 °C, 113 mm at 0 °C, 185 mm at 10 °C, 265 mm at 20 °C, 410 mm at 30 °C, 575 mm at 40 °C, 760 mm at 47.7 °C [16].

FIRE HAZARDS

Flash Point: 2 °C [19].

Lower Explosive Limit (LEL): 9.7% [19].

Upper Explosive Limit (UEL): 12.8% [19].

HEALTH HAZARD DATA

Immediately Dangerous to Life or Health (IDLH): No data found, however, a level of 4,000 ppm has been established for 1,2-dichloroethylene containing both the *cis*- and *trans*- isomers [20].

Permissible Exposure Limits (PEL) in Air: No standards set, however, for the compound containing both isomers the TLV is 200 ppm ($\approx$790 mg/m^3) [21].

MANUFACTURING

Selected Manufacturers:

Fluka Chemical Corp.
980 South Second St.
Ronkonkoma, NY 11779

Pfaltz and Bauer, Inc.
126-04 Northern Blvd.
Flushing, NY 11368

Uses: A mixture of *cis*- and *trans*- isomers is used as a solvent for fats, phenols, camphor, etc.; ingredient in perfumes; low temperature solvent for sensitive substances such as caffeine; refrigerant; organic synthesis.

REFERENCES

1. Weast, R.C., Ed. *CRC Handbook of Chemistry and Physics*, 67th ed. (Boca Raton, FL: CRC Press, Inc., 1986), 2406 p.
2. Gossett, J.M. "Measurement of Henry's Law Constants for C_1 and C_2 Chlorinated Hydrocarbons," *Environ. Sci. Technol.*, 21(2):202-208 (1987).
3. Pankow, J.F., and M.E. Rosen. "Determination of Volatile Compounds in Water by Purging Directly to a Capillary Column with Whole Column Cryotrapping," *Environ. Sci. Technol.*, 22(4):398-405 (1988).
4. "Treatability Manual - Volume 1: Treatability Data," Office of Research and Development, U.S. EPA Report-600/8-80-042a (1980), 1035 p.
5. Hine, J., and P.K. Mookerjee. "The Intrinsic Hydrophilic Character of Organic Compounds. Correlations in Terms of Structural Contributions," *J. Org. Chem.*, 40(3):292-298 (1975).
6. Sato, A., and T. Nakajima. "A Structure-Activity Relationship of Some Chlorinated Hydrocarbons," *Arch. Environ. Health*, 34(2):69-75 (1979).
7. Franklin, J.L., Dillard, J.G., Rosenstock, H.M., Herron, J.T., Draxl K., and F.H. Field. "Ionization Potentials, Appearance Potentials and Heats of Formation of Gaseous Positive Ions," National Bureau of Standards Report NSRDS-NBS 26, U.S. Government Printing Office (1969), 289 p.
8. Schwille, F. *Dense Chlorinated Solvents* (Chelsea, MI: Lewis

238   *trans*-1,2-Dichloroethylene

Publishers, Inc., 1988), 146 p.

9. Mabey, W.R., Smith, J.H., Podoll, R.T., Johnson, H.L., Mill, T., Chou, T.-W., Gates, J., Partridge, I.W., Jaber, H., and D. Vandenberg. "Aquatic Fate Process Data for Organic Priority Pollutants - Final Report," Office of Regulations and Standards, U.S. EPA Report-440/4-81-014 (1982), 407 p.

10. "Chemical, Physical, and Biological Properties of Compounds Present at Hazardous Waste Sites," U.S. EPA Report-530/SW-89-010 (1985), 619 p.

11. Verschueren, K. *Handbook of Environmental Data on Organic Chemicals* (New York: Van Nostrand Reinhold Co., 1983), 1310 p.

12. Dilling, W.L. "Interphase Transfer Processes. II. Evaporation Rates of Chloro Methanes, Ethanes, Ethylenes, Propanes, and Propylenes from Dilute Aqueous Solutions. Comparisons with Theoretical Predictions," *Environ. Sci. Technol.*, 11(4):405-409 (1977).

13. Riddick, J.A., Bunger, W.B., and T.K. Sakano. *Organic Solvents - Physical Properties and Methods of Purification. Volume II* (New York: John Wiley and Sons, Inc., 1986), 1325 p.

14. Kamlet, M.J., Doherty, R.M., Abraham, M.H., Carr, P.W., Doherty, R.F., and R.W. Taft. "Linear Solvation Energy Relationships. 41. Important Differences Between Aqueous Solubility Relationships for Aliphatic and Aromatic Solutes," *J. Phys. Chem.*, 91(7):1996-2004 (1987).

15. Weiss, G. *Hazardous Chemicals Data Book* (Park Ridge, NJ: Noyes Data Corp., 1986), 1069 p.

16. Standen, A., Ed. *Kirk-Othmer Encyclopedia of Chemical Technology, Volume 4*, 2nd ed. (New York: John Wiley and Sons, Inc., 1964), 937 p.

17. Wilson, B.H., Smith, G.B., and J.F. Rees. "Biotransformations of Selected Alkylbenzenes and Halogenated Aliphatic Hydrocarbons in Methanogenic Aquifer Material: A Microcosm Study," *Environ. Sci. Technol.*, 20(10):997-1002 (1986).

18. Barrio-Lage, G., Parsons, F.Z., Nassar, R.S., and P.A. Lorenzo. "Sequential Dehalogenation of Chlorinated Ethenes," *Environ. Sci. Technol.*, 20(1):96-99 (1986).

19. Sax, N.I. *Dangerous Properties of Industrial Materials* (New York: Van Nostrand Reinhold Co., 1984), 3124 p.

20. "NIOSH Pocket Guide to Chemical Hazards," U.S. Department of Health and Human Services, U.S. Government Printing Office (1987), 241 p.

21. *Documentation of the Threshold Limit Values and Biological Exposure Indices for 1986-1987* (Cincinnati, OH: American Conference of Governmental Industrial Hygienists, 1986), 111 p.

2,4-DICHLOROPHENOL

Synonyms: 3-Chloro-4-hydroxychlorobenzene; DCP; 2,4-DCP; 2,4-Dichlorohydroxybenzene; 4,6-Dichlorohydroxybenzene; NCI-C55345; RCRA waste number U081.

Structural Formula:

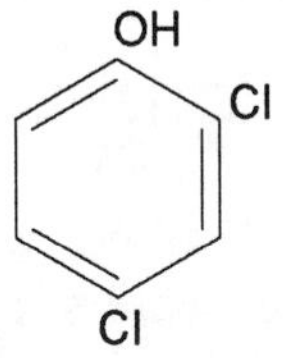

CHEMICAL DESIGNATIONS

CAS Registry Number: 120-83-2

DOT Designation: None assigned.

Empirical Formula: $C_6H_4Cl_2O$

Formula Weight: 163.00

RTECS Number: SK 8575000

PHYSICAL AND CHEMICAL PROPERTIES

Appearance and Odor: Colorless to yellow crystals with a sweet, musty or medicinal odor.

Boiling Point: 210 °C [1]; 216 °C [2].

Dissociation Constant: 7.85 [3].

Henry's Law Constant: 6.66 x 10^{-6} atm·m³/mol (calculated) [4]; 3.23 x 10^{-6} atm·m³/mol at 25 °C (estimated) [5].

Ionization Potential: No data found.

Log K_{oc}: 2.94 using method of Karickhoff and others [6].

Log K_{ow}: 3.15 [7]; 3.06 [8]; 3.08 [9]; 3.23 [10].

Melting Point: 45 °C [1].

Solubility in Organics: Soluble in ethanol, benzene, ether, chloroform [11], and carbon tetrachloride [12].

Solubility in Water: 4,600 mg/L at 20 °C, 4,500 mg/L at 25 °C [13]; 4,500 mg/L at 20 °C [9]; 0.092 M at 25 °C [14].

Specific Density: 1.383 at 60/25 °C [13]; 1.40 at 15/4 °C [2].

Transformation Products: In distilled water, photolysis occurs at a slower rate than in estuarine waters containing humic substances. Photolysis products identified in distilled water were the three isomers of chlorocyclopentadienic acid [15]. In freshwater lake sediments, anaerobic reductive dechlorination produced 4-chlorophenol [16].

Vapor Pressure: 0.12 mm at 20 °C (calculated) [17]; 1 mm at 53 °C, 10 mm at 92.8 °C, 40 mm at 123.4 °C, 100 mm at 146 °C, 400 mm at 187.5 °C, 760 mm at 210 °C [1]; 0.015 mm at 8 °C, 0.089 mm at 25 °C [5].

FIRE HAZARDS

Flash Point: 113.9 °C, 93.3 °C (open cup) [2].

Lower Explosive Limit (LEL): No data found.

Upper Explosive Limit (UEL): No data found.

HEALTH HAZARD DATA

Immediately Dangerous to Life or Health (IDLH): No data found.

Permissible Exposure Limits (PEL) in Air: No standards set.

MANUFACTURING

Selected Manufacturers:

American Aniline Products, Inc.
25 McLean Blvd.
Paterson, NJ 07509

Fluka Chemical Corp.
980 South Second St.
Ronkonkoma, NY 11779

Martin Marietta Corp.
Southern Dyestuff Co., Division
Charlotte, NC 28201

Uses: A chemical intermediate in the manufacture of the pesticide 2,4-dichlorophenoxyacetic acid (2,4-D), and other compounds for use as germicides, antiseptics, and seed disinfectants.

REFERENCES

1. Weast, R.C., Ed. *CRC Handbook of Chemistry and Physics*, 67th ed. (Boca Raton, FL: CRC Press, Inc., 1986), 2406 p.
2. Weiss, G. *Hazardous Chemicals Data Book* (Park Ridge, NJ: Noyes Data Corp., 1986), 1069 p.
3. Dean, J.A., Ed., *Lange's Handbook of Chemistry*, 11th ed. (New York: McGraw-Hill, Inc., 1973), 1570 p.
4. "Treatability Manual - Volume 1: Treatability Data," Office of Research and Development, U.S. EPA Report-600/8-80-042a (1980), 1035 p.
5. Leuenberger, C., Ligocki, M.P., and J.F. Pankow. "Trace Organic Compounds in Rain. 4. Identities, Concentrations, and Scavenging Mechanisms for Phenols in Urban Air and Rain," *Environ. Sci. Technol.*, 19(11):1053-1058 (1985).
6. Karickhoff, S.W., Brown, D.S., and T.A. Scott. "Sorption of Hydrophobic Pollutants on Natural Sediments," *Water Res.*, 13:241-248 (1979).
7. Roberts, P.V. "Nature of Organic Contaminants in Groundwater and Approaches to Treatment," Proceedings from the American Water Works Association Seminar on Organic Chemical Contaminants in Groundwater: Transport and Removal, St. Louis, MO (1981), pp 47-65.
8. Banerjee, S., Howard, P.H., Rosenburg, A.M., Dombrowski, A.E., Sikka, H., and D.L. Tullis. "Development of a General Kinetic Model for Biodegradation and its Application to Chlorophenols and Related Compounds," *Environ. Sci. Technol.*, 18(6):416-422 (1984).
9. Krijgsheld, K.R., and A. van der Gen. "Assessment of the Impact of the Emission of Certain Organochlorine Compounds on the Aquatic Environment - Part I: Monochlorophenols and

2,4-Dichlorophenol," *Chemosphere*, 15(7):825-860 (1986).

10. Schellenberg, K., Leuenberger, C., and R.P. Schwarzenbach. "Sorption of Chlorinated Phenols by Natural Sediments and Aquifer Materials," *Environ. Sci. Technol.*, 18(9):652-657 (1984).

11. "Chemical, Physical, and Biological Properties of Compounds Present at Hazardous Waste Sites," U.S. EPA Report-530/SW-89-010 (1985), 619 p.

12. *Toxic and Hazardous Industrial Chemicals Safety Manual for Handling and Disposal with Toxicity and Hazard Data* (Tokyo, Japan: International Technical Information Institute, 1986), 700 p.

13. Verschueren, K. *Handbook of Environmental Data on Organic Chemicals* (New York: Van Nostrand Reinhold Co., 1983), 1310 p.

14. Caturla, F., Martin-Martinez, J.M., Molina-Sabio, M., Rodriguez-Reinoso, F., and R. Torregrosa. "Adsorption of Substituted Phenols on Activated Carbon," *J. Colloid Interface Sci.*, 124(2):528-534 (1988).

15. Hwang, H.-M., Hodson, R.E., and R.F. Lee. "Degradation of Phenol and Chlorophenols by Sunlight and Microbes in Estuarine Water," *Environ. Sci. Technol.*, 20(10):1002-1007 (1986).

16. Kohring, G.-W., Rogers, J.E., and J. Wiegel. "Anaerobic Biodegradation of 2,4-Dichlorophenol in Freshwater Lake Sediments at Different Temperatures," *Appl. Environ. Microbiol.*, 55(2):348-353 (1989).

17. Callahan, M.A., Slimak, M.W., Gable, N.W., May, I.P., Fowler, C.F., Freed, J.R., Jennings, P., Durfee, R.L., Whitmore, F.C., Maestri, B., Mabey, W.R., Holt, B.R., and C. Gould. "Water-Related Environmental Fate of 129 Priority Pollutants Volumes I and II," National Technical Information Service, U.S. EPA Report-440/4-79-029 (1979), 1160 p.

1,2-DICHLOROPROPANE

Synonyms: α,β-Dichloropropane; ENT 15,406; NCI-C55141; Propylene chloride; Propylene dichloride; α,β-Propylene dichloride; RCRA waste number U083.

Structural Formula:

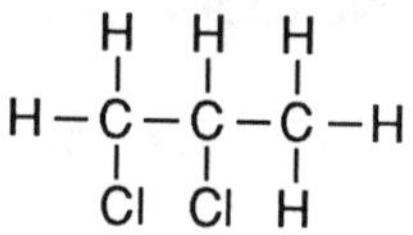

$$
\begin{array}{ccc}
H & H & H \\
| & | & | \\
H-C-C-C-H \\
| & | & | \\
Cl & Cl & H
\end{array}
$$

CHEMICAL DESIGNATIONS

CAS Registry Number: 78-87-5

DOT Designation: 1279

Empirical Formula: $C_3H_6Cl_2$

Formula Weight: 112.99

RTECS Number: TX 9625000

PHYSICAL AND CHEMICAL PROPERTIES

Appearance and Odor: Colorless liquid with a sweet, chloroform-like odor.

Boiling Point: 96.4 °C [1].

Henry's Law Constant: 0.0023 atm·m^3/mol [2]; 0.00294 atm·m^3/mol at 25 °C [3]; 0.00471 atm·m^3/mol at 37 °C [4].

Ionization Potential: 10.87 eV [5].

Log K_{oc}: 1.71 [6]; 1.431 [7].

Log K_{ow}: 2.28 [8].

Melting Point: -100.4 °C [1]; -70 °C [9].

Solubility in Organics: Miscible with organic solvents [10].

Solubility in Water: 2,700 mg/L at 20 °C [11]; 0.26 wt% at 20 °C [5]; 0.280 wt% at 25 °C [12]; 2,800 mg/L at 25 °C [13].

Specific Density: 1.560 at 20/4 °C [1]; 1.1593 at 20/20 °C [14]; 1.158 at 20/4 °C [15].

Transformation Products: No data found.

Vapor Density: 4.62 g/L at 25 °C, 3.90 (air = 1).

Vapor Pressure: 42 mm at 20 °C, 50 mm at 25 °C, 66 mm at 30 °C [16]; 1 mm at -38.5 °C, 10 mm at -6.1 °C, 40 mm at 19.4 °C, 100 mm at 39.4 °C, 400 mm at 76.0 °C, 760 mm at 96.8 °C [1].

FIRE HAZARDS

Flash Point: 15.6 °C [5].

Lower Explosive Limit (LEL): 3.4% [5].

Upper Explosive Limit (UEL): 14.5% [5].

HEALTH HAZARD DATA

Immediately Dangerous to Life or Health (IDLH): 2,000 ppm [5].

Permissible Exposure Limits (PEL) in Air: 75 ppm ($\approx$350 mg/m^3) [17]; 75 ppm TLV, 110 ppm ($\approx$510 mg/m^3) STEL [18].

MANUFACTURING

Selected Manufacturers:

> Jefferson Chemical Co., Inc.
> 3336 Richmond Ave.
> Houston, TX 77052

> Union Carbide Corp.
> 270 Park Ave.
> New York, NY 10017

Uses: Preparation of tetrachloroethylene and carbon tetrachloride; lead scavenger for antiknock fluids; metal cleanser; soil fumigant for nematodes; solvent for oils, fats, gums, waxes, and resins; spotting agent.

REFERENCES

1. Weast, R.C., Ed. *CRC Handbook of Chemistry and Physics*, 67th ed. (Boca Raton, FL: CRC Press, Inc., 1986), 2406 p.
2. Pankow, J.F., and M.E. Rosen. "Determination of Volatile Compounds in Water by Purging Directly to a Capillary Column with Whole Column Cryotrapping," *Environ. Sci. Technol.*, 22(4):398-405 (1988).
3. Hine, J., and P.K. Mookerjee. "The Intrinsic Hydrophilic Character of Organic Compounds. Correlations in Terms of Structural Contributions," *J. Org. Chem.*, 40(3):292-298 (1975).
4. Sato, A., and T. Nakajima. "A Structure-Activity Relationship of Some Chlorinated Hydrocarbons," *Arch. Environ. Health*, 34(2):69-75 (1979).
5. "NIOSH Pocket Guide to Chemical Hazards," U.S. Department of Health and Human Services, U.S. Government Printing Office (1987), 241 p.
6. Schwille, F. *Dense Chlorinated Solvents* (Chelsea, MI: Lewis Publishers, Inc., 1988), 146 p.
7. Chiou, C.T., Peters, L.J., and V.H. Freed. "A Physical Concept of Soil-Water Equilibria for Nonionic Organic Compounds," *Science*, 206:831-832 (1979).
8. Mills, W.B., Porcella, D.B., Ungs, M.J., Gherini, S.A., Summers, K.V., Mok, L., Rupp, G.L., and G.L. Bowie. "Water Quality Assessment: A Screening Procedure for Toxic and Conventional Pollutants in Surface and Groundwater-Part I," Office of Research and Development, U.S. EPA Report-600/6-85-002a (1985), 638 p.
9. Melnikov, N.N. *Chemistry of Pesticides* (New York: Springer-Verlag, Inc., 1971), 480 p.
10. "Chemical, Physical, and Biological Properties of Compounds Present at Hazardous Waste Sites," U.S. EPA Report-530/SW-89-010 (1985), 619 p.
11. Gunther, F.A., Westlake, W.E., and P.S. Jaglan. "Reported Solubilities of 738 Pesticide Chemicals in Water," *Res. Rev.*, 20:1-148 (1968).
12. Stephen, H., and T. Stephen. *Solubilities of Inorganic and Organic Compounds - Part 1, Volume 1* (London: Pergamon Printing and Art Services, Ltd., 1963), 960 p.
13. Gross, P. "The Determination of the Solubility of Slightly Soluble

Liquids in Water and the Solubilities of the Dichloro- Ethanes and -Propanes," *J. Am. Chem. Soc.*, 51(8):2362-2366 (1929).

14. Sax, N.I. *Dangerous Properties of Industrial Materials* (New York: Van Nostrand Reinhold Co., 1984), 3124 p.

15. *Fluka Catalog 1988/89 - Chemika-Biochemika* (Ronkonkoma, NY: Fluka Chemical Corp., 1988), 1536 p.

16. Verschueren, K. *Handbook of Environmental Data on Organic Chemicals* (New York: Van Nostrand Reinhold Co., 1983), 1310 p.

17. "General Industry Standards for Toxic and Hazardous Substances," U.S. Code of Federal Regulations 1910, Subpart Z, Section 1910.1000 (July 1982).

18. *Documentation of the Threshold Limit Values and Biological Exposure Indices for 1986-1987* (Cincinnati, OH: American Conference of Governmental Industrial Hygienists, 1986), 111 p.

cis-1,3-DICHLOROPROPYLENE

Synonyms: *cis*-1,3-Dichloropropene; *cis*-1,3-Dichloro-1-propene; (Z)-1,3-Dichloropropene; **(Z)-1,3-Dichloro-1-propene;** 1,3-Dichloroprop-1-ene; *cis*-1,3-Dichloro-1-propylene.

Structural Formula:

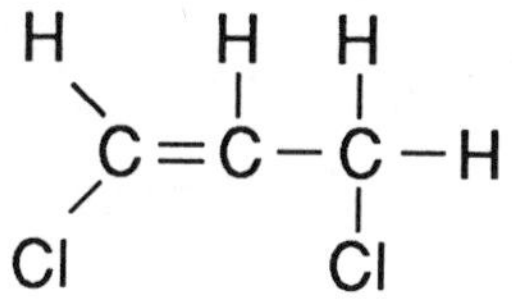

$$H_2C=CH-CH_2-H$$

Cl   Cl

CHEMICAL DESIGNATIONS

CAS Registry Number: 10061-01-5

DOT Designation: 2047 for a mixture containing *cis*- and *trans*- isomers.

Empirical Formula: $C_3H_4Cl_2$

Formula Weight: 110.97

RTECS Number: UC 8325000

PHYSICAL AND CHEMICAL PROPERTIES

Appearance and Odor: Colorless liquid with a chloroform-like odor.

Boiling Point: 104.3 °C [1].

Henry's Law Constant: 0.0013 atm·m^3/mol [2]; 0.00355 atm·m^3/mol [3].

Ionization Potential: No data found.

Log K_{oc}: 1.68 [4]; 1.36 [5].

Log K_{ow}: 1.41 [6].

Melting Point: -84 °C for the mixture containing *cis*- and *trans*- isomers [6].

Solubility in Organics: Soluble in benzene, chloroform, and ether [7].

Solubility in Water: 2,700 mg/L at 25 °C [8]; 1,000 mg/L at 20 °C [4]; 2,700 mg/L at 20 °C [9].

Specific Density: 1.224 at 20/4 °C [10].

Transformation Products: No data found.

Vapor Density: 4.54 g/L at 25 °C, 3.83 (air = 1).

Vapor Pressure: 43 mm at 25 °C [8]; 25 mm at 20 °C [4].

FIRE HAZARDS

Flash Point: 35 °C includes *trans*- isomer [11].

Lower Explosive Limit (LEL): 5.3% includes *trans*- isomer [11].

Upper Explosive Limit (UEL): 14.5% includes *trans*- isomer [11].

HEALTH HAZARD DATA

Immediately Dangerous to Life or Health (IDLH): No data found.

Permissible Exposure Limits (PEL) in Air: No standards set.

MANUFACTURING

Selected Manufacturer:

> Pfaltz & Bauer, Inc.
> 172 East Aurora St.
> Waterbury, CT 06708

Uses: A mixture containing *cis*- and *trans*- isomers is used as a soil fumigant and a nematocide.

REFERENCES

1. Weast, R.C., Ed. *CRC Handbook of Chemistry and Physics*, 67th ed.

(Boca Raton, FL: CRC Press, Inc., 1986), 2406 p.

2. Pankow, J.F., and M.E. Rosen. "Determination of Volatile Compounds in Water by Purging Directly to a Capillary Column with Whole Column Cryotrapping," *Environ. Sci. Technol.*, 22(4):398-405 (1988).

3. "Treatability Manual - Volume 1: Treatability Data," Office of Research and Development, U.S. EPA Report-600/8-80-042a (1980), 1035 p.

4. Schwille, F. *Dense Chlorinated Solvents* (Chelsea, MI: Lewis Publishers, Inc., 1988), 146 p.

5. Kenaga, E.E. "Predicted Bioconcentration Factors and Soil Sorption Coefficients of Pesticides and other Chemicals," *Ecotoxicol. Environ. Safety*, 4(1):26-38 (1980).

6. Krijgsheld, K.R., and A. van der Gen. "Assessment of the Impact of the Emission of Certain Organochlorine Compounds on the Aquatic Environment - Part II: Allylchloride, 1,3- and 2,3-Dichloropropene," *Chemosphere*, 15(7):861-880 (1986).

7. "Chemical, Physical, and Biological Properties of Compounds Present at Hazardous Waste Sites," U.S. EPA Report-530/SW-89-010 (1985), 619 p.

8. Verschueren, K. *Handbook of Environmental Data on Organic Chemicals* (New York: Van Nostrand Reinhold Co., 1983), 1310 p.

9. Dilling, W.L. "Interphase Transfer Processes. II. Evaporation Rates of Chloro Methanes, Ethanes, Ethylenes, Propanes, and Propylenes from Dilute Aqueous Solutions. Comparisons with Theoretical Predictions," *Environ. Sci. Technol.*, 11(4):405-409 (1977).

10. Melnikov, N.N. *Chemistry of Pesticides* (New York: Springer-Verlag, Inc., 1971), 480 p.

11. *Fire Protection Guide on Hazardous Materials* (Quincy, MA: National Fire Protection Association, 1984), 443 p.

trans-1,3-DICHLOROPROPYLENE

Synonyms: (*e*)-1,3-Dichloropropene; *trans*-1,3-Dichloropropene; (*E*)-1,3-Dichloro-1-propene; *trans*-1,3-Dichloro-1-propene; 1,3-Dichloroprop-1-ene; *trans*-1,3-Dichloro-1-propylene.

Structural Formula:

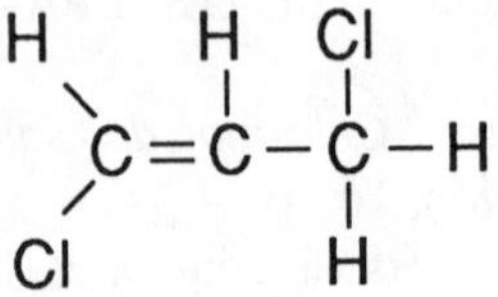

$$\begin{array}{ccccc} H & & H & Cl & \\ \diagdown & & | & | & \\ C & = & C & - C & - H \\ \diagup & & & | & \\ Cl & & & H & \end{array}$$

CHEMICAL DESIGNATIONS

CAS Registry Number: 10061-02-6

DOT Designation: 2047 for the mixture *cis-* and *trans-* isomers.

Empirical Formula: $C_3H_4Cl_2$

Formula Weight: 110.97

RTECS Number: UC 8320000

PHYSICAL AND CHEMICAL PROPERTIES

Appearance and Odor: Liquid with a chloroform-like odor.

Boiling Point: 77 °C at 757 mm [1]; 112.1 °C [2].

Henry's Law Constant: 0.0013 atm·m³/mol [3]; 0.00355 atm·m³/mol [4].

Ionization Potential: No data found.

Log K_{oc}: 1.68 [5]; 1.415 [6].

Log K_{ow}: 1.41 [7].

Melting Point: -84 °C for the mixture containing *cis-* and *trans-* isomers [7].

Solubility in Organics: Soluble in benzene, chloroform, and ether [8].

Solubility in Water: 2,800 mg/L at 25 °C [9]; 1,000 mg/L at 20 °C [5]; 2,800 mg/L at 20 °C [10].

Specific Density: 1.1818 at 20/4 °C [1]; 1.217 at 20/4 °C [7].

Transformation Products: No data found.

Vapor Density: 4.54 g/L at 25 °C, 3.83 (air = 1).

Vapor Pressure: 34 mm at 25 °C [9]; 25 mm at 20 °C [5].

FIRE HAZARDS

Flash Point: 5.3 °C includes *cis*- isomer [11].

Lower Explosive Limit (LEL): 5.3% includes *cis*- isomer [11].

Upper Explosive Limit (UEL): 14.5% includes *cis*- isomer [11].

HEALTH HAZARD DATA

Immediately Dangerous to Life or Health (IDLH): No data found.

Permissible Exposure Limits (PEL) in Air: No standards set.

MANUFACTURING

Selected Manufacturer:

> Pfaltz & Bauer, Inc.
> 172 East Aurora St.
> Waterbury, CT 06708

Uses: A mixture containing *cis*- and *trans*- isomers is used as a soil fumigant and a nematocide.

REFERENCES

1.   Weast, R.C., Ed. *CRC Handbook of Chemistry and Physics*, 67th ed.

(Boca Raton, FL: CRC Press, Inc., 1986), 2406 p.

2. Melnikov, N.N. *Chemistry of Pesticides* (New York: Springer-Verlag, Inc., 1971), 480 p.

3. Pankow, J.F., and M.E. Rosen. "Determination of Volatile Compounds in Water by Purging Directly to a Capillary Column with Whole Column Cryotrapping," *Environ. Sci. Technol.*, 22(4):398-405 (1988).

4. "Treatability Manual - Volume 1: Treatability Data," Office of Research and Development, U.S. EPA Report-600/8-80-042a (1980), 1035 p.

5. Schwille, F. *Dense Chlorinated Solvents* (Chelsea, MI: Lewis Publishers, Inc., 1988), 146 p.

6. Kenaga, E.E. "Predicted Bioconcentration Factors and Soil Sorption Coefficients of Pesticides and other Chemicals," *Ecotoxicol. Environ. Safety*, 4(1):26-38 (1980).

7. Krijgsheld, K.R., and A. van der Gen. "Assessment of the Impact of the Emission of Certain Organochlorine Compounds on the Aquatic Environment - Part II: Allylchloride, 1,3- and 2,3-Dichloropropene," *Chemosphere*, 15(7):861-880 (1986).

8. "Chemical, Physical, and Biological Properties of Compounds Present at Hazardous Waste Sites," U.S. EPA Report-530/SW-89-010 (1985), 619 p.

9. Verschueren, K. *Handbook of Environmental Data on Organic Chemicals* (New York: Van Nostrand Reinhold Co., 1983), 1310 p.

10. Dilling, W.L. "Interphase Transfer Processes. II. Evaporation Rates of Chloro Methanes, Ethanes, Ethylenes, Propanes, and Propylenes from Dilute Aqueous Solutions. Comparisons with Theoretical Predictions," *Environ. Sci. Technol.*, 11(4):405-409 (1977).

11. *Fire Protection Guide on Hazardous Materials* (Quincy, MA: National Fire Protection Association, 1984), 443 p.

DIELDRIN

Synonyms: Alvit; Compound 497; Dieldrite; Dieldrix; ENT 16,225; HEOD; Hexachloroepoxyoctahydro-*endo-exo*-dimethanonaphthalene; 1,2,3,4,10,10-Hexachloro-6,7-epoxy-1,4,4a,5,6,7,8,8a-octahydro-1,4-*endo-exo*-5,8-di-methanonaphthalene; **3,4,5,6,9,9-Hexachloro-1a,2,2a,3,6,6a,7,7a-octahydro-2,7:3,6-dimethanonaphth[2,3-*b*]oxirene**; Illoxol; Insecticide number 497; NA 2,761; NCI-C00124; Octalox; Panoram D-31; Quintox; RCRA waste number P037.

Structural Formula:

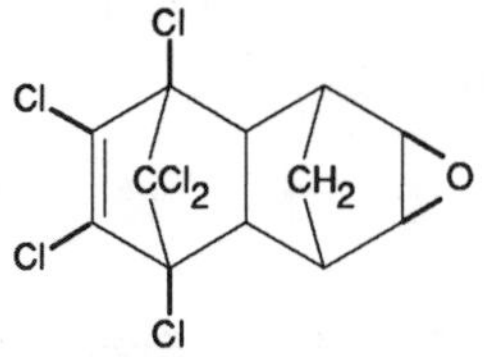

CHEMICAL DESIGNATIONS

CAS Registry Number: 60-57-1

DOT Designation: 2761

Empirical Formula: $C_{12}H_8Cl_6O$

Formula Weight: 380.91

RTECS Number: IO 1750000

PHYSICAL AND CHEMICAL PROPERTIES

Appearance and Odor: White crystals or pale tan flakes with an odorless to mild chemical odor.

Boiling Point: Decomposes [1].

Henry's Law Constant: 2 x 10^{-7} atm·m^3/mol [2]; 5.8 x 10^{-5} atm·m^3/mol [3]; 0.807-3.18 x 10^{-5} atm·m^3/mol [4].

Ionization Potential: No data found.

Log K_{oc}: 4.55 [5]; 4.08 [6].

Log K_{ow}: 5.16 [7]; 5.48 [8]; 4.32 [9]; 4.49, 4.51, 4.55, 4.66 [10]; 3.692 [11].

Melting Point: 175-176 °C [1]; 150 °C [12].

Solubility in Organics: Soluble in ethanol and benzene [1].

Solubility in Water: 0.1 mg/L [13]; 200 ppb at 20 °C [14]; 0.186 mg/L at 20 °C [15]; 0.05 mg/L at 26 °C [16]; 0.20 mg/L at 25 °C [17]; 0.022 ppm [18]; 90 ppb at 15 °C, 195 ppb at 25 °C, 400 ppb at 35 °C, 650 ppb at 45 °C (particle sizes $\leq$ 5μ) [19]; at room temperature, solubilities of 180 ppb (maximum particle size 5μ) and 140 ppb (maximum particle size 0.04μ) were reported [20]; 0.05 mg/L [21].

Specific Density: 1.75 at 20/4 °C [22].

Transformation Products: Identified metabolites of dieldrin from solution cultures containing *Pseudomonas* sp. in soils include aldrin and dihydroxydihydroaldrin. Other unidentified byproducts included a ketone, an aldehyde, and an acid [23].

Vapor Density: 54 ng/L at 20 °C, 202 ng/L at 30 °C, 676 ng/L at 40 °C [24]; 210 ng/L at 30 °C [25].

Vapor Pressure: 3.1 x 10^{-6} mm at 20 °C [26]; 1.8 x 10^{-7} mm at 25 °C [13]; 1.78 x 10^{-7} mm at 20 °C [27]; 2.8 x 10^{-6} mm at 20 °C [24]; 1.0 x 10^{-5} mm at 30 °C [25]; 7.78 x 10^{-7} at 20.25 °C [21].

FIRE HAZARDS

Flash Point: Not flammable [22].

Lower Explosive Limit (LEL): Not flammable [22].

Upper Explosive Limit (UEL): Not flammable [22].

HEALTH HAZARD DATA

Immediately Dangerous to Life or Health (IDLH): 450 mg/m^3 (carcinogen) [22].

Permissible Exposure Limits (PEL) in Air: 0.25 mg/m^3 [28]; lowest detectable limit (0.15 mg/m^3 time weighted average by NIOSH validated method) [29]; 0.25 mg/m^3 TWA [30].

MANUFACTURING

Selected Manufacturers:

> Shell Chemical Co.
> Agricultural Division
> 2401 Crow Canyon Rd.
> San Ramon, CA 94583

Uses: Insecticide; wool processing industry.

REFERENCES

1. Weast, R.C., Ed. *CRC Handbook of Chemistry and Physics*, 67th ed. (Boca Raton, FL: CRC Press, Inc., 1986), 2406 p.
2. Eisenreich, S.J., Looney, B.B., and J.D. Thornton. "Airborne Organic Contaminants in the Great Lakes Ecosystem," *Environ. Sci. Technol.*, 15(1):30-38 (1981).
3. Warner, H.P., Cohen, J.M., and J.C. Ireland. "Determination of Henry's Law Constants of Selected Priority Pollutants," Office of Science and Development, U.S. EPA Report-600/D-87/229 (1987), 14 p.
4. Jury, W.A., Spencer, W.F., and W.J. Farmer. "Behavior Assessment Model for Trace Organics in Soil: IV. Review of Experimental Evidence," *J. Environ. Qual.*, 13(4):580-586 (1984).
5. Kenaga, E.E. "Correlation of Bioconcentration Factors of Chemicals in Aquatic and Terrestrial Organisms with Their Physical and Chemical Properties," *Environ. Sci. Technol.*, 14(5):553-556 (1980).
6. Jury, W.A., Spencer, W.F., and W.J. Farmer. "Behavior Assessment Model for Trace Organics in Soil: III. Application of Screening Model," *J. Environ. Qual.*, 13(4):573-579 (1984).
7. Travis, C.C., and A.D. Arms. "Bioconcentration of Organics in Beef, Milk and Vegetation," *Environ. Sci. Technol.*, 22(3):271-274 (1988).
8. Mackay, D. "Correlation of Bioconcentration Factors," *Environ. Sci. Technol.*, 16(5):274-278 (1982).
9. Geyer, H.J., Scheunert, I., and F. Korte. "Correlation Between the Bioconcentration Potential of Organic Environmental Chemicals in Humans and Their *n*-Octanol/Water Partition Coefficients,"

Chemosphere, 16(1):239-252 (1987).

10. Brooke, D.N., Dobbs, A.J., and N. Williams. "Octanol:Water Partition Coefficients (P): Measurement, Estimation, and Interpretation, Particularly for Chemicals with P > 10^5," *Ecotoxicol. Environ. Safety*, 11(3):251-260 (1986).

11. Rao, P.S.C., and J.M. Davidson. "Estimation of Pesticide Retention and Transformation Parameters Required in Nonpoint Source Pollution Models," in *Environmental Impact of Nonpoint Source Pollution*, Overcash, M.R., and J.M. Davidson, Eds. (Ann Arbor, MI: Ann Arbor Science Publishers, Inc., 1980), pp 23-67.

12. Sax, N.I. *Dangerous Properties of Industrial Materials* (New York: Van Nostrand Reinhold Co., 1984), 3124 p.

13. Verschueren, K. *Handbook of Environmental Data on Organic Chemicals* (New York: Van Nostrand Reinhold Co., 1983), 1310 p.

14. Weil, L., Dure, G., and K.E. Quentin. "Solubility in Water of Insecticide Chlorinated Hydrocarbons and Polychlorinated Biphenyls in View of Water Pollution," *Z. Wasser Forsch.*, 7(6):169-175 (1974).

15. Worthing, C.R., and S.B. Walker, Eds. *The Pesticide Manual - A World Compendium*, 7th ed. (Lavenham, Suffolk, Great Britain: The Lavenham Press Ltd., 1983), 695 p.

16. Melnikov, N.N. *Chemistry of Pesticides* (New York: Springer-Verlag, Inc., 1971), 480 p.

17. Walton, W.C. *Practical Aspects of Ground Water Modeling* (Worthington, OH: National Water Well Association, 1985), 587 p.

18. Kenaga, E.E., and C.A.I. Goring. "Relationship Between Water Solubility, Soil Sorption, Octanol-Water Partitioning and Concentration of Chemicals in Biota," in *Aquatic Toxicology, ASTM STP 707*, Eaton, J.G., Parrish, P.R., and A.C. Hendricks, Eds. (Philadelphia, PA: American Society for Testing and Materials, 1980), pp 78-115.

19. Biggar, J.W., and I.R. Riggs. "Apparent Solubility of Organochlorine Insecticides in Water at Various Temperatures," *Hilgardia*, 42(10):383-391 (1974).

20. Robeck, G.G., Dostal, K.A., Cohen, J.M., and J.F. Kreissl. "Effectiveness of Water Treatment Processes in Pesticide Removal," *J. Am. Water Works Assoc.*, 57(2):181-200 (1965).

21. Gile, J.D., and J.W. Gillett. "Fate of Selected Fungicides in a Terrestrial Laboratory Ecosystem," *J. Agric. Food Chem.*, 27(6):1159-1164 (1979).

22. Weiss, G. *Hazardous Chemicals Data Book* (Park Ridge, NJ: Noyes Data Corp., 1986), 1069 p.

23. Kearney, P.C., and D.D. Kaufman. *Herbicides: Chemistry,*

Degradation and Mode of Action (New York: Marcel Dekker, Inc., 1976), 1036 p.

24. Spencer, W.F., and M.M. Cliath. "Vapor Density of Dieldrin," *Environ. Sci. Technol.*, 3(7):670-674 (1969).

25. Tinsley, L.J. *Chemical Concepts in Pollutant Behavior* (New York: John Wiley and Sons, Inc., 1979), 265 p.

26. Windholz, M., Budavari, S., Blumetti, R.F., and E.S. Otterbein, Eds. *The Merck Index*, 10th ed., (Rahway, NJ: Merck and Co., 1983), 1463 p.

27. "Treatability Manual - Volume 1: Treatability Data," Office of Research and Development, U.S. EPA Report-600/8-80-042a (1980), 1035 p.

28. "General Industry Standards for Toxic and Hazardous Substances," U.S. Code of Federal Regulations 1910, Subpart Z Section 1910.1000 (July 1982).

29. "NIOSH Pocket Guide to Chemical Hazards," U.S. Department of Health and Human Services, U.S. Government Printing Office (1987), 241 p.

30. *Documentation of the Threshold Limit Values and Biological Exposure Indices for 1986-1987* (Cincinnati, OH: American Conference of Governmental Industrial Hygienists, 1986), 111 p.

DIETHYL PHTHALATE

Synonyms: Anozol; **1,2-Benzenedicarboxylic acid, diethyl ester**; DEP; Diethyl-*o*-phthalate; Estol 1550; Ethyl phthalate; NCI-C60048; Neantine; Palatinol A; Phthalol; Placidol E; RCRA waste number U088; Solvanol.

Structural Formula:

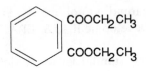

CHEMICAL DESIGNATIONS

CAS Registry Number: 84-66-2

DOT Designation: None assigned.

Empirical Formula: $C_{12}H_{14}O_4$

Formula Weight: 222.24

RTECS Number: TI 1050000

PHYSICAL AND CHEMICAL PROPERTIES

Appearance and Odor: Clear, colorless liquid with a mild chemical odor.

Boiling Point: 298 °C [1]; 302 °C [2]; 296 °C [3].

Henry's Law Constant: 8.46 x 10^{-7} atm·m³/mol [4].

Ionization Potential: No data found.

Log K_{oc}: 1.84 [5].

Log K_{ow}: 3.00 [6]; 2.35 [7]; 2.47 [8]; 2.24 [9]; 1.40 [10].

Melting Point: -40.5 °C [11].

Solubility in Organics: Soluble in acetone and benzene; miscible with ethanol, ether, esters, and ketones [12].

Solubility in Water: 210 mg/L [11]; 896 mg/L at 25 °C [13]; 600 mg/L at 20 °C [14]; 1,000 mg/L at 25 °C [15]; 928 mg/L at 20 °C [16]; 1,080 mg/L at 25 °C [9]; 1,200 ppm at 25 °C [17]; 0.1 wt% at 20 °C [18]; 680 mg/L at 25 °C [5].

Specific Density: 1.1175 at 20/4 °C [1]; 1.123 at 25/4 °C [18].

Transformation Products: A proposed microbial degradation mechanism is as follows: 4-hydroxy-3-methylbenzyl alcohol, 4-hydroxy-3-methyl-benzaldehyde, 3-methyl-4-hydroxybenzoic acid, 4-hydroxy-isophthalic acid, protocatechuic acid, and β-ketoadipic acid [19]. In anaerobic sludge, diethyl phthalate degraded as follows: monoethyl phthalate to phthalic acid to protocatechuic acid followed by ring cleavage and mineralization [20].

Vapor Density: 9.08 g/L at 25 °C, 7.67 (air = 1).

Vapor Pressure: 14 mm at 163 °C, 30 mm at 182 °C, 734 mm at 295 °C [11]; 1 mm at 108.8 °C, 10 mm at 156.0 °C, 40 mm at 192.1 °C, 100 mm at 219.5 °C, 400 mm at 267.5 °C, 760 mm at 294.0 °C [1]; 0.05 mm at 70 °C [18]; 0.22 (± 0.7) Pa at 25 °C [9].

FIRE HAZARDS

Flash Point: 140 °C [21]; 163 °C (open cup) [2].

Lower Explosive Limit (LEL): 0.7% at 186 °C [22].

Upper Explosive Limit (UEL): No data found.

HEALTH HAZARD DATA

Immediately Dangerous to Life or Health (IDLH): No data found.

Permissible Exposure Limits (PEL) in Air: 5 mg/m^3 TWA [23].

MANUFACTURING

Selected Manufacturers:

Aldrich Chemical Co.
940 West Saint Paul Ave.
Milwaukee, WI 53233

Monsanto Industrial Chemicals Co.
800 North Lindbergh Blvd.
St. Louis, MO 63166

Uses: Plasticizer; plastic manufacturing and processing; denaturant for ethyl alcohol; ingredient in insecticidal sprays and explosives (propellant); dye application agent; wetting agent; perfumery as fixative and solvent; solvent for nitrocellulose and cellulose acetate; camphor substitute.

REFERENCES

1. Weast, R.C., Ed. *CRC Handbook of Chemistry and Physics*, 67th ed. (Boca Raton, FL: CRC Press, Inc., 1986), 2406 p.
2. Sax, N.I. *Dangerous Properties of Industrial Materials* (New York: Van Nostrand Reinhold Co., 1984), 3124 p.
3. Standen, A., Ed. *Kirk-Othmer Encyclopedia of Chemical Technology, Volume 15*, 2nd ed. (New York: John Wiley and Sons, Inc., 1968), 923 p.
4. "Treatability Manual - Volume 1: Treatability Data," Office of Research and Development, U.S. EPA Report-600/8-80-042a (1980), 1035 p.
5. Russell, D.J., and B. McDuffie. "Chemodynamic Properties of Phthalate Esters: Partitioning and Soil Migration," *Chemosphere*, 15(8):1003-1021 (1986).
6. Isnard, S., and S. Lambert. "Estimating Bioconcentration Factors from Octanol-Water Partition Coefficient and Aqueous Solubility," *Chemosphere*, 17(1):21-34 (1988).
7. Leyder, F., and P. Boulanger. "Ultraviolet Absorption, Aqueous Solubility and Octanol-Water Partition for Several Phthalates," *Bull. Environ. Contam. Toxicol.*, 30(2):152-157 (1983).
8. Mabey, W.R., Smith, J.H., Podoll, R.T., Johnson, H.L., Mill, T., Chou, T.-W., Gates, J., Partridge, I.W., Jaber, H., and D. Vandenberg. "Aquatic Fate Process Data for Organic Priority Pollutants - Final

Report," Office of Regulations and Standards, U.S. EPA Report-440/4-81-014 (1982), 407 p.

9. Howard, P.H., Banerjee, S., and K.H. Robillard. "Measurement of Water Solubilities, Octanol/Water Partition Coefficients and Vapor Pressures of Commercial Phthalate Esters," *Environ. Toxicol. Chem.*, 4:653-661 (1985).

10. Veith, G.D., Macek, K.J., Petrocelli, S.R., and J. Carroll. "An Evaluation of Using Partition Coefficients and Water Solubility to Estimate Bioconcentration Factors for Organic Chemicals in Fish," in *Aquatic Toxicology, ASTM STP 707*, Eaton, J.G., Parrish, P.R., and A.C. Hendricks, Eds. (Philadelphia, PA: American Society for Testing and Materials, 1980), pp 116-129.

11. Verschueren, K. *Handbook of Environmental Data on Organic Chemicals* (New York: Van Nostrand Reinhold Co., 1983), 1310 p.

12. "Chemical, Physical, and Biological Properties of Compounds Present at Hazardous Waste Sites," U.S. EPA Report-530/SW-89-010 (1985), 619 p.

13. Callahan, M.A., Slimak, M.W., Gable, N.W., May, I.P., Fowler, C.F., Freed, J.R., Jennings, P., Durfee, R.L., Whitmore, F.C., Maestri, B., Mabey, W.R., Holt, B.R., and C. Gould. "Water-Related Environmental Fate of 129 Priority Pollutants Volumes I and II," National Technical Information Service, U.S. EPA Report-440/4-79-029 (1979), 1160 p.

14. Meites, L., Ed. *Handbook of Analytical Chemistry*, 1st ed. (New York: McGraw-Hill, Inc., 1963), 1782 p.

15. Walton, W.C. *Practical Aspects of Ground Water Modeling* (Worthington, OH: National Water Well Association, 1985), 587 p.

16. Leyder, F., and P. Boulanger. "Ultraviolet Absorption, Aqueous Solubility and Octanol-Water Partition for Several Phthalates," *Bull. Environ. Contam. Toxicol.*, 30(2):152-157 (1983).

17. Fukano, I., and Y. Obata. "Solubility of Phthalates in Water," [Chemical abstract 120601u, 86(17):486 (1977)]: *Purasuchikkusu*, 27(7):48-49 (1976).

18. Fishbein, L., and P.W. Albro. "Chromatographic and Biological Aspects of the Phthalate Esters," *J. Chromatogr.*, 70(2):365-412 (1972).

19. Chapman, P.J. "An Outline of Reaction Sequences Used for the Bacterial Degradation of Phenolic Compounds" in *Degradation of Synthetic Organic Molecules in the Biosphere: Natural, Pesticidal, and Various Other Man-Made Compounds* (Washington, DC: National Academy of Sciences, 1972), pp 17-55.

20. Shelton, D.R., Boyd, S.A., and J.M. Tiedje. "Anaerobic Biodegradation of Phthalic Acid Esters in Sludge," *Environ. Sci.*

Technol., 18(2):93-97 (1984).

21. Windholz, M., Budavari, S., Blumetti, R.F., and E.S. Otterbein, Eds. *The Merck Index,* 10th ed., (Rahway, NJ: Merck and Co., 1983), 1463 p.

22. *Fire Protection Guide on Hazardous Materials* (Quincy, MA: National Fire Protection Association, 1984), 443 p.

23. *Documentation of the Threshold Limit Values and Biological Exposure Indices for 1986-1987* (Cincinnati, OH: American Conference of Governmental Industrial Hygienists, 1986), 111 p.

2,4-DIMETHYLPHENOL

Synonyms: 4,6-Dimethylphenol; 2,4-DMP; 1-Hydroxy-2,4-dimethyl-benzene; 4-Hydroxy-1,3-dimethylbenzene; RCRA waste number U101; 1,3,4-Xylenol; 2,4-Xylenol; *m*-Xylenol.

Structural Formula:

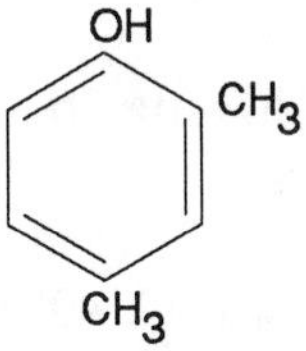

CHEMICAL DESIGNATIONS

CAS Registry Number: 105-67-9

DOT Designation: None assigned.

Empirical Formula: $C_8H_{10}O$

Formula Weight: 122.17

RTECS Number: ZE 5600000

PHYSICAL AND CHEMICAL PROPERTIES

Appearance: Colorless solid, slowly turning brown on exposure to air.

Boiling Point: 210 °C [1].

Dissociation Constant: 10.63 [2].

Henry's Law Constant: 6.55 x 10^{-6} atm·m³/mol at 25 °C (estimated) [3].

Ionization Potential: No data found.

Log K_{oc}: 2.07 (estimated) [4].

Log K_{ow}: 2.42 [5]; 2.50 [6]; 2.30 [7].

Melting Point: 27-28 °C [1].

Solubility in Organics: Freely soluble in ethanol, chloroform, ether, and benzene [8].

Solubility in Water: 4,200 mg/L at 20 °C [9]; 7,868 mg/L at 25 °C [10]; 6.2 g/L at 25 °C [3].

Specific Density: 0.9650 at 20/4 °C [1].

Transformation Products: No data found.

Vapor Pressure: As a supercooled liquid - 0.0621 mm at 20 °C [11]; 1 mm at 51.8 °C, 10 mm at 92.3 °C, 40 mm at 121.5 °C, 100 mm at 143.0 °C, 400 mm at 184.2 °C, 760 mm at 211.5 °C [1]; 0.098 mm at 25.0 °C [3].

FIRE HAZARDS

Flash Point: > 110 °C [12].

Lower Explosive Limit (LEL): No data found.

Upper Explosive Limit (UEL): No data found.

HEALTH HAZARD DATA

Immediately Dangerous to Life or Health (IDLH): No data found.

Permissible Exposure Limits (PEL) in Air: No standards set.

MANUFACTURING

Selected Manufacturers:

Aldrich Chemical Co.
940 West Saint Paul Ave.
Milwaukee, WI 53233

Fluka Chemical Corp.
980 South Second St.
Ronkonkoma, NY 11779

Pfaltz & Bauer, Inc.
172 East Aurora St.
Waterbury, CT 06708

Uses: Wetting agent; dyestuffs; preparation of phenolic antioxidants; plastics, resins, solvent, disinfectant, pharmaceuticals, insecticides, fungicides, and rubber chemicals manufacturing; lubricant and gasoline additive; possibly used a pesticide; plasticizers.

REFERENCES

1. Weast, R.C., Ed. *CRC Handbook of Chemistry and Physics*, 67th ed. (Boca Raton, FL: CRC Press, Inc., 1986), 2406 p.
2. Howard, P.H. *Handbook of Environmental Fate and Exposure Data for Organic Chemicals* (Chelsea, MI: Lewis Publishers, Inc., 1989), 574 p.
3. Leuenberger, C., Ligocki, M.P., and J.F. Pankow. "Trace Organic Compounds in Rain. 4. Identities, Concentrations, and Scavenging Mechanisms for Phenols in Urban Air and Rain," *Environ. Sci. Technol.*, 19(11):1053-1058 (1985).
4. Montgomery, J.H. Unpublished results (1989).
5. Veith, G.D., Macek, K.J., Petrocelli, S.R., and J. Carroll. "An Evaluation of Using Partition Coefficients and Water Solubility to Estimate Bioconcentration Factors for Organic Chemicals in Fish," in *Aquatic Toxicology, ASTM STP 707*, Eaton, J.G., Parrish, P.R., and A.C. Hendricks, Eds. (Philadelphia, PA: American Society for Testing and Materials, 1980), pp 116-129.
6. "Treatability Manual - Volume 1: Treatability Data," Office of Research and Development, U.S. EPA Report-600/8-80-042a (1980), 1035 p.
7. Mabey, W.R., Smith, J.H., Podoll, R.T., Johnson, H.L., Mill, T., Chou, T.-W., Gates, J., Partridge, I.W., Jaber, H., and D. Vandenberg. "Aquatic Fate Process Data for Organic Priority Pollutants - Final Report," Office of Regulations and Standards, U.S. EPA Report-440/4-81-014 (1982), 407 p.
8. "Chemical, Physical, and Biological Properties of Compounds Present at Hazardous Waste Sites," U.S. EPA Report-530/SW-89-010 (1985), 619 p.
9. Verschueren, K. *Handbook of Environmental Data on Organic Chemicals* (New York: Van Nostrand Reinhold Co., 1983), 1310 p.
10. Banerjee, S., Yalkowsky, S.H., and S.C. Valvani. "Water Solubility and Octanol/Water Partition Coefficients of Organics. Limitations

of the Solubility-Partition Coefficient Correlation," *Environ. Sci. Technol.*, 14(10):1227-1229 (1980).

11. Andon, R.J.L., Biddiscombe, D.P., Cox, F.D., Handley, R., Harrop, D., Herington, E.F.G., and J.F. Martin. "Thermodynamic Properties of Organic Oxygen Compounds. Part 1. Preparation of Physical Properties of Pure Phenol, Cresols and Xylenols," *J. Chem. Soc. (London)*, (1960), pp 5246-5254.

12. *Catalog Handbook of Fine Chemicals* (Milwaukee, WI: Aldrich Chemical Co., 1988), 2212 p.

DIMETHYL PHTHALATE

Synonyms: Avolin; **1,2-Benzenedicarboxylic acid, dimethyl ester;** Dimethyl-1,2-benzenedicarboxylate; Dimethylbenzeneorthodicarboxylate; DMP; ENT 262; Fermine; Methyl phthalate; Mipax; NTM; Palatinol M; Phthalic acid, dimethyl ester; Phthalic acid, methyl ester; RCRA waste number U102; Solvanom; Solvarone.

Structural Formula:

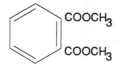

CHEMICAL DESIGNATIONS

CAS Registry Number: 131-11-3

DOT Designation: None assigned.

Empirical Formula: $C_{10}H_{10}O_4$

Formula Weight: 194.19

RTECS Number: TI 1575000

PHYSICAL AND CHEMICAL PROPERTIES

Appearance and Odor: Colorless, odorless, moderately viscous liquid.

Boiling Point: 283.8 °C [1]; 282 °C [2].

Henry's Law Constant: 4.2 x 10^{-7} atm·m^3/mol [3].

Ionization Potential: 9.75 eV [4].

Log K_{oc}: A value of 1.63 was found using an agricultural soil having a pH of 7.4. Values 0.88 and 1.84 were determined using two forest soils with a pH of 5.6 and 4.2, respectively [5]; 2.28 [6].

Log K_{ow}: 2.00 [7]; 1.53 [8]; 1.56 [9]; 1.47 [10]; 1.61 [11]; 1.86 [12].

Melting Point: 0 °C [13]; 5.5 °C [14].

Solubility in Organics: Soluble in ethanol, ether, and benzene [1].

Solubility in Water: 1,744 mg/L [15]; 4,320 mg/L at 25 °C [16]; 4,290 mg/L at 20 °C [8]; 4,000 mg/L at 25 °C, 3,960 mg/L in well water at 25 °C, 3,160 mg/L in natural seawater at 25 °C [10]; 45,000 ppm at 25 °C [17]; 0.5 wt% at 20 °C [13].

Specific Density: 1.1905 at 20/4 °C [1]; 1.189 at 25/25 °C [18]; 1.192 at 20/4 °C [2].

Transformation Products: In anaerobic sludge, degradation occurred as follows: monomethyl phthalate to phthalic acid to protocatechuic acid followed by ring cleavage and mineralization [19].

Vapor Density: 7.94 g/L at 25 °C, 6.70 (air = 1).

Vapor Pressure: < 0.01 mm at 20 °C [13]; 1 mm at 100.3 °C, 10 mm at 147.6 °C, 40 mm at 182.8 °C, 100 mm at 210.0 °C, 400 mm at 257.8 °C, 760 mm at 283.7 °C [1]; 0.22 (± 0.7) Pa at 25 °C [10].

FIRE HAZARDS

Flash Point: 146 °C [4].

Lower Explosive Limit (LEL): 1.2% at 146 °C [4].

Upper Explosive Limit (UEL): Unknown [4].

HEALTH HAZARD DATA

Immediately Dangerous to Life or Health (IDLH): 9,300 mg/m^3 [4].

Permissible Exposure Limits (PEL) in Air: 5 mg/m^3 [20]; 5 mg/m^3 TWA [21].

MANUFACTURING

Selected Manufacturers:

Aldrich Chemical Co.
940 West Saint Paul Ave.
Milwaukee, WI 53233

Fluka Chemical Corp.
980 South Second St.
Ronkonkoma, NY 11779

Uses: Plasticizer for cellulose acetate, nitrocellulose, resins, rubber, elastomers; ingredient in lacquers; coating agents; safety glass; insect repellant; molding powders; perfumes.

REFERENCES

1. Weast, R.C., Ed. *CRC Handbook of Chemistry and Physics*, 67th ed. (Boca Raton, FL: CRC Press, Inc., 1986), 2406 p.
2. Standen, A., Ed. *Kirk-Othmer Encyclopedia of Chemical Technology, Volume 15*, 2nd ed. (New York: John Wiley and Sons, Inc., 1968), 923 p.
3. Petrasek, A.C., Kugelman, I.J., Austern, B.M., Pressley, T.A., Winslow, L.A., and R.H. Wise. "Fate of Toxic Organic Compounds in Wastewater Treatment Plants," *J. Water Poll. Control Fed.*, 55(10):1286-1296 (1983).
4. "NIOSH Pocket Guide to Chemical Hazards," U.S. Department of Health and Human Services, U.S. Government Printing Office (1987), 241 p.
5. Seip, H.M., Alstad, J., Carlberg, G.E., Matinsen, K., and R. Skaane. "Measurement of Mobility of Organic Compounds in Soils," *Sci. Tot. Environ.*, 50:87-101 (1986).
6. Banerjee, P., Piwoni, M.D., and K. Ebeid. "Sorption of Organic Contaminants to a Low Carbon Subsurface Core," *Chemosphere*, 14(8):1057-1067 (1985).
7. Isnard, S., and S. Lambert. "Estimating Bioconcentration Factors from Octanol-Water Partition Coefficient and Aqueous Solubility," *Chemosphere*, 17(1):21-34 (1988).
8. Leyder, F., and P. Boulanger. "Ultraviolet Absorption, Aqueous Solubility and Octanol-Water Partition for Several Phthalates," *Bull. Environ. Contam. Toxicol.*, 30(2):152-157 (1983).

9. Mabey, W.R., Smith, J.H., Podoll, R.T., Johnson, H.L., Mill, T., Chou, T.-W., Gates, J., Partridge, I.W., Jaber, H., and D. Vandenberg. "Aquatic Fate Process Data for Organic Priority Pollutants - Final Report," Office of Regulations and Standards, U.S. EPA Report-440/4-81-014 (1982), 407 p.

10. Howard, P.H., Banerjee, S., and K.H. Robillard. "Measurement of Water Solubilities, Octanol/Water Partition Coefficients and Vapor Pressures of Commercial Phthalate Esters," *Environ. Toxicol. Chem.,* 4:653-661 (1985).

11. Veith, G.D., Macek, K.J., Petrocelli, S.R., and J. Carroll. "An Evaluation of Using Partition Coefficients and Water Solubility to Estimate Bioconcentration Factors for Organic Chemicals in Fish," in *Aquatic Toxicology, ASTM STP 707*, Eaton, J.G., Parrish, P.R., and A.C. Hendricks, Eds. (Philadelphia, PA: American Society for Testing and Materials, 1980), pp 116-129.

12. Eadsforth, C.V. "Application of Reverse-Phase H.P.L.C. for the Determination of Partition Coefficients," *Pestic. Sci.,* 17(3):311-325 (1986).

13. Fishbein, L., and P.W. Albro. "Chromatographic and Biological Aspects of the Phthalate Esters," *J. Chromatogr.,* 70(2):365-412 (1972).

14. "Treatability Manual - Volume 1: Treatability Data," Office of Research and Development, U.S. EPA Report-600/8-80-042a (1980), 1035 p.

15. Verschueren, K. *Handbook of Environmental Data on Organic Chemicals* (New York: Van Nostrand Reinhold Co., 1983), 1310 p.

16. Callahan, M.A., Slimak, M.W., Gable, N.W., May, I.P., Fowler, C.F., Freed, J.R., Jennings, P., Durfee, R.L., Whitmore, F.C., Maestri, B., Mabey, W.R., Holt, B.R., and C. Gould. "Water-Related Environmental Fate of 129 Priority Pollutants Volumes I and II," National Technical Information Service, U.S. EPA Report-440/4-79-029 (1979), 1160 p.

17. Fukano, I., and Y. Obata. "Solubility of Phthalates in Water," [Chemical abstract 120601u, 86(17):486 (1977)]: *Purasuchikkusu,* 27(7):48-49 (1976).

18. Sax, N.I. *Dangerous Properties of Industrial Materials* (New York: Van Nostrand Reinhold Co., 1984), 3124 p.

19. Shelton, D.R., Boyd, S.A., and J.M. Tiedje. "Anaerobic Biodegradation of Phthalic Acid Esters in Sludge," *Environ. Sci. Technol.,* 18(2):93-97 (1984).

20. "General Industry Standards for Toxic and Hazardous Substances," U.S. Code of Federal Regulations 1910, Subpart Z, Section 1910.1000 (July 1982).

21. *Documentation of the Threshold Limit Values and Biological Exposure Indices for 1986-1987* (Cincinnati, OH: American Conference of Governmental Industrial Hygienists, 1986), 111 p.

4,6-DINITRO-o-CRESOL

Synonyms: Antinonin; Antinonnon; Arborol; Capsine; Chemsect DNOC; Degrassan; Dekrysil; Detal; Dinitrocresol; Dinitro-*o*-cresol; 2,4-Dinitro-*o*-cresol; 3,5-Dinitro-*o*-cresol; Dinitrodendtroxal; 3,5-Dinitro-2-hydroxytoluene; Dinitrol; Dinitromethyl cyclohexyltrienol; 2,4-Dinitro-2-methylphenol; 2,4-Dinitro-6-methylphenol; 4,6-Dinitro-2-methylphenol; Dinitrosol; Dinoc; Dinurania; DN; DNC; DN-dry mix no. 2; DNOC; Effusan; Effusan 3,436; Elgetol; Elgetol 30; Elipol; ENT 154; Extrar; Hedolit; Hedolite; K III; K IV; Kresamone; Krezotol 50; Lipan; **2-Methyl-4,6-dinitrophenol**; 6-Methyl-2,4-dinitrophenol; Nitrador; Nitrofan; Prokarbol; Rafex; Rafex 35; Raphatox; RCRA waste number P047; Sandolin; Sandolin A; Selinon; Sinox; Trifina; Trifocide; Winterwash.

Structural Formula:

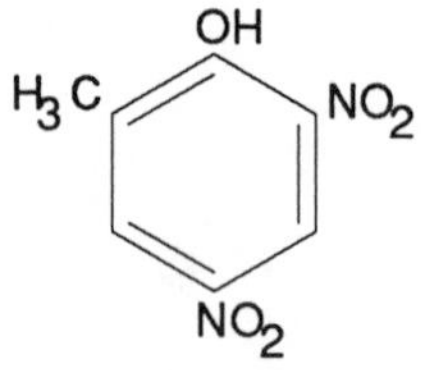

CHEMICAL DESIGNATIONS

CAS Registry Number: 534-52-1

DOT Designation: 1598

Empirical Formula: $C_7H_6N_2O_5$

Formula Weight: 198.14

RTECS Number: GO 9625000

PHYSICAL AND CHEMICAL PROPERTIES

Appearance and Odor: Yellow, odorless crystals.

272

Boiling Point: 312 °C [1].

Dissociation Constant: 4.35 [2].

Henry's Law Constant: 1.4 x 10^{-6} atm·m^3/mol [3]; 4.3 x 10^{-4} atm·m^3/mol at 20 °C (calculated) [4].

Ionization Potential: No data found.

Log K$_{oc}$: 2.64 using method of Karickhoff and others [5].

Log K$_{ow}$: 2.85 [6]; 2.12 [4].

Melting Point: 86.5 °C [7]; 85.8 °C [8].

Solubility in Organics: Soluble in acetone, ethanol, ether [7], and slightly soluble in petroleum ether [9].

Solubility in Water: 0.01 wt% at 20 °C [10]; 0.013% at 15 °C [11]; 128 mg/L at 20 °C [12]; 250 mg/L at 25 °C [13]; 0.198 g/L at 20 °C [4].

Specific Density: No data found.

Transformation Products: No data found.

Vapor Pressure: 5.2 x 10^{-5} mm at 25 °C [14]; 5 x 10^{-5} mm at 20 °C [1]; 3.2 x 10^{-4} mm at 20 °C [4].

FIRE HAZARDS

Flash Point: none [10].

Lower Explosive Limit (LEL): No data found.

Upper Explosive Limit (UEL): No data found.

HEALTH HAZARD DATA

Immediately Dangerous to Life or Health (IDLH): 5 mg/m^3 [10].

Permissible Exposure Limits (PEL) in Air: 0.2 mg/m^3 [15]; 0.2 mg/m^3 10-hour TWA [10].

MANUFACTURING

Selected Manufacturers:

J.T. Baker Chemical Co.
Phillipsburg, NJ 08865

Pfaltz and Bauer, Inc.
375 Fairfield Ave.
Stamford, CT 06902

Sherwin Williams Chemicals
P.O. Box 6520
Cleveland, OH 44101

Uses: Dormant ovicidal spray for fruit trees (highly phototoxic and cannot be used successfully on actively growing plants); herbicide; insecticide.

REFERENCES

1. *Documentation of the Threshold Limit Values and Biological Exposure Indices* (Cincinnati, OH: American Conference of Governmental Industrial Hygienists, 1986), 744 p.
2. Dean, J.A., Ed., *Lange's Handbook of Chemistry*, 11th ed. (New York: McGraw-Hill, Inc., 1973), 1570 p.
3. Warner, H.P., Cohen, J.M., and J.C. Ireland. "Determination of Henry's Law Constants of Selected Priority Pollutants," Office of Science and Development, U.S. EPA Report-600/D-87/229 (1987), 14 p.
4. Schwarzenbach, R.P., Stierli, R., Folsom, B.R., and J. Zeyer. "Compound Properties Relevant for Assessing Environmental Partitioning of Nitrophenols," *Environ. Sci. Technol.*, 22(1):83-92 (1988).
5. Karickhoff, S.W., Brown, D.S., and T.A. Scott. "Sorption of Hydrophobic Pollutants on Natural Sediments," *Water Res.*, 13:241-248 (1979).
6. Mills, W.B., Porcella, D.B., Ungs, M.J., Gherini, S.A., Summers, K.V., Mok, L., Rupp, G.L., and G.L. Bowie. "Water Quality Assessment: A Screening Procedure for Toxic and Conventional Pollutants in Surface and Groundwater-Part I," Office of Research and Development, U.S. EPA Report-600/6-85-002a (1985), 638 p.

7. Weast, R.C., Ed. *CRC Handbook of Chemistry and Physics*, 67th ed. (Boca Raton, FL: CRC Press, Inc., 1986), 2406 p.

8. Sax, N.I. *Dangerous Properties of Industrial Materials* (New York: Van Nostrand Reinhold Co., 1984), 3124 p.

9. *Toxic and Hazardous Industrial Chemicals Safety Manual for Handling and Disposal with Toxicity and Hazard Data* (Tokyo, Japan: International Technical Information Institute, 1986), 700 p.

10. "NIOSH Pocket Guide to Chemical Hazards," U.S. Department of Health and Human Services, U.S. Government Printing Office (1987), 241 p.

11. Berg, G.L., Ed., *The Farm Book* (Willoughby, OH: Meister Publishing Co., 1983), 440 p.

12. Meites, L., Ed. *Handbook of Analytical Chemistry*, 1st ed. (New York: McGraw-Hill, Inc., 1963), 1782 p.

13. Walton, W.C. *Practical Aspects of Ground Water Modeling* (Worthington, OH: National Water Well Association, 1985), 587 p.

14. Melnikov, N.N. *Chemistry of Pesticides* (New York: Springer-Verlag, Inc., 1971), 480 p.

15. "General Industry Standards for Toxic and Hazardous Substances," U.S. Code of Federal Regulations 1910, Subpart Z, Section 1910.1000 (July 1982).

2,4-DINITROPHENOL

Synonyms: Aldifen; Chemox PE; α-Dinitrophenol; DNP; 2,4-DNP; Fenoxyl carbon n; 1-Hydroxy-2,4-dinitrobenzene; Maroxol-50; Nitro kleenup; NSC 1,532; RCRA waste number P048; Solfo black B; Solfo black BB; Solfo black 2B supra; Solfo black G; Solfo black SB; Tetrasulphur black PB; Tetrosulphur PBR.

Structural Formula:

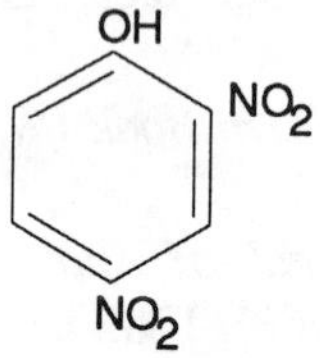

CHEMICAL DESIGNATIONS

CAS Registry Number: 51-28-5

DOT Designation: 0076

Empirical Formula: $C_6H_4N_2O_5$

Formula Weight: 184.11

RTECS Number: SL 2800000

PHYSICAL AND CHEMICAL PROPERTIES

Appearance and Odor: Yellow crystals with a sweet, musty odor.

Boiling Point: sublimes [1].

Dissociation Constant: 4.09 [2].

Henry's Law Constant: 1.57×10^{-8} atm·m³/mol at 18-20 °C (calculated).

Ionization Potential: No data found.

Log K_{oc}: 1.25 (estimated) [3].

Log K_{ow}: 1.53 [4]; 1.51, 1.54 [5]; 1.67 [6].

Melting Point: 115-116 °C [1]; 106-108 °C [7]; 111-113 °C [8].

Solubility in Organics: Soluble in acetone, ethanol, ether, benzene, and chloroform [1].

Solubility in Water: 5,600 mg/L at 18 °C; 43,000 mg/L at 100 °C [9]; 6,000 mg/L at 25 °C [10]; 0.172 g/L at 5 °C, 0.207 g/L at 10 °C, 0.335 g/L at 20 °C, 0.473 g/L at 30 °C [6]; 0.00346 M at 25 °C [11].

Specific Density: 1.683 at 24/4 °C [1]; 1.68 at 20/4 °C [12].

Transformation Products: No data found.

Vapor Pressure: 0.0000149 mm at 18 °C [13]; 0.00039 mm at 20 °C [6].

FIRE HAZARDS

Flash Point: No data found.

Lower Explosive Limit (LEL): Not pertinent [12].

Upper Explosive Limit (UEL): Not pertinent [12].

HEALTH HAZARD DATA

Immediately Dangerous to Life or Health (IDLH): 5 mg/m^3 [12].

Permissible Exposure Limits (PEL) in Air: 0.2 mg/m^3 [12].

MANUFACTURING

Selected Manufacturers:

American Aniline Products
25 McLean Blvd.
Paterson, NJ 07509

Fluka Chemical Corp.
980 South Second St.
Ronkonkoma, NY 11779

Martin Marietta Corp.
Southern Dyestuff Co., Division
Charlotte, NC 28201

Uses: Organic synthesis; photographic agent; manufacturing of pesticides, herbicides, explosives, and wood preservatives; yellow dyes; preparation of picric acid and diaminophenol (photographic developer); indicator; reagent for potassium (K^+) and ammonium (NH_4^+) ions.

REFERENCES

1. Weast, R.C., Ed. *CRC Handbook of Chemistry and Physics*, 67th ed. (Boca Raton, FL: CRC Press, Inc., 1986), 2406 p.
2. Dean, J.A., Ed., *Lange's Handbook of Chemistry*, 11th ed. (New York: McGraw-Hill, Inc., 1973), 1570 p.
3. Montgomery, J.H. Unpublished results (1989).
4. "Treatability Manual - Volume 1: Treatability Data," Office of Research and Development, U.S. EPA Report-600/8-80-042a (1980), 1035 p.
5. Leo, A., Hansch, C., and D. Elkins. "Partition Coefficients and Their Uses," *Chem. Rev.*, 71(6):525-616 (1971).
6. Schwarzenbach, R.P., Stierli, R., Folsom, B.R., and J. Zeyer. "Compound Properties Relevant for Assessing Environmental Partitioning of Nitrophenols," *Environ. Sci. Technol.*, 22(1):83-92 (1988).
7. *Catalog Handbook of Fine Chemicals* (Milwaukee, WI: Aldrich Chemical Co., 1988), 2212 p.
8. *Fluka Catalog 1988/89 - Chemika-Biochemika* (Ronkonkoma, NY: Fluka Chemical Corp., 1988), 1536 p.
9. Verschueren, K. *Handbook of Environmental Data on Organic Chemicals* (New York: Van Nostrand Reinhold Co., 1983), 1310 p.
10. Morrison, R.T., and R.N. Boyd. *Organic Chemistry* (Boston: Allyn and Bacon, Inc., 1971), 1258 p.
11. Caturla, F., Martin-Martinez, J.M., Molina-Sabio, M., Rodriguez-Reinoso, F., and R. Torregrosa. "Adsorption of Substituted Phenols on Activated Carbon," *J. Colloid Interface Sci.*, 124(2):528-534 (1988).
12. Weiss, G. *Hazardous Chemicals Data Book* (Park Ridge, NJ: Noyes Data Corp., 1986), 1069 p.
13. Mabey, W.R., Smith, J.H., Podoll, R.T., Johnson, H.L., Mill, T., Chou, T.-W., Gates, J., Partridge, I.W., Jaber, H., and D. Vandenberg. "Aquatic Fate Process Data for Organic Priority Pollutants - Final

Report," Office of Regulations and Standards, U.S. EPA Report-440/4-81-014 (1982), 407 p.

2,4-DINITROTOLUENE

Synonyms: 2,4-Dinitromethylbenzene; Dinitrotoluol; 2,4- Dinitrotoluol; DNT; 2,4-DNT; 1-Methyl-2,4-dinitrobenzene; NCI-C01865; RCRA waste number U105.

Structural Formula:

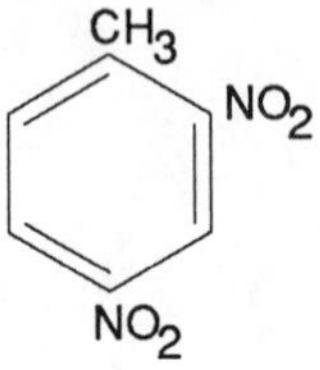

CHEMICAL DESIGNATIONS

CAS Registry Number: 121-14-2

DOT Designation: 1600 (liquid); 2038 (solid).

Empirical Formula: $C_7H_6N_2O_4$

Formula Weight: 182.14

RTECS Number: XT 1575000

PHYSICAL AND CHEMICAL PROPERTIES

Appearance and Odor: Yellow to red needles or yellow liquid with a slight odor.

Boiling Point: 250 °C [1]; 300 °C with slight decomposition [2].

Henry's Law Constant: 8.67 x 10^{-7} atm·m^3/mol [2].

Ionization Potential: No data found.

Log K_{oc}: 1.79 using method of Karickhoff and others [3].

Log K_{ow}: 1.98 [4].

Melting Point: 71.1 °C [5]; 67-70 °C [6].

Solubility in Organics: Soluble in acetone, ethanol, benzene, ether, and pyrimidine [7].

Solubility in Water: 270 mg/L at 22 °C [8].

Specific Density: 1.3208 at 71/4 °C [8]; 1.521 at 15/4 °C [10]; 1.379 at 20/4 °C [11].

Transformation Products: No data found.

Vapor Pressure: 1 mm at 20 °C [1]; 0.001298 mm at 58.8 °C, 0.003902 mm at 69.1 °C [5]; 0.0051 mm at 20 °C [4]; 1.1×10^{-4} mm at 20 °C [2].

FIRE HAZARDS

Flash Point: 206.7 °C [11].

Lower Explosive Limit (LEL): No data found.

Upper Explosive Limit (UEL): No data found.

HEALTH HAZARD DATA

Immediately Dangerous to Life or Health (IDLH): 200 mg/m^3 (carcinogen) [11].

Permissible Exposure Limits (PEL) in Air: 1.5 mg/m^3 TWA [12].

MANUFACTURING

Selected Manufacturers:

> Air Products and Chemicals, Inc.
> Chemicals Group.
> 5 Executive Mall
> Wayne, PA 19087

> Fluka Chemical Corp.
> 980 South Second St.
> Ronkonkoma, NY 11779

282   2,4-Dinitrotoluene

Rubicon Chemicals, Inc.
Gusmar, LA 70734

Uses: Organic synthesis; intermediate for toluidine, dyes, and explosives.

REFERENCES

1. "NIOSH Pocket Guide to Chemical Hazards," U.S. Department of Health and Human Services, U.S. Government Printing Office (1987), 241 p.
2. Howard, P.H. *Handbook of Environmental Fate and Exposure Data for Organic Chemicals* (Chelsea, MI: Lewis Publishers, Inc., 1989), 574 p.
3. Karickhoff, S.W., Brown, D.S., and T.A. Scott. "Sorption of Hydrophobic Pollutants on Natural Sediments," *Water Res.*, 13:241-248 (1979).
4. Mabey, W.R., Smith, J.H., Podoll, R.T., Johnson, H.L., Mill, T., Chou, T.-W., Gates, J., Partridge, I.W., Jaber, H., and D. Vandenberg. "Aquatic Fate Process Data for Organic Priority Pollutants - Final Report," Office of Regulations and Standards, U.S. EPA Report-440/4-81-014 (1982), 407 p.
5. Lenchitz, C., and R.W. Velicky. "Vapor Pressure and Heat of Sublimation of three Nitrotoluenes," *J. Chem. Eng. Data*, 15(3):401-403 (1970).
6. *Catalog Handbook of Fine Chemicals* (Milwaukee, WI: Aldrich Chemical Co., 1988), 2212 p.
7. Weast, R.C., Ed. *CRC Handbook of Chemistry and Physics*, 67th ed. (Boca Raton, FL: CRC Press, Inc., 1986), 2406 p.
8. Verschueren, K. *Handbook of Environmental Data on Organic Chemicals* (New York: Van Nostrand Reinhold Co., 1983), 1310 p.
9. Windholz, M., Budavari, S., Blumetti, R.F., and E.S. Otterbein, Eds. *The Merck Index*, 10th ed., (Rahway, NJ: Merck and Co., 1983), 1463 p.
10. Sax, N.I. *Dangerous Properties of Industrial Materials* (New York: Van Nostrand Reinhold Co., 1984), 3124 p.
11. Weiss, G. *Hazardous Chemicals Data Book* (Park Ridge, NJ: Noyes Data Corp., 1986), 1069 p.
12. *Documentation of the Threshold Limit Values and Biological Exposure Indices for 1986-1987* (Cincinnati, OH: American Conference of Governmental Industrial Hygienists, 1986), 111 p.

2,6-DINITROTOLUENE

Synonyms: 2,6-Dinitromethylbenzene; 2,6-Dinitrotoluol; 2,6-DNT; 2-Methyl-1,3-dinitrobenzene; RCRA waste number U106.

Structural Formula:

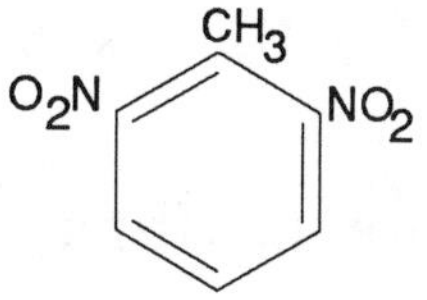

$$O_2N \qquad \overset{CH_3}{\bigcirc} \qquad NO_2$$

CHEMICAL DESIGNATIONS

CAS Registry Number: 606-20-2

DOT Designation: 1600 (liquid); 2038 (solid).

Empirical Formula: $C_7H_6N_2O_4$

Formula Weight: 182.14

RTECS Number: XT 1925000

PHYSICAL AND CHEMICAL PROPERTIES

Appearance: Yellow crystals.

Boiling Point: 285 °C [1].

Henry's Law Constant: 2.17 x 10^{-7} atm·m^3/mol [2].

Ionization Potential: No data found.

Log K_{oc}: 1.79 using method of Karickhoff and others [3].

Log K_{ow}: 2.00 [4].

Melting Point: 66 °C [5]; 60.5 °C [6].

Solubility in Organics: Soluble in ethanol [5].

Solubility in Water: ≈300 mg/L [4].

Specific Density: 1.2833 at 111/4 °C [5].

Transformation Products: No data found.

Vapor Pressure: 0.018 mm at 20 °C [7]; 3.5 x 10^{-4} mm at 20 °C [2].

FIRE HAZARDS

Flash Point: 206.7 °C (calculated) [6].

Lower Explosive Limit (LEL): No data found.

Upper Explosive Limit (UEL): No data found.

HEALTH HAZARD DATA

Immediately Dangerous to Life or Health (IDLH): 200 mg/m^3 [6].

Permissible Exposure Limits (PEL) in Air: 1.5 mg/m^3 [6].

MANUFACTURING

Selected Manufacturers:

Aldrich Chemical Co.
940 West Saint Paul Ave.
Milwaukee, WI 53233

Fluka Chemical Corp.
980 South Second St.
Ronkonkoma, NY 11779

Pfaltz & Bauer, Inc.
172 East Aurora St.
Waterbury, CT 06708

Uses: Organic synthesis; manufacture of explosives.

REFERENCES

1. Maksimov, Y.Y. "Vapor Pressures of Aromatic Nitro Compounds at Various Temperatures," *Zh. Fiz. Khim.*, 42(11):2921-2925 (1968).
2. Howard, P.H. *Handbook of Environmental Fate and Exposure Data for Organic Chemicals* (Chelsea, MI: Lewis Publishers, Inc., 1989), 574 p.
3. Karickhoff, S.W., Brown, D.S., and T.A. Scott. "Sorption of Hydrophobic Pollutants on Natural Sediments," *Water Res.*, 13:241-248 (1979).
4. Mills, W.B., Porcella, D.B., Ungs, M.J., Gherini, S.A., Summers, K.V., Mok, L., Rupp, G.L., and G.L. Bowie. "Water Quality Assessment: A Screening Procedure for Toxic and Conventional Pollutants in Surface and Groundwater-Part I," Office of Research and Development, U.S. EPA Report-600/6-85-002a (1985), 638 p.
5. Weast, R.C., Ed. *CRC Handbook of Chemistry and Physics*, 67th ed. (Boca Raton, FL: CRC Press, Inc., 1986), 2406 p.
6. Weiss, G. *Hazardous Chemicals Data Book* (Park Ridge, NJ: Noyes Data Corp., 1986), 1069 p.
7. Mabey, W.R., Smith, J.H., Podoll, R.T., Johnson, H.L., Mill, T., Chou, T.-W., Gates, J., Partridge, I.W., Jaber, H., and D. Vandenberg. "Aquatic Fate Process Data for Organic Priority Pollutants - Final Report," Office of Regulations and Standards, U.S. EPA Report-440/4-81-014 (1982), 407 p.

DI-n-OCTYL PHTHALATE

Synonyms: 1,2-Benzenedicarboxylic acid, dioctyl ester; 1,2-Benzenedicarboxylic acid, di-*n*-octyl ester; *o*-Benzenedicarboxylic acid, dioctyl ester; Celluflex DOP; Dinopol NOP; Dioctyl-*o*-benzenedicarboxylate; Dioctyl phthalate; *n*-Dioctyl phthalate; DNOP; DOP; Octyl phthalate; *n*-Octyl phthalate; Polycizer 162; PX-138; RCRA waste number U107; Vinicizer 85.

Structural Formula:

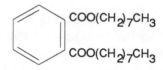

CHEMICAL DESIGNATIONS

CAS Registry Number: 117-84-0

DOT Designation: None assigned.

Empirical Formula: $C_{24}H_{38}O_4$

Formula Weight: 390.57

RTECS Number: TI 1925000

PHYSICAL AND CHEMICAL PROPERTIES

Appearance and Odor: Clear, viscous liquid with a slight odor.

Boiling Point: 220 °C at 4 mm [1]; 386 °C [2].

Henry's Law Constant: 1.41 x 10^{-12} atm·m³/mol at 25 °C (calculated).

Ionization Potential: No data found.

Log K_{oc}: 8.99 using method of Karickhoff and others [3].

Log K_{ow}: 9.2 (calculated) [4].

Melting Point: -30 °C [5]; -25 °C [6].

Solubility in Organics: No data found.

Solubility in Water: 0.285 at 24 °C [7]; 3 mg/L at 25 °C [8].

Specific Density: 0.99 at 20/20 °C [9]; 0.978 at 25/4 °C [10]; 0.978 at 20/4 °C [6].

Transformation Product: *o*-Phthalic acid was tentatively idenitfied as the major degradation product of di-*n*-octyl phthalate produced by the bacterium *Serratia marcescens* [11].

Vapor Density: 16.00 g/L at 25 °C, 13.52 (air = 1).

Vapor Pressure: 1.2 mm at 200 °C [7]; 5×10^{-8} mm at 82 °C, 5×10^{-6} mm at 132 °C [12]; a value of 0.00014 mm at 25 °C was assigned by analogy [13].

FIRE HAZARDS

Flash Point: 218.3 °C [2].

Lower Explosive Limit (LEL): Not pertinent [2].

Upper Explosive Limit (UEL): Not pertinent [2].

HEALTH HAZARD DATA

Immediately Dangerous to Life or Health (IDLH): No data found.

Permissible Exposure Limits (PEL) in Air: No standards set.

MANUFACTURING

Use: Plasticizer for polyvinyl chloride (PVC) and other vinyl polymers.

REFERENCES

1. Sax, N.I. *Dangerous Properties of Industrial Materials* (New York: Van Nostrand Reinhold Co., 1984), 3124 p.

2. Weiss, G. *Hazardous Chemicals Data Book* (Park Ridge, NJ: Noyes Data Corp., 1986), 1069 p.

3. Karickhoff, S.W., Brown, D.S., and T.A. Scott. "Sorption of Hydrophobic Pollutants on Natural Sediments," *Water Res.*, 13:241-248 (1979).

4. Leo, A., Hansch, C., and D. Elkins. "Partition Coefficients and Their Uses," *Chem. Rev.*, 71(6):525-616 (1971).

5. Clayton, G.D., and F.E. Clayton. Eds. *Patty's Industrial Hygiene and Toxicology*, 3rd ed. (New York, John Wiley and Sons, Inc., 1981), 2878 p.

6. Fishbein, L., and P.W. Albro. "Chromatographic and Biological Aspects of the Phthalate Esters," *J. Chromatogr.*, 70(2):365-412 (1972).

7. Verschueren, K. *Handbook of Environmental Data on Organic Chemicals* (New York: Van Nostrand Reinhold Co., 1983), 1310 p.

8. Callahan, M.A., Slimak, M.W., Gable, N.W., May, I.P., Fowler, C.F., Freed, J.R., Jennings, P., Durfee, R.L., Whitmore, F.C., Maestri, B., Mabey, W.R., Holt, B.R., and C. Gould. "Water-Related Environmental Fate of 129 Priority Pollutants Volumes I and II," National Technical Information Service, U.S. EPA Report-440/4-79-029 (1979), 1160 p.

9. Weast, R.C., Ed. *CRC Handbook of Chemistry and Physics*, 67th ed. (Boca Raton, FL: CRC Press, Inc., 1986), 2406 p.

10. Standen, A., Ed. *Kirk-Othmer Encyclopedia of Chemical Technology, Volume 15*, 2nd ed. (New York: John Wiley and Sons, Inc., 1968), 923 p.

11. Mathur, S.P., and J.W. Rouatt. "Utilization of the Pollutant Di-2-ethylhexyl Phthalate by a Bacterium," *J. Environ. Qual.*, 4(2):273-275 (1975).

12. Gross, F.C., and J.A. Colony. "The Ubiquitous Nature and Objectionable Characteristics of Phthalate Esters in Aerospace Industry," *Environ. Health Perspect.*, (January 1973), pp 37-48.

13. Mabey, W.R., Smith, J.H., Podoll, R.T., Johnson, H.L., Mill, T., Chou, T.-W., Gates, J., Partridge, I.W., Jaber, H., and D. Vandenberg. "Aquatic Fate Process Data for Organic Priority Pollutants - Final Report," Office of Regulations and Standards, U.S. EPA Report-440/4-81-014 (1982), 407 p.

1,2-DIPHENYLHYDRAZINE

Synonyms: *N,N'*-Bianiline; *N,N'*-Diphenylhydrazine; *sym*-Diphenyl-hydrazine; DPH; Hydrazobenzene; 1,1'-Hydrazobenzene; Hydrazodi-benzene; NCI-C01854; RCRA waste number U109.

Structural Formula:

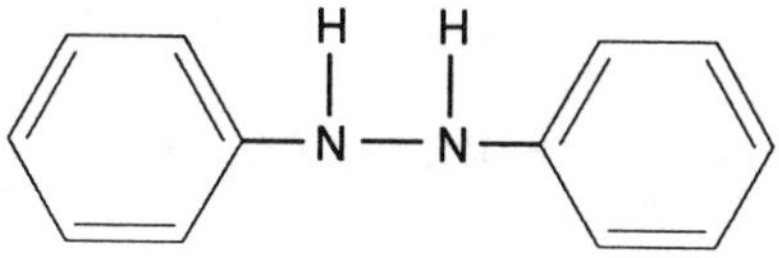

CHEMICAL DESIGNATIONS

CAS Registry Number: 122-66-7

DOT Designation: None assigned.

Empirical Formula: $C_{12}H_{12}N_2$

Formula Weight: 184.24

RTECS Number: MW 2625000

PHYSICAL AND CHEMICAL PROPERTIES

Appearance: Colorless to pale yellow crystals.

Boiling Point: Decomposes near the melting point [1].

Henry's Law Constant: 4.11 x 10^{-11} atm·m^3/mol at 25 °C (calculated).

Ionization Potential: No data found.

Log K_{oc}: 2.82 using method of Karickhoff and others [2].

Log K_{ow}: 2.94 [3].

Melting Point: 131 °C [4]; 126-128 °C [5].

Solubility in Organics: Soluble in ethanol [4].

Solubility in Water: 221 mg/L at 25 °C [1].

Specific Density: 1.158 at 16/4 °C [4].

Transformation Products: No data found.

Vapor Pressure: 2.6 x 10^{-5} mm at 25 °C [3].

FIRE HAZARDS

Flash Point: No data found.

Lower Explosive Limit (LEL): No data found.

Upper Explosive Limit (UEL): No data found.

HEALTH HAZARD DATA

Immediately Dangerous to Life or Health (IDLH): No data found.

Permissible Exposure Limits (PEL) in Air: No standards set.

MANUFACTURING

Selected Manufacturers:

> Aldrich Chemical Co.
> 940 West Saint Paul Ave.
> Milwaukee, WI 53233

> Fluka Chemical Corp.
> 980 South Second St.
> Ronkonkoma, NY 11779

> Pfaltz & Bauer, Inc.
> 172 East Aurora St.
> Waterbury, CT 06708

Uses: Manufacture of benzidine and starting material for pharmaceutical drugs.

REFERENCES

1. "Treatability Manual - Volume 1: Treatability Data," Office of Research and Development, U.S. EPA Report-600/8-80-042a (1980), 1035 p.
2. Karickhoff, S.W., Brown, D.S., and T.A. Scott. "Sorption of Hydrophobic Pollutants on Natural Sediments," *Water Res.*, 13:241-248 (1979).
3. Mabey, W.R., Smith, J.H., Podoll, R.T., Johnson, H.L., Mill, T., Chou, T.-W., Gates, J., Partridge, I.W., Jaber, H., and D. Vandenberg. "Aquatic Fate Process Data for Organic Priority Pollutants - Final Report," Office of Regulations and Standards, U.S. EPA Report-440/4-81-014 (1982), 407 p.
4. Weast, R.C., Ed. *CRC Handbook of Chemistry and Physics*, 67th ed. (Boca Raton, FL: CRC Press, Inc., 1986), 2406 p.
5. *Fluka Catalog 1988/89 - Chemika-Biochemika* (Ronkonkoma, NY: Fluka Chemical Corp., 1988), 1536 p.

α–ENDOSULFAN

Synonyms: Benzoepin; Beosit; Bio 5,462; Chlorthiepin; Crisulfan; Cyclodan; Endocel; Endosol; Endosulfan; Endosulfan I; Endosulphan; ENT 23,979; FMC 5,462; 1,2,3,7,7-Hexachlorobicyclo[2.2.1]-2-heptene-5,6-bisoxymethylene sulfite; α,β-1,2,3,7,7-Hexachlorobicyclo[2.2.1]-2-heptene-5,6-bisoxymethylene sulfite; Hexachlorohexahydromethano-2,4,3-benzodioxathiepin-3-oxide; **(3α,5aβ,6α,9α,9aβ)-6,7,8,9,10,10-Hexachloro-1,5,5a,6,9,9a-hexahydro-6,9-methano-2,4,3-benzodioxathiepin-3-oxide;** 1,4,5,6,7,7-Hexachloro-5-norborene-2,3-dimethanol cyclic sulfite; Hildan; HOE 2,671; Insectophene; KOP-thiodan; Malix; NCI-C00566; NIA 5,462; Niagara 5,462; OMS-570; RCRA waste number P050; Thifor; Thimul; Thiodan; Thiofor; Thiomul; Thionex; Thiosulfan; Tionel; Tiovel.

Structural Formula:

CHEMICAL DESIGNATIONS

CAS Registry Number: 959-98-8

DOT Designation: 2761

Empirical Formula: $C_9H_6Cl_6O_3S$

Formula Weight: 406.92

RTECS Number: RB 9275000

PHYSICAL AND CHEMICAL PROPERTIES

Appearance and Odor: Colorless to brown crystals with a sulfur dioxide odor.

Boiling Point: No data found.

Henry's Law Constant: 1.01×10^{-4} atm·m^3/mol at 25 °C (calculated).

Ionization Potential: No data found.

Log K$_{oc}$: 3.31 using method of Kenaga and Goring [1].

Log K$_{ow}$: 3.55 [2].

Melting Point: 108-110 °C [2]; 106 °C [3].

Solubility in Organics: No data found.

Solubility in Water: 0.530 ppm at 25 °C [4].

Specific Density: 1.745 at 20/4 °C [3].

Transformation Products: Endosulfan sulfate was the major biodegradation product in soils under aerobic conditions. Endosulfan diol and endosulfan-α-hydroxyether were produced under flooded conditions [5].

Vapor Pressure: 10^{-5} mm at 25 °C [6].

FIRE HAZARDS

Flash Point: No data found.

Lower Explosive Limit (LEL): No data found.

Upper Explosive Limit (UEL): No data found.

HEALTH HAZARD DATA

Immediately Dangerous to Life or Health (IDLH): No data found.

Permissible Exposure Limits (PEL) in Air: 0.1 mg/m^3 on skin [7].

MANUFACTURING

Use: Insecticide for vegetable crops.

REFERENCES

1. Kenaga, E.E., and C.A.I. Goring. "Relationship Between Water Solubility, Soil Sorption, Octanol-Water Partitioning and Concentration of Chemicals in Biota," in *Aquatic Toxicology, ASTM STP 707*, Eaton, J.G., Parrish, P.R., and A.C. Hendricks, Eds. (Philadelphia, PA: American Society for Testing and Materials, 1980), pp 78-115.

2. Ali, S. "Degradation and Environmental Fate of Endosulfan Isomers and Endosulfan Sulfate in Mouse, Insect, and Laboratory Ecosystem," PhD Thesis, University of Illinois, Ann Arbor, MI (1978).

3. Sax, N.I. *Dangerous Properties of Industrial Materials* (New York: Van Nostrand Reinhold Co., 1984), 3124 p.

4. Weil, L., Dure, G., and K.E. Quentin. "Solubility in Water of Insecticide Chlorinated Hydrocarbons and Polychlorinated Biphenyls in View of Water Pollution," *Z. Wasser Forsch.*, 7(6):169-175 (1974).

5. Verschueren, K. *Handbook of Environmental Data on Organic Chemicals* (New York: Van Nostrand Reinhold Co., 1983), 1310 p.

6. Martens, R. "Degradation of $[8,9-^{14}C]$Endosulfan by Soil Microorganisms," *Appl. Environ. Microbiol.*, 31(6):853-858 (1975).

7. Weiss, G. *Hazardous Chemicals Data Book* (Park Ridge, NJ: Noyes Data Corp., 1986), 1069 p.

β-ENDOSULFAN

Synonyms: Benzoepin; Beosit; Bio 5,462; Chlorthiepin; Crisulfan; Cyclodan; Endocel; Endosol; Endosulfan; Endosulfan II; Endosulphan; ENT 23,979; FMC 5,462; 1,2,3,7,7-Hexachlorobicyclo[2.2.1]-2-heptene-5,6-bisoxymethylene sulfite; α,β-1,2,3,7,7-Hexachlorobicyclo[2.2.1]-2-heptene-5,6-bisoxymethylene sulfite; Hexachlorohexahydromethano-2,4,3-benzodioxathiepin-3-oxide; **(3α,5aα,6β,9β,9aα)-6,7,8,9,10,10-Hexachloro-1,5,5a,6,9,9a-hexahydro-6,9-methano-2,4,3-benzodioxathiepin-3-oxide;** 1,4,5,6,7,7-Hexachloro-5-norborene-2,3-dimethanol cyclic sulfite; Hildan; HOE 2,671; Insectophene; KOP-thiodan; Malix; NCI-C00566; NIA 5,462; Niagara 5,462; OMS-570; RCRA waste number P050; Thifor; Thimul; Thiomul; Thiodan; Thiofor; Thionex; Thiosulfan; Tionel; Tiovel.

Structural Formula:

CHEMICAL DESIGNATIONS

CAS Registry Number: 33213-65-9

DOT Designation: 2761

Empirical Formula: $C_9H_6Cl_6O_3S$

Formula Weight: 406.92

RTECS Number: RB 9275000

PHYSICAL AND CHEMICAL PROPERTIES

Appearance and Odor: Colorless to brown crystals with a sulfur dioxide odor.

Boiling Point: No data found.

Henry's Law Constant: 1.91×10^{-5} atm·m^3/mol at 25 °C (calculated).

Ionization Potential: No data found.

Log K$_{oc}$: 3.37 using method of Kenaga and Goring [1].

Log K$_{ow}$: 3.62 [2].

Melting Point: 207-209 °C [2].

Solubility in Organics: No data found.

Solubility in Water: 0.280 ppm at 25 °C [3].

Specific Density: 1.745 at 20/20 °C [4].

Transformation Products: Endosulfan sulfate was the major biodegradation product in soils under aerobic conditions. Endosulfan diol and endosulfan-α-hydroxyether were produced under flooded conditions [5].

Vapor Pressure: 10^{-5} mm at 25 °C [6].

FIRE HAZARDS

Flash Point: No data found.

Lower Explosive Limit (LEL): No data found.

Upper Explosive Limit (UEL): No data found.

HEALTH HAZARD DATA

Immediately Dangerous to Life or Health (IDLH): No data found.

Permissible Exposure Limits (PEL) in Air: No standards set.

MANUFACTURING

Use: Insecticide for vegetable crops.

REFERENCES

1. Kenaga, E.E., and C.A.I. Goring. "Relationship Between Water Solubility, Soil Sorption, Octanol-Water Partitioning and Concentration of Chemicals in Biota," in *Aquatic Toxicology, ASTM STP 707*, Eaton, J.G., Parrish, P.R., and A.C. Hendricks, Eds. (Philadelphia, PA: American Society for Testing and Materials, 1980), pp 78-115.

2. Ali, S. "Degradation and Environmental Fate of Endosulfan Isomers and Endosulfan Sulfate in Mouse, Insect, and Laboratory Ecosystem," PhD Thesis, University of Illinois, Ann Arbor, MI (1978).

3. Weil, L., Dure, G., and K.E. Quentin. "Solubility in Water of Insecticide Chlorinated Hydrocarbons and Polychlorinated Biphenyls in View of Water Pollution," *Z. Wasser Forsch.*, 7(6):169-175 (1974).

4. Sax, N.I. *Dangerous Properties of Industrial Materials* (New York: Van Nostrand Reinhold Co., 1984), 3124 p.

5. Verschueren, K. *Handbook of Environmental Data on Organic Chemicals* (New York: Van Nostrand Reinhold Co., 1983), 1310 p.

6. Martens, R. "Degradation of [8,9-^{14}C]Endosulfan by Soil Microorganisms," *Appl. Environ. Microbiol.*, 31(6):853-858 (1975).

ENDOSULFAN SULFATE

Synonyms: 6,7,8,9,10,10-Hexachloro-1,5,5a,6,9,9a-hexahydro-3,3-dioxide; 6,9-Methano-2,4,3-Benzodioxathiepin.

Structural Formula:

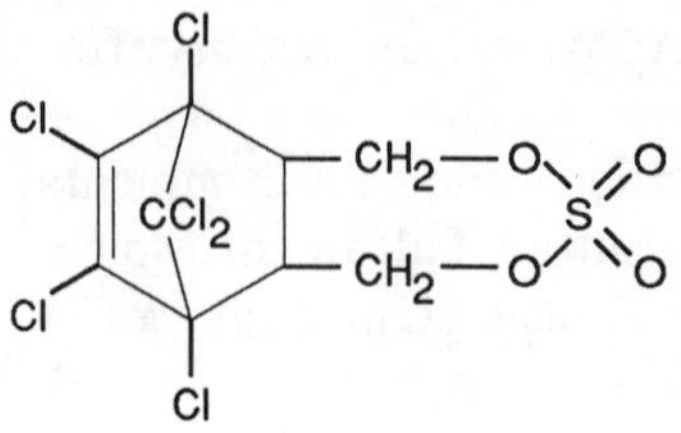

CHEMICAL DESIGNATIONS

CAS Registry Number: 1031-07-8

DOT Designation: 2761

Empirical Formula: $C_9H_6Cl_6O_4S$

Formula Weight: 422.92

RTECS Number: Not assigned.

PHYSICAL AND CHEMICAL PROPERTIES

Appearance: Solid.

Boiling Point: No data found.

Henry's Law Constant: Insufficient vapor pressure data for calculation.

Ionization Potential: No data found.

Log K_{oc}: 3.37 using method of Kenaga and Goring [1].

Log K_{ow}: 3.66 [2].

Melting Point: 198-201 °C [2].

Solubility in Organics: No data found.

Solubility in Water: 0.117 ppm [2].

Specific Density: No data found.

Transformation Products: A mixed culture of soil microorganisms biodegradated endosulfan sulfate into endosulfan ether, endosulfan-α-hydroxy ether, and endosulfan lactone [3].

Vapor Pressure: No data found.

FIRE HAZARDS

Flash Point: No data found.

Lower Explosive Limit (LEL): No data found.

Upper Explosive Limit (UEL): No data found.

HEALTH HAZARD DATA

Immediately Dangerous to Life or Health (IDLH): No data found.

Permissible Exposure Limits (PEL) in Air: No standards set.

MANUFACTURING

Uses: No data found.

REFERENCES

1. Kenaga, E.E., and C.A.I. Goring. "Relationship Between Water Solubility, Soil Sorption, Octanol-Water Partitioning and Concentration of Chemicals in Biota," in *Aquatic Toxicology, ASTM STP 707*, Eaton, J.G., Parrish, P.R., and A.C. Hendricks, Eds. (Philadelphia, PA: American Society for Testing and Materials, 1980), pp 78-115.
2. Ali, S. "Degradation and Environmental Fate of Endosulfan Isomers and Endosulfan Sulfate in Mouse, Insect, and Laboratory Ecosystem," PhD Thesis, University of Illinois, Ann Arbor, MI

(1978).

3. Verschueren, K. *Handbook of Environmental Data on Organic Chemicals* (New York: Van Nostrand Reinhold Co., 1983), 1310 p.

ENDRIN

Synonyms: Compound 269; Endrex; ENT 17,251; Experimental insecticide no. 269; Hexachloroepoxyoctahydro-*endo-endo*-dimethanonaphthalene; 1,2,3,4,10,10-Hexachloro-6,7-epoxy-1,4,4a,5,6,7,8,8a-octahydro-*endo,endo*-1,4:5,8-dimethanonaphthalene; **3,4,5,6,9,9-Hexachloro-1a,2,2a,3,6,6a,7,7a-octahydro-2,7:3,6-dimethanonaphth[2,3-*b*]oxirene**; Hexadrin; Isodrin epoxide; Mendrin; NA 2,761; NCI-C00157; Nendrin; RCRA waste number P051.

Structural Formula:

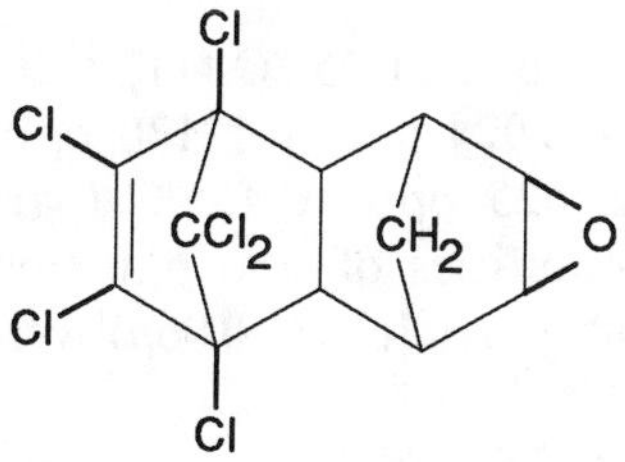

CHEMICAL DESIGNATIONS

CAS Registry Number: 72-20-8

DOT Designation: 2761

Empirical Formula: $C_{12}H_8Cl_6O$

Formula Weight: 380.92

RTECS Number: IO 1575000

PHYSICAL AND CHEMICAL PROPERTIES

Appearance and Odor: White, odorless, crystalline solid when pure; light tan color with faint chemical odor for technical grade.

Boiling Point: Decomposes at 245 °C [1].

Henry's Law Constant: 5.0 x 10^{-7} atm·m^3/mol [2].

301

Ionization Potential: No data found.

Log K_{oc}: 3.92 [3].

Log K_{ow}: 4.56 [4]; 5.16 [5]; 5.339 [6]; 3.209 [7].

Melting Point: 200 °C [8].

Solubility in Organics: Soluble in acetone, benzene, carbon tetrachloride, hexane, xylene [9], aromatic hydrocarbons, esters, and ketones [10].

Solubility in Water: 220 ppb at 25 °C [11]; 0.26 mg/L at 25 °C [2]; 0.230 mg/L at 20-25 °C [12]; 0.024 ppm [6]; 130 ppb at 15 °C, 250 ppb at 25 °C, 420 ppb at 35 °C, 625 ppb at 45 °C (particle sizes $\leq$ 5μ) [13]; at room temperature, solubilities of 260 ppb (maximum particle size 5μ) and 190 ppb (maximum particle size 0.06μ) were reported [14].

Specific Density: 1.65 at 25/4 °C [15].

Transformation Products: Microbial degradation in soil formed several ketones and aldehydes of which *keto*-endrin was the only metabolite identified [16].

Vapor Pressure: 7 x 10^{-7} mm at 25 °C [17]; 2 x 10^{-7} mm at 25 °C [1].

FIRE HAZARDS

Flash Point: Non-flammable as a solid, but in a xylene solution a value > 26.6 °C was reported [15].

Lower Explosive Limit (LEL): 1.1% in xylene [15].

Upper Explosive Limit (UEL): 7.0% in xylene [15].

HEALTH HAZARD DATA

Immediately Dangerous to Life or Health (IDLH): 200 mg/m^3 [18].

Permissible Exposure Limits (PEL) in Air: 0.1 mg/m^3 [19]; 0.1 mg/m^3 TWA [20].

MANUFACTURING

Selected Manufacturers:

Velsicol Chemical Corp.
341 East Ohio St.
Chicago, IL 60611

Shell Chemical Co.
1 Shell Plaza
Houston, TX 77002

Use: Insecticide.

REFERENCES

1. *Documentation of the Threshold Limit Values and Biological Exposure Indices* (Cincinnati, OH: American Conference of Governmental Industrial Hygienists, 1986), 744 p.
2. "Treatability Manual - Volume 1: Treatability Data," Office of Research and Development, U.S. EPA Report-600/8-80-042a (1980), 1035 p.
3. Sharom, M.S., Miles, J.R.W., Harris C.R., and F.L. McEwen. "Behaviour of 12 Pesticides in Soil and Aqueous Suspensions of Soil and Sediment," *Water Res.*, 14:1095-1100 (1980).
4. Isnard, S., and S. Lambert. "Estimating Bioconcentration Factors from Octanol-Water Partition Coefficient and Aqueous Solubility," *Chemosphere*, 17(1):21-34 (1988).
5. Travis, C.C., and A.D. Arms. "Bioconcentration of Organics in Beef, Milk and Vegetation," *Environ. Sci. Technol.*, 22(3):271-274 (1988).
6. Kenaga, E.E., and C.A.I. Goring. "Relationship Between Water Solubility, Soil Sorption, Octanol-Water Partitioning and Concentration of Chemicals in Biota," in *Aquatic Toxicology, ASTM STP 707*, Eaton, J.G., Parrish, P.R., and A.C. Hendricks, Eds. (Philadelphia, PA: American Society for Testing and Materials, 1980), pp 78-115.
7. Rao, P.S.C., and J.M. Davidson. "Estimation of Pesticide Retention and Transformation Parameters Required in Nonpoint Source Pollution Models," in *Environmental Impact of Nonpoint Source Pollution*, Overcash, M.R., and J.M. Davidson, Eds. (Ann Arbor, MI: Ann Arbor Science Publishers, Inc., 1980), pp 23-67.
8. Caswell, R.L., DeBold, K.J., and L.S. Gilbert, Eds. *Pesticide*

Handbook, 29th ed. (College Park, MD: The Entomological Society of America, 1981), 286 p.

9. "Chemical, Physical, and Biological Properties of Compounds Present at Hazardous Waste Sites," U.S. EPA Report-530/SW-89-010 (1985), 619 p.

10. *Toxic and Hazardous Industrial Chemicals Safety Manual for Handling and Disposal with Toxicity and Hazard Data* (Tokyo, Japan: International Technical Information Institute, 1986), 700 p.

11. Mills, W.B., Porcella, D.B., Ungs, M.J., Gherini, S.A., Summers, K.V., Mok, L., Rupp, G.L., and G.L. Bowie. "Water Quality Assessment: A Screening Procedure for Toxic and Conventional Pollutants in Surface and Groundwater-Part I," Office of Research and Development, U.S. EPA Report-600/6-85-002a (1985), 638 p.

12. Geyer, H., Kraus, A.S., Klein, W., Richter, E., and F. Korte. "Relationship Between Water Solubility and Bioaccumulation Potential of Organic Chemicals in Rats," *Chemosphere*, 9(5/6):277-294 (1980).

13. Biggar, J.W., and I.R. Riggs. "Apparent Solubility of Organochlorine Insecticides in Water at Various Temperatures," *Hilgardia*, 42(10):383-391 (1974).

14. Robeck, G.G., Dostal, K.A., Cohen, J.M., and J.F. Kreissl. "Effectiveness of Water Treatment Processes in Pesticide Removal," *J. Am. Water Works Assoc.*, 57(2):181-200 (1965).

15. Weiss, G. *Hazardous Chemicals Data Book* (Park Ridge, NJ: Noyes Data Corp., 1986), 1069 p.

16. Kearney, P.C., and D.D. Kaufman. *Herbicides: Chemistry, Degradation and Mode of Action* (New York: Marcel Dekker, Inc., 1976), 1036 p.

17. Verschueren, K. *Handbook of Environmental Data on Organic Chemicals* (New York: Van Nostrand Reinhold Co., 1983), 1310 p.

18. "NIOSH Pocket Guide to Chemical Hazards," U.S. Department of Health and Human Services, U.S. Government Printing Office (1987), 241 p.

19. "General Industry Standards for Toxic and Hazardous Substances," U.S. Code of Federal Regulations 1910, Subpart Z, Section 1910.1000 (July 1982).

20. *Documentation of the Threshold Limit Values and Biological Exposure Indices for 1986-1987* (Cincinnati, OH: American Conference of Governmental Industrial Hygienists, 1986), 111 p.

ENDRIN ALDEHYDE

Synonym: 2,2a,3,3,4,7-Hexachlorodecahydro-1,2,4-methenocyclo-penta[c,d]pentalene-5-carboxaldehyde.

Structural Formula:

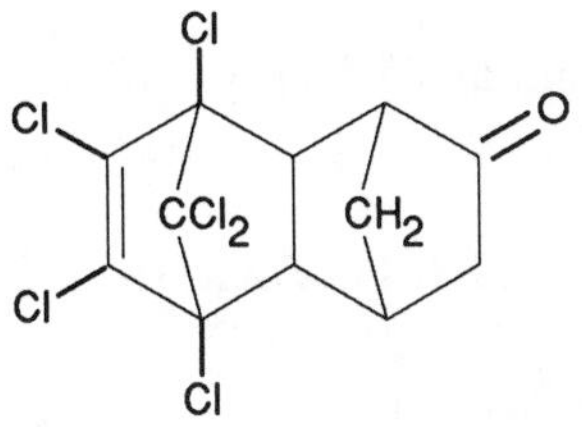

CHEMICAL DESIGNATIONS

CAS Registry Number: 7421-93-4

DOT Designation: 2761

Empirical Formula: $C_{12}H_8Cl_6O$

Formula Weight: 380.92

RTECS Number: Not assigned.

PHYSICAL AND CHEMICAL PROPERTIES

Appearance: Solid.

Boiling Point: Decomposes at 235 °C [1].

Henry's Law Constant: 3.86 x 10^{-7} atm·m³/mol 25 °C (calculated).

Ionization Potential: No data found.

Log K_{oc}: 4.43 using method of Kenaga and Goring [2].

Log K_{ow}: 5.6 (calculated) [3].

Melting Point: 145-149 °C [4].

Solubility in Organics: No data found.

Solubility in Water: 0.26 ppm at 25 °C [5].

Specific Density: No data found.

Transformation Products: No data found.

Vapor Pressure: 2×10^{-7} mm at 25 °C [6].

FIRE HAZARDS

Flash Point: No data found.

Lower Explosive Limit (LEL): No data found.

Upper Explosive Limit (UEL): No data found.

HEALTH HAZARD DATA

Immediately Dangerous to Life or Health (IDLH): No data found.

Permissible Exposure Limits (PEL) in Air: No standards set.

MANUFACTURING

Uses: No data found.

REFERENCES

1. Callahan, M.A., Slimak, M.W., Gable, N.W., May, I.P., Fowler, C.F., Freed, J.R., Jennings, P., Durfee, R.L., Whitmore, F.C., Maestri, B., Mabey, W.R., Holt, B.R., and C. Gould. "Water-Related Environmental Fate of 129 Priority Pollutants Volumes I and II," National Technical Information Service, U.S. EPA Report-440/4-79-029 (1979), 1160 p.
2. Kenaga, E.E., and C.A.I. Goring. "Relationship Between Water Solubility, Soil Sorption, Octanol-Water Partitioning and Concentration of Chemicals in Biota," in *Aquatic Toxicology, ASTM STP 707*, Eaton, J.G., Parrish, P.R., and A.C. Hendricks, Eds. (Philadelphia, PA: American Society for Testing and Materials,

1980), pp 78-115.

3. Neely, W.B., Branson, D.R., and G.E. Blau. "Partition Coefficient to Measure Bioconcentration Potential of Organic Pesticides in Fish," *Environ. Sci. Technol.*, 6(7):629-632 (1974).

4. "Treatability Manual - Volume 1: Treatability Data," Office of Research and Development, U.S. EPA Report-600/8-80-042a (1980), 1035 p.

5. Weil, L., Dure, G., and K.E. Quentin. "Solubility in Water of Insecticide Chlorinated Hydrocarbons and Polychlorinated Biphenyls in View of Water Pollution," *Z. Wasser Forsch.*, 7(6):169-175 (1974).

6. Martin, H., Ed. *Pesticide Manual*, 3rd ed. (Worcester, England: British Crop Protection Council, 1972).

ETHYLBENZENE

Synonyms: EB; Ethylbenzol; NCI-C56393; Phenylethane; UN 1775.

Structural Formula:

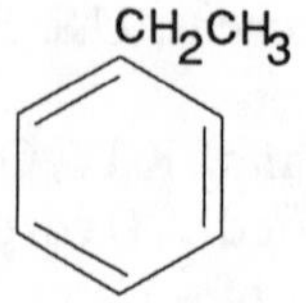

CHEMICAL DESIGNATIONS

CAS Registry Number: 100-41-4

DOT Designation: 1175

Empirical Formula: C_8H_{10}

Formula Weight: 106.17

RTECS Number: DA 0700000

PHYSICAL AND CHEMICAL PROPERTIES

Appearance and Odor: Clear, colorless liquid with a sweet gasoline-like odor.

Boiling Point: 136.2 °C [1].

Henry's Law Constant: 0.0066 atm·m³/mol [2]; 0.00644 atm·m³/mol [3]; 0.00868 atm·m³/mol at 25 °C [4].

Ionization Potential: 8.76 eV [5]; 9.12 eV [6].

Log K_{oc}: 1.98 [7]; 2.41 [8].

Log K_{ow}: 3.13 [9]; 3.15 [10]; 3.05 [11].

Melting Point: -95.0 °C [12]; -94.4 °C [13].

Solubility in Organics: Freely soluble in most solvents [14].

Solubility in Water: 140 mg/L at 15 °C, 152 mg/L at 20 °C, 206 mg/L at 25 °C [15]; 187 mg/L at 25 °C [16]; 152 mg/L at 25 °C [17]; 168 mg/L at 25 °C [18]; 208 mg/L at 25 °C [19]; 0.014 wt% at 15 °C [20]; 0.00185 M at 10 °C, 0.001770 M at 20 °C, 0.001811 M at 25 °C, 0.001777 M at 30 °C [21]; 197 ppm at 0 °C, 177 ppm at 25 °C [22]; 161.2 ppm at 25 °C, 111.0 ppm in artificial seawater at 25 °C [23]; 131.0 ppm at 25 °C [24]; 181 mg/L at 20 °C [25]; 77 mg/L in fresh water at 25 °C, 70 mg/L in salt water at 25 °C [26]; 0.00200 M at 25 °C [27].

Specific Density: 0.8670 at 20/4 °C [1]; 0.86690 at 20/4 °C, 0.86250 at 25/4 °C [13].

Transformation Products: No data found.

Vapor Density: 4.34 g/L at 25 °C, 3.66 (air = 1).

Vapor Pressure: 12 mm at 30 °C [15]; 1 mm at -9.8 °C, 10 mm at 25.9 °C, 40 mm at 52.8 °C, 100 mm at 74.1 °C, 400 mm at 113.8 °C, 760 mm at 136.2 °C [1]; 7.08 mm at 20 °C [25].

FIRE HAZARDS

Flash Point: 15 °C [5].

Lower Explosive Limit (LEL): 1.0% [5].

Upper Explosive Limit (UEL): 6.7% [5].

HEALTH HAZARD DATA

Immediately Dangerous to Life or Health (IDLH): 2,000 ppm [5].

Permissible Exposure Limits (PEL) in Air: 100 ppm ($\approx$435 mg/m^3) [28]; 100 ppm TWA, 125 ppm ($\approx$545 mg/m^3) STEL [29].

MANUFACTURING

Selected Manufacturers:

Dow Chemical Co.
Midland, MI 48640

Amoco Chemical Corp.
130 East Randolph Drive
Chicago, IL 60601

Monsanto Industrial Chemicals Co.
800 North Lindbergh Blvd.
St. Louis, MO 63166

Uses: Intermediate in production of styrene and acetophenone; solvent; organic synthesis.

REFERENCES

1. Weast, R.C., Ed. *CRC Handbook of Chemistry and Physics*, 67th ed. (Boca Raton, FL: CRC Press, Inc., 1986), 2406 p.
2. Pankow, J.F., and M.E. Rosen. "Determination of Volatile Compounds in Water by Purging Directly to a Capillary Column with Whole Column Cryotrapping," *Environ. Sci. Technol.*, 22(4):398-405 (1988).
3. Valsaraj, K.T. "On the Physio-Chemical Aspects of Partitioning of Non-Polar Hydrophobic Organics at the Air-Water Interface," *Chemosphere*, 17(5):875-887 (1988).
4. Hine, J., and P.K. Mookerjee. "The Intrinsic Hydrophilic Character of Organic Compounds. Correlations in Terms of Structural Contributions," *J. Org. Chem.*, 40(3):292-298 (1975).
5. "NIOSH Pocket Guide to Chemical Hazards," U.S. Department of Health and Human Services, U.S. Government Printing Office (1987), 241 p.
6. Yoshida, K., Shigeoka, T., and F. Yamauchi. "Non-Steady State Equilibrium Model for the Preliminary Prediction of the Fate of Chemicals in the Environment," *Ecotoxicol. Environ. Safety*, 7(2):179-190 (1983).
7. Abdul, S.A., Gibson, T.L., and D.N. Rai. "Statistical Correlations for Predicting the Partition Coefficient for Nonpolar Organic Contaminants Between Aquifer Organic Carbon and Water," *Haz. Waste Haz. Mater.*, 4(3):211-222 (1987).
8. Hodson, J., and N.A. Williams. "The Estimation of the Adsorption Coefficient (K_{oc}) for Soils by High Performance Liquid Chromatography," *Chemosphere*, 19(1):67-77 (1988).
9. Yalkowsky, S.H., Valvani, S.C., and D. Mackay. "Estimation of the Aqueous Solubility of Some Aromatic Compounds," *Res. Rev.*, 85:43-55 (1983).

10. Hansch, C., Quinlan, J.E., and G.L. Lawrence. "The Linear Free-Energy Relationship Between Partition Coefficients and the Aqueous Solubility of Organic Liquids," *J. Org. Chem.*, 33(1):347-350 (1968).

11. Walton, W.C. *Practical Aspects of Ground Water Modeling* (Worthington, OH: National Water Well Association, 1985), 587 p.

12. Dean, J.A., Ed., *Lange's Handbook of Chemistry*, 11th ed. (New York: McGraw-Hill, Inc., 1973), 1570 p.

13. Huntress, E.H., and S.P. Mulliken. *Identification of Pure Organic Compounds - Tables of Data on Selected Compounds of Order I* (New York: John Wiley and Sons, Inc., 1941), 691 p.

14. "Chemical, Physical, and Biological Properties of Compounds Present at Hazardous Waste Sites," U.S. EPA Report-530/SW-89-010 (1985), 619 p.

15. Verschueren, K. *Handbook of Environmental Data on Organic Chemicals* (New York: Van Nostrand Reinhold Co., 1983), 1310 p.

16. Miller, M.M., Wasik, S.P., Huang, G.-L., Shiu, W.-Y., and D. Mackay. "Relationships Between Octanol-Water Partition Coefficient and Aqueous Solubility," *Environ. Sci. Technol.*, 19(6):522-529 (1985).

17. McAuliffe, C. "Solubility in Water of Paraffin, Cycloparaffin, Olefin, Acetylene, Cycloolefin, and Aromatic Compounds," *J. Phys. Chem.*, 70(4):1267-1275 (1966).

18. Andrews, L.J., and R.M. Keefer. "Cation Complexes of Compounds Containing Carbon-Carbon Double Bonds. VII. Further Studies on the Argentation of Substituted Benzenes," *J. Am. Chem. Soc.*, 72(11):5034-5037 (1950).

19. Bohon, R.L., and W.F. Claussen. "The Solubility of Aromatic Hydrocarbons in Water," *J. Am. Chem. Soc.*, 73(4):1571-1578 (1951).

20. Stephen, H., and T. Stephen. *Solubilities of Inorganic and Organic Compounds - Part 1, Volume 1* (London: Pergamon Printing and Art Services, Ltd., 1963), 960 p.

21. Owens, J.W., Wasik, S.P., and H. DeVoe. "Aqueous Solubilities and Enthalpies of Solution of *n*-Alkylbenzenes," *J. Chem. Eng. Data*, 31(1):47-51 (1986).

22. Polak, J., and B.C.-Y. Lu. "Mutual Solubilities of Hydrocarbons and Water at 0 and 25 °C," *Can. J. Chem.*, 51(24):4018-4023 (1973).

23. Sutton, C., and J.A. Calder. "Solubility of Alkylbenzenes in Distilled Water and Seawater at 25 °C," *J. Chem. Eng. Data*, 20(3):320-322 (1975).

24. Price, L.C. "Aqueous Solubility of Petroleum as Applied to its Origin and Primary Migration," *Am. Assoc. Pet. Geol. Bull.*, 60(2):213-244 (1976).

25. Burris, D.R., and W.G. MacIntyre. "A Thermodynamic Study of

Solutions of Liquid Hydrocarbon Mixtures in Water," *Geochim. Cosmochim. Acta*, 50(7):1545-1549 (1986).

26. Krasnoshchekova, R.Y., and M. Gubergrits. "Solubility of *n*-Alkylbenzene in Fresh and Salt Waters," [Chemical Abstracts 83(16):136583p]: *Vodn. Resur.*, 2:170-173 (1975).

27. Ben-Naim, A., and J. Wilf. "Solubilities and Hydrophobic Interactions in Aqueous Solutions of Monoalkylbenzene Molecules," *J. Phys. Chem.*, 84(6):583-586 (1980).

28. "General Industry Standards for Toxic and Hazardous Substances," U.S. Code of Federal Regulations 1910, Subpart Z, Section 1910.1000 (July 1982).

29. *Documentation of the Threshold Limit Values and Biological Exposure Indices for 1986-1987* (Cincinnati, OH: American Conference of Governmental Industrial Hygienists, 1986), 111 p.

FLUORANTHENE

Synonyms: 1,2-Benzacenaphthene; Benzo[*jk*]fluorene; Idryl; 1,2-(1,8-Naphthylene)benzene; 1,2-(1,8-Naphthalenediyl)benzene; RCRA waste number U120.

Structural Formula:

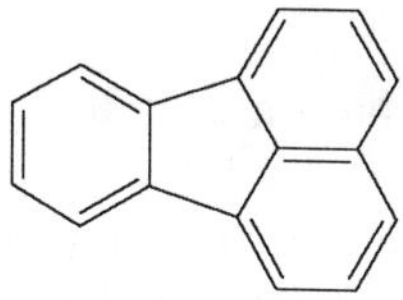

CHEMICAL DESIGNATIONS

CAS Registry Number: 206-44-0

DOT Designation: None assigned.

Empirical Formula: $C_{16}H_{10}$

Formula Weight: 202.26

RTECS Number: LL 4025000

PHYSICAL AND CHEMICAL PROPERTIES

Appearance: Colorless crystals.

Boiling Point: 375 °C [1]; 384 °C [2]; 367 °C [3].

Henry's Law Constant: 0.0169 atm·m^3/mol at 25 °C [4].

Ionization Potential: 8.54 eV [5].

Log K_{oc}: 4.62 [6].

Log K_{ow}: 5.22 [6].

Melting Point: 107 °C [7]; 111 °C [8]; 120 °C [3]; 109-110 °C [9].

Solubility in Organics: Soluble in acetic acid, benzene, chloroform, carbon disulfide, ethanol, and ether [10].

Solubility in Water: 0.265 mg/L at 25 °C [11]; 0.120 mg/L at 24 °C, 0.1 ppm in seawater at 22 °C [7]; 0.26 mg/L at 25 °C [12]; 0.206 mg/L at 25 °C; 0.264 mg/L at 29 °C [13]; 0.236 mg/L at 25 °C [14]; 0.240 mg/L [15]; 0.133, 0.275 mg/L at 15 °C, 0.166 mg/L at 20 °C, 0.222, 0.373 mg/L at 25 °C [16].

Specific Density: 1.252 at 0/4 °C [1].

Transformation Products: No data found.

Vapor Pressure: 0.01 mm at 20 °C [3]; estimated as 10^{-6}-10^{-4} mm at 20 °C based on data for structurally similar compounds [8]; 5.0 x 10^{-6} mm at 25 °C [17].

FIRE HAZARDS

Flash Point: No data found.

Lower Explosive Limit (LEL): No data found.

Upper Explosive Limit (UEL): No data found.

HEALTH HAZARD DATA

Immediately Dangerous to Life or Health (IDLH): Potential human carcinogen [18].

Permissible Exposure Limits (PEL) in Air: No individual standards have been set, however, as a constituent in coal tar pitch volatiles, the following exposure limits have been established: 0.2 mg/m^3 (benzene-soluble fraction) [19]; 0.1 mg/m^3 10-hour TWA (cyclohexane-extractable fraction) [18]; 0.2 mg/m^3 TWA (benzene solubles) [20].

MANUFACTURING

Selected Manufacturers:

Aldrich Chemical Co.
940 West Saint Paul Ave.
Milwaukee, WI 53233

Fluka Chemical Corp.
980 South Second St.
Ronkonkoma, NY 11779

Use: Research chemical.

REFERENCES

1. Weast, R.C., Ed. *CRC Handbook of Chemistry and Physics*, 67th ed. (Boca Raton, FL: CRC Press, Inc., 1986), 2406 p.
2. *Catalog Handbook of Fine Chemicals* (Milwaukee, WI: Aldrich Chemical Co., 1988), 2212 p.
3. Sax, N.I. *Dangerous Properties of Industrial Materials* (New York: Van Nostrand Reinhold Co., 1984), 3124 p.
4. Hine, J., and P.K. Mookerjee. "The Intrinsic Hydrophilic Character of Organic Compounds. Correlations in Terms of Structural Contributions," *J. Org. Chem.*, 40(3):292-298 (1975).
5. Franklin, J.L., Dillard, J.G., Rosenstock, H.M., Herron, J.T., Draxl K., and F.H. Field. "Ionization Potentials, Appearance Potentials and Heats of Formation of Gaseous Positive Ions," National Bureau of Standards Report NSRDS-NBS 26, U.S. Government Printing Office (1969), 289 p.
6. Abdul, S.A., Gibson, T.L., and D.N. Rai. "Statistical Correlations for Predicting the Partition Coefficient for Nonpolar Organic Contaminants Between Aquifer Organic Carbon and Water," *Haz. Waste Haz. Mater.*, 4(3):211-222 (1987).
7. Verschueren, K. *Handbook of Environmental Data on Organic Chemicals* (New York: Van Nostrand Reinhold Co., 1983), 1310 p.
8. "Treatability Manual - Volume 1: Treatability Data," Office of Research and Development, U.S. EPA Report-600/8-80-042a (1980), 1035 p.
9. *Fluka Catalog 1988/89 - Chemika-Biochemika* (Ronkonkoma, NY: Fluka Chemical Corp., 1988), 1536 p.
10. "Chemical, Physical, and Biological Properties of Compounds Present at Hazardous Waste Sites," U.S. EPA Report-530/SW-89-010 (1985), 619 p.
11. Harrison, R.M., Perry, R., and R.A. Wellings. "Polynuclear Aromatic Hydrocarbons in Raw, Potable and Waste Waters," *Water Res.*, 9:331-346 (1975).
12. Mackay, D., and W.Y. Shiu. "Aqueous Solubility of Polynuclear Aromatic Hydrocarbons," *J. Chem. Eng. Data*, 22(4):399-402 (1977).
13. May, W.E., Wasik, S.P., and D.H. Freeman. "Determination of the

Aqueous Solubility of Polynuclear Aromatic Hydrocarbons by a Coupled Column Liquid Chromatographic Technique," *Anal. Chem.*, 50(1):175-179 (1978).

14. Schwarz, F.P., and S.P. Wasik. "Fluorescence Measurements of Benzene, Naphthalene, Anthracene, Pyrene, Fluoranthene, and Benzo[e]pyrene in Water," *Anal. Chem.*, 48(3):524-528 (1976).

15. Davis, W.W., Krahl, M.E., and G.H.A. Clowes. "Solubility of Carcinogenic and Related Hydrocarbons in Water," *J. Am. Chem. Soc.*, 64(1):108-110 (1942).

16. Kishi, H., and Y. Hashimoto. "Evaluation of the Procedures for the Measurement of Water Solubility and n-Octanol/Water Partition Coefficient of Chemicals Results of a Ring Test in Japan," *Chemosphere*, 18(9/10):1749-1749 (1989).

17. Mabey, W.R., Smith, J.H., Podoll, R.T., Johnson, H.L., Mill, T., Chou, T.-W., Gates, J., Partridge, I.W., Jaber, H., and D. Vandenberg. "Aquatic Fate Process Data for Organic Priority Pollutants - Final Report," Office of Regulations and Standards, U.S. EPA Report-440/4-81-014 (1982), 407 p.

18. "NIOSH Pocket Guide to Chemical Hazards," U.S. Department of Health and Human Services, U.S. Government Printing Office (1987), 241 p.

19. "General Industry Standards for Toxic and Hazardous Substances," U.S. Code of Federal Regulations 1910, Subpart Z Section 1910.1000 (July 1982).

20. *Documentation of the Threshold Limit Values and Biological Exposure Indices for 1986-1987* (Cincinnati, OH: American Conference of Governmental Industrial Hygienists, 1986), 111 p.

FLUORENE

Synonyms: 2,3-Benzindene; *o*-Biphenylenemethane; *o*-Biphenylmethane; Diphenylenemethane; *o*-Diphenylenemethane; **9*H*-Fluorene**; 2,2'-Methylenebiphenyl.

Structural Formula:

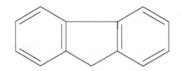

CHEMICAL DESIGNATIONS

CAS Registry Number: 86-73-7

DOT Designation: None assigned.

Empirical Formula: $C_{13}H_{10}$

Formula Weight: 166.22

RTECS Number: LL 5670000

PHYSICAL AND CHEMICAL PROPERTIES

Appearance: Small white leaflets or flakes. Fluorescent when impure.

Boiling Point: 298 °C [1]; 294 °C [2].

Henry's Law Constant: 2.1 x 10^{-4} atm·m^3/mol [3].

Ionization Potential: 8.63 eV [4]; 8.56 eV [5].

Log K_{oc}: 3.70 [6].

Log K_{ow}: 4.12 [7]; 4.18 [8]; 4.38 [9].

Melting Point: 116-117 °C [10]; 116 °C [2]; 113-115 °C [11].

Solubility in Organics: Soluble in most solvents [12].

Solubility in Water: 0.8 ppm [13]; 1.69 mg/L at 25 °C [14]; 1.98 mg/L at 25 °C [15]; 0.190 mg/L at 25 °C [16]; 1.6622 mg/L at 25 °C [17].

Specific Density: 1.203 at 0/4 °C [10].

Transformation Products: No data found.

Vapor Pressure: 10 mm at 146 °C, 40 mm at 185.2 °C, 100 mm at 214.7 °C, 400 mm at 268.6 °C, 760 mm at 295 °C [10]; estimated as 0.001-0.01 mm at 20 °C based on data for structurally similar compounds [18]; 7.1 x 10^{-4} mm [19]; 0.113 mm at 75 °C, 0.167 mm at 80 °C, 0.242 mm at 85 °C, 0.356 mm at 90 °C, 0.512 mm at 95 °C, 0.744 mm at 100 °C, 1.061 mm at 105 °C, 1.500 mm at 110 °C, 1.943 mm at 114 °C [20]; 0.000195 mm at 34.85 °C, 0.000250 mm at 37.20 °C, 0.000164 mm at 33.30 °C, 0.000281 mm at 38.45 °C, 0.000343 mm at 40.30 °C, 0.000543 mm at 45.00 °C, 0.000708 mm at 47.75 °C, 0.000818 mm at 49.25 °C, 0.000833 mm at 49.55 °C [21].

FIRE HAZARDS

Flash Point: No data found.

Lower Explosive Limit (LEL): No data found.

Upper Explosive Limit (UEL): No data found.

HEALTH HAZARD DATA

Immediately Dangerous to Life or Health (IDLH): Potential human carcinogen [22].

Permissible Exposure Limits (PEL) in Air: No individual standards have been set, however, as a constituent in coal tar pitch volatiles, the following exposure limits have been established: 0.2 mg/m^3 (benzene-soluble fraction) [23]; 0.1 mg/m^3 10-hour TWA (cyclohexane-extractable fraction) [22]; 0.2 mg/m^3 TWA (benzene solubles) [24].

MANUFACTURING

Selected Manufacturers:

Aldrich Chemical Co.
940 West Saint Paul Ave.
Milwaukee, WI 53233

Fluka Chemical Corp.
980 South Second St.
Ronkonkoma, NY 11779

Uses: Chemical intermediate in numerous applications and in the formation of polyradicals for resins; insecticides and dyestuffs. Derived from industrial and experimental coal gasification operations where maximum concentrations detected in gas, liquid, and coal tar streams were 9.1 mg/m^3, 0.057 mg/m^3, and 8.0 mg/m^3, respectively [25].

REFERENCES

1. *Catalog Handbook of Fine Chemicals* (Milwaukee, WI: Aldrich Chemical Co., 1988), 2212 p.
2. Huntress, E.H., and S.P. Mulliken. *Identification of Pure Organic Compounds - Tables of Data on Selected Compounds of Order I* (New York: John Wiley and Sons, Inc., 1941), 691 p.
3. Petrasek, A.C., Kugelman, I.J., Austern, B.M., Pressley, T.A., Winslow, L.A., and R.H. Wise. "Fate of Toxic Organic Compounds in Wastewater Treatment Plants," *J. Water Poll. Control Fed.*, 55(10):1286-1296 (1983).
4. Franklin, J.L., Dillard, J.G., Rosenstock, H.M., Herron, J.T., Draxl K., and F.H. Field. "Ionization Potentials, Appearance Potentials and Heats of Formation of Gaseous Positive Ions," National Bureau of Standards Report NSRDS-NBS 26, U.S. Government Printing Office (1969), 289 p.
5. Yoshida, K., Shigeoka, T., and F. Yamauchi. "Non-Steady State Equilibrium Model for the Preliminary Prediction of the Fate of Chemicals in the Environment," *Ecotoxicol. Environ. Safety*, 7(2):179-190 (1983).
6. Abdul, S.A., Gibson, T.L., and D.N. Rai. "Statistical Correlations for Predicting the Partition Coefficient for Nonpolar Organic Contaminants Between Aquifer Organic Carbon and Water," *Haz. Waste Haz. Mater.*, 4(3):211-222 (1987).

7. Chou, J.T., and P.C. Jurs. "Computer Assisted Computation of Partition Coefficients from Molecular Structures using Fragment Constants," *J. Chem. Info. Comp. Sci.*, 19:172-178 (1979).

8. Hansch, C., Leo, A., and D. Nikaitani. "On the Additive-Constitutive Character of Partition Coefficients," *J. Org. Chem.*, 37(20):3090-3092 (1972).

9. Isnard, S., and S. Lambert. "Estimating Bioconcentration Factors from Octanol-Water Partition Coefficient and Aqueous Solubility," *Chemosphere*, 17(1):21-34 (1988).

10. Weast, R.C., Ed. *CRC Handbook of Chemistry and Physics*, 67th ed. (Boca Raton, FL: CRC Press, Inc., 1986), 2406 p.

11. *Fluka Catalog 1988/89 - Chemika-Biochemika* (Ronkonkoma, NY: Fluka Chemical Corp., 1988), 1536 p.

12. "Chemical, Physical, and Biological Properties of Compounds Present at Hazardous Waste Sites," U.S. EPA Report-530/SW-89-010 (1985), 619 p.

13. Verschueren, K. *Handbook of Environmental Data on Organic Chemicals* (New York: Van Nostrand Reinhold Co., 1983), 1310 p.

14. May, W.E., Wasik, S.P., and D.H. Freeman. "Determination of the Aqueous Solubility of Polynuclear Aromatic Hydrocarbons by a Coupled Column Liquid Chromatographic Technique," *Anal. Chem.*, 50(1):175-179 (1978).

15. Mackay, D., and W.Y. Shiu. "Aqueous Solubility of Polynuclear Aromatic Hydrocarbons," *J. Chem. Eng. Data*, 22(4):399-402 (1977).

16. Wauchope, R.D., and F.W. Getzen. "Temperature Dependence of Solubilities in Water and Heats of Fusion of Solid Aromatic Hydrocarbons," *J. Chem. Eng. Data*, 17:38-41 (1972).

17. Sahyun, M.R.V. "Binding of Aromatic Compounds to Bovine Serum Albumin," *Nature*, 209(5023):613-614 (1966).

18. "Treatability Manual - Volume 1: Treatability Data," Office of Research and Development, U.S. EPA Report-600/8-80-042a (1980), 1035 p.

19. Mabey, W.R., Smith, J.H., Podoll, R.T., Johnson, H.L., Mill, T., Chou, T.-W., Gates, J., Partridge, I.W., Jaber, H., and D. Vandenberg. "Aquatic Fate Process Data for Organic Priority Pollutants - Final Report," Office of Regulations and Standards, U.S. EPA Report-440/4-81-014 (1982), 407 p.

20. Osborn, A.G., and D.R. Douslin. "Vapor Pressures and Derived Enthalpies of Vaporization of Some Condensed-Ring Hydrocarbons," *J. Chem. Eng. Data*, 20(3):229-231 (1975).

21. Bradley, R.S., and T.G. Cleasby. "The Vapour Pressure and Lattice Energy of Some Aromatic Ring Compounds," *J. Chem. Soc. (London)*, pp 1690-1692 (1953).

22. "NIOSH Pocket Guide to Chemical Hazards," U.S. Department of Health and Human Services, U.S. Government Printing Office (1987), 241 p.

23. "General Industry Standards for Toxic and Hazardous Substances," U.S. Code of Federal Regulations 1910, Subpart Z Section 1910.1000 (July 1982).

24. *Documentation of the Threshold Limit Values and Biological Exposure Indices for 1986-1987* (Cincinnati, OH: American Conference of Governmental Industrial Hygienists, 1986), 111 p.

25. Cleland, J.G. "Project Summary - Environmental Hazard Rankings of Pollutants Generated in Coal Gasification Processes," Office of Research and Development, U.S. EPA Report-600/S7-81-101 (1981), 19 p.

HEPTACHLOR

Synonyms: Aahepta; Agroceres; Basaklor; 3-Chlorochlordene; Drinox; Drinox H-34; E 3,314; ENT 15,152; GPKh; H-34; Heptachlorane; 3,4,5,6,7,8,8-Heptachlorodicyclopentadiene; 3,4,5,6,7,8,8a-Heptachlorodicyclopentadiene; 1(3a),4,5,6,7,8,8-Heptachloro-3a(1),4,7,7a-tetrahydro-4,7-methanoindene; 1,4,5,6,7,8,8-Heptachloro-3a,4,7,7a-tetrahydro-4,7-methanoindene; **1,4,5,6,7,8,8-Heptachloro-3a,4,7,7a-tetrahydro-4,7-methanol-1*H*-indene**; 1,4,5,6,7,8,8-Heptachloro-3a,4,7,7a-tetrahydro-4,7-*endo*-methanoindene; 1,4,5,6,7,8,8a-Heptachloro-3a,4,7,7a-tetrahydro-4,7-methanoindene; 1,4,5,6,7,8,8-Heptachloro-3a,4,7,7a-tetrahydro-4,7-methyleneindene; 1,4,5,6,7,10,10-Heptachloro-4,7,8,9-tetrahydro-4,7-methyleneindene; 1,4,5,6,7,10,10-Heptachloro-4,7,8,9-tetrahydro-4,7-*endo*-methyleneindene; 3,4,5,6,7,8,8a-Heptachloro-α-dicyclopentadiene; Heptadichlorocyclopentadiene; Heptagran; Heptagranox; Heptamak; Heptamul; Heptasol; Heptox; NA 2,761; NCI-C00180; Soleptax; RCRA waste number P059; Rhodiachlor; Velsicol 104; Velsicol heptachlor.

Structural Formula:

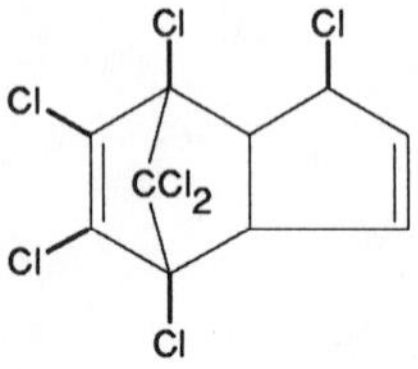

CHEMICAL DESIGNATIONS

CAS Registry Number: 76-44-8

DOT Designation: 2761

Empirical Formula: $C_{10}H_5Cl_7$

Formula Weight: 373.32

RTECS Number: PC 0700000

PHYSICAL AND CHEMICAL PROPERTIES

Appearance and Odor: Crystalline white to light tan, waxy solid with a camphor-like odor.

322

Boiling Point: Decomposes [1]; 135-145 °C at 1-1.5 mm [2].

Henry's Law Constant: 0.0023 atm·m^3/mol [3].

Ionization Potential: No data found.

Log K$_{oc}$: 4.34 using method of Kenaga and Goring [4].

Log K$_{ow}$: 5.44 [5]; 4.40 [6].

Melting Point: 95°-96 °C [1]; 46-74 °C [7].

Solubility in Organics: Soluble in acetone, benzene, ethanol, ether, carbon tetrachloride, xylene, kerosene, cyclohexanone, and ligroin [8].

Solubility in Water: 100 ppb at 15 °C, 180 ppb at 25 °C, 315 ppb at 35 °C, 490 ppb at 45 °C (particle sizes ≤ 5μ) [9]; 0.056 mg/L at 25 °C [2]; 0.030 ppm [4].

Specific Density: 1.57 at 9/4 °C [1]; 1.66 at 20/4 °C [7]; 1.65-1.67 at 25/4 °C [10]; 1.60 at 71/4 °C (technical grade) [11].

Transformation Products: Many soil microorganisms were found to oxidize heptachlor to heptachlor epoxide. In addition, hydrolysis produced hydroxychlordene with subsequent epoxidation yielding 1-hydroxy-2,3-epoxychlordene [12].

Vapor Pressure: 3 x 10^{-4} mm at 20 °C [13]; 4 x 10^{-4} mm at 25 °C [10]; technical grade has vapor pressures of 0.004 mm, 0.0016 mm, and 0.0280 mm at 25 °C, 30 °C, and 71 °C, respectively.

FIRE HAZARDS

Flash Point: Not flammable [7].

Lower Explosive Limit (LEL): Not flammable [7].

Upper Explosive Limit (UEL): Not flammable [7].

HEALTH HAZARD DATA

Immediately Dangerous to Life or Health (IDLH): 100 mg/m^3 [14].

Permissible Exposure Limits (PEL) in Air: 0.5 mg/m^3 [15]; 0.5 mg/m^3 TWA [16].

MANUFACTURING

Selected Manufacturer:

Velsicol Chemical Corp.
341 East Ohio St.
Chicago, IL 60611

Use: Insecticide for termite control.

REFERENCES

1. Weast, R.C., Ed. *CRC Handbook of Chemistry and Physics*, 67th ed. (Boca Raton, FL: CRC Press, Inc., 1986), 2406 p.
2. *IARC Monographs on the Evaluation of Carcinogenic Risk of Chemicals to Man. Some Halogenated Hydrocarbons, Volume 20* (Lyon, France: International Agency for Research on Cancer, 1979), 609 p.
3. Petrasek, A.C., Kugelman, I.J., Austern, B.M., Pressley, T.A., Winslow, L.A., and R.H. Wise. "Fate of Toxic Organic Compounds in Wastewater Treatment Plants," *J. Water Poll. Control Fed.*, 55(10):1286-1296 (1983).
4. Kenaga, E.E., and C.A.I. Goring. "Relationship Between Water Solubility, Soil Sorption, Octanol-Water Partitioning and Concentration of Chemicals in Biota," in *Aquatic Toxicology, ASTM STP 707*, Eaton, J.G., Parrish, P.R., and A.C. Hendricks, Eds. (Philadelphia, PA: American Society for Testing and Materials 1980), pp 78-115.
5. Travis, C.C., and A.D. Arms. "Bioconcentration of Organics in Beef, Milk and Vegetation," *Environ. Sci. Technol.*, 22(3):271-274 (1988).
6. "Treatability Manual - Volume 1: Treatability Data," Office of Research and Development, U.S. EPA Report-600/8-80-042a (1980), 1035 p.
7. Weiss, G. *Hazardous Chemicals Data Book* (Park Ridge, NJ: Noyes Data Corp., 1986), 1069 p.
8. "Chemical, Physical, and Biological Properties of Compounds Present at Hazardous Waste Sites," U.S. EPA Report-530/SW-89-010 (1985), 619 p.

9. Biggar, J.W., and I.R. Riggs. "Apparent Solubility of Organochlorine Insecticides in Water at Various Temperatures," *Hilgardia*, 42(10):383-391 (1974).

10. *Environmental Health Criteria 38: Heptachlor* (Geneva: World Health Organization, 1984), 81 p.

11. Standen, A., Ed. *Kirk-Othmer Encyclopedia of Chemical Technology, Volume 4*, 2nd ed. (New York: John Wiley and Sons, Inc., 1964), 937 p.

12. Kearney, P.C., and D.D. Kaufman. *Herbicides: Chemistry, Degradation and Mode of Action* (New York: Marcel Dekker, Inc., 1976), 1036 p.

13. Sims, R.C., Doucette, W.C., McLean, J.E., Grenney, W.J., and R.R. Dupont. "Treatment Potential for 56 EPA Listed Hazardous Chemicals in Soil," U.S. EPA Report-600/6-88-001 (1988), 105 p.

14. "NIOSH Pocket Guide to Chemical Hazards," U.S. Department of Health and Human Services, U.S. Government Printing Office (1987), 241 p.

15. "General Industry Standards for Toxic and Hazardous Substances," U.S. Code of Federal Regulations 1910, Subpart Z Section 1910.1000 (July 1982).

16. *Documentation of the Threshold Limit Values and Biological Exposure Indices for 1986-1987* (Cincinnati, OH: American Conference of Governmental Industrial Hygienists, 1986), 111 p.

HEPTACHLOR EPOXIDE

Synonyms: ENT 25,584; Epoxy heptachlor; HCE; 1,4,5,6,7,8,8-Heptachloro-2,3-epoxy-2,3,3a,4,7,7a-hexahydro-4,7-methanoindene; 1,2,3,4,5,6,7,8,8-Heptachloro-2,3-epoxy-3a,4,7,7a-tetrahydro-4,7-methanoindene; **2,3,4,5,6,7,7-Heptachloro-1a,1b,5,5a,6,6a-hexahydro-2,5-methano-2H-indeno[1,2-b]oxirene**; 2,3,4,5,6,7,7- Heptachloro-1a,1b,5,5a,6,6a-hexahydro-2,5-methano-2H-oxireno[a]indene; Velsicol 53-CS-17.

Structural Formula:

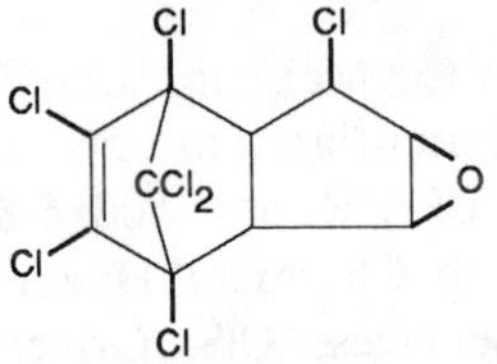

CHEMICAL DESIGNATIONS

CAS Registry Number: 1024-57-3

DOT Designation: 2761

Empirical Formula: $C_{10}H_5Cl_7O$

Formula Weight: 389.32

RTECS Number: PB 9450000

PHYSICAL AND CHEMICAL PROPERTIES

Appearance: Liquid.

Boiling Point: No data found.

Henry's Law Constant: 3.2×10^{-5} atm·m^3/mol [1].

Ionization Potential: No data found.

Log K_{oc}: 4.32 using method of Kenaga and Goring [2].

Log K_{ow}: 5.40 [3]; 3.65 [4].

Melting Point: 157-160 °C [5].

Solubility in Organics: No data found.

Solubility in Water: 0.350 ppm at 25 °C [6]; 0.275 mg/L at 25 °C [1]; 110 ppb at 15 °C, 200 ppb at 25 °C, 350 ppb at 35 °C, 600 ppb at 45 °C (particle sizes ≤ 5μ) [7].

Specific Density: No data found.

Transformation Products: No data found.

Vapor Pressure: 2.6 x 10^{-6} mm at 20 °C [8].

FIRE HAZARDS

Flash Point: No data found.

Lower Explosive Limit (LEL): No data found.

Upper Explosive Limit (UEL): No data found.

HEALTH HAZARD DATA

Immediately Dangerous to Life or Health (IDLH): No data found.

Permissible Exposure Limits (PEL) in Air: No standards set.

MANUFACTURING

Uses: No data found.

REFERENCES

1. Warner, H.P., Cohen, J.M., and J.C. Ireland. "Determination of Henry's Law Constants of Selected Priority Pollutants," Office of Science and Development, U.S. EPA Report-600/D-87/229 (1987), 14 p.
2. Kenaga, E.E., and C.A.I. Goring. "Relationship Between Water

Solubility, Soil Sorption, Octanol-Water Partitioning and Concentration of Chemicals in Biota," in *Aquatic Toxicology, ASTM STP 707*, Eaton, J.G., Parrish, P.R., and A.C. Hendricks, Eds. (Philadelphia, PA: American Society for Testing and Materials 1980), pp 78-115.

3.   Travis, C.C., and A.D. Arms. "Bioconcentration of Organics in Beef, Milk and Vegetation," *Environ. Sci. Technol.*, 22(3):271-274 (1988).

4.   "Treatability Manual - Volume 1: Treatability Data," Office of Research and Development, U.S. EPA Report-600/8-80-042a (1980), 1035 p.

5.   Singh, J. "Conversion of Heptachloride to its Epoxide," *Bull. Environ. Contam. Toxicol.*, 4(2):77-79 (1969).

6.   Weil, L., Dure, G., and K.E. Quentin. "Solubility in Water of Insecticide Chlorinated Hydrocarbons and Polychlorinated Biphenyls in View of Water Pollution," *Z. Wasser Forsch.*, 7(6):169-175 (1974).

7.   Biggar, J.W., and I.R. Riggs. "Apparent Solubility of Organochlorine Insecticides in Water at Various Temperatures," *Hilgardia*, 42(10):383-391 (1974).

8.   *IARC Monographs on the Evaluation of Carcinogenic Risk of Chemicals to Man. Some Organochlorine Pesticides, Volume 4* (Lyon, France: International Agency for Research on Cancer, 1974), 241 p.

HEXACHLOROBENZENE

Synonyms: Amatin; Anticarie; Bunt-cure; Bunt-no-more; Co-op hexa; Granox NM; HCB; Hexa C.B.; Julin's carbon chloride; No bunt; No bunt 40; No bunt 80; No bunt liquid; Pentachlorophenyl chloride; Perchlorobenzene; Phenyl perchloryl; RCRA waste number U127; Sanocide; Smut-go; Snieciotox; UN 2729.

Structural Formula:

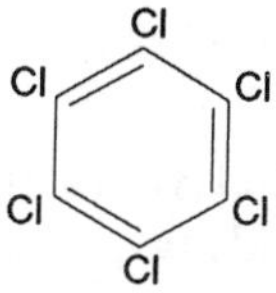

CHEMICAL DESIGNATIONS

CAS Registry Number: 118-74-1

DOT Designation: 2729

Empirical Formula: C_6Cl_6

Formula Weight: 284.78

RTECS Number: DA 2975000

PHYSICAL AND CHEMICAL PROPERTIES

Appearance: Monoclinic white crystals.

Boiling Point: 332 °C [1].

Henry's Law Constant: 0.0017 atm·m³/mol [2]; 0.0013 atm·m³/mol at 23 °C [3].

Ionization Potential: No data found.

Log K_{oc}: 3.59 [4]; 4.01 [5]; 2.56, 2.70, 4.32, and 4.54 [6]; 2.56, 2.70 [7].

Log K_{ow}: 5.45 [8]; 5.50 [9]; 5.66 [10]; 5.47 [11]; 6.18 [12]; 5.23 [13]; 5.31 [14]; 5.57 [15]; 5.20, 5.55 [16]; 5.44 [5]; 3.93 [17]; 5.2 [18]; 6.06 [19]; 5.68 at 13 °C, 5.70 at 19 °C, 5.58 at 28 °C, 5.17 at 33 °C [20]; 6.22 [21]; 5.00 [22].

Melting Point: 230 °C [23]; 227 °C [24].

Solubility in Organics: Soluble in acetone, benzene, ether, and chloroform [25].

Solubility in Water: 0.11 mg/L at 24 °C, 0.006 mg/L at 25 °C [26]; 0.0047 mg/L at 25 °C [27]; 0.005 mg/L at 25 °C [28]; 0.035 ppm [29]; 0.02 mg/L [30].

Specific Density: 1.5691 at 23.6/4 °C [23]; 2.049 at 20/4 °C [31].

Transformation Products: Reductive monodechlorination occurred in an anaerobic sewage sludge yielding principally 1,3,5-trichlorobenzene. Other compounds identified included pentachlorobenzene, 1,2,3,5-tetrachlorobenzene, and dichlorobenzenes [32].

Vapor Pressure: 1.089×10^{-5} mm at 20 °C [33]; 1 mm at 114.4 °C; 10 mm at 166.4 °C, 40 mm at 206.0 °C, 100 mm at 235.5 °C, 400 mm at 283.5 °C, 760 mm at 309.4 °C [23].

FIRE HAZARDS

Flash Point: 242 °C [34].

Lower Explosive Limit (LEL): No data found.

Upper Explosive Limit (UEL): No data found.

HEALTH HAZARD DATA

Immediately Dangerous to Life or Health (IDLH): No data found.

Permissible Exposure Limits (PEL) in Air: No standards set.

MANUFACTURING

Selected Manufacturer:

Aldrich Chemical Co.
940 West Saint Paul Ave.
Milwaukee, WI 53233

Uses: Organic synthesis; seed fungicide; wood preservative.

REFERENCES

1. *Catalog Handbook of Fine Chemicals* (Milwaukee, WI: Aldrich Chemical Co., 1988), 2212 p.
2. Warner, H.P., Cohen, J.M., and J.C. Ireland. "Determination of Henry's Law Constants of Selected Priority Pollutants," Office of Science and Development, U.S. EPA Report-600/D-87/229 (1987), 14 p.
3. Howard, P.H. *Handbook of Environmental Fate and Exposure Data for Organic Chemicals* (Chelsea, MI: Lewis Publishers, Inc., 1989), 574 p.
4. Kenaga, E.E. "Predicted Bioconcentration Factors and Soil Sorption Coefficients of Pesticides and other Chemicals," *Ecotoxicol. Environ. Safety*, 4(1):26-38 (1980).
5. Briggs, G.G. "Theoretical and Experimental Relationships Between Soil Adsorption, Octanol-Water Partition Coefficients, Water Solubilities, Bioconcentration Factors, and the Parachor," *J. Agric. Food Chem.*, 29(5):1050-1059 (1981).
6. Hodson, J., and N.A. Williams. "The Estimation of the Adsorption Coefficient (K_{oc}) for Soils by High Performance Liquid Chromatography," *Chemosphere*, 19(1):67-77 (1988).
7. Rippen, G., Ilgenstein, M., and W. Klöpffer. "Screening of the Adsorption Behavior of New Chemicals: Natural Soils and Model Adsorbents," *Ecotoxicol. Environ. Safety*, 6(3):236-245 (1982).
8. Travis, C.C., and A.D. Arms. "Bioconcentration of Organics in Beef, Milk and Vegetation," *Environ. Sci. Technol.*, 22(3):271-274 (1988).
9. Chiou, C.T., Schmedding D.W., and M. Manes. "Partitioning of Organic Compounds in Octanol-Water Systems," *Environ. Sci. Technol.*, 16(1):4-10 (1982).
10. Isnard, S., and S. Lambert. "Estimating Bioconcentration Factors from Octanol-Water Partition Coefficient and Aqueous Solubility," *Chemosphere*, 17(1):21-34 (1988).
11. Miller, M.M., Ghodbane, S., Wasik, S.P., Tewari, Y.B., and D.E. Martire. "Aqueous Solubilities, Octanol/Water Partition Coefficients, and Entropies of Melting of Chlorinated Benzenes and Biphenyls," *J. Chem. Eng. Data*, 29(2):184-190 (1984).
12. "Treatability Manual - Volume 1: Treatability Data," Office of Research and Development, U.S. EPA Report-600/8-80-042a (1980), 1035 p.
13. Mackay, D. "Correlation of Bioconcentration Factors," *Environ. Sci.*

Technol., 16(5):274-278 (1982).

14. Watarai, H., Tanaka, M., and N. Suzuki. "Determination of Partition Coefficients of Halobenzenes in Heptane/Water and 1-Octanol/Water Systems and Comparison with the Scaled Particle Calculation," *Anal. Chem.*, 54(4):702-705 (1982).

15. Brooke, D.N., Dobbs, A.J., and N. Williams. "Octanol:Water Partition Coefficients (P): Measurement, Estimation, and Interpretation, Particularly for Chemicals with P > 10^5," *Ecotoxicol. Environ. Safety*, 11(3):251-260 (1986).

16. Geyer, H., Politzki, G., and D. Freitag. "Prediction of Ecotoxicological Behaviour of Chemicals: Relationship Between *n*-Octanol/Water Partition Coefficient and Bioaccumulation of Organic Chemicals by Alga *Chlorella*," *Chemosphere*, 13(2):269-284 (1984).

17. Veith, G.D., Macek, K.J., Petrocelli, S.R., and J. Carroll. "An Evaluation of Using Partition Coefficients and Water Solubility to Estimate Bioconcentration Factors for Organic Chemicals in Fish," in *Aquatic Toxicology, ASTM STP 707*, Eaton, J.G., Parrish, P.R., and A.C. Hendricks, Eds. (Philadelphia, PA: American Society for Testing and Materials 1980), pp 116-129.

18. Platford, R.F., Carey, J.H., and E.J. Hale. "The Environmental Significance of Surface Films: Part 1. Octanol-Water Partition Coefficients for DDT and Hexachlorobenzene," *Environ. Poll. (Series B)*, 3(2):125-128 (1982).

19. Schwarzenbach, R.P., Giger, W., Hoehn, E., and J.K. Schneider. "Behavior of Organic Compounds during Infiltration of River Water to Groundwater. Field Studies, " *Environ. Sci. Technol.*, 17(8):472-479 (1983).

20. Opperhuizen, A., Serné, P., and J.M.D. Van der Steen. "Thermodynamics of Fish/Water Octan-1-ol/Water Partitioning of Some Chlorinated Benzenes," *Environ. Sci. Technol.*, 22(3):286-298 (1988).

21. Rao, P.S.C., and J.M. Davidson. "Estimation of Pesticide Retention and Transformation Parameters Required in Nonpoint Source Pollution Models," in *Environmental Impact of Nonpoint Source Pollution*, Overcash, M.R., and J.M. Davidson, Eds. (Ann Arbor, MI: Ann Arbor Science Publishers, Inc., 1980), pp 23-67.

22. Könemann, H., Zelle, R., and F. Busser. "Determination of Log P_{oct} Values of Chloro-Substituted Benzenes, Toluenes, and Anilines by High-Performance Liquid Chromatography on ODS-Silica," *J. Chromatogr.*, 178:559-565 (1979).

23. Weast, R.C., Ed. *CRC Handbook of Chemistry and Physics*, 67th ed. (Boca Raton, FL: CRC Press, Inc., 1986), 2406 p.

24. Standen, A., Ed. *Kirk-Othmer Encyclopedia of Chemical Technology, Volume 4*, 2nd ed. (New York: John Wiley and Sons, Inc., 1964), 937 p.

25. "Chemical, Physical, and Biological Properties of Compounds Present at Hazardous Waste Sites," U.S. EPA Report-530/SW-89-010 (1985), 619 p.

26. Metcalf, R.L., Kapoor, I.P., Lu, P.-Y., Schuth, C.K., and P. Sherman. "Model Ecosystem Studies of the Environmental Fate of Six Organochlorine Pesticides," *Environ. Health Perspect.*, (June 1973), pp 35-44.

27. Miller, M.M., Wasik, S.P., Huang, G.-L., Shiu, W.-Y., and D. Mackay. "Relationships Between Octanol-Water Partition Coefficient and Aqueous Solubility," *Environ. Sci. Technol.*, 19(6):522-529 (1985).

28. Yalkowsky, S.H., Orr, R.J., and S.C. Valvani. "Solubility and Partitioning. 3. The Solubility of Halobenzenes in Water," *Indust. Eng. Chem. Fund.*, 18(4):351-353 (1979).

29. Kenaga, E.E., and C.A.I. Goring. "Relationship Between Water Solubility, Soil Sorption, Octanol-Water Partitioning and Concentration of Chemicals in Biota," in *Aquatic Toxicology, ASTM STP 707*, Eaton, J.G., Parrish, P.R., and A.C. Hendricks, Eds. (Philadelphia, PA: American Society for Testing and Materials 1980), pp 78-115.

30. Gile, J.D., and J.W. Gillett. "Fate of Selected Fungicides in a Terrestrial Laboratory Ecosystem," *J. Agric. Food Chem.*, 27(6):1159-1164 (1979).

31. Melnikov, N.N. *Chemistry of Pesticides* (New York: Springer-Verlag, Inc., 1971), 480 p.

32. Fathepure, B.Z., Tiedje, J.M., and S.A. Boyd. "Reductive Dechlorination of Hexachlorobenzene to Tri- and Dichlorobenzenes in Anaerobic Sewage Sludge," *Appl. Environ. Microbiol.*, 54(2):327-330 (1988).

33. Isensee, A.R., Holden, E.R., Woolson, E.A., and G.E. Jones. "Soil Persistence and Aquatic Bioaccumulation Potential of Hexachlorobenzene," *J. Agric. Food Chem.*, 24(6):1210-1214 (1976).

34. Hawley, G.G. *The Condensed Chemical Dictionary* (New York: Van Nostrand Reinhold Co., 1981), 1135 p.

HEXACHLOROBUTADIENE

Synonyms: Dolen-pur; GP-40-66:120; HCBD; Hexachlorbutadiene; 1,1,2,3,4,4-Hexachlorobutadiene; 1,3-Hexachlorobutadiene; Hexachloro-1,3-butadiene; **1,1,2,3,4,4-Hexachloro-1,3-butadiene**; Perchlorobutadiene; RCRA waste number U128; UN 2279.

Structural Formula:

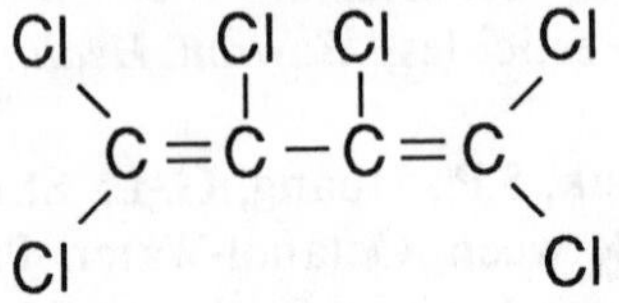

$$\underset{Cl}{\overset{Cl}{\underset{\diagup}{\overset{\diagdown}{C}}}} = \underset{}{\overset{Cl}{\underset{|}{C}}} - \underset{}{\overset{Cl}{\underset{|}{C}}} = \underset{Cl}{\overset{Cl}{\underset{\diagdown}{\overset{\diagup}{C}}}}$$

CHEMICAL DESIGNATIONS

CAS Registry Number: 87-68-3

DOT Designation: 2279

Empirical Formula: C_4Cl_6

Formula Weight: 260.76

RTECS Number: EJ 0700000

PHYSICAL AND CHEMICAL PROPERTIES

Appearance and Odor: Clear, yellowish-green liquid with a mild to pungent odor.

Boiling Point: 215 °C [1].

Henry's Law Constant: 0.026 atm·m³/mol [2]; 0.00102 atm·m³/mol [3]; 0.0103 atm·m³/mol [4].

Ionization Potential: No data found.

Log K_{oc}: 3.67 using method of Chiou and others [5].

Log K_{ow}: 4.78 [6].

Melting Point: -21 °C [1].

Solubility in Organics: Soluble in ethanol and ether [7].

Solubility in Water: 3.23 ppm at 25 °C [6]; 200 mg/L [8]; 2 mg/L at 20 °C [9]; 4 mg/L at 20-25 °C [10]; 2.55 mg/L at 20 °C [4].

Specific Density: 1.5542 at 20/4 °C [1]; 1.6820 at 20/4 °C [8].

Transformation Products: No data found.

Vapor Density: 10.66 g/L at 25 °C, 9.00 (air = 1).

Vapor Pressure: 22 mm at 100 °C, 500 mm at 200 °C [11]; 0.15 mm at 20 °C [12].

FIRE HAZARDS

Flash Point: Non-combustible [13].

Lower Explosive Limit (LEL): No data found.

Upper Explosive Limit (UEL): No data found.

HEALTH HAZARD DATA

Immediately Dangerous to Life or Health (IDLH): Suspected carcinogen [14].

Permissible Exposure Limits (PEL) in Air: 0.02 ppm ($\approx$0.24 mg/m^3) [15].

MANUFACTURING

Selected Manufacturers:

Aldrich Chemical Co.
940 West Saint Paul Ave.
Milwaukee, WI 53233

Fluka Chemical Corp.
980 South Second St.
Ronkonkoma, NY 11779

Pfaltz & Bauer, Inc.
172 East Aurora St.
Waterbury, CT 06708

Uses: Solvent for elastomers, natural rubber, synthetic rubber; heat-transfer liquid; transformer and hydraulic fluid; wash liquor for C_4 and higher hydrocarbons; sniff gas recovery agent in chlorine plants; chemical intermediate for fluorinated lubricants and rubber compounds; fluid for gyroscopes; fumigant for grapes.

REFERENCES

1. Weast, R.C., Ed. *CRC Handbook of Chemistry and Physics*, 67th ed. (Boca Raton, FL: CRC Press, Inc., 1986), 2406 p.
2. Pankow, J.F., and M.E. Rosen. "Determination of Volatile Compounds in Water by Purging Directly to a Capillary Column with Whole Column Cryotrapping," *Environ. Sci. Technol.*, 22(4):398-405 (1988).
3. Warner, H.P., Cohen, J.M., and J.C. Ireland. "Determination of Henry's Law Constants of Selected Priority Pollutants," Office of Science and Development, U.S. EPA Report-600/D-87/229 (1987), 14 p.
4. Howard, P.H. *Handbook of Environmental Fate and Exposure Data for Organic Chemicals* (Chelsea, MI: Lewis Publishers, Inc., 1989), 574 p.
5. Chiou, C.T., Peters, L.J., and V.H. Freed. "A Physical Concept of Soil-Water Equilibria for Nonionic Organic Compounds," *Science*, 206:831-832 (1979).
6. Banerjee, S., Yalkowsky, S.H., and S.C. Valvani. "Water Solubility and Octanol/Water Partition Coefficients of Organics. Limitations of the Solubility-Partition Coefficient Correlation," *Environ. Sci. Technol.*, 14(10):1227-1229 (1980).
7. "Chemical, Physical, and Biological Properties of Compounds Present at Hazardous Waste Sites," U.S. EPA Report-530/SW-89-010 (1985), 619 p.
8. Melnikov, N.N. *Chemistry of Pesticides* (New York: Springer-Verlag, Inc., 1971), 480 p.
9. Callahan, M.A., Slimak, M.W., Gable, N.W., May, I.P., Fowler, C.F., Freed, J.R., Jennings, P., Durfee, R.L., Whitmore, F.C., Maestri, B., Mabey, W.R., Holt, B.R., and C. Gould. "Water-Related Environmental Fate of 129 Priority Pollutants Volumes I and II," National Technical Information Service, U.S. EPA

Report-440/4-79-029 (1979), 1160 p.

10. Geyer, H., Kraus, A.S., Klein, W., Richter, E., and F. Korte. "Relationship Between Water Solubility and Bioaccumulation Potential of Organic Chemicals in Rats," *Chemosphere*, 9(5/6):277-294 (1980).

11. Verschueren, K. *Handbook of Environmental Data on Organic Chemicals* (New York: Van Nostrand Reinhold Co., 1983), 1310 p.

12. McConnell, G., Ferguson, D.M., and C.R. Pearson. "Chlorinated Hydrocarbons and the Environment," *Endeavour*, 34(121):13-18 (1975).

13. "NIOSH Pocket Guide to Chemical Hazards," U.S. Department of Health and Human Services, U.S. Government Printing Office (1987), 241 p.

14. Weiss, G. *Hazardous Chemicals Data Book* (Park Ridge, NJ: Noyes Data Corp., 1986), 1069 p.

15. *Documentation of the Threshold Limit Values and Biological Exposure Indices for 1986-1987* (Cincinnati, OH: American Conference of Governmental Industrial Hygienists, 1986), 111 p.

HEXACHLOROCYCLOPENTADIENE

Synonyms: C-56; Graphlox; HCCP; HCCPD; HCPD; Hex; **1,2,3,4,5,5-Hexachloro-1,3-cyclopentadiene**; HRS 1,655; NCI-C55607; PCL; Perchlorocyclopentadiene; RCRA waste number U130; UN 2646.

Structural Formula:

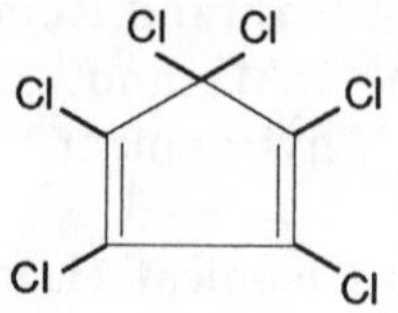

CHEMICAL DESIGNATIONS

CAS Registry Number: 77-47-4

DOT Designation: 2646

Empirical Formula: C_5Cl_6

Formula Weight: 272.77

RTECS Number: GY 1225000

PHYSICAL AND CHEMICAL PROPERTIES

Appearance and Odor: Greenish-yellow liquid with a harsh, unpleasant odor.

Boiling Point: 236-238 °C [1]; 239 °C at 753 mm [2].

Henry's Law Constant: 0.016 atm·m^3/mol [3].

Ionization Potential: No data found.

Log K_{oc}: 3.63 [4].

Log K_{ow}: 4.00 [5]; 5.51 [6]; 5.04 [7].

Melting Point: -9 °C [2]; 9.9 °C [8].

Solubility in Organics: Based on structurally similar compounds,

338

hexachlorocyclopentadiene is expected to be soluble in benzene, ethanol, chloroform, and other liquid halogenated solvents.

Solubility in Water: 0.805 mg/L [9]; 1.8 mg/L at 25 °C [10]; at 22 °C the following solubilities were reported: 1.11 mg/L in distilled water, 1.14 mg/L in deionized water, 1.08 mg/L in tap water, and 1.08 mg/L in Sugar Creek water [4].

Specific Density: 1.7019 at 25/4 °C [2]; 1.715 at 15/15 °C [8]; 1.7119 at 20/4 °C [11].

Transformation Products: No data found.

Vapor Density: 11.15 g/L at 25 °C, 9.42 (air = 1).

Vapor Pressure: 0.081 mm at 25 °C [11]; 0.4 mm at 50 °C, 0.54 mm at 54.2 °C, 1.90 mm at 75 °C, 7.50 mm at 100 °C, 24.0 mm at 125 °C, 44.0 mm at 140 °C, 64.0 mm at 150 °C, 162.0 mm at 175 °C, 760.0 mm at 239 °C [8]; 1 mm at 78-79 °C [4].

FIRE HAZARDS

Flash Point: Not flammable [12].

Lower Explosive Limit (LEL): Not flammable [12].

Upper Explosive Limit (UEL): Not flammable [12].

HEALTH HAZARD DATA

Immediately Dangerous to Life or Health (IDLH): No data found.

Permissible Exposure Limits (PEL) in Air: 0.01 ppm ($\approx$0.1 mg/m^3) [13].

MANUFACTURING

Selected Manufacturers:

Hooker Chemical Corp.
Industrial Chemicals Division
Niagara Falls, NY 14302

Velsicol Chemical Corp.
341 East Ohio St.
Chicago, IL 60611

Uses: Intermediate in the synthesis of dyes, cyclodiene pesticides, fungicides, and pharmaceuticals.

REFERENCES

1. Melnikov, N.N. *Chemistry of Pesticides* (New York: Springer-Verlag, Inc., 1971), 480 p.
2. Weast, R.C., Ed. *CRC Handbook of Chemistry and Physics*, 67th ed. (Boca Raton, FL: CRC Press, Inc., 1986), 2406 p.
3. Pankow, J.F., and M.E. Rosen. "Determination of Volatile Compounds in Water by Purging Directly to a Capillary Column with Whole Column Cryotrapping," *Environ. Sci. Technol.*, 22(4):398-405 (1988).
4. Chou, S.F.J., and R.A. Griffin. "Soil, Clay and Caustic Soda Effects on Solubility, Sorption and Mobility of Hexachlorocyclopentadiene," Environmental Geology Notes 104, Illinois Department of Energy and Natural Resources (1983), 54 p.
5. Mills, W.B., Porcella, D.B., Ungs, M.J., Gherini, S.A., Summers, K.V., Mok, L., Rupp, G.L., and G.L. Bowie. "Water Quality Assessment: A Screening Procedure for Toxic and Conventional Pollutants in Surface and Groundwater-Part I," Office of Research and Development, U.S. EPA Report-600/6-85-002a (1985), 638 p.
6. Mackay, D. "Correlation of Bioconcentration Factors," *Environ. Sci. Technol.*, 16(5):274-278 (1982).
7. Geyer, H., Politzki, G., and D. Freitag. "Prediction of Ecotoxicological Behaviour of Chemicals: Relationship Between *n*-Octanol/Water Partition Coefficient and Bioaccumulation of Organic Chemicals by Alga *Chlorella*," *Chemosphere*, 13(2):269-284 (1984).
8. Standen, A., Ed. *Kirk-Othmer Encyclopedia of Chemical Technology, Volume 4*, 2nd ed. (New York: John Wiley and Sons, Inc., 1964), 937 p.
9. Lu, P.-Y., Metcalf, R.L., Hirwe, A.S., and J.W. Williams. "Evaluation of Environmental Distribution and Fate of Hexachlorocyclopentadiene, Chlordene, Heptachlor, and Heptachlor Epoxide in a Laboratory Model Ecosystem," *J. Agric. Food Chem.*, 23(5):967-973 (1975).
10. Zepp, R.G., Wolfe, N.L., Baughman, G.L., Schlotzhauer, P.F., and

J.N. MacAllister. "Dynamics of Processes Influencing the Behavior of Hexachlorocyclopentadiene in the Aquatic Environment," 178th Meeting of the American Chemical Society, Washington, D.C. (September 1979).

11. Ungnade, H.E., and E.T. McBee. "The Chemistry of Perchlorocyclopentadienes and Cyclopentadienes," *Chem. Rev.*, 58:249-320 (1957).

12. Weiss, G. *Hazardous Chemicals Data Book* (Park Ridge, NJ: Noyes Data Corp., 1986), 1069 p.

13. *Documentation of the Threshold Limit Values and Biological Exposure Indices for 1986-1987* (Cincinnati, OH: American Conference of Governmental Industrial Hygienists, 1986), 111 p.

HEXACHLOROETHANE

Synonyms: Avlothane; Carbon hexachloride; Carbon trichloride; Distokal; Distopan; Distopin; Egitol; Ethane hexachloride; Ethylene hexachloride; Falkitol; Fasciolin; HCE; 1,1,1,2,2,2-Hexachloroethane; Hexachloroethylene; Mottenhexe; NA 9,037; NCI-C04604; Perchloroethane; Phenohep; RCRA waste number U131.

Structural Formula:

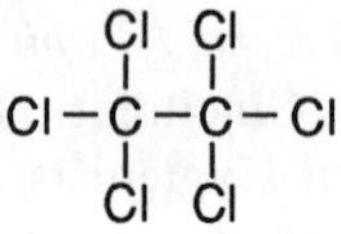

$$Cl - \underset{\underset{Cl}{|}}{\overset{\overset{Cl}{|}}{C}} - \underset{\underset{Cl}{|}}{\overset{\overset{Cl}{|}}{C}} - Cl$$

CHEMICAL DESIGNATIONS

CAS Registry Number: 67-72-1

DOT Designation: 9037

Empirical Formula: C_2Cl_6

Formula Weight: 236.74

RTECS Number: KI 4025000

PHYSICAL AND CHEMICAL PROPERTIES

Appearance and Odor: Rhombic, triclinic or cubic colorless crystals with a camphor-like odor.

Boiling Point: 184.4 °C [1]; 186.8 °C at 811 mm [2].

Henry's Law Constant: 0.0025 atm·m^3/mol [3]; 0.00218 atm·m^3/mol at 25 °C [4].

Ionization Potential: No data found.

Log K_{oc}: 3.34 [5].

Log K_{ow}: 4.62 [5]; 3.93 [6]; 3.58 [7].

Melting Point: 186.6 °C [8].

342

Solubility in Organics: Soluble in ethanol, benzene, chloroform, and ether [9].

Solubility in Water: 49.9 mg/L at 22 °C [10]; 27.2 mg/L at 25 °C [11]; 0.005 wt% at 22.3 °C [12]; 50 ppm at 20 °C [13]; 50 ppm at 25 °C [14].

Specific Density: 2.091 at 20/4 °C [8].

Transformation Products: No data found.

Vapor Density: 6.3 kg/m^3 at the sublimation point and 1 bar (0.987 atm) [13].

Vapor Pressure: 0.8 mm at 30 °C [15]; 1 mm at 32.7 °C, 10 mm at 73.5 °C, 40 mm at 102.3 °C, 100 mm at 124.2 °C, 400 mm at 163.8 °C, 760 mm at 185.6 °C [8]; 0.18 mm at 20 °C [10].

FIRE HAZARDS

Flash Point: Non-combustible [16].

Lower Explosive Limit (LEL): No data found.

Upper Explosive Limit (UEL): No data found.

HEALTH HAZARD DATA

Immediately Dangerous to Life or Health (IDLH): 300 ppm (potential human carcinogen) [16].

Permissible Exposure Limits (PEL) in Air: 1 ppm ($\approx$10 mg/m^3) [17]; lowest feasible level [16]; 10 ppm ($\approx$100 mg/m^3) TWA [18].

MANUFACTURING

Selected Manufacturers:

Aldrich Chemical Co.
940 West Saint Paul Ave.
Milwaukee, WI 53233

Fluka Chemical Corp.
980 South Second St.
Ronkonkoma, NY 11779

ICI Australia Ltd.
ICI House - 1 Nicholson St.
P.O. Box 4311
Melbourne Victoria 3001 Australia

Uses: Plasticizer for cellulose resins; moth repellant; camphor substitute in cellulose solvent; manufacturing of smoke candles and explosives; solvent; rubber vulcanization accelerator.

REFERENCES

1. Dean, J.A., Ed., *Lange's Handbook of Chemistry*, 11th ed. (New York: McGraw-Hill, Inc., 1973), 1570 p.
2. Standen, A., Ed. *Kirk-Othmer Encyclopedia of Chemical Technology, Volume 4*, 2nd ed. (New York: John Wiley and Sons, Inc., 1964), 937 p.
3. Pankow, J.F., and M.E. Rosen. "Determination of Volatile Compounds in Water by Purging Directly to a Capillary Column with Whole Column Cryotrapping," *Environ. Sci. Technol.*, 22(4):398-405 (1988).
4. Hine, J., and P.K. Mookerjee. "The Intrinsic Hydrophilic Character of Organic Compounds. Correlations in Terms of Structural Contributions," *J. Org. Chem.*, 40(3):292-298 (1975).
5. Abdul, S.A., Gibson, T.L., and D.N. Rai. "Statistical Correlations for Predicting the Partition Coefficient for Nonpolar Organic Contaminants Between Aquifer Organic Carbon and Water," *Haz. Waste Haz. Mater.*, 4(3):211-222 (1987).
6. Isnard, S., and S. Lambert. "Estimating Bioconcentration Factors from Octanol-Water Partition Coefficient and Aqueous Solubility," *Chemosphere*, 17(1):21-34 (1988).
7. Oliver, B.G., and A.J. Niimi. "Bioconcentration of Chlorobenzenes from Water by Rainbow Trout: Correlations with Partition Coefficients and Environmental Residues," *Environ. Sci. Technol.*, 17(5):287-291 (1983).
8. Weast, R.C., Ed. *CRC Handbook of Chemistry and Physics*, 67th ed. (Boca Raton, FL: CRC Press, Inc., 1986), 2406 p.
9. "Chemical, Physical, and Biological Properties of Compounds Present at Hazardous Waste Sites," U.S. EPA Report-530/SW-89-010

(1985), 619 p.

10. Munz, C., and P.V. Roberts. "Air-Water Phase Equilibria of Volatile Organic Solutes," *J. Am. Water Works Assoc.*, 79(5):62-69 (1987).

11. Lyman, W.J., Reehl, W.F., and D.H. Rosenblatt. *Handbook of Chemical Property Estimation Methods: Environmental Behavior of Organic Compounds* (New York: McGraw-Hill, Inc., 1982).

12. Stephen, H., and T. Stephen. *Solubilities of Inorganic and Organic Compounds - Part 1, Volume 1* (London: Pergamon Printing and Art Services, Ltd., 1963), 960 p.

13. Konietzko, H. "Chlorinated Ethanes: Sources, Distribution, Environmental Impact, and Health Effects," in *Hazard Assessment of Chemicals, Volume 3*, J. Saxena, Ed. (New York: Academic Press, Inc., 1984), pp 401-448.

14. Amoore, J.E., and E. Hautala. "Odor as an Aide to Chemical Safety: Odor Thresholds Compared with Threshold Limit Values and Volatilities for 214 Industrial Chemicals in Air and Water Dilution," *J. Appl. Toxicol.*, 3(6):272-290 (1983).

15. Verschueren, K. *Handbook of Environmental Data on Organic Chemicals* (New York: Van Nostrand Reinhold Co., 1983), 1310 p.

16. "NIOSH Pocket Guide to Chemical Hazards," U.S. Department of Health and Human Services, U.S. Government Printing Office (1987), 241 p.

17. "General Industry Standards for Toxic and Hazardous Substances," U.S. Code of Federal Regulations 1910, Subpart Z Section 1910.1000 (July 1982).

18. *Documentation of the Threshold Limit Values and Biological Exposure Indices for 1986-1987* (Cincinnati, OH: American Conference of Governmental Industrial Hygienists, 1986), 111 p.

2-HEXANONE

Synonyms: Butyl ketone; Butyl methyl ketone; *n*-Butyl methyl ketone; Hexanone-2; MBK; Methyl *n*-butyl ketone; MNBK; Propylacetone; RCRA waste number U159.

Structural Formula:

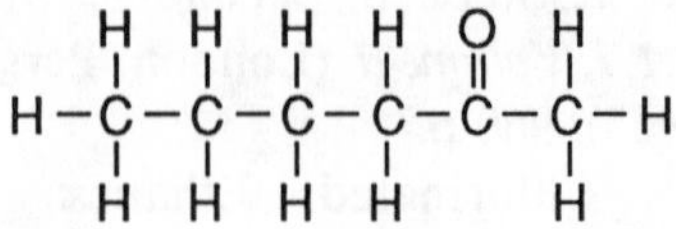

$$
\begin{array}{c}
\text{H} \quad \text{H} \quad \text{H} \quad \text{H} \quad \text{O} \quad \text{H} \\
| \quad\;\; | \quad\;\; | \quad\;\; | \quad\;\; || \quad\;\; | \\
\text{H}-\text{C}-\text{C}-\text{C}-\text{C}-\text{C}-\text{C}-\text{H} \\
| \quad\;\; | \quad\;\; | \quad\;\; | \qquad\;\; | \\
\text{H} \quad \text{H} \quad \text{H} \quad \text{H} \qquad \text{H}
\end{array}
$$

CHEMICAL DESIGNATIONS

CAS Registry Number: 591-78-6

DOT Designation: None assigned.

Empirical Formula: $C_6H_{12}O$

Formula Weight: 100.16

RTECS Number: MP 1400000

PHYSICAL AND CHEMICAL PROPERTIES

Appearance: Colorless liquid.

Boiling Point: 128 °C [1].

Henry's Law Constant: 0.00175 atm·m³/mol at 25 °C (calculated).

Ionization Potential: 9.35 eV [2].

Log K_{oc}: 2.13 using method of Kenaga and Goring [3].

Log K_{ow}: 1.38 [4].

Melting Point: -56.9 °C [5].

Solubility in Organics: Soluble in acetone, ethanol, and ether [1].

Solubility in Water: 35,000 mg/L at 25 °C [6]; 1.4 wt% at 20 °C [7];

34,400 mg/L at 25 °C [8]; 1.75 wt% at 20 °C, 1.64 wt% at 25 °C, 1.52 wt% at 30 °C [9].

Specific Density: 0.8113 at 20/4 °C [1]; 0.8300 at 20/4 °C [10].

Transformation Products: No data found.

Vapor Density: 4.09 g/L at 25 °C, 3.46 (air = 1).

Vapor Pressure: 2 mm at 20 °C [6]; 1 mm at 7.7 °C, 10 mm at 38.8 °C, 40 mm at 62.0 °C, 100 mm at 79.8 °C, 400 mm at 110.0 °C, 760 mm at 127.5 °C [1]; 3.8 mm at 25 °C [11].

FIRE HAZARDS

Flash Point: -9 °C [12].

Lower Explosive Limit (LEL): 1.2% [7].

Upper Explosive Limit (UEL): 8.0% [7].

HEALTH HAZARD DATA

Immediately Dangerous to Life or Health (IDLH): 5,000 ppm [7].

Permissible Exposure Limits (PEL) in Air: 100 ppm ($\approx$400 mg/m^3) [13]; 1 ppm ($\approx$4 mg/m^3) 10-hour TWA [7]; 5 ppm ($\approx$20 mg/m^3) TWA [11].

MANUFACTURING

Selected Manufacturers:

Aldrich Chemical Co.
940 West Saint Paul Ave.
Milwaukee, WI 53233

Fluka Chemical Corp.
980 South Second St.
Ronkonkoma, NY 11779

Pfaltz & Bauer, Inc.
172 East Aurora St.
Waterbury, CT 06708

Uses: Solvent of paints, varnishes, nitrocellulose lacquers; denaturant for ethyl alcohol; organic synthesis.

REFERENCES

1. Weast, R.C., Ed. *CRC Handbook of Chemistry and Physics*, 67th ed. (Boca Raton, FL: CRC Press, Inc., 1986), 2406 p.
2. Franklin, J.L., Dillard, J.G., Rosenstock, H.M., Herron, J.T., Draxl K., and F.H. Field. "Ionization Potentials, Appearance Potentials and Heats of Formation of Gaseous Positive Ions," National Bureau of Standards Report NSRDS-NBS 26, U.S. Government Printing Office (1969), 289 p.
3. Kenaga, E.E., and C.A.I. Goring. "Relationship Between Water Solubility, Soil Sorption, Octanol-Water Partitioning and Concentration of Chemicals in Biota," in *Aquatic Toxicology, ASTM STP 707*, Eaton, J.G., Parrish, P.R., and A.C. Hendricks, Eds. (Philadelphia, PA: American Society for Testing and Materials 1980), pp 78-115.
4. Leo, A., Hansch, C., and D. Elkins. "Partition Coefficients and Their Uses," *Chem. Rev.*, 71(6):525-616 (1971).
5. Dean, J.A., Ed., *Lange's Handbook of Chemistry*, 11th ed. (New York: McGraw-Hill, Inc., 1973), 1570 p.
6. Verschueren, K. *Handbook of Environmental Data on Organic Chemicals* (New York: Van Nostrand Reinhold Co., 1983), 1310 p.
7. "NIOSH Pocket Guide to Chemical Hazards," U.S. Department of Health and Human Services, U.S. Government Printing Office (1987), 241 p.
8. Meites, L., Ed. *Handbook of Analytical Chemistry*, 1st ed. (New York: McGraw-Hill, Inc., 1963), 1782 p.
9. Ginnings, P.M., Plonk, D., and E. Carter. "Aqueous Solubilities of Some Aliphatic Ketones," *J. Am. Chem. Soc.*, 62(8):1923-1924 (1940).
10. Standen, A., Ed. *Kirk-Othmer Encyclopedia of Chemical Technology, Volume 12*, 2nd ed. (New York: John Wiley and Sons, Inc., 1967), 905 p.
11. *Documentation of the Threshold Limit Values and Biological Exposure Indices for 1986-1987* (Cincinnati, OH: American Conference of Governmental Industrial Hygienists, 1986), 111 p.
12. *Fire Protection Guide on Hazardous Materials* (Quincy, MA: National

Fire Protection Association, 1984), 443 p.

13. "General Industry Standards for Toxic and Hazardous Substances," U.S. Code of Federal Regulations 1910, Subpart Z Section 1910.1000 (July 1982).

INDENO[1,2,3-cd]PYRENE

Synonyms: Indenopyren; IP; 1,10-(*o*-Phenylene)pyrene; 2,3-Phenylene-pyrene; 2,3-*o*-Phenylenepyrene; 3,4-(*o*-Phenylene)pyrene; 2,3-Phenylene-*o*-pyrene; 1,10-(1,2-Phenylene)pyrene; *o*-Phenylpyrene; RCRA waste number U137.

Structural Formula:

CHEMICAL DESIGNATIONS

CAS Registry Number: 193-39-5

DOT Designation: None assigned.

Empirical Formula: $C_{22}H_{12}$

Formula Weight: 276.34

RTECS Number: NK 9300000

PHYSICAL AND CHEMICAL PROPERTIES

Appearance: Solid.

Boiling Point: 536 °C [1].

Henry's Law Constant: 2.96 x 10^{-20} atm·m³/mol at 25 °C (calculated).

Ionization Potential: No data found.

Log K_{oc}: 7.49 using method of Karickhoff and others [2].

Log K_{ow}: 7.70 [3]; 5.97 [4].

Melting Point: 160-163 °C [1].

Solubility in Organics: Soluble in most solvents [5].

Solubility in Water: 0.062 mg/L [4].

Specific Density: No data found.

Transformation Products: No data found.

Vapor Pressure: 10^{-10} mm at 25 °C (calculated) [6].

FIRE HAZARDS

Flash Point: No data found.

Lower Explosive Limit (LEL): No data found.

Upper Explosive Limit (UEL): No data found.

HEALTH HAZARD DATA

Immediately Dangerous to Life or Health (IDLH): Potential human carcinogen [7].

Permissible Exposure Limits (PEL) in Air: No individual standards have been set, however, as a constituent in coal tar pitch volatiles, the following exposure limits have been established: 0.2 mg/m^3 (benzene-soluble fraction) [8]; 0.1 mg/m^3 10-hour TWA (cyclohexane-extractable fraction) [7]; 0.2 mg/m^3 TWA (benzene solubles) [9].

MANUFACTURING

Uses: Derived from industrial and experimental coal gasification operations where the maximum concentration detected in coal tar streams was 1.7 mg/m^3 [10].

REFERENCES

1. Verschueren, K. *Handbook of Environmental Data on Organic*

Chemicals (New York: Van Nostrand Reinhold Co., 1983), 1310 p.

2. Karickhoff, S.W., Brown, D.S., and T.A. Scott. "Sorption of Hydrophobic Pollutants on Natural Sediments," *Water Res.*, 13:241-248 (1979).

3. Mills, W.B., Porcella, D.B., Ungs, M.J., Gherini, S.A., Summers, K.V., Mok, L., Rupp, G.L., and G.L. Bowie. "Water Quality Assessment: A Screening Procedure for Toxic and Conventional Pollutants in Surface and Groundwater-Part I," Office of Research and Development, U.S. EPA Report-600/6-85-002a (1985), 638 p.

4. Sims, R.C., Doucette, W.C., McLean, J.E., Grenney, W.J., and R.R. Dupont. "Treatment Potential for 56 EPA Listed Hazardous Chemicals in Soil," U.S. EPA Report-600/6-88-001 (1988), 105 p.

5. "Chemical, Physical, and Biological Properties of Compounds Present at Hazardous Waste Sites," U.S. EPA Report-530/SW-89-010 (1985), 619 p.

6. "Treatability Manual - Volume 1: Treatability Data," Office of Research and Development, U.S. EPA Report-600/8-80-042a (1980), 1035 p.

7. "NIOSH Pocket Guide to Chemical Hazards," U.S. Department of Health and Human Services, U.S. Government Printing Office (1987), 241 p.

8. "General Industry Standards for Toxic and Hazardous Substances," U.S. Code of Federal Regulations 1910, Subpart Z Section 1910.1000 (July 1982).

9. *Documentation of the Threshold Limit Values and Biological Exposure Indices* (Cincinnati, OH: American Conference of Governmental Industrial Hygienists, 1986), 744 p.

10. Cleland, J.G. "Project Summary - Environmental Hazard Rankings of Pollutants Generated in Coal Gasification Processes," Office of Research and Development, U.S. EPA Report-600/S7-81-101 (1981), 19 p.

ISOPHORONE

Synonyms: Isoacetophorone; Isoforon; Isoforone; Isooctaphenone; Isophoron; NCI-C55618; 1,1,3-Trimethyl-3-cyclohexene-5-one; Trimethylcyclohexenone; **3,5,5-Trimethyl-2-cyclohexen-1-one.**

Structural Formula:

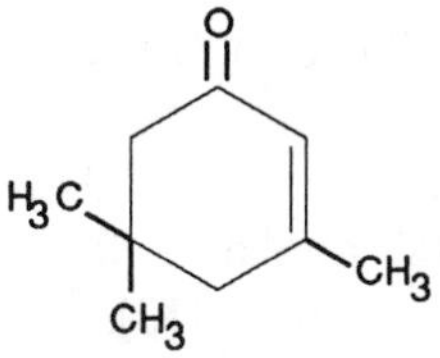

CHEMICAL DESIGNATIONS

CAS Registry Number: 78-59-1

DOT Designation: 1224

Empirical Formula: $C_9H_{14}O$

Formula Weight: 138.21

RTECS Number: GW 7700000

PHYSICAL AND CHEMICAL PROPERTIES

Appearance and Odor: Colorless liquid with a camphor-like odor.

Boiling Point: 215.2 °C [1]; 208-212 °C [2].

Henry's Law Constant: 5.8 x 10^{-6} atm·m³/mol (calculated) [3].

Ionization Potential: No data found.

Log K_{oc}: 1.49 using method of Karickhoff and others [4].

Log K_{ow}: 1.70 [5]; 1.67 [6].

Melting Point: -8.1 °C [7].

Solubility in Organics: Soluble in acetone, ethanol, and ether [8].

353

354   Isophorone

Solubility in Water: 12,000 mg/L at 25 °C [9].

Specific Density: 0.9229 at 20/4 °C [8]; 0.921 at 25/4 °C [10]; 0.9225 at 20.5/4 °C [11]; 0.921 at 20/4 °C [2].

Transformation Products: No data found.

Vapor Density: 5.65 g/L at 25 °C, 4.77 (air = 1).

Vapor Pressure: 0.38 mm at 20 °C [12]; 1 mm at 38.0 °C, 10 mm at 81.2 °C, 40 mm at 114.5 °C, 100 mm at 140.6 °C, 400 mm at 188.7 °C, 760 mm at 215.2 °C [8]; 0.2 mm at 20 °C [13].

FIRE HAZARDS

Flash Point: 84.4 °C [14].

Lower Explosive Limit (LEL): 0.8% [14].

Upper Explosive Limit (UEL): 3.8% [14].

HEALTH HAZARD DATA

Immediately Dangerous to Life or Health (IDLH): 800 ppm [14].

Permissible Exposure Limits (PEL) in Air: 25 ppm ($\approx$140 mg/m^3) [15]; 4 ppm ($\approx$23 mg/m^3) 10-hour TWA [14]; 5 ppm ($\approx$25 mg/m^3) TLV-C [13].

MANUFACTURING

Selected Manufacturers:

Exxon Chemical Corp.
P.O. Box 3272
Houston, TX 77001

Fluka Chemical Corp.
980 South Second St.
Ronkonkoma, NY 11779

Pfaltz & Bauer, Inc.
172 East Aurora St.
Waterbury, CT 06708

Uses: Solvent for paints, tin coatings, agricultural chemicals, and synthetic resins; excellent solvent for vinyl resins, cellulose esters, and ethers; pesticides; storing lacquers; pesticide manufacturing.

REFERENCES

1. Sax, N.I. *Dangerous Properties of Industrial Materials* (New York: Van Nostrand Reinhold Co., 1984), 3124 p.
2. *Fluka Catalog 1988/89 - Chemika-Biochemika* (Ronkonkoma, NY: Fluka Chemical Corp., 1988), 1536 p.
3. "Treatability Manual - Volume 1: Treatability Data," Office of Research and Development, U.S. EPA Report-600/8-80-042a (1980), 1035 p.
4. Karickhoff, S.W., Brown, D.S., and T.A. Scott. "Sorption of Hydrophobic Pollutants on Natural Sediments," *Water Res.*, 13:241-248 (1979).
5. Mills, W.B., Porcella, D.B., Ungs, M.J., Gherini, S.A., Summers, K.V., Mok, L., Rupp, G.L., and G.L. Bowie. "Water Quality Assessment: A Screening Procedure for Toxic and Conventional Pollutants in Surface and Groundwater-Part I," Office of Research and Development, U.S. EPA Report-600/6-85-002a (1985), 638 p.
6. Veith, G.D., Macek, K.J., Petrocelli, S.R., and J. Carroll. "An Evaluation of Using Partition Coefficients and Water Solubility to Estimate Bioconcentration Factors for Organic Chemicals in Fish," in *Aquatic Toxicology, ASTM STP 707*, Eaton, J.G., Parrish, P.R., and A.C. Hendricks, Eds. (Philadelphia, PA: American Society for Testing and Materials 1980), pp 116-129.
7. Dean, J.A., Ed., *Lange's Handbook of Chemistry*, 11th ed. (New York: McGraw-Hill, Inc., 1973), 1570 p.
8. Weast, R.C., Ed. *CRC Handbook of Chemistry and Physics*, 67th ed. (Boca Raton, FL: CRC Press, Inc., 1986), 2406 p.
9. Amoore, J.E., and E. Hautala. "Odor as an Aide to Chemical Safety: Odor Thresholds Compared with Threshold Limit Values and Volatilities for 214 Industrial Chemicals in Air and Water Dilution," *J. Appl. Toxicol.*, 3(6):272-290 (1983).
10. Weiss, G. *Hazardous Chemicals Data Book* (Park Ridge, NJ: Noyes Data Corp., 1986), 1069 p.
11. Huntress, E.H., and S.P. Mulliken. *Identification of Pure Organic*

Compounds - Tables of Data on Selected Compounds of Order I (New York: John Wiley and Sons, Inc., 1941), 691 p.

12. Verschueren, K. *Handbook of Environmental Data on Organic Chemicals* (New York: Van Nostrand Reinhold Co., 1983), 1310 p.

13. *Documentation of the Threshold Limit Values and Biological Exposure Indices* (Cincinnati, OH: American Conference of Governmental Industrial Hygienists, 1986), 744 p.

14. "NIOSH Pocket Guide to Chemical Hazards," U.S. Department of Health and Human Services, U.S. Government Printing Office (1987), 241 p.

15. "General Industry Standards for Toxic and Hazardous Substances," U.S. Code of Federal Regulations 1910, Subpart Z Section 1910.1000 (July 1982).

LINDANE

Synonyms: Aalindan; Aficide; Agrisol G-20; Agrocide; Agrocide 2; Agrocide 6G; Agrocide 7; Agrocide III; Agrocide WP; Agronexit; Ameisenatod; Ameisenmittel merck; Aparasin; Aphtiria; Aplidal; Arbitex; BBX; Ben-hex; Bentox 10; Benzene hexachloride; Benzene-γ-hexachloride; γ-Benzene hexachloride; Bexol; BHC; γ-BHC; Celanex; Chloran; Chloresene; Codechine; DBH; Detmol-extrakt; Detox 25; Devoran; Dol granule; Drill tox-spezial aglukon; ENT 7,796; Entomoxan; Exagama; Forlin; Gallogama; Gamacid; Gamaphex; Gamene; Gamiso; Gammahexa; Gammalin; Gammexene; Gammopaz; Gexane; HCCH; HCH; γ-HCH; Heclotox; Hexa; γ-Hexachlor; Hexachloran; γ-Hexachloran; Hexachlorane; γ-Hexachlorane; γ-Hexachlorobenzene; 1,2,3,4,5,6-Hexachlorocyclohexane; 1α,2α,3β,4α,5α,6-β-Hexachlorocyclohexane; 1,2,3,4,5,6-Hexachloro-γ-cyclohexane; γ-Hexachlorocyclohexane; γ-1,2,3,4,5,6-Hexachlorocyclohexane; Hexatox; Hexaverm; Hexicide; Hexyclan; HGI; Hortex; Inexit; γ-Isomer; Isotox; Jacutin; Kokotine; Kwell; Lendine; Lentox; Lidenal; Lindafor; Lindagam; Lindagrain; Lindagranox; γ-Lindane; Lindapoudre; Lindatox; Lindosep; Lintox; Lorexane; Milbol 49; Mszycol; NA 2,761; NCI-C00204; Neo-scabicidol; Nexen FB; Nexit; Nexit-stark; Nexol-E; Nicochloran; Novigam; Omnitox; Ovadziak; Owadziak; Pedraczak; Pflanzol; Quellada; RCRA waste number U129; Silvanol; Spritz-rapidin; Spruehpflanzol; Streunex; Tap 85; TBH; Tri-6; Viton.

Structural Formula:

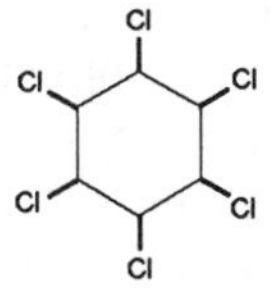

CHEMICAL DESIGNATIONS

CAS Registry Number: 58-89-9

DOT Designation: 2761

Empirical Formula: $C_6H_6Cl_6$

Formula Weight: 290.83

RTECS Number: GV 4900000

PHYSICAL AND CHEMICAL PROPERTIES

Appearance and Odor: Colorless and odorless solid.

Boiling Point: 323.4 °C [1].

Henry's Law Constant: 4.8 x 10^{-7} atm·m^3/mol [2]; 2.45 x 10^{-5} atm·m^3/mol [3]; 3.25 x 10^{-6} [4].

Ionization Potential: No data found.

Log K$_{oc}$: 3.03 [5]; 2.87 [6]; 2.96 [7]; 3.11 [3]; 3.52 [8]; 2.93 [9]; 3.42 [10]; 3.238 [11]; 3.2885 [12]; 2.96 [13]; 3.00 [14]; 2.81 [15].

Log K$_{ow}$: 3.70 [16]; 3.66 [17]; 3.85 [18]; 3.72 [19]; 3.30 [20]; 3.20 [21]; 3.89 [13]; 3.57 [22].

Melting Point: 112.5 °C [1]; 112.9 °C [23].

Solubility in Organics: Soluble in benzene, chloroform, and ether [1].

Solubility in Water: 1.4 mg/L in salt water [24]; 7.52 ppm at 25 °C [25]; 7.8 ppm at 25 °C [26]; 7.3 ppm at 25 °C, 12 ppm at 35 °C, 14 ppm at 45 °C [27]; 10 mg/L [28]; 6.98 ppm [29]; 7.5 g/m^3 at 25 °C [3]; 7.87 mg/L at 24 °C [16]; 17.0 mg/L at 24 °C [30]; 7.8 mg/L at 25 °C [11]; 0.150 ppm [31]; 10 ppm at 20 °C [32]; 5.75-7.40 ppm at 28 °C [19]; 12 ppm at 26.5 °C [33]; 2,150 ppb at 15 °C, 6,800 ppb at 25 °C, 11,400 ppb at 35 °C, 15,200 ppb at 45 °C (particle sizes ≤ 5μ) [34]; at room temperature, solubilities of 6,600 ppb (maximum particle size 5μ) and 500 ppb (maximum particle size 0.04μ) were reported [35].

Specific Density: 1.5691 at 23.6/4 °C [1]; 1.891 at 19/4 °C [36].

Transformation Products: In moist soils, microbial degradation of lindane produced γ-pentachlorocyclohexene with trace amounts of benzene and chlorobenzene as possible trace contaminants [37]. Under anaerobic conditions, degradation by soil bacteria yielded γ-3,4,5,6-tetrachloro-1-cyclohexane and α-BHC [38]. Photolysis produces β-BHC [39].

Vapor Density: 518 ng/L at 20 °C, 1,971 ng/L at 30 °C, 6,784 ng/L at 40 °C [40].

Vapor Pressure: 9.4 x 10^{-6} mm at 20 °C [24]; 2.69 x 10^{-4} mm [29]; 6.7 x 10^{-5} mm at 25 °C [3]; 3.225 x 10^{-5} mm at 20 °C [41]; 1.28 x 10^{-4} mm at 30 °C [42].

FIRE HAZARDS

Flash Point: Non-combustible [43].

Lower Explosive Limit (LEL): Not flammable [36].

Upper Explosive Limit (UEL): Not flammable [36].

HEALTH HAZARD DATA

Immediately Dangerous to Life or Health (IDLH): 1,000 mg/m^3 [43].

Permissible Exposure Limits (PEL) in Air: 0.5 mg/m^3 [44]; 0.5 mg/m^3 TWA [45].

MANUFACTURING

Selected Manufacturers:

Aldrich Chemical Co.
940 West Saint Paul Ave.
Milwaukee, WI 53233

Pfaltz & Bauer, Inc.
172 East Aurora St.
Waterbury, CT 06708

Uses: Pesticide and insecticide.

REFERENCES

1. Weast, R.C., Ed. *CRC Handbook of Chemistry and Physics*, 67th ed. (Boca Raton, FL: CRC Press, Inc., 1986), 2406 p.
2. Eisenreich, S.J., Looney, B.B., and J.D. Thornton. "Airborne Organic Contaminants in the Great Lakes Ecosystem," *Environ. Sci. Technol.*,

15(1):30-38 (1981).

3. Spencer, W.F., Cliath, M.M., Jury, W.A., and L.-Z. Zhang. "Volatilization of Organic Chemicals from Soil as Related to Their Henry's Law Constants," *J. Environ. Qual.*, 17(3):504-509 (1988).

4. Jury, W.A., Spencer, W.F., and W.J. Farmer. "Use of Models for Assessing Relative Volatility, Mobility, and Persistence of Pesticides and other Trace Organics in Soil Systems," in *Hazard Assessment of Chemicals, Volume 2*, J. Saxena, Ed. (New York: Academic Press, Inc., 1983), pp 1-43.

5. Wahid, P.A., and N. Sethunathan. "Sorption-Desorption of Alpha, Beta and Gamma Isomers of BHC in Soils," *J. Agric. Food Chem.*, 27(5):1050-1053 (1979).

6. Chiou, C.T., Shoup, T.D., and P.E. Porter. "Mechanistic Roles of Soil Humus and Minerals in the Sorption of Nonionic Organic Compounds from Aqueous and Organic Solutions," *Org. Geochem.*, 8(1):9-14 (1985).

7. Kenaga, E.E. "Correlation of Bioconcentration Factors of Chemicals in Aquatic and Terrestrial Organisms with Their Physical and Chemical Properties," *Environ. Sci. Technol.*, 14(5):553-556 (1980).

8. Ramamoorthy, S. "Competition of Fate Processes in the Bioconcentration of Lindane," *Bull. Environ. Contam. Toxicol.*, 34(3):349-358 (1985).

9. Sharom, M.S., Miles, J.R.W., Harris C.R., and F.L. McEwen. "Behaviour of 12 Pesticides in Soil and Aqueous Suspensions of Soil and Sediment," *Water Res.*, 14:1095-1100 (1980).

10. Chin, Y.-P., Peven, C.S., and W.J. Weber. "Estimating Soil/Sediment Partition Coefficients for Organic Compounds by High Performance Reverse Phase Liquid Chromatography," *Water Res.*, 22(7):873-881 (1988).

11. Chiou, C.T., Peters, L.J., and V.H. Freed. "A Physical Concept of Soil-Water Equilibria for Nonionic Organic Compounds," *Science*, 206:831-832 (1979).

12. Reinbold, K.A., Hassett, J.J., Means, J.C., and W.L. Banwart. "Adsorption of Energy-Related Organic Pollutants: A Literature Review," Office of Research and Development, U.S. EPA Report-600/3-79-086 (1979), 180 p.

13. Hodson, J., and N.A. Williams. "The Estimation of the Adsorption Coefficient (K_{oc}) for Soils by High Performance Liquid Chromatography," *Chemosphere*, 19(1):67-77 (1988).

14. Rippen, G., Ilgenstein, M., and W. Klöpffer. "Screening of the Adsorption Behavior of New Chemicals: Natural Soils and Model Adsorbents," *Ecotoxicol. Environ. Safety*, 6(3):236-245 (1982).

15. Rao, P.S.C., and J.M. Davidson. "Estimation of Pesticide Retention

and Transformation Parameters Required in Nonpoint Source Pollution Models," in *Environmental Impact of Nonpoint Source Pollution*, Overcash, M.R., and J.M. Davidson, Eds. (Ann Arbor, MI: Ann Arbor Science Publishers, Inc., 1980), pp 23-67.

16. Chiou, C.T., Malcolm, R.L., Brinton, T.I., and D.E. Kile. "Water Solubility Enhancement of Some Organic Pollutants and Pesticides by Dissolved Humic and Fulvic Acids," *Environ. Sci. Technol.*, 20(5):502-508 (1986).

17. Travis, C.C., and A.D. Arms. "Bioconcentration of Organics in Beef, Milk and Vegetation," *Environ. Sci. Technol.*, 22(3):271-274 (1988).

18. Isnard, S., and S. Lambert. "Estimating Bioconcentration Factors from Octanol-Water Partition Coefficient and Aqueous Solubility," *Chemosphere*, 17(1):21-34 (1988).

19. Kurihara, N., Uchida, M., Fujita, T., and M. Nakajima. "Studies on BHC Isomers and Related Compounds. V. Some Physicochemical Properties of BHC isomers," *Pestic. Biochem. Physiol.*, 2(4):383-390 (1973).

20. Geyer, H.J., Scheunert, I., and F. Korte. "Correlation Between the Bioconcentration Potential of Organic Environmental Chemicals in Humans and Their *n*-Octanol/Water Partition Coefficients," *Chemosphere*, 16(1):239-252 (1987).

21. Geyer, H., Politzki, G., and D. Freitag. "Prediction of Ecotoxicological Behaviour of Chemicals: Relationship Between *n*-Octanol/Water Partition Coefficient and Bioaccumulation of Organic Chemicals by Alga *Chlorella*," *Chemosphere*, 13(2):269-284 (1984).

22. Kishi, H., and Y. Hashimoto. "Evaluation of the Procedures for the Measurement of Water Solubility and *n*-Octanol/Water Partition Coefficient of Chemicals Results of a Ring Test in Japan," *Chemosphere*, 18(9/10):1749-1749 (1989).

23. Sunshine, I., Ed. *Handbook of Analytical Toxicology* (Cleveland, OH: The Chemical Rubber Co., 1969), 1081 p.

24. Verschueren, K. *Handbook of Environmental Data on Organic Chemicals* (New York: Van Nostrand Reinhold Co., 1983), 1310 p.

25. Masterton, W.L., and T.P. Lee. "Effect of Dissolved Salts on Water Solubility of Lindane," *Environ. Sci. Technol.*, 6(10):919-921 (1972).

26. Weil, L., Dure, G., and K.E. Quentin. "Solubility in Water of Insecticide Chlorinated Hydrocarbons and Polychlorinated Biphenyls in View of Water Pollution," *Z. Wasser Forsch.*, 7(6):169-175 (1974).

27. Berg, G.L., Ed., *The Farm Book* (Willoughby, OH: Meister Publishing Co., 1983), 440 p.

28. Murty, A.S. *Toxicity of Pesticides to Fish, Volume. 1* (Boca Raton, FL:

CRC Press, Inc., 1986), 178 p.

29. Caron, G., Suffet, I.H., and T. Belton. "Effect of Dissolved Organic Carbon on the Environmental Distribution of Nonpolar Organic Compounds," *Chemosphere*, 14(8):993-1000 (1985).

30. Hollifield, H.C. "Rapid Nephelometric Estimate of Water Solubility of Highly Insoluble Organic Chemicals of Environmental Interest," *Bull. Environ. Contam. Toxicol.*, 23(4/5):579-586 (1979).

31. Kenaga, E.E., and C.A.I. Goring. "Relationship Between Water Solubility, Soil Sorption, Octanol-Water Partitioning and Concentration of Chemicals in Biota," in *Aquatic Toxicology, ASTM STP 707*, Eaton, J.G., Parrish, P.R., and A.C. Hendricks, Eds. (Philadelphia, PA: American Society for Testing and Materials 1980), pp 78-115.

32. Gunther, F.A., Westlake, W.E., and P.S. Jaglan. "Reported Solubilities of 738 Pesticide Chemicals in Water," *Res. Rev.*, 20:1-148 (1968).

33. Bhavnagary, H.M., and M. Jayaram. "Determination of Water Solubilities of Lindane and Dieldrin at Different Temperatures," *Bull. Grain Technol.*, 12(2):95-99 (1974).

34. Biggar, J.W., and I.R. Riggs. "Apparent Solubility of Organochlorine Insecticides in Water at Various Temperatures," *Hilgardia*, 42(10):383-391 (1974).

35. Robeck, G.G., Dostal, K.A., Cohen, J.M., and J.F. Kreissl. "Effectiveness of Water Treatment Processes in Pesticide Removal," *J. Am. Water Works Assoc.*, 57(2):181-200 (1965).

36. Weiss, G. *Hazardous Chemicals Data Book* (Park Ridge, NJ: Noyes Data Corp., 1986), 1069 p.

37. Kearney, P.C., and D.D. Kaufman. *Herbicides: Chemistry, Degradation and Mode of Action* (New York: Marcel Dekker, Inc., 1976), 1036 p.

38. Kobayashi, H., and B.E. Rittman. "Microbial Removal of Hazardous Organic Compounds," *Environ. Sci. Technol.*, 16(3):170A-183A (1982).

39. Hazardous Substances Data Bank. Lindane, National Library of Medicine, Toxicology Information Program (1989).

40. Spencer, W.F., and M.M. Cliath. "Vapor Density and Apparent Vapor Pressure of Lindane (γ-BHC)," *J. Agric. Food Chem.*, 18(3):529-530 (1970).

41. Dobbs, A.J., and M.R. Cull. "Volatilisation of Chemicals - Relative Loss Rates and the Estimation of Vapour Pressures," *Environ. Poll. (Series B)*, 3(4):289-298 (1982).

42. Tinsley, L.J. *Chemical Concepts in Pollutant Behavior* (New York: John Wiley and Sons, Inc., 1979), 265 p.

43. "NIOSH Pocket Guide to Chemical Hazards," U.S. Department of Health and Human Services, U.S. Government Printing Office

(1987), 241 p.

44. "General Industry Standards for Toxic and Hazardous Substances," U.S. Code of Federal Regulations 1910, Subpart Z Section 1910.1000 (July 1982).

45. *Documentation of the Threshold Limit Values and Biological Exposure Indices for 1986-1987* (Cincinnati, OH: American Conference of Governmental Industrial Hygienists, 1986), 111 p.

METHOXYCHLOR

Synonyms: 2,2-Bis(*p*-anisyl)-1,1,1-trichloroethane; 1,1-Bis(*p*-methoxyphenyl)-2,2,2-trichloroethane; 2,2-Bis(*p*-methoxyphenyl)-1,1,1-trichloroethane; Chemform; 2,2-Di-*p*-anisyl-1,1,1-trichloro-ethane; Dimethoxy-DDT; *p,p'*-Dimethoxydiphenyltrichloroethane; Dimethoxy-DT; 2,2-Di-(*p*-methoxyphenyl)-1,1,1-trichloroethane; Di(*p*-methoxyphenyl)trichloromethyl methane; DMDT; 4,4'-DMDT; *p,p'*-DMDT; DMTD; ENT 1,716; Maralate; Marlate; Marlate 50; Methoxcide; Methoxo; 4,4'-Methoxychlor; *p,p'*-Methoxychlor; Methoxy-DDT; Metox; Moxie; NCI-C00497; RCRA waste number U247; 1,1,1-Trichloro-2,2-bis(*p*-anisyl)ethane; 1,1,1-Trichloro-2,2-bis(*p*-methoxyphenol)ethanol; 1,1,1-Trichloro-2,2-bis(*p*-methoxy-phenyl)ethane; 1,1,1-Trichloro-2,2-di(4-methoxy-phenyl)ethane; **1,1'-(2,2,2-Trichloroethylidene)bis(4-methoxybenzene)**.

Structural Formula:

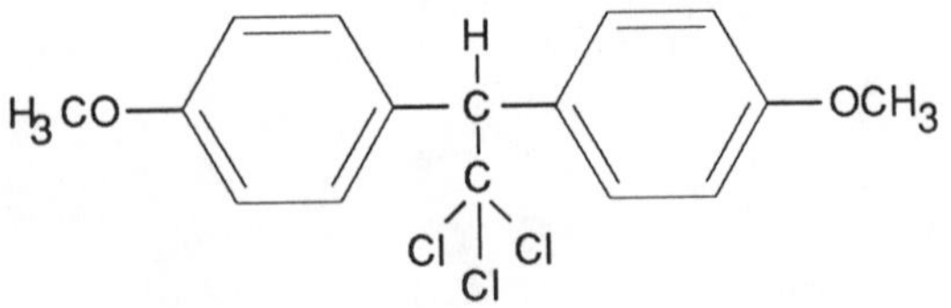

CHEMICAL DESIGNATIONS

CAS Registry Number: 72-43-5

DOT Designation: 2761

Empirical Formula: $C_{16}H_{15}Cl_3O_2$

Formula Weight: 345.66

RTECS Number: KJ 3675000

PHYSICAL AND CHEMICAL PROPERTIES

Appearance and Odor: White, gray or pale yellow crystals or powder. May be dissolved in an organic solvent or petroleum distillate for application. Pungent to mild fruity odor.

Boiling Point: Decomposes [1].

Henry's Law Constant: Insufficient vapor pressure data for calculation at 25 °C.

Ionization Potential: No data found.

Log K_{oc}: 4.90 [2]; 4.95 [3].

Log K_{ow}: 4.30 [4]; 4.68 [2]; 4.40 [5]; 3.40 [6]; 3.31, 5.08 [7].

Melting Point: 98 °C [8]; 89 °C [9]; 92 °C [10]; 77 °C (technical grade) [11].

Solubility in Organics: Soluble in ethanol [12].

Solubility in Water: 0.040 mg/L at 24 °C [8]; 0.1 mg/L at 25 °C [9]; 0.003 ppm [13]; 0.62 ppm [10]; 20 ppb at 15 °C, 45 ppb at 25 °C, 95 ppb at 35 °C, 185 ppb at 45 °C (particle sizes ≤ 5μ) [14].

Specific Density: 1.41 at 25/4 °C [8].

Transformation Products: Degradation by *Aerobacter aerogenes* under aerobic or anaerobic conditions yielded 1,1-dichloro-2,2-bis(*p*-methoxyphenyl)ethylene and 1,1-dichloro-2,2-bis(*p*-methoxyphenyl)ethane [15]. Hydrolysis at common aquatic pH's produced anisoin, anisil, and 2,2-bis(*p*-methoxy-phenyl)-1,1-dichloroethylene [6].

Vapor Pressure: No data found.

FIRE HAZARDS

Flash Point: Burns only at high temperatures [16].

Lower Explosive Limit (LEL): Not pertinent [16].

Upper Explosive Limit (UEL): Not pertinent [16].

HEALTH HAZARD DATA

Immediately Dangerous to Life or Health (IDLH): 7,500 mg/m^3 [17].

Permissible Exposure Limits (PEL) in Air: 15 mg/m^3 [18]; 10 mg/m^3 TWA [19].

MANUFACTURING

Selected Manufacturers:

Ansul Co.
1 Stanton St.
Marinette, WI 54143

Chemical Formulators, Inc.
P.O. Box 26
Nitro, WV 25143

E.I. duPont de Nemours and Co.
Biochemical Department
308 East Lancaster Ave.
Wynnewood, PA 19096

Uses: Insecticide to control mosquito larvae and house flies; to control ectoparasites on cattle, sheep, and goats; recommended for use in dairy barns.

REFERENCES

1. Weast, R.C., Ed. *CRC Handbook of Chemistry and Physics*, 67th ed. (Boca Raton, FL: CRC Press, Inc., 1986), 2406 p.
2. Kenaga, E.E. "Correlation of Bioconcentration Factors of Chemicals in Aquatic and Terrestrial Organisms with Their Physical and Chemical Properties," *Environ. Sci. Technol.*, 14(5):553-556 (1980).
3. Karickhoff, S.W., Brown, D.S., and T.A. Scott. "Sorption of Hydrophobic Pollutants on Natural Sediments," *Water Res.*, 13:241-248 (1979).
4. Mackay, D. "Correlation of Bioconcentration Factors," *Environ. Sci. Technol.*, 16(5):274-278 (1982).
5. Garten, C.T., and J.R. Trabalka. "Evaluation of Models for Predicting Terrestrial Food Chain Behavior of Xenobiotics," *Environ. Sci. Technol.*, 17(10):590-595 (1983).
6. Wolfe, N.L., Zepp, R.G., Paris, D.F., Baughman, G.L., and R.C. Hollis. "Methoxychlor and DDT Degradation in Water: Rates and

Products," *Environ. Sci. Technol.*, 11(12):1077-1081 (1977).

7. Rao, P.S.C., and J.M. Davidson. "Estimation of Pesticide Retention and Transformation Parameters Required in Nonpoint Source Pollution Models," in *Environmental Impact of Nonpoint Source Pollution*, Overcash, M.R., and J.M. Davidson, Eds. (Ann Arbor, MI: Ann Arbor Science Publishers, Inc., 1980), pp 23-67.

8. Verschueren, K. *Handbook of Environmental Data on Organic Chemicals* (New York: Van Nostrand Reinhold Co., 1983), 1310 p.

9. *IARC Monographs on the Evaluation of Carcinogenic Risk of Chemicals to Man. Some Halogenated Hydrocarbons, Volume 20* (Lyon, France: International Agency for Research on Cancer, 1979), 609 p.

10. Kapoor, I.P., Metcalf, R.L., Nystrom, R.F., and G.K. Sanghua. "Comparative Metabolism of Methoxychlor, Methiochlor, and DDT in Mouse, Insects, and in a Model Ecosystem," *J. Agric. Food Chem.*, 18(6):1145-1152 (1970).

11. Sunshine, I., Ed. *Handbook of Analytical Toxicology* (Cleveland, OH: The Chemical Rubber Co., 1969), 1081 p.

12. Windholz, M., Budavari, S., Blumetti, R.F., and E.S. Otterbein, Eds. *The Merck Index*, 10th ed., (Rahway, NJ: Merck and Co., 1983), 1463 p.

13. Kenaga, E.E., and C.A.I. Goring. "Relationship Between Water Solubility, Soil Sorption, Octanol-Water Partitioning and Concentration of Chemicals in Biota," in *Aquatic Toxicology, ASTM STP 707*, Eaton, J.G., Parrish, P.R., and A.C. Hendricks, Eds. (Philadelphia, PA: American Society for Testing and Materials 1980), pp 78-115.

14. Biggar, J.W., and I.R. Riggs. "Apparent Solubility of Organochlorine Insecticides in Water at Various Temperatures," *Hilgardia*, 42(10):383-391 (1974).

15. Kobayashi, H., and B.E. Rittman. "Microbial Removal of Hazardous Organic Compounds," *Environ. Sci. Technol.*, 16(3):170A-183A (1982).

16. Weiss, G. *Hazardous Chemicals Data Book* (Park Ridge, NJ: Noyes Data Corp., 1986), 1069 p.

17. "NIOSH Pocket Guide to Chemical Hazards," U.S. Department of Health and Human Services, U.S. Government Printing Office (1987), 241 p.

18. "General Industry Standards for Toxic and Hazardous Substances," U.S. Code of Federal Regulations 1910, Subpart Z Section 1910.1000 (July 1982).

19. *Documentation of the Threshold Limit Values and Biological Exposure Indices for 1986-1987* (Cincinnati, OH: American Conference of Governmental Industrial Hygienists, 1986), 111 p.

METHYL BROMIDE

Synonyms: Brom-o-gas; Brom-o-gaz; **Bromomethane**; Celfume; Dawson 100; Dowfume; Dowfume MC-2; Dowfume MC-2 soil fumigant; Dowfume MC-33; Edco; Embafume; Fumigant-1; Halon 1001; Iscobrome; Kayafume; MB; M-B-C Fumigant; MBX; MEBR; Metafume; Methogas; Monobromomethane; Pestmaster; Profume; R 40B1; RCRA waste number U029; Rotox; Terabol; Terr-o-gas 100; UN 1062; Zytox.

Structural Formula:

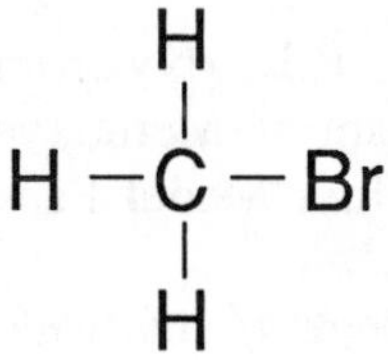

$$H - \overset{\displaystyle H}{\underset{\displaystyle H}{\overset{\displaystyle |}{\underset{\displaystyle |}{C}}}} - Br$$

CHEMICAL DESIGNATIONS

CAS Registry Number: 74-83-9

DOT Designation: 1062

Empirical Formula: CH_3Br

Formula Weight: 94.94

RTECS Number: PA 4900000

PHYSICAL AND CHEMICAL PROPERTIES

Appearance and Odor: Colorless liquid with an odor similar to chloroform at high concentrations.

Boiling Point: 3.55 °C [1].

Henry's Law Constant: 0.20 atm·m^3/mol [2]; 0.0318 atm·m^3/mol [3].

368

Ionization Potential: 10.54 eV [4].

Log K_{oc}: 1.92 using method of Kenaga and Goring [5].

Log K_{ow}: 1.00 [6]; 1.10 [7]; 1.19 [8].

Melting Point: -93.6 °C [9].

Solubility in Organics: Soluble in ethanol, ether [9], chloroform, carbon disulfide, carbon tetrachloride, and benzene [10].

Solubility in Water: 900 mg/L at 20 °C [11]; 13,000 mg/L at 25 °C [12]; 13,200 mg/L at 25 °C [13]; 17,500 mg/L at 20 °C under 748 mm atmosphere consisting of methyl bromide and water vapor [14]; 13,400 mg/L at 25 °C [15].

Specific Density: 1.6755 at 20/4 °C [9]; 1.732 at 0/0 °C [16].

Transformation Products: No data found.

Vapor Density: 3.88 g/L at 25 °C, 3.28 (air = 1).

Vapor Pressure: 760 mm at 3.6 °C, 1,520 mm at 23.3 °C, 3,800 mm at 54.8 °C, 7,600 mm at 84.0 °C, 15,200 mm at 121.7 °C, 30,400 mm at 170.2 °C [9]; 1,824 mm at 25 °C [16]; 1,633 mm at 25 °C [17].

FIRE HAZARDS

Flash Point: Unknown [4].

Lower Explosive Limit (LEL): 10% [18].

Upper Explosive Limit (UEL): 16% [18].

HEALTH HAZARD DATA

Immediately Dangerous to Life or Health (IDLH): 2,000 ppm (carcinogen) [19].

Permissible Exposure Limits (PEL) in Air: 20 ppm ($\approx$40 mg/m^3) ceiling [20]; lowest feasible limit [4]; 5 ppm ($\approx$20 mg/m^3) TLV [21].

MANUFACTURING

Selected Manufacturers:

Aldrich Chemical Co.
940 West Saint Paul Ave.
Milwaukee, WI 53233

Northwest Industries, Inc.
Michigan Chemicals Corp.
351 East Ohio St.
Chicago, IL 60611

Pfaltz & Bauer, Inc.
172 East Aurora St.
Waterbury, CT 06708

Uses: Soil, space, and food fumigant; organic synthesis; fire extinguishing agent; refrigerant; disinfestation of potatoes, tomatoes, and other crops; solvent for extracting vegetable oils.

REFERENCES

1. Kudchadker, A.P., Kudchadker, S.A., Shukla, R.P., and P.R. Patnaik. "Vapor Pressures and Boiling Points of Selected Halomethanes," *J. Phys. Chem. Ref. Data*, 8(2):499-517 (1979).
2. Pankow, J.F., and M.E. Rosen. "Determination of Volatile Compounds in Water by Purging Directly to a Capillary Column with Whole Column Cryotrapping," *Environ. Sci. Technol.*, 22(4):398-405 (1988).
3. Jury, W.A., Spencer, W.F., and W.J. Farmer. "Behavior Assessment Model for Trace Organics in Soil: III. Application of Screening Model," *J. Environ. Qual.*, 13(4):573-579 (1984).
4. "NIOSH Pocket Guide to Chemical Hazards," U.S. Department of Health and Human Services, U.S. Government Printing Office (1987), 241 p.
5. Kenaga, E.E., and C.A.I. Goring. "Relationship Between Water Solubility, Soil Sorption, Octanol-Water Partitioning and Concentration of Chemicals in Biota," in *Aquatic Toxicology, ASTM STP 707*, Eaton, J.G., Parrish, P.R., and A.C. Hendricks, Eds. (Philadelphia, PA: American Society for Testing and Materials 1980), pp 78-115.

6. Mills, W.B., Porcella, D.B., Ungs, M.J., Gherini, S.A., Summers, K.V., Mok, L., Rupp, G.L., and G.L. Bowie. "Water Quality Assessment: A Screening Procedure for Toxic and Conventional Pollutants in Surface and Groundwater-Part I," Office of Research and Development, U.S. EPA Report-600/6-85-002a (1985), 638 p.
7. Walton, W.C. *Practical Aspects of Ground Water Modeling* (Worthington, OH: National Water Well Association, 1985), 587 p.
8. Hansch, C., and A. Leo. *Substituent Constants for Correlation Analysis in Chemistry and Biology* (New York: John Wiley and Sons, Inc., 1979), 339 p.
9. Weast, R.C., Ed. *CRC Handbook of Chemistry and Physics*, 67th ed. (Boca Raton, FL: CRC Press, Inc., 1986), 2406 p.
10. *Toxic and Hazardous Industrial Chemicals Safety Manual for Handling and Disposal with Toxicity and Hazard Data* (Tokyo, Japan: International Technical Information Institute, 1986), 700 p.
11. Verschueren, K. *Handbook of Environmental Data on Organic Chemicals* (New York: Van Nostrand Reinhold Co., 1983), 1310 p.
12. Lyman, W.J., Reehl, W.F., and D.H. Rosenblatt. *Handbook of Chemical Property Estimation Methods: Environmental Behavior of Organic Compounds* (New York: McGraw-Hill, Inc., 1982).
13. Gordon, A.J., and R.A. Ford. *The Chemist's Companion* (New York: John Wiley and Sons, Inc., 1972), 551 p.
14. Standen, A., Ed. *Kirk-Othmer Encyclopedia of Chemical Technology, Volume 3*, 2nd ed. (New York: John Wiley and Sons, Inc., 1964), 927 p.
15. Gunther, F.A., Westlake, W.E., and P.S. Jaglan. "Reported Solubilities of 738 Pesticide Chemicals in Water," *Res. Rev.*, 20:1-148 (1968).
16. Sax, N.I. *Dangerous Properties of Industrial Materials* (New York: Van Nostrand Reinhold Co., 1984), 3124 p.
17. Howard, P.H. *Handbook of Environmental Fate and Exposure Data for Organic Chemicals* (Chelsea, MI: Lewis Publishers, Inc., 1989), 574 p.
18. *Fire Protection Guide on Hazardous Materials* (Quincy, MA: National Fire Protection Association, 1984), 443 p.
19. Weiss, G. *Hazardous Chemicals Data Book* (Park Ridge, NJ: Noyes Data Corp., 1986), 1069 p.
20. "General Industry Standards for Toxic and Hazardous Substances," U.S. Code of Federal Regulations 1910, Subpart Z Section 1910.1000 (July 1982).
21. *Documentation of the Threshold Limit Values and Biological Exposure Indices for 1986-1987* (Cincinnati, OH: American Conference of Governmental Industrial Hygienists, 1986), 111 p.

METHYL CHLORIDE

Synonyms: Artic; **Chloromethane;** Monochloromethane; RCRA waste number U045; UN 1063.

Structural Formula:

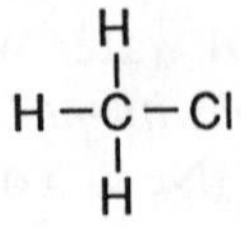

$$H-\overset{\displaystyle H}{\underset{\displaystyle H}{C}}-Cl$$

CHEMICAL DESIGNATIONS

CAS Registry Number: 74-87-3

DOT Designation: 1063

Empirical Formula: CH_3Cl

Formula Weight: 50.48

RTECS Number: PA 6300000

PHYSICAL AND CHEMICAL PROPERTIES

Appearance and Odor: Liquified compressed gas, colorless, odorless or sweet ethereal odor.

Boiling Point: -24.2 °C [1]; -23.73 °C [2].

Henry's Law Constant: 0.00882 atm·m^3/mol [3]; 0.0066 atm·m^3/mol [4]; 0.010 atm·m^3/mol at 25 °C [5].

Ionization Potential: 11.26 eV [6]; 11.33 eV [7].

Log K_{oc}: 1.40 using method of Chiou and others [8].

Log K_{ow}: 0.90 [9]; 0.91 [10].

Melting Point: -97.1 °C [1]; -97.7 °C [2].

Solubility in Organics: Miscible with chloroform, ether, and glacial acetic acid; soluble in ethanol [11].

Solubility in Water: 4,000 mg/L [12]; 7,400 mg/L at 25 °C [13]; 6,450-7,250 mg/L at 20 °C [14]; 7,250 mg/L at 20 °C [15]; 0.648 wt% at 30 °C [16]; 4,800 mg/L at 25 °C [2].

Specific Density: 0.9159 at 20/4 °C [1]; 0.997 at -24/4 °C [17].

Transformation Products: No data found.

Vapor Density: 2.06 g/L at 25 °C, 1.74 (air = 1).

Vapor Pressure: 3,789 mm at 20 °C [18]; 5,092 mm at 30 °C [12]; 760 mm at -24.0 °C, 1,520 mm at -6.4 °C, 3,800 mm at 27.0 °C, 7,600 mm at 47.3 °C, 15,200 mm at 77.3 °C, 30,400 mm at 113.8 °C, 45,600 mm at 137.5 °C [1]; 3,765 mm at 20 °C [14]; 3,756 mm at 20 °C [19]; 6.529 atm at 30 °C, 9.623 atm at 50 °C [20]; 4,309.7 mm at 30 °C [21].

FIRE HAZARDS

Flash Point: -50 °C [22].

Lower Explosive Limit (LEL): 7.6% [6].

Upper Explosive Limit (UEL): 19% [6].

HEALTH HAZARD DATA

Immediately Dangerous to Life or Health (IDLH): 10,000 ppm (carcinogen) [17].

Permissible Exposure Limits (PEL) in Air: 100 ppm, 200 ppm ceiling, 300 ppm 5 minute 3-hour peak [23]; lowest feasible limit [6]; 50 ppm ($\approx$105 mg/m^3) TWA, 100 ppm ($\approx$205 mg/m^3) STEL [24].

MANUFACTURING

Selected Manufacturers:

Aldrich Chemical Co.
940 West Saint Paul Ave.
Milwaukee, WI 53233

Dow Chemical Co.
Midland, MI 48640

Ethyl Corp.
Industrial Chemicals Division
451 Florida St.
Baton Rouge, LA 70801

Uses: Coolant and refrigerant; herbicide and fumigant; organic synthesis-methylating agent; manufacturing of silicone polymers pharmaceuticals, tetraethyllead, synthetic rubber, methyl cellulose, agricultural chemicals, and non-flammable films; preparation of methylene chloride, carbon tetrachloride, chloroform; low temperature solvent and extractant; catalytic carrier for butyl rubber polymerization; topical anesthetic; fluid for thermometric and thermostatic equipment.

REFERENCES

1. Weast, R.C., Ed. *CRC Handbook of Chemistry and Physics*, 67th ed. (Boca Raton, FL: CRC Press, Inc., 1986), 2406 p.
2. Standen, A., Ed. *Kirk-Othmer Encyclopedia of Chemical Technology, Volume 4*, 2nd ed. (New York: John Wiley and Sons, Inc., 1964), 937 p.
3. Gossett, J.M. "Measurement of Henry's Law Constants for C_1 and C_2 Chlorinated Hydrocarbons," *Environ. Sci. Technol.*, 21(2):202-208 (1987).
4. Pankow, J.F., and M.E. Rosen. "Determination of Volatile Compounds in Water by Purging Directly to a Capillary Column with Whole Column Cryotrapping," *Environ. Sci. Technol.*, 22(4):398-405 (1988).
5. Hine, J., and P.K. Mookerjee. "The Intrinsic Hydrophilic Character of Organic Compounds. Correlations in Terms of Structural Contributions," *J. Org. Chem.*, 40(3):292-298 (1975).
6. "NIOSH Pocket Guide to Chemical Hazards," U.S. Department of Health and Human Services, U.S. Government Printing Office (1987), 241 p.
7. Yoshida, K., Shigeoka, T., and F. Yamauchi. "Non-Steady State Equilibrium Model for the Preliminary Prediction of the Fate of Chemicals in the Environment," *Ecotoxicol. Environ. Safety*, 7(2):179-190 (1983).
8. Chiou, C.T., Peters, L.J., and V.H. Freed. "A Physical Concept of

Soil-Water Equilibria for Nonionic Organic Compounds," *Science*, 206:831-832 (1979).

9. Mills, W.B., Porcella, D.B., Ungs, M.J., Gherini, S.A., Summers, K.V., Mok, L., Rupp, G.L., and G.L. Bowie. "Water Quality Assessment: A Screening Procedure for Toxic and Conventional Pollutants in Surface and Groundwater-Part I," Office of Research and Development, U.S. EPA Report-600/6-85-002a (1985), 638 p.

10. Hansch, C., and A. Leo. *Substituent Constants for Correlation Analysis in Chemistry and Biology* (New York: John Wiley and Sons, Inc., 1979), 339 p.

11. "Chemical, Physical, and Biological Properties of Compounds Present at Hazardous Waste Sites," U.S. EPA Report-530/SW-89-010 (1985), 619 p.

12. Verschueren, K. *Handbook of Environmental Data on Organic Chemicals* (New York: Van Nostrand Reinhold Co., 1983), 1310 p.

13. Lyman, W.J., Reehl, W.F., and D.H. Rosenblatt. *Handbook of Chemical Property Estimation Methods: Environmental Behavior of Organic Compounds* (New York: McGraw-Hill, Inc., 1982).

14. "Treatability Manual - Volume 1: Treatability Data," Office of Research and Development, U.S. EPA Report-600/8-80-042a (1980), 1035 p.

15. Pearson, C.R., and G. McConnell. "Chlorinated C_1 and C_2 Hydrocarbons in the Marine Environment," in *Proc. R. Soc. London*, B189(1096):305-322 (1975).

16. Riddick, J.A., Bunger, W.B., and T.K. Sakano. *Organic Solvents - Physical Properties and Methods of Purification. Volume II* (New York: John Wiley and Sons, Inc., 1986), 1325 p.

17. Weiss, G. *Hazardous Chemicals Data Book* (Park Ridge, NJ: Noyes Data Corp., 1986), 1069 p.

18. Melnikov, N.N. *Chemistry of Pesticides* (New York: Springer-Verlag, Inc., 1971), 480 p.

19. McConnell, G., Ferguson, D.M., and C.R. Pearson. "Chlorinated Hydrocarbons and the Environment," *Endeavour*, 34(121):13-18 (1975).

20. Hsu, C.C., and J.J. McKetta. "Pressure-Volume-Temperature Properties of Methyl Chloride," *J. Chem. Eng. Data*, 9(1):45-51 (1964).

21. Howard, P.H. *Handbook of Environmental Fate and Exposure Data for Organic Chemicals* (Chelsea, MI: Lewis Publishers, Inc., 1989), 574 p.

22. *Fire Protection Guide on Hazardous Materials* (Quincy, MA: National Fire Protection Association, 1984), 443 p.

23. "General Industry Standards for Toxic and Hazardous Substances,"

U.S. Code of Federal Regulations 1910, Subpart Z Section 1910.1000 (July 1982).

24. *Documentation of the Threshold Limit Values and Biological Exposure Indices for 1986-1987* (Cincinnati, OH: American Conference of Governmental Industrial Hygienists, 1986), 111 p.

METHYLENE CHLORIDE

Synonyms: Aerothene MM; DCM; **Dichloromethane**; Freon 30; Methane dichloride; Methylene bichloride; Methylene dichloride; Narcotil; NCI-C50102; RCRA waste number U080; Solaesthin; Solmethine; UN 1593.

Structural Formula:

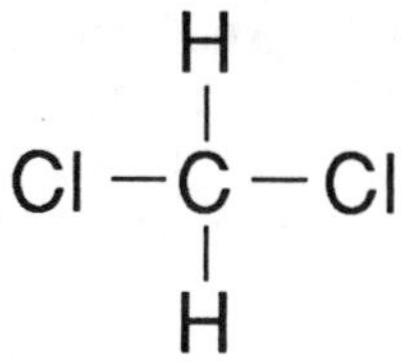

$$
\begin{array}{c}
\text{H} \\
| \\
\text{Cl} - \text{C} - \text{Cl} \\
| \\
\text{H}
\end{array}
$$

CHEMICAL DESIGNATIONS

CAS Registry Number: 75-09-2

DOT Designation: 1593

Empirical Formula: CH_2Cl_2

Formula Weight: 84.93

RTECS Number: PA 8050000

PHYSICAL AND CHEMICAL PROPERTIES

Appearance and Odor: Colorless liquid with a sweet, penetrating ethereal odor.

Boiling Point: 40.2 °C [1].

Henry's Law Constant: 0.0020 atm·m³/mol [2]; 0.00319 atm·m³/mol [3]; 0.00269 atm·m³/mol at 25 °C [4]; 0.00218 atm·m³/mol at 25 °C [5]; 0.00353 atm·m³/mol at 37 °C [6].

Ionization Potential: 11.35 eV [7].

Log K_{oc}: 0.94 [8].

Log K$_{ow}$: 1.30 [9]; 1.25 [10].

Melting Point: -95.1 °C [11]; -96.7 °C [12].

Solubility in Organics: Miscible with ethanol and ether [13].

Solubility in Water: 20,000 mg/L at 20 °C, 16,700 mg/L at 25 °C [14]; 1.3 wt% at 20 °C [7]; 13,000 mg/L at 25 °C [15]; 19,400 ppm at 25 °C [4]; 13,200 mg/L at 25 °C [16]; 23,600 mg/L at 0 °C, 21,200 mg/L at 10 °C, 19,700 at 30 °C [12]; 19,000 ppm at 25 °C [17].

Specific Density: 1.3266 at 20/4 °C [11]; 1.378 at 0/4 °C [12].

Transformation Products: Under laboratory conditions, methylene chloride hydrolyzed with subsequent oxidation and reduction to produce methyl chloride, methanol, formic acid, and formaldehyde [18].

Vapor Density: 3.47 g/L at 25 °C, 2.93 (air = 1).

Vapor Pressure: 147.4 mm 0 °C, 229.7 mm at 10 °C, 348.9 mm at 20 °C, 511.4 mm at 30 °C, 600 mm at 35 °C [12]; 1 mm at -70.0 °C, 10 mm at -43.3 °C, 40 mm at -22.3 °C, 100 mm at -6.3 °C, 400 mm at 24.1 °C, 760 mm at 40.7 °C [11]; 440 mm at 25 °C [19]; 362.4 mm at 20 °C [20]; 455 mm at 25 °C [21]; 380 mm at 22 °C [22].

FIRE HAZARDS

Flash Point: None [23]; $\geq$ 30 °C [24].

Lower Explosive Limit (LEL): 12% [7].

Upper Explosive Limit (UEL): 19% [7].

HEALTH HAZARD DATA

Immediately Dangerous to Life or Health (IDLH): 5,000 ppm (carcinogen) [25].

Permissible Exposure Limits (PEL) in Air: 500 ppm, 1,000 ppm ceiling, 2,000 ppm ceiling 5 minute/2-hour peak [26]; lowest feasible limit [7]; 50 ppm ($\approx$175 mg/m^3) [19].

MANUFACTURING

Selected Manufacturers:

Aldrich Chemical Co.
940 West Saint Paul Ave.
Milwaukee, WI 53233

Fluka Chemical Corp.
980 South Second St.
Ronkonkoma, NY 11779

Pfaltz & Bauer, Inc.
172 East Aurora St.
Waterbury, CT 06708

Uses: Low temperature solvent; ingredient in paint and varnish removers; cleaning, degreasing and drying metal parts; fumigant; manufacturing of aerosols; refrigerant; dewaxing; blowing agent in foams; solvent for cellulose acetate; organic synthesis

REFERENCES

1. Dean, J.A., Ed., *Lange's Handbook of Chemistry*, 11th ed. (New York: McGraw-Hill, Inc., 1973), 1570 p.
2. Pankow, J.F., and M.E. Rosen. "Determination of Volatile Compounds in Water by Purging Directly to a Capillary Column with Whole Column Cryotrapping," *Environ. Sci. Technol.*, 22(4):398-405 (1988).
3. Warner, H.P., Cohen, J.M., and J.C. Ireland. "Determination of Henry's Law Constants of Selected Priority Pollutants," Office of Science and Development, U.S. EPA Report-600/D-87/229 (1987), 14 p.
4. Dilling, W.L. "Interphase Transfer Processes. II. Evaporation Rates of Chloro Methanes, Ethanes, Ethylenes, Propanes, and Propylenes from Dilute Aqueous Solutions. Comparisons with Theoretical Predictions," *Environ. Sci. Technol.*, 11(4):405-409 (1977).
5. Hine, J., and P.K. Mookerjee. "The Intrinsic Hydrophilic Character of Organic Compounds. Correlations in Terms of Structural Contributions," *J. Org. Chem.*, 40(3):292-298 (1975).
6. Sato, A., and T. Nakajima. "A Structure-Activity Relationship of Some Chlorinated Hydrocarbons," *Arch. Environ. Health*, 34(2):69-75

(1979).

7. "NIOSH Pocket Guide to Chemical Hazards," U.S. Department of Health and Human Services, U.S. Government Printing Office (1987), 241 p.

8. Schwille, F. *Dense Chlorinated Solvents* (Chelsea, MI: Lewis Publishers, Inc., 1988), 146 p.

9. Mills, W.B., Porcella, D.B., Ungs, M.J., Gherini, S.A., Summers, K.V., Mok, L., Rupp, G.L., and G.L. Bowie. "Water Quality Assessment: A Screening Procedure for Toxic and Conventional Pollutants in Surface and Groundwater-Part I," Office of Research and Development, U.S. EPA Report-600/6-85-002a (1985), 638 p.

10. Hansch, C., Vittoria, A., Silipo, C., and P.Y.C. Jow. "Partition Coefficients and the Structure-Activity Relationship of the Anesthetic Gases," *J. Med. Chem.*, 18(6):546-548 (1975).

11. Weast, R.C., Ed. *CRC Handbook of Chemistry and Physics*, 67th ed. (Boca Raton, FL: CRC Press, Inc., 1986), 2406 p.

12. Standen, A., Ed. *Kirk-Othmer Encyclopedia of Chemical Technology, Volume 4*, 2nd ed. (New York: John Wiley and Sons, Inc., 1964), 937 p.

13. "Chemical, Physical, and Biological Properties of Compounds Present at Hazardous Waste Sites," U.S. EPA Report-530/SW-89-010 (1985), 619 p.

14. Verschueren, K. *Handbook of Environmental Data on Organic Chemicals* (New York: Van Nostrand Reinhold Co., 1983), 1310 p.

15. Haque, R., Ed., *Dynamics, Exposure and Hazardous Assessment of Toxic Chemicals* (Ann Arbor, MI: Ann Arbor Science Publishers, Inc., 1980), 496 p.

16. Pearson, C.R., and G. McConnell. "Chlorinated C_1 and C_2 Hydrocarbons in the Marine Environment," in *Proc. R. Soc. London*, B189(1096):305-322 (1975).

17. Amoore, J.E., and E. Hautala. "Odor as an Aide to Chemical Safety: Odor Thresholds Compared with Threshold Limit Values and Volatilities for 214 Industrial Chemicals in Air and Water Dilution," *J. Appl. Toxicol.*, 3(6):272-290 (1983).

18. Smith, L.R., and J. Dragun. "Degradation of Volatile Chlorinated Aliphatic Priority Pollutants in Groundwater," *Environ. Int.*, 19(4):291-298 (1984).

19. *Documentation of the Threshold Limit Values and Biological Exposure Indices for 1986-1987* (Cincinnati, OH: American Conference of Governmental Industrial Hygienists, 1986), 111 p.

20. McConnell, G., Ferguson, D.M., and C.R. Pearson. "Chlorinated Hydrocarbons and the Environment," *Endeavour*, 34(121):13-18 (1975).

21. Valsaraj, K.T. "On the Physio-Chemical Aspects of Partitioning of Non-Polar Hydrophobic Organics at the Air-Water Interface," *Chemosphere*, 17(5):875-887 (1988).

22. Sax, N.I. *Dangerous Properties of Industrial Materials* (New York: Van Nostrand Reinhold Co., 1984), 3124 p.

23. *Fire Protection Guide on Hazardous Materials* (Quincy, MA: National Fire Protection Association, 1984), 443 p.

24. Kuchta, J.M., Furno, A.L., Bartkowiak, A., and G.H. Martindill. "Effect of Pressure and Temperature on Flammability Limits of Chlorinated Hydrocarbons in Oxygen-Nitrogen and Nitrogen Tetroxide-Nitrogen Atmospheres," *J. Chem. Eng. Data*, 13(3):421-428 (1968).

25. Weiss, G. *Hazardous Chemicals Data Book* (Park Ridge, NJ: Noyes Data Corp., 1986), 1069 p.

26. "General Industry Standards for Toxic and Hazardous Substances," U.S. Code of Federal Regulations 1910, Subpart Z Section 1910.1000 (July 1982).

2-METHYLNAPHTHALENE

Synonym: β-Methylnaphthalene.

Structural Formula:

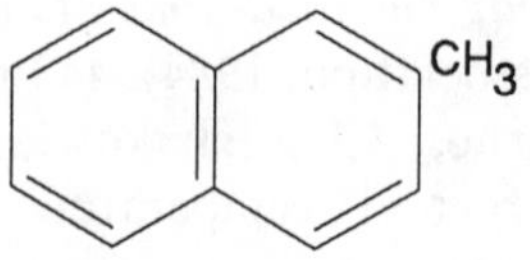

CHEMICAL DESIGNATIONS

CAS Registry Number: 91-57-6

DOT Designation: None assigned.

Empirical Formula: $C_{11}H_{10}$

Formula Weight: 142.20

RTECS Number: QJ 9635000

PHYSICAL AND CHEMICAL PROPERTIES

Appearance: Solid.

Boiling Point: 241.052 °C [1].

Henry's Law Constant: Insufficient vapor pressure data for calculation at 25 °C.

Ionization Potential: 7.955 eV [2]; 8.48 eV [3].

Log K_{oc}: 3.93 [4]; 3.87 [5].

Log K_{ow}: 4.11 [4]; 3.86 [3].

Melting Point: 34.6 °C [6].

Solubility in Organics: Soluble in most solvents [7].

Solubility in Water: 24.6 mg/L at 25 °C [8]; 25.4 mg/L at 25 °C [9].

Specific Density: 1.0058 at 20/4 °C [6].

Transformation Products: No data found.

Vapor Pressure: No data found.

FIRE HAZARDS

Flash Point: 97 °C [10].

Lower Explosive Limit (LEL): No data found.

Upper Explosive Limit (UEL): No data found.

HEALTH HAZARD DATA

Immediately Dangerous to Life or Health (IDLH): No data found.

Permissible Exposure Limits (PEL) in Air: No standards set.

MANUFACTURING

Selected Manufacturers:

Aldrich Chemical Co.
940 West Saint Paul Ave.
Milwaukee, WI 53233

Fluka Chemical Corp.
980 South Second St.
Ronkonkoma, NY 11779

Pfaltz & Bauer, Inc.
172 East Aurora St.
Waterbury, CT 06708

Uses: Organic synthesis; insecticides. Derived from industrial and experimental coal gasification operations where the maximum concentration detected in gas, liquid, and coal tar streams were 2.1 mg/m^3, 0.22 mg/m^3, and 10 mg/m^3, respectively [11].

REFERENCES

1. Zwolinski, B.J., and R.C. Wilhoit. *Handbook of Vapor Pressure and Heats of Vaporization of Hydrocarbons and Related Compounds*, Publication 101 (College Station, TX: Thermodynamics Research Station, 1971), 329 p.

2. *Instruction Manual - Model ISP1 101: Intrinsically Safe Portable Photoionization Analyzer* (Newton, MA: HNU Systems, Inc., 1986), 86 p.

3. Yoshida, K., Shigeoka, T., and F. Yamauchi. "Non-Steady State Equilibrium Model for the Preliminary Prediction of the Fate of Chemicals in the Environment," *Ecotoxicol. Environ. Safety*, 7(2):179-190 (1983).

4. Abdul, S.A., Gibson, T.L., and D.N. Rai. "Statistical Correlations for Predicting the Partition Coefficient for Nonpolar Organic Contaminants Between Aquifer Organic Carbon and Water," *Haz. Waste Haz. Mater.*, 4(3):211-222 (1987).

5. Hodson, J., and N.A. Williams. "The Estimation of the Adsorption Coefficient (K_{oc}) for Soils by High Performance Liquid Chromatography," *Chemosphere*, 19(1):67-77 (1988).

6. Weast, R.C., Ed. *CRC Handbook of Chemistry and Physics*, 67th ed. (Boca Raton, FL: CRC Press, Inc., 1986), 2406 p.

7. "Chemical, Physical, and Biological Properties of Compounds Present at Hazardous Waste Sites," U.S. EPA Report-530/SW-89-010 (1985), 619 p.

8. Eganhouse, R.P., and J.A. Calder. "The Solubility of Medium Weight Aromatic Hydrocarbons and the Effect of Hydrocarbon Co-solutes and Salinity," *Geochim. Cosmochim. Acta*, 40(5):555-561 (1976).

9. Mackay, D., and W.Y. Shiu. "Aqueous Solubility of Polynuclear Aromatic Hydrocarbons," *J. Chem. Eng. Data*, 22(4):399-402 (1977).

10. *Catalog Handbook of Fine Chemicals* (Milwaukee, WI: Aldrich Chemical Co., 1988), 2212 p.

11. Cleland, J.G. "Project Summary - Environmental Hazard Rankings of Pollutants Generated in Coal Gasification Processes," Office of Research and Development, U.S. EPA Report-600/S7-81-101 (1981), 19 p.

4-METHYL-2-PENTANONE

Synonyms: Hexanone; Hexone; Isobutyl methyl ketone; Isopropyl-acetone; Methyl isobutyl ketone; 2-Methyl-4-pentanone; MIBK; MIK; RCRA waste number U161; Shell MIBK; UN 1245.

Structural Formula:

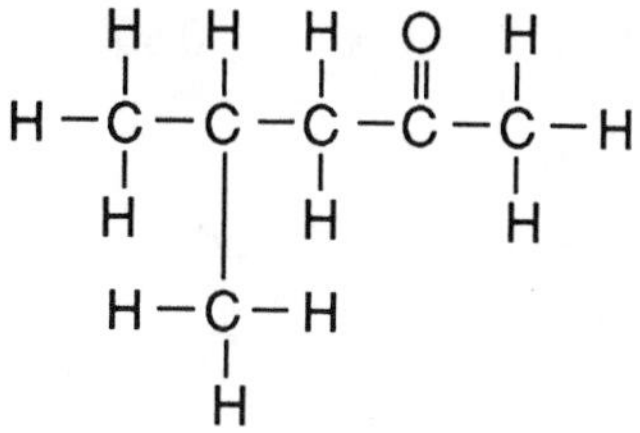

$$
\begin{array}{ccccccccc}
 & H & & H & & H & & O & & H \\
 & | & & | & & | & & \| & & | \\
H-& C & -& C & -& C & -& C & -& C & -H \\
 & | & & | & & | & & & & | \\
 & H & & | & & H & & & & H \\
 & & & H-C-H & & & & & & \\
 & & & | & & & & & & \\
 & & & H & & & & & & \\
\end{array}
$$

CHEMICAL DESIGNATIONS

CAS Registry Number: 108-10-1

DOT Designation: 1245

Empirical Formula: $C_6H_{12}O$

Formula Weight: 100.16

RTECS Number: SA 9275000

PHYSICAL AND CHEMICAL PROPERTIES

Appearance and Odor: Clear, colorless, watery liquid with a mild pleasant odor.

Boiling Point: 116.8 °C [1]; 118 °C [2].

Henry's Law Constant: 1.49×10^{-5} atm·m^3/mol at 25 °C (calculated).

Ionization Potential: 9.30 eV [3].

Log K_{oc}: 0.79 (estimated) [4].

Log K_{ow}: 1.09 [5].

Melting Point: -84.7 °C [1].

Solubility in Organics: Soluble in acetone, ethanol, benzene, chloroform, ether, and many other solvents [6].

Solubility in Water: 17,000 mg/L at 20 °C [7]; 1.9 wt% at 20 °C [8]; 2.1 vol% at 20 °C [9]; 16,800 ppm [10]; 2.04 wt% at 20 °C, 1.91 wt% at 25 °C, 1.78 wt% at 30 °C [11].

Specific Density: 0.7978 at 20/4 °C [1]; 0.8008 at 20/4 °C [12]; 0.800 at 20/4 °C [13].

Transformation Products: No data found.

Vapor Density: 4.09 g/L at 25 °C, 3.46 (air = 1).

Vapor Pressure: 15 mm at 20 °C [8]; 6 mm at 20 °C [7]; 1 mm at -1.4 °C, 10 mm at 30 °C, 40 mm at 52.8 °C, 100 mm at 70.4 °C, 400 mm at 102.0 °C, 760 mm at 119.0 °C [1]; 16 mm at 20 °C [2].

FIRE HAZARDS

Flash Point: 22.8 °C [8].

Lower Explosive Limit (LEL): 1.4% [8].

Upper Explosive Limit (UEL): 7.5% [8].

HEALTH HAZARD DATA

Immediately Dangerous to Life or Health (IDLH): 3,000 ppm [8].

Permissible Exposure Limits (PEL) in Air: 100 ppm ($\approx$410 mg/m^3) [14]; 50 ppm ($\approx$200 mg/m^3) 10-hour TWA [8]; 50 ppm ($\approx$205 mg/m^3) TWA, 75 ppm ($\approx$300 mg/m^3) STEL [15].

MANUFACTURING

Selected Manufacturers:

Exxon Chemical Co.
Houston, TX 77001

Shell Chemical Co.
Industrial Chemicals Division
Houston, TX 77001

Union Carbide Corp.
Chemicals and Plastics Division
270 Park Ave.
New York, NY 10017

Uses: Denaturant for ethyl alcohol; solvent for paints, varnishes nitrocellulose lacquers; preparation of methyl amyl alcohol; extraction of uranium from fission products; organic synthesis.

REFERENCES

1. Weast, R.C., Ed. *CRC Handbook of Chemistry and Physics*, 67th ed. (Boca Raton, FL: CRC Press, Inc., 1986), 2406 p.
2. Sax, N.I. *Dangerous Properties of Industrial Materials* (New York: Van Nostrand Reinhold Co., 1984), 3124 p.
3. Franklin, J.L., Dillard, J.G., Rosenstock, H.M., Herron, J.T., Draxl K., and F.H. Field. "Ionization Potentials, Appearance Potentials and Heats of Formation of Gaseous Positive Ions," National Bureau of Standards Report NSRDS-NBS 26, U.S. Government Printing Office (1969), 289 p.
4. Montgomery, J.H. Unpublished results (1989).
5. Hansch, C., Quinlan, J.E., and G.L. Lawrence. "The Linear Free-Energy Relationship Between Partition Coefficients and the Aqueous Solubility of Organic Liquids," *J. Org. Chem.*, 33(1):347-350 (1968).
6. "Chemical, Physical, and Biological Properties of Compounds Present at Hazardous Waste Sites," U.S. EPA Report-530/SW-89-010 (1985), 619 p.
7. Verschueren, K. *Handbook of Environmental Data on Organic Chemicals* (New York: Van Nostrand Reinhold Co., 1983), 1310 p.
8. "NIOSH Pocket Guide to Chemical Hazards," U.S. Department of Health and Human Services, U.S. Government Printing Office (1987), 241 p.
9. Meites, L., Ed. *Handbook of Analytical Chemistry*, 1st ed. (New York: McGraw-Hill, Inc., 1963), 1782 p.
10. Amidon, G.L., Yalkowsky, S.H., Anik, S.T., and S.C. Valvani. "Solubility of Nonelectrolytes in Polar Solvents. V. Estimation of the Solubility of Aliphatic Monofunctional Compounds in Water

using a Molecular Surface Area Approach," *J. Phys. Chem.*, 79(21):2239-2246 (1975).

11. Ginnings, P.M., Plonk, D., and E. Carter. "Aqueous Solubilities of Some Aliphatic Ketones," *J. Am. Chem. Soc.*, 62(8):1923-1924 (1940).

12. Huntress, E.H., and S.P. Mulliken. *Identification of Pure Organic Compounds - Tables of Data on Selected Compounds of Order I* (New York: John Wiley and Sons, Inc., 1941), 691 p.

13. *Fluka Catalog 1988/89 - Chemika-Biochemika* (Ronkonkoma, NY: Fluka Chemical Corp., 1988), 1536 p.

14. "General Industry Standards for Toxic and Hazardous Substances," U.S. Code of Federal Regulations 1910, Subpart Z Section 1910.1000 (July 1982).

15. *Documentation of the Threshold Limit Values and Biological Exposure Indices for 1986-1987* (Cincinnati, OH: American Conference of Governmental Industrial Hygienists, 1986), 111 p.

2-METHYLPHENOL

Synonyms: 2-Cresol; *o*-Cresol; *o*-Cresylic acid; 1-Hydroxy-2-methyl-benzene; 2-Hydroxytoluene; *o*-Hydroxytoluene; 2-Methyl-hydroxy-benzene; *o*-Methylhydroxybenzene; *o*-Methylphenol; *o*-Methyl-phenylol; Orthocresol; *o*-Oxytoluene; RCRA waste number U052; 2-Toluol; *o*-Toluol; UN 2076.

Structural Formula:

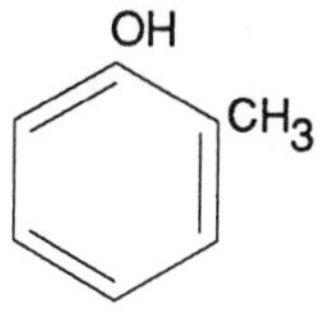

CHEMICAL DESIGNATIONS

CAS Registry Number: 95-48-7

DOT Designation: 2076

Empirical Formula: C_7H_8O

Formula Weight: 108.14

RTECS Number: GO 6300000

PHYSICAL AND CHEMICAL PROPERTIES

Appearance and Odor: Colorless solid or liquid with a phenolic odor; darkens on exposure to air.

Boiling Point: 191.0 °C [1].

Dissociation Constant: 10.26 [1].

Henry's Law Constant: 1.23×10^{-6} atm·m^3/mol at 25 °C [2].

Ionization Potential: 8.98 eV [3].

Log K_{oc}: 1.34 [4].

Log K$_{ow}$: 1.93 [4]; 1.95 [5]; 1.99 [6].

Melting Point: 30.9 °C [7].

Solubility in Organics: Miscible with ethanol, benzene, ether, and glycerol [8].

Solubility in Water: 31,000 mg/L at 40 °C, 56,000 mg/L at 100 °C [9]; 25,000 mg/L at 25 °C [10]; 24,500 mg/L at 20 °C [11]; ≈3% at 35 °C [12]; 3.08 wt% at 40 °C [13]; 13,000 mg/L at 0 °C, 29,000 mg/L at 46.2 °C, 45,000 mg/L at 104.5 °C, 69,000 mg/L at 212 °C, 109.3 g/L at 155.35 °C, 355.1 g/L at 169.25 °C [14]; 30.8 g/L at 40 °C [4]; 23 g/L at 8 °C, 26 g/L at 25 °C [15].

Specific Density: 1.0273 at 20/4 °C [7]; 1.047 at 20/4 °C [16]; 1.0465 at 20/4 °C [14].

Transformation Product: Bacterial degradation of 2-methylphenol may introduce a hydroxyl group to produce *m*-methylcatechol [17].

Vapor Pressure: 0.24 mm at 25 °C, 5 mm at 64 °C [9]; 1 mm at 38.2 °C, 10 mm at 76.7 °C, 40 mm at 105.8 °C, 100 mm at 127.4 °C, 400 mm at 168.4 °C, 760 mm at 190.8 °C [7]; 1 mm at 36.6 °C, 20 mm at 90.6 °C, 40 mm at 105.2 °C [14]; 0.31 mm at 25 °C [4]; 0.045 mm at 8 °C, 0.29 mm at 25 °C [15].

FIRE HAZARDS

Flash Point: 81 °C [3].

Lower Explosive Limit (LEL): 1.35% [18].

Upper Explosive Limit (UEL): Unknown [3].

HEALTH HAZARD DATA

Immediately Dangerous to Life or Health (IDLH): 250 ppm [3].

Permissible Exposure Limits (PEL) in Air: 5 ppm (≈22 mg/m^3) [18]; 2.3 ppm (≈10 mg/m^3) 10-hour TWA [3].

MANUFACTURING

Selected Manufacturers:

Aldrich Chemical Co.
940 West Saint Paul Ave.
Milwaukee, WI 53233

Fluka Chemical Corp.
980 South Second St.
Ronkonkoma, NY 11779

Uses: Disinfectant; phenolic resins; tricresyl phosphate; ore flotation; textile scouring agent; organic intermediate; manufacturing of salicylaldehyde, coumarin, and herbicides; surfactant; synthetic food flavors (*para* isomer only); food antioxidant; dye, perfume, plastics, and resins manufacturing.

REFERENCES

1. Dean, J.A., Ed., *Lange's Handbook of Chemistry*, 11th ed. (New York: McGraw-Hill, Inc., 1973), 1570 p.
2. Hine, J., and P.K. Mookerjee. "The Intrinsic Hydrophilic Character of Organic Compounds. Correlations in Terms of Structural Contributions," *J. Org. Chem.*, 40(3):292-298 (1975).
3. "NIOSH Pocket Guide to Chemical Hazards," U.S. Department of Health and Human Services, U.S. Government Printing Office (1987), 241 p.
4. Howard, P.H. *Handbook of Environmental Fate and Exposure Data for Organic Chemicals* (Chelsea, MI: Lewis Publishers, Inc., 1989), 574 p.
5. Leo, A., Hansch, C., and D. Elkins. "Partition Coefficients and Their Uses," *Chem. Rev.*, 71(6):525-616 (1971).
6. Dearden, J.C. "Partitioning and Lipophilicity in Quantitative Structure-Activity Relationships," *Environ. Health Perspect.*, (September 1985), pp 203-228.
7. Weast, R.C., Ed. *CRC Handbook of Chemistry and Physics*, 67th ed. (Boca Raton, FL: CRC Press, Inc., 1986), 2406 p.
8. "Chemical, Physical, and Biological Properties of Compounds Present at Hazardous Waste Sites," U.S. EPA Report-530/SW-89-010 (1985), 619 p.
9. Verschueren, K. *Handbook of Environmental Data on Organic*

Chemicals (New York: Van Nostrand Reinhold Co., 1983), 1310 p.

10. Morrison, R.T., and R.N. Boyd. *Organic Chemistry* (Boston: Allyn and Bacon, Inc., 1971), 1258 p.

11. Meites, L., Ed. *Handbook of Analytical Chemistry*, 1st ed. (New York: McGraw-Hill, Inc., 1963), 1782 p.

12. Huntress, E.H., and S.P. Mulliken. *Identification of Pure Organic Compounds - Tables of Data on Selected Compounds of Order I* (New York: John Wiley and Sons, Inc., 1941), 691 p.

13. Riddick, J.A., Bunger, W.B., and T.K. Sakano. *Organic Solvents - Physical Properties and Methods of Purification. Volume II* (New York: John Wiley and Sons, Inc., 1986), 1325 p.

14. Standen, A., Ed. *Kirk-Othmer Encyclopedia of Chemical Technology, Volume 5*, 2nd ed. (New York: John Wiley and Sons, Inc., 1965), 884 p.

15. Leuenberger, C., Ligocki, M.P., and J.F. Pankow. "Trace Organic Compounds in Rain. 4. Identities, Concentrations, and Scavenging Mechanisms for Phenols in Urban Air and Rain," *Environ. Sci. Technol.*, 19(11):1053-1058 (1985).

16. Sax, N.I. *Dangerous Properties of Industrial Materials* (New York: Van Nostrand Reinhold Co., 1984), 3124 p.

17. Chapman, P.J. "An Outline of Reaction Sequences Used for the Bacterial Degradation of Phenolic Compounds" in *Degradation of Synthetic Organic Molecules in the Biosphere: Natural, Pesticidal, and Various Other Man-Made Compounds* (Washington, DC: National Academy of Sciences, 1972), pp 17-55.

18. Weiss, G. *Hazardous Chemicals Data Book* (Park Ridge, NJ: Noyes Data Corp., 1986), 1069 p.

19. "General Industry Standards for Toxic and Hazardous Substances," U.S. Code of Federal Regulations 1910, Subpart Z Section 1910.1000 (July 1982).

4–METHYLPHENOL

Synonyms: 4-Cresol; *p*-Cresol; *p*-Cresylic acid; 1-Hydroxy-4-methylbenzene; *p*-Hydroxytoluene; 4-Hydroxytoluene; *p*-Kresol; 1-Methyl-4-hydroxybenzene; 4-Methylhydroxybenzene; *p*-Methyl-hydroxybenzene; *p*-Methylphenol; 4-Oxytoluene; *p*-Oxytoluene; Para-cresol; Paramethylphenol; RCRA waste number U052; 4-Toluol; *p*-Toluol; *p*-Tolyl alcohol; UN 2076.

Structural Formula:

CHEMICAL DESIGNATIONS

CAS Registry Number: 106-44-5

DOT Designation: 2076

Empirical Formula: C_7H_8O

Formula Weight: 108.14

RTECS Number: GO 6475000

PHYSICAL AND CHEMICAL PROPERTIES

Appearance and Odor: Colorless solid with a phenolic odor.

Boiling Point: 201.9 °C [1].

Dissociation Constant: 10.26 [2].

Henry's Law Constant: 7.92 x 10^{-7} atm·m³/mol at 25 °C [3].

Ionization Potential: 8.97 eV [4].

Log K_{oc}: 1.69 [5].

Log K_{ow}: 1.67 [6]; 3.01 [7]; 1.92 [8]; 1.94 [9].

Melting Point: 34.8 °C [1]; 36 °C [10].

Solubility in Organics: Miscible with ethanol, benzene, ether, and glycerol [11].

Solubility in Water: 24,000 mg/L at 40 °C, 53,000 mg/L at 100 °C [12]; 23,000 mg/L at 25 °C [13]; 1.9 wt% at 20 °C [4]; 19,400 mg/L at 20 °C [14]; ≈2.3% at 40 °C [10]; 2.26 wt% at 40 °C [15]; 22,100 mg/L at 30 °C, 54,000 mg/L at 105 °C, 164.0 g/L at 138 °C [16]; 13 g/L at 8 °C, 18 g/L at 25 °C [17].

Specific Density: 1.0178 at 20/4 °C [1]; 1.0341 at 20/4 °C [16].

Transformation Products: Protocatechuic acid (3,4-dihydroxybenzoic acid) is the central metabolite in the bacterial degradation of 4-methylphenol. Intermediate byproducts included *p*-hydroxybenzyl alcohol, *p*-hydroxybenzaldehyde, and *p*-hydroxybenzoic acid. In addition, 4-methylphenol may undergo hydroxylation to form *p*-methylcatechol [18]. Oxidation products reported include Pummerer's ketone and 2,2'-dihydroxy-4,4'-dimethylbiphenyl [19]. A species of *Pseudomonas*, isolated from creosote-contaminated soil, degraded 4-methylphenol into *p*-hydroxybenzaldehyde and *p*-hydroxybenzoate. Both metabolites were then converted into protocatechuate [20].

Vapor Pressure: 0.04 mm at 20 °C [12]; 1 mm at 53.0 °C, 10 mm at 88.6 °C, 40 mm at 117.7 °C, 100 mm at 140.0 °C, 400 mm at 179.4 °C, 760 mm at 201.8 °C [1]; 0.08 mm at 25 °C [21]; 1 mm at 55.7 °C, 20 mm at 101.8 °C, 40 mm at 116.4 °C [16]; 0.108 mm at 25 °C [19]; 0.13 mm at 25 °C [5]; 0.020 mm at 8°, 0.12 mm at 25 °C [17].

FIRE HAZARDS

Flash Point: 86 °C [4].

Lower Explosive Limit (LEL): 1.06% [22].

Upper Explosive Limit (UEL): 1.4% [22].

HEALTH HAZARD DATA

Immediately Dangerous to Life or Health (IDLH): 250 ppm [4].

Permissible Exposure Limits (PEL) in Air: 5 ppm ($\approx$22 mg/m^3) [23]; 2.3 ppm ($\approx$10 mg/m^3) 10-hour TWA [4].

MANUFACTURING

Selected Manufacturers:

> Aldrich Chemical Co.
> 940 West Saint Paul Ave.
> Milwaukee, WI 53233

> Fluka Chemical Corp.
> 980 South Second St.
> Ronkonkoma, NY 11779

> Pfaltz & Bauer, Inc.
> 172 East Aurora St.
> Waterbury, CT 06708

Uses: Disinfectant; phenolic resins; tricresyl phosphate; ore flotation; textile scouring agent; organic intermediate; manufacturing of salicylaldehyde, coumarin, and herbicides; surfactant; synthetic food flavors.

REFERENCES

1. Weast, R.C., Ed. *CRC Handbook of Chemistry and Physics*, 67th ed. (Boca Raton, FL: CRC Press, Inc., 1986), 2406 p.
2. Dean, J.A., Ed., *Lange's Handbook of Chemistry*, 11th ed. (New York: McGraw-Hill, Inc., 1973), 1570 p.
3. Hine, J., and P.K. Mookerjee. "The Intrinsic Hydrophilic Character of Organic Compounds. Correlations in Terms of Structural Contributions," *J. Org. Chem.*, 40(3):292-298 (1975).
4. "NIOSH Pocket Guide to Chemical Hazards," U.S. Department of Health and Human Services, U.S. Government Printing Office (1987), 241 p.
5. Howard, P.H. *Handbook of Environmental Fate and Exposure Data for Organic Chemicals* (Chelsea, MI: Lewis Publishers, Inc., 1989), 574 p.
6. Neely, W.B., and G.E. Blau, Eds. *Environmental Exposure from Chemicals. Volume 1* (Boca Raton, FL: CRC Press, Inc. 1985), 245 p.

7. Mackay, D., and S. Paterson. "Calculating Fugacity," *Environ. Sci. Technol.*, 15(9):1006-1014 (1981).

8. Leo, A., Hansch, C., and D. Elkins. "Partition Coefficients and Their Uses," *Chem. Rev.*, 71(6):525-616 (1971).

9. Fujita, T., Iwasa, J., and C. Hansch. "A New Substituent Constant, π, Derived from Partition Coefficients," *J. Am. Chem. Soc.*, 86(23):5175-5180 (1964).

10. Huntress, E.H., and S.P. Mulliken. *Identification of Pure Organic Compounds - Tables of Data on Selected Compounds of Order I* (New York: John Wiley and Sons, Inc., 1941), 691 p.

11. "Chemical, Physical, and Biological Properties of Compounds Present at Hazardous Waste Sites," U.S. EPA Report-530/SW-89-010 (1985), 619 p.

12. Verschueren, K. *Handbook of Environmental Data on Organic Chemicals* (New York: Van Nostrand Reinhold Co., 1983), 1310 p.

13. Morrison, R.T., and R.N. Boyd. *Organic Chemistry* (Boston: Allyn and Bacon, Inc., 1971), 1258 p.

14. Meites, L., Ed. *Handbook of Analytical Chemistry*, 1st ed. (New York: McGraw-Hill, Inc., 1963), 1782 p.

15. Riddick, J.A., Bunger, W.B., and T.K. Sakano. *Organic Solvents - Physical Properties and Methods of Purification. Volume II* (New York: John Wiley and Sons, Inc., 1986), 1325 p.

16. Standen, A., Ed. *Kirk-Othmer Encyclopedia of Chemical Technology, Volume 5*, 2nd ed. (New York: John Wiley and Sons, Inc., 1965), 884 p.

17. Leuenberger, C., Ligocki, M.P., and J.F. Pankow. "Trace Organic Compounds in Rain. 4. Identities, Concentrations, and Scavenging Mechanisms for Phenols in Urban Air and Rain," *Environ. Sci. Technol.*, 19(11):1053-1058 (1985).

18. Chapman, P.J. "An Outline of Reaction Sequences Used for the Bacterial Degradation of Phenolic Compounds" in *Degradation of Synthetic Organic Molecules in the Biosphere: Natural, Pesticidal, and Various Other Man-Made Compounds* (Washington, DC: National Academy of Sciences, 1972), pp 17-55.

19. Smith, J.H., Mabey, W.R., Bohonos, N., Holt, B.R., Lee, S.S., Chou, T.-W., Bomberger, D.C., and T. Mill. "Environmental Pathways of Selected Chemicals in Freshwater Systems. Part II: Laboratory Studies," U.S. EPA Report-600/7-78-074 (1978), 406 p.

20. O'Reilly, K.T., and R.L. Crawford. "Kinetics of *p*-Cresol Degradation by an Immobilized *Pseudomonas* sp.," *Appl. Environ. Microbiol.*, 55(4):866-870 (1989).

21. Valsaraj, K.T. "On the Physio-Chemical Aspects of Partitioning of Non-Polar Hydrophobic Organics at the Air-Water Interface,"

Chemosphere, 17(5):875-887 (1988).

22. Weiss, G. *Hazardous Chemicals Data Book* (Park Ridge, NJ: Noyes Data Corp., 1986), 1069 p.

23. "General Industry Standards for Toxic and Hazardous Substances," U.S. Code of Federal Regulations 1910, Subpart Z Section 1910.1000 (July 1982).

NAPHTHALENE

Synonyms: Camphor tar; Mighty 150; Mighty RD1; Moth balls; Moth flakes; Naphthalin; Naphthaline; Naphthene; NCI-C52904; RCRA waste number U165; Tar camphor; UN 1334; White tar.

Structural Formula:

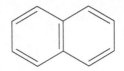

CHEMICAL DESIGNATIONS

CAS Registry Number: 91-20-3

DOT Designation: 1334

Empirical Formula: $C_{10}H_8$

Formula Weight: 128.18

RTECS Number: QJ 0525000

PHYSICAL AND CHEMICAL PROPERTIES

Appearance and Odor: White, crystalline flakes with an aromatic odor resembling coal-tar.

Boiling Point: 217.942 °C [1].

Henry's Law Constant: 4.6 x 10^{-4} atm·m^3/mol [2]; 4.8 x 10^{-4} atm·m^3/mol [3]; 0.00122 atm·m^3/mol [4]; 3.6 x 10^{-4} atm·m^3/mol [5]; 5.53 x 10^{-4} atm·m^3/mol [6].

Ionization Potential: 8.14 eV [7]; 8.26 eV [8].

Log K_{oc}: 2.74 [9]; 3.11 [10]; 3.11, 3.52 [11]; 2.96, 3.04 [12]; 3.11, 3.16, 3.21, 3.50 [13]; 3.04 [14].

Log K_{ow}: 3.36 [10]; 4.70 [15]; 3.59 [16]; 3.23, 3.24, 3.26, 3.28 [17]; 3.30 [18]; 3.31 [19]; 3.01, 3.20, 3.45 [20]; 3.37 [21].

Melting Point: 80.5 °C [22]; 80.28 °C [23].

Solubility in Organics: Soluble in acetone, ethanol, benzene, and ether [24].

Solubility in Water: 30 mg/L at 25 °C [25]; 31-34 mg/L at 25 °C, 20 mg/L in seawater at 22 °C [26]; 40 mg/L at 25 °C [27]; 34.4 mg/L at 25 °C [28]; 31.7 mg/L at 25 °C [29]; 31.5 mg/L at 25 °C [30]; 22 mg/L mg/L at 25 °C [31]; 33.6 mg/L at 25 °C [32]; 31.3 mg/L at 25 °C, 22.0 mg/L in artificial seawater at 25 °C [33]; 31.2 mg/L at 25 °C [34]; 37.7 mg/L at 20-25 °C [35]; 0.019 g/L at 0 °C, 0.030-0.0344 g/L at 25 °C [36]; 0.003169 wt% at 25 °C [37]; 20.315 mg/L at 25 °C [38]; 0.00022 M at 21 °C [39]; 0.000140 M at 8.4 °C, 0.000149 M at 11.1 °C, 0.000166 M at 14.0 °C, 0.000188 M at 17.5 °C, 0.000207 M at 20.2 °C, 0.000222 M at 23.2 °C, 0.000236 M at 25.0 °C, 0.000248 M at 26.3 °C, 0.000268 M at 29.2 °C, 0.000283 M at 31.8 °C [40]; 13.66 mg/L at 1 °C, 29.41 mg/L at 23 °C, 53.90 mg/L at 40 °C [41]; 12.07 mg/L at 1.9 °C, 17.19 mg/L at 10.7 °C, 21.64 mg/L at 15.4 °C, 26.72 mg/L at 21.7 °C, 30.72 mg/L at 25.2 °C, 40.09 mg/L at 30.7 °C, 46.31 mg/L at 35.1 °C, 54.80 mg/L at 39.3 °C, 68.90 mg/L at 44.9 °C [42].

Specific Density: 0.9625 at 100/4 °C [22]; 1.162 at 20/4 °C [43]; 1.145 at 20/4 °C [44]; 1.01813 at 30/4 °C, 0.9752 at 85/4 °C [45].

Transformation Products: Under certain conditions, *Pseudomonas* sp. oxidized naphthalene to *cis*-1,2-dihydro-1,2-dihydroxynaphthalene [46]. Under aerobic conditions, *Cuninghamella elegans* biodegraded naphthalene to α-naphthol, β-naphthol, *trans*-1,2-dihydroxy-1,2-dihydronaphthalene, 4-hydroxy-1-tetralene, and 1,4-naphthoquinone. Also under aerobic conditions, *Agnenellum, Oscillatoria* and *Anabaena* reportedly biodegraded naphthalene into 1-naphthol, *cis*-1,2-dihydroxyl-1,2-dihydronaphthalene and 4-hydroxy-1-tetralene [47].

Vapor Pressure: 0.005 mm at 0 °C, 0.021 mm at 10 °C, 0.054 mm at 20 °C, 0.320 mm at 40 °C [48]; 0.23 mm at 25 °C [49]; 1 mm at 52.6 °C, 10 mm at 85.8 °C, 40 mm at 119.3 °C, 100 mm at 145.5 °C, 400 mm at 193.2 °C, 760 mm at 217.9 °C [22]; 0.87 mm at 25 °C [43]; 0.0492 mm at 20 °C [50]; 0.0132 mm at 7.15 °C, 0.0232 mm at 12.80 °C, 0.0419 mm at 18.40 °C, 0.0445 mm at 18.85 °C, 0.0944 mm at 26.40 °C, 0.154 mm at 31.85 °C [51]; 0.00122 mm at 6.70 °C, 0.00141 mm at 8.10 °C, 0.00222 mm at 12.30 °C, 0.00235 mm at 12.70 °C, 0.00263 mm at 13.85 °C, 0.00320 mm at 15.65 °C, 0.00350 mm at 16.85 °C, 0.00382 mm at 17.35 °C, 0.00383 mm at 17.55 °C, 0.00438 mm at 18.70 °C, 0.00534 mm at

20.70 °C [52]; 0.0062 mm at 1 °C, 0.071 mm at 23 °C, 0.34 mm at 40 °C [53]; 0.354 mm at 40.33 °C, 1.836 mm at 60.23 °C, 6.995 mm at 79.34 °C, 21.257 mm at 102.93 °C, 41.03 mm at 119.22 °C, 89.64 mm at 140.83 °C, 290.95 mm at 179.51 °C [23].

FIRE HAZARDS

Flash Point: 79 °C [7].

Lower Explosive Limit (LEL): 0.9% [7].

Upper Explosive Limit (UEL): 5.9% [7].

HEALTH HAZARD DATA

Immediately Dangerous to Life or Health (IDLH): 500 ppm [7].

Permissible Exposure Limits (PEL) in Air: 10 ppm ($\approx$50 mg/m^3) [54]; 15 ppm ($\approx$75 mg/m^3) STEL [43].

MANUFACTURING

Selected Manufacturers:

Aldrich Chemical Co.
940 West Saint Paul Ave.
Milwaukee, WI 53233

Fluka Chemical Corp.
980 South Second St.
Ronkonkoma, NY 11779

Pfaltz & Bauer, Inc.
172 East Aurora St.
Waterbury, CT 06708

Uses: Intermediate for phthalic anhydride, naphthol, Tetralin, Decalin, chlorinated naphthalenes, naphthyl, naphthol derivatives, and dyes; mothballs manufacturing; preparation of pesticides, fungicides, dyes, detergents, and wetting agents, synthetic resins, celluloids, and lubricants; synthetic tanning; preservative; textile chemicals; emulsion breakers; scintillation counters; smokeless powders.

REFERENCES

1. Zwolinski, B.J., and R.C. Wilhoit. *Handbook of Vapor Pressure and Heats of Vaporization of Hydrocarbons and Related Compounds*, Publication 101 (College Station, TX: Thermodynamics Research Station, 1971), 329 p.
2. Pankow, J.F., and M.E. Rosen. "Determination of Volatile Compounds in Water by Purging Directly to a Capillary Column with Whole Column Cryotrapping," *Environ. Sci. Technol.*, 22(4):398-405 (1988).
3. Valsaraj, K.T. "On the Physio-Chemical Aspects of Partitioning of Non-Polar Hydrophobic Organics at the Air-Water Interface," *Chemosphere*, 17(5):875-887 (1988).
4. Jury, W.A., Spencer, W.F., and W.J. Farmer. "Behavior Assessment Model for Trace Organics in Soil: III. Application of Screening Model," *J. Environ. Qual.*, 13(4):573-579 (1984).
5. Petrasek, A.C., Kugelman, I.J., Austern, B.M., Pressley, T.A., Winslow, L.A., and R.H. Wise. "Fate of Toxic Organic Compounds in Wastewater Treatment Plants," *J. Water Poll. Control Fed.*, 55(10):1286-1296 (1983).
6. Southworth, G.R. "The Role Volatilization in Removing Polycyclic Aromatic Hydrocarbons from Aquatic Environments," *Bull. Environ. Contam. Toxicol.*, 21(4/5):507-514 (1979).
7. "NIOSH Pocket Guide to Chemical Hazards," U.S. Department of Health and Human Services, U.S. Government Printing Office (1987), 241 p.
8. Yoshida, K., Shigeoka, T., and F. Yamauchi. "Non-Steady State Equilibrium Model for the Preliminary Prediction of the Fate of Chemicals in the Environment," *Ecotoxicol. Environ. Safety*, 7(2):179-190 (1983).
9. Abdul, S.A., Gibson, T.L., and D.N. Rai. "Statistical Correlations for Predicting the Partition Coefficient for Nonpolar Organic Contaminants Between Aquifer Organic Carbon and Water," *Haz. Waste Haz. Mater.*, 4(3):211-222 (1987).
10. Karickhoff, S.W., Brown, D.S., and T.A. Scott. "Sorption of Hydrophobic Pollutants on Natural Sediments," *Water Res.*, 13:241-248 (1979).
11. Chin, Y.-P., Peven, C.S., and W.J. Weber. "Estimating Soil/Sediment Partition Coefficients for Organic Compounds by High

Performance Reverse Phase Liquid Chromatography," *Water Res.*, 22(7):873-881 (1988).

12. Hodson, J., and N.A. Williams. "The Estimation of the Adsorption Coefficient (K_{oc}) for Soils by High Performance Liquid Chromatography," *Chemosphere*, 19(1):67-77 (1988).

13. Rippen, G., Ilgenstein, M., and W. Klöpffer. "Screening of the Adsorption Behavior of New Chemicals: Natural Soils and Model Adsorbents," *Ecotoxicol. Environ. Safety*, 6(3):236-245 (1982).

14. Løkke, H. "Sorption of Selected Organic Pollutants in Danish Soils," *Ecotoxicol. Environ. Safety*, 8(5):395-409 (1984).

15. Isnard, S., and S. Lambert. "Estimating Bioconcentration Factors from Octanol-Water Partition Coefficient and Aqueous Solubility," *Chemosphere*, 17(1):21-34 (1988).

16. Mackay, D. "Correlation of Bioconcentration Factors," *Environ. Sci. Technol.*, 16(5):274-278 (1982).

17. Brooke, D.N., Dobbs, A.J., and N. Williams. "Octanol:Water Partition Coefficients (P): Measurement, Estimation, and Interpretation, Particularly for Chemicals with $P > 10^5$," *Ecotoxicol. Environ. Safety*, 11(3):251-260 (1986).

18. Geyer, H., Politzki, G., and D. Freitag. "Prediction of Ecotoxicological Behaviour of Chemicals: Relationship Between *n*-Octanol/Water Partition Coefficient and Bioaccumulation of Organic Chemicals by Alga *Chlorella*," *Chemosphere*, 13(2):269-284 (1984).

19. Kenaga, E.E., and C.A.I. Goring. "Relationship Between Water Solubility, Soil Sorption, Octanol-Water Partitioning and Concentration of Chemicals in Biota," in *Aquatic Toxicology, ASTM STP 707*, Eaton, J.G., Parrish, P.R., and A.C. Hendricks, Eds. (Philadelphia, PA: American Society for Testing and Materials 1980), pp 78-115.

20. Leo, A., Hansch, C., and D. Elkins. "Partition Coefficients and Their Uses," *Chem. Rev.*, 71(6):525-616 (1971).

21. Hansch, C., and T. Fujita. "ρ-σ-π Analysis. A Method for the Correlation of Biological Activity and Chemical Structure," *J. Am. Chem. Soc.*, 86(8):1616-1617 (1964).

22. Weast, R.C., Ed. *CRC Handbook of Chemistry and Physics*, 67th ed. (Boca Raton, FL: CRC Press, Inc., 1986), 2406 p.

23. Fowler, L., Trump, W.N., and C.E. Vogler. "Vapor Pressure of Naphthalene - New Measurements Between 40° and 180 °C," *J. Chem. Eng. Data*, 13(2):209-210 (1968).

24. "Chemical, Physical, and Biological Properties of Compounds Present at Hazardous Waste Sites," U.S. EPA Report-530/SW-89-010 (1985), 619 p.

25. Dean, J.A., Ed., *Lange's Handbook of Chemistry*, 11th ed. (New York: McGraw-Hill, Inc., 1973), 1570 p.
26. Verschueren, K. *Handbook of Environmental Data on Organic Chemicals* (New York: Van Nostrand Reinhold Co., 1983), 1310 p.
27. Meites, L., Ed. *Handbook of Analytical Chemistry*, 1st ed. (New York: McGraw-Hill, Inc., 1963), 1782 p.
28. Bohon, R.L., and W.F. Claussen. "The Solubility of Aromatic Hydrocarbons in Water," *J. Am. Chem. Soc.*, 73(4):1571-1578 (1951).
29. Mackay, D., and W.Y. Shiu. "Aqueous Solubility of Polynuclear Aromatic Hydrocarbons," *J. Chem. Eng. Data*, 22(4):399-402 (1977).
30. Andrews, L.J., and R.M. Keefer. "Cation Complexes of Compounds Containing Carbon-Carbon Double Bonds. IV. The Argentation of Aromatic Hydrocarbons," *J. Am. Chem. Soc.*, 71(11):3644-3647 (1949).
31. Schwarz, F.P., and S.P. Wasik. "Fluorescence Measurements of Benzene, Naphthalene, Anthracene, Pyrene, Fluoranthene, and Benzo[e]pyrene in Water," *Anal. Chem.*, 48(3):524-528 (1976).
32. Gordon, J.E., and R.L. Thorne. "Salt Effects on the Activity Coefficient of Naphthalene in Mixed Aqueous Electrolyte Solutions. I. Mixtures of Two Salts," *J. Phys. Chem.*, 71(13):4390-4399 (1967).
33. Eganhouse, R.P., and J.A. Calder. "The Solubility of Medium Weight Aromatic Hydrocarbons and the Effect of Hydrocarbon Co-solutes and Salinity," *Geochim. Cosmochim. Acta*, 40(5):555-561 (1976).
34. Wauchope, R.D., and F.W. Getzen. "Temperature Dependence of Solubilities in Water and Heats of Fusion of Solid Aromatic Hydrocarbons," *J. Chem. Eng. Data*, 17:38-41 (1972).
35. Geyer, H., Sheehan, P., Kotzias, D., Freitag, D., and Friedhelm Korte. "Prediction of Ecotoxicological Behaviour of Chemicals: Relationship Between Physico-Chemical Properties and Bioaccumulation of Organic Chemicals in the Mussell *Mytilus edulis*," *Chemosphere*, 11(11):1121-1134 (1982).
36. Stephen, H., and T. Stephen. *Solubilities of Inorganic and Organic Compounds - Part 1, Volume 1* (London: Pergamon Printing and Art Services, Ltd., 1963), 960 p.
37. Riddick, J.A., Bunger, W.B., and T.K. Sakano. *Organic Solvents - Physical Properties and Methods of Purification. Volume II* (New York: John Wiley and Sons, Inc., 1986), 1325 p.
38. Sahyun, M.R.V. "Binding of Aromatic Compounds to Bovine Serum Albumin," *Nature*, 209(5023):613-614 (1966).
39. Almgren, M., Grieser, F., Powell, J.R., and J.K. Thomas. "A Correlation Between the Solubility of Aromatic Hydrocarbons in Water and Micellar Solutions, with Their Normal Boiling Points," *J. Chem. Eng. Data*, 24(4):285-287 (1979).

40. Schwarz, F.P. "Determination of Temperature Dependence of Solubilities of Polycyclic Aromatic Hydrocarbons in Aqueous Solutions by a Fluorescence Method," *J. Chem. Eng. Data,* 22(3):273-277 (1977).

41. Klöpffer, W., Kaufman, G., Rippen, G., and H.-P. Poremski. "A Laboratory Method for Testing the Volatility from Aqueous Solution: First Results and Comparison with Theory," *Ecotoxicol. Environ. Safety,* 6(6):545-559 (1982).

42. Bennett, D., and W.J. Canady. "Thermodynamics of Solution of Naphthalene in Various Water-Ethanol Mixtures," *J. Am. Chem. Soc.,* 106(4):910-915 (1984).

43. *Documentation of the Threshold Limit Values and Biological Exposure Indices* (Cincinnati, OH: American Conference of Governmental Industrial Hygienists, 1986), 744 p.

44. Weiss, G. *Hazardous Chemicals Data Book* (Park Ridge, NJ: Noyes Data Corp., 1986), 1069 p.

45. Standen, A., Ed. *Kirk-Othmer Encyclopedia of Chemical Technology, Volume 13,* 2nd ed. (New York: John Wiley and Sons, Inc., 1967), 994 p.

46. Dagley, S. "Microbial Degradation of Stable Chemical Structures: General Features of Metabolic Pathways" in *Degradation of Synthetic Organic Molecules in the Biosphere: Natural, Pesticidal, and Various Other Man-Made Compounds* (Washington, DC: National Academy of Sciences, 1972), pp 1-16.

47. Kobayashi, H., and B.E. Rittman. "Microbial Removal of Hazardous Organic Compounds," *Environ. Sci. Technol.,* 16(3):170A-183A (1982).

48. Standen, A., Ed. *Kirk-Othmer Encyclopedia of Chemical Technology, Volume 12,* 2nd ed. (New York: John Wiley and Sons, Inc., 1967), 905 p.

49. Mackay, D., and A.W. Wolkoff. "Rate of Evaporation of Low-Solubility Contaminants from Water Bodies to Atmosphere," *Environ. Sci. Technol.,* 7(7):611-614 (1973).

50. "Treatability Manual - Volume 1: Treatability Data," Office of Research and Development, U.S. EPA Report-600/8-80-042a (1980), 1035 p.

51. Macknick, A.B., and J.M. Prausnitz. "Vapor Pressures of High Molecular Weight Hydrocarbons," *J. Chem. Eng. Data,* 24(3):175-178 (1979).

52. Bradley, R.S., and T.G. Cleasby. "The Vapour Pressure and Lattice Energy of Some Aromatic Ring Compounds," *J. Chem. Soc. (London),* pp 1690-1692 (1953).

53. Klöppfer, W., Kaufman, G., Rippen, G., and H.-P. Poremski. "A Laboratory Method for Testing the Volatility from Aqueous

Solution: First Results and Comparison with Theory," *Ecotoxicol. Environ. Safety*, 6(6):545-559 (1982).

54. "General Industry Standards for Toxic and Hazardous Substances," U.S. Code of Federal Regulations 1910, Subpart Z Section 1910.1000 (July 1982).

2-NITROANILINE

Synonyms: 1-Amino-2-nitrobenzene; Azoene fast orange GR base; Azoene fast orange GR salt; Azofix orange GR salt; Azogene fast orange GR; Azoic diazo component 6; Brentamine fast orange GR base; Brentamine fast orange GR salt; C.I. 37,025; C.I. azoic diazo component 6; Devol orange B; Devol orange salt B; Diazo fast orange GR; Fast orange base GR; Fast orange base GR salt; Fast orange base JR; Fast orange GR base; Fast orange O base; Fast orange O salt; Fast orange salt JR; Hiltonil fast orange GR base; Hiltosal fast orange GR salt; Hindasol orange GR salt; Natasol fast orange GR salt; *o*-Nitraniline; *o*-Nitroaniline; **3-Nitrobenzenamine**; ONA; Orange base CIBA II; Orange base IRGA II; Orange GRS salt; Orange salt CIBA II; Orange salt IRGA II; Orthonitroaniline; UN 1661.

Structural Formula:

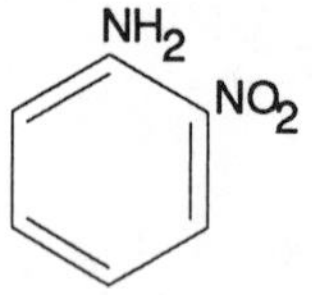

CHEMICAL DESIGNATIONS

CAS Registry Number: 88-74-4

DOT Designation: 1661

Empirical Formula: $C_6H_6N_2O_2$

Formula Weight: 138.13

RTECS Number: BY 6650000

PHYSICAL AND CHEMICAL PROPERTIES

Appearance and Odor: Orange-yellow crystals with a musty odor.

Boiling Point: 284.1 °C [1].

Dissociation Constant: -0.26 [1].

Henry's Law Constant: 9.72×10^{-5} atm·m³/mol at 25 °C (calculated).

Ionization Potential: 8.66 eV [2].

Log K_{oc}: 1.23-1.62 using method of Karickhoff and others [3].

Log K_{ow}: 1.44, 1.79, 1.83 [4].

Melting Point: 71.5 °C [5]; 69-71 °C [6].

Solubility in Organics: Soluble in acetone, ethanol, benzene, and ether [5].

Solubility in Water: 1,260 mg/L at 25 °C [7]; 1.47 g/L at 30 °C [8].

Specific Density: 1.442 at 15/4 °C [5]; 0.9015 at 25/4 °C [6]; 1.44 at 20/4 °C [9].

Transformation Products: No data found.

Vapor Pressure: 8.1 mm at 25 °C [10].

FIRE HAZARDS

Flash Point: 168 °C [11].

Lower Explosive Limit (LEL): Not pertinent [9].

Upper Explosive Limit (UEL): No data found.

HEALTH HAZARD DATA

Immediately Dangerous to Life or Health (IDLH): No data found.

Permissible Exposure Limits (PEL) in Air: No standards set.

MANUFACTURING

Selected Manufacturers:

American Aniline Products, Inc.
Mount Vernon Street
Lock Haven, PA 17745

Monsanto Industrial Chemicals Co.
800 North Lindbergh Blvd.
St. Louis, MO 63166

Use: Organic synthesis.

REFERENCES

1. Dean, J.A., Ed., *Lange's Handbook of Chemistry*, 11th ed. (New York: McGraw-Hill, Inc., 1973), 1570 p.
2. Franklin, J.L., Dillard, J.G., Rosenstock, H.M., Herron, J.T., Draxl K., and F.H. Field. "Ionization Potentials, Appearance Potentials and Heats of Formation of Gaseous Positive Ions," National Bureau of Standards Report NSRDS-NBS 26, U.S. Government Printing Office (1969), 289 p.
3. Karickhoff, S.W., Brown, D.S., and T.A. Scott. "Sorption of Hydrophobic Pollutants on Natural Sediments," *Water Res.*, 13:241-248 (1979).
4. Leo, A., Hansch, C., and D. Elkins. "Partition Coefficients and Their Uses," *Chem. Rev.*, 71(6):525-616 (1971).
5. Weast, R.C., Ed. *CRC Handbook of Chemistry and Physics*, 67th ed. (Boca Raton, FL: CRC Press, Inc., 1986), 2406 p.
6. Sax, N.I. *Dangerous Properties of Industrial Materials* (New York: Van Nostrand Reinhold Co., 1984), 3124 p.
7. Verschueren, K. *Handbook of Environmental Data on Organic Chemicals* (New York: Van Nostrand Reinhold Co., 1983), 1310 p.
8. Gross, P.M., Saylor, J.H., and M.A. Gorman. "Solubility Studies. IV. The Solubilities of Certain Slightly Soluble Organic Compounds in Water," *J. Am. Chem. Soc.*, 55(2):650-652 (1933).
9. Weiss, G. *Hazardous Chemicals Data Book* (Park Ridge, NJ: Noyes Data Corp., 1986), 1069 p.
10. Mabey, W.R., Smith, J.H., Podoll, R.T., Johnson, H.L., Mill, T., Chou, T.-W., Gates, J., Partridge, I.W., Jaber, H., and D. Vandenberg. "Aquatic Fate Process Data for Organic Priority Pollutants - Final Report," Office of Regulations and Standards, U.S. EPA Report-440/4-81-014 (1982), 407 p.
11. Hawley, G.G. *The Condensed Chemical Dictionary* (New York: Van Nostrand Reinhold Co., 1981), 1135 p.

3-NITROANILINE

Synonyms: Amarthol fast orange R base; 1-Amino-3-nitrobenzene; 3-Aminonitrobenzene; *m*-Aminonitrobenzene; Azobase MNA; C.I. 37,030; C.I. azoic diazo component 7; Daito orange base R; Devol orange R; Diazo fast orange R; Fast orange base R; Fast orange M base; Fast orange MM base; Fast orange R base; Fast orange R salt; Hiltonil fast orange R base; MNA; Naphtoelan orange R base; Nitranilin; *m*-Nitraniline; 3-Nitroaminobenzene; *m*-Nitro-aminobenzene; *m*-Nitroaniline; **3-Nitrobenzenamine**; *m*-Nitro-benzenamine; *m*-Nitrophenylamine; Orange base IRGA I; UN 1661.

Structural Formula:

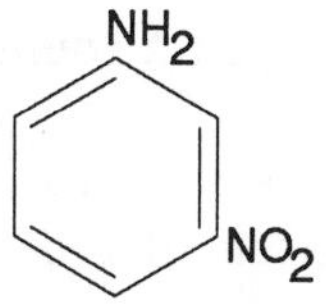

CHEMICAL DESIGNATIONS

CAS Registry Number: 99-09-2

DOT Designation: 1661

Empirical Formula: $C_6H_6N_2O_2$

Formula Weight: 138.13

RTECS Number: BY 6825000

PHYSICAL AND CHEMICAL PROPERTIES

Appearance: Yellow, rhombic crystals.

Boiling Point: 306.4 °C [1].

Dissociation Constant: 2.46 [1].

Henry's Law Constant: Insufficient vapor pressure data for calculation.

Ionization Potential: 8.80 eV [2].

Log K$_{oc}$: 1.26 using method of Karickhoff and others [3].

Log K$_{ow}$: 1.37 [4].

Melting Point: 114 °C [5].

Solubility in Organics: Soluble in acetone, ethanol, and ether [5].

Solubility in Water: 890 mg/L at 25 °C [6]; 1,100 mg/L at 20 °C [1]; 1.21 g/L at 30 °C [7].

Specific Density: 1.1747 at 160/4 °C [5]; 0.9011 at 25/4 °C [8].

Transformation Products: No data found.

Vapor Pressure: 1 mm at 119.3 °C, 10 mm at 167.8 °C, 40 mm at 204.2 °C, 100 mm at 213.0 °C, 400 mm at 280.2 °C, 760 mm at 305.7 °C [5].

FIRE HAZARDS

Flash Point: 306 °C (calculated) [9].

Lower Explosive Limit (LEL): No data found.

Upper Explosive Limit (UEL): No data found.

HEALTH HAZARD DATA

Immediately Dangerous to Life or Health (IDLH): No data found.

Permissible Exposure Limits (PEL) in Air: No standards set.

MANUFACTURING

Selected Manufacturers:

Aldrich Chemical Co.
940 West Saint Paul Ave.
Milwaukee, WI 53233

Fluka Chemical Corp.
980 South Second St.
Ronkonkoma, NY 11779

Use: Organic synthesis.

REFERENCES

1. Dean, J.A., Ed., *Lange's Handbook of Chemistry*, 11th ed. (New York: McGraw-Hill, Inc., 1973), 1570 p.
2. Franklin, J.L., Dillard, J.G., Rosenstock, H.M., Herron, J.T., Draxl K., and F.H. Field. "Ionization Potentials, Appearance Potentials and Heats of Formation of Gaseous Positive Ions," National Bureau of Standards Report NSRDS-NBS 26, U.S. Government Printing Office (1969), 289 p.
3. Karickhoff, S.W., Brown, D.S., and T.A. Scott. "Sorption of Hydrophobic Pollutants on Natural Sediments," *Water Res.*, 13:241-248 (1979).
4. Fujita, T., Iwasa, J., and C. Hansch. "A New Substituent Constant, π, Derived from Partition Coefficients," *J. Am. Chem. Soc.*, 86(23):5175-5180 (1964).
5. Weast, R.C., Ed. *CRC Handbook of Chemistry and Physics*, 67th ed. (Boca Raton, FL: CRC Press, Inc., 1986), 2406 p.
6. Verschueren, K. *Handbook of Environmental Data on Organic Chemicals* (New York: Van Nostrand Reinhold Co., 1983), 1310 p.
7. Gross, P.M., Saylor, J.H., and M.A. Gorman. "Solubility Studies. IV. The Solubilities of Certain Slightly Soluble Organic Compounds in Water," *J. Am. Chem. Soc.*, 55(2):650-652 (1933).
8. Sax, N.I. *Dangerous Properties of Industrial Materials* (New York: Van Nostrand Reinhold Co., 1984), 3124 p.
9. "NIOSH Pocket Guide to Chemical Hazards," U.S. Department of Health and Human Services, U.S. Government Printing Office (1987), 241 p.

4–NITROANILINE

Synonyms: 1-Amino-4-nitrobenzene; 4-Aminonitrobenzene; *p*-Aminonitrobenzene; Azoamine red ZH; Azofix Red GG salt; Azoic diazo compound 37; C.I. 37,035; C.I. azoic diazo component 37; C.I. developer 17; Developer P; Devol red GG; Diazo fast red GG; Fast red base GG; Fast red base 2J; Fast red 2G base; Fast red 2G salt; Fast red GG base; Fast red GG salt; Fast red MP base; Fast red P base; Fast red P salt; Fast red salt GG; Fast red salt 2J; IG base; Naphtolean red GG base; NCI-C60786; 4-Nitraniline; *p*-Nitraniline; Nitrazol 2F extra; *p*-Nitroaniline; **4-Nitrobenzenamine**; *p*-Nitrobenzenamine; *p*-Nitrophenylamine; PNA; RCRA waste number P077; Red 2G base; Shinnippon fast red GG base; UN 1661.

Structural Formula:

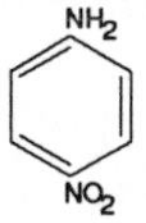

CHEMICAL DESIGNATIONS

CAS Registry Number: 100-01-6

DOT Designation: 1661

Empirical Formula: $C_6H_6N_2O_2$

Formula Weight: 138.13

RTECS Number: BY 7000000

PHYSICAL AND CHEMICAL PROPERTIES

Appearance: Bright yellow powder.

Boiling Point: 331.7 °C [1]; 336 °C [2].

Dissociation Constant: 0.99 [3].

Henry's Law Constant: 1.14 x 10^{-8} atm·m^3/mol at 25 °C (calculated).

Ionization Potential: 8.85 eV [4].

Log K_{oc}: 1.08 (estimated) [5].

Log K_{ow}: 1.39 [6].

Melting Point: 148-149 °C [1].

Solubility in Organics: Soluble in acetone, ethanol, chloroform, and ether [1].

Solubility in Water: 800 mg/L at 18.5 °C [3]; 22,000 mg/L at 100 °C [7]; 728 mg/L at 30 °C [8].

Specific Density: 1.424 at 20/4 °C [1].

Transformation Products: No data found.

Vapor Pressure: 0.0015 mm at 20 °C, 0.007 mm at 30 °C [7]; 1 mm at 142.4 °C, 10 mm at 194.4 °C, 40 mm at 234.2 °C, 100 mm at 261.8 °C, 400 mm at 310.2 °C, 760 mm at 336.0 °C [1].

FIRE HAZARDS

Flash Point: 165 °C [9]; 198.9 °C [10].

Lower Explosive Limit (LEL): Not pertinent [2].

Upper Explosive Limit (UEL): Not pertinent [2].

HEALTH HAZARD DATA

Immediately Dangerous to Life or Health (IDLH): 300 mg/m^3 [4].

Permissible Exposure Limits (PEL) in Air: 1 ppm ($\approx$6 mg/m^3) [11]; 3 mg/m^3 [4].

MANUFACTURING

Selected Manufacturers:

> Eastman Organic Chemicals
> Rochester, NY 14650

American Aniline Products, Inc.
Mount Vernon Street
Lock Haven, PA 17745

Monsanto Industrial Chemicals Co.
800 North Lindbergh Blvd.
St. Louis, MO 63166

Uses: Intermediate for dyes and antioxidants; inhibits gum formation in gasoline; corrosion inhibiter; organic synthesis.

REFERENCES

1. Weast, R.C., Ed. *CRC Handbook of Chemistry and Physics*, 67th ed. (Boca Raton, FL: CRC Press, Inc., 1986), 2406 p.
2. Weiss, G. *Hazardous Chemicals Data Book* (Park Ridge, NJ: Noyes Data Corp., 1986), 1069 p.
3. Dean, J.A., Ed., *Lange's Handbook of Chemistry*, 11th ed. (New York: McGraw-Hill, Inc., 1973), 1570 p.
4. "NIOSH Pocket Guide to Chemical Hazards," U.S. Department of Health and Human Services, U.S. Government Printing Office (1987), 241 p.
5. Montgomery, J.H. Unpublished results (1989).
6. Fujita, T., Iwasa, J., and C. Hansch. "A New Substituent Constant, π, Derived from Partition Coefficients," *J. Am. Chem. Soc.*, 86(23):5175-5180 (1964).
7. Verschueren, K. *Handbook of Environmental Data on Organic Chemicals* (New York: Van Nostrand Reinhold Co., 1983), 1310 p.
8. Gross, P.M., and J.H. Saylor. "The Solubilities of Certain Slightly Soluble Organic Compounds in Water," *J. Am. Chem. Soc.*, 53(5):1744-1751 (1931).
9. *Catalog Handbook of Fine Chemicals* (Milwaukee, WI: Aldrich Chemical Co., 1988), 2212 p.
10. Sax, N.I. *Dangerous Properties of Industrial Materials* (New York: Van Nostrand Reinhold Co., 1984), 3124 p.
11. "General Industry Standards for Toxic and Hazardous Substances," U.S. Code of Federal Regulations 1910, Subpart Z Section 1910.1000 (July 1982).

NITROBENZENE

Synonyms: Essence of mirbane; Essence of myrbane; Mirbane oil; NCI-C60082; Nitrobenzol; Oil of bitter almonds; Oil of mirbane; Oil of myrbane; RCRA waste number U169; UN 1662.

Structural Formula:

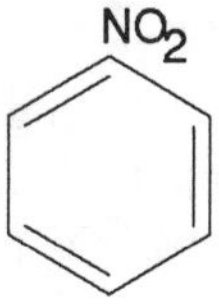

CHEMICAL DESIGNATIONS

CAS Registry Number: 98-95-3

DOT Designation: 1662

Empirical Formula: $C_6H_5NO_2$

Formula Weight: 123.11

RTECS Number: DA 6475000

PHYSICAL AND CHEMICAL PROPERTIES

Appearance and Odor: Clear, light yellow to brown, oily liquid with an almond or shoe polish odor.

Boiling Point: 210.8 °C [1].

Henry's Law Constant: 2.45×10^{-5} atm·m^3/mol [2].

Ionization Potential: 9.92 eV [3].

Log K_{oc}: 2.36 [4]; a value of 1.95 was found using an agricultural soil having a pH of 7.4, values 1.49 and 2.01 were determined using two forest soils with a pH of 5.6 and 4.2, respectively [5].

Log K_{ow}: 1.85 [6]; 1.83 [7]; 1.84 [8]; 1.792 [9]; 1.88 [10].

Melting Point: 5.7 °C [1].

Solubility in Organics: Soluble in acetone, ethanol, benzene, and ether [1].

Solubility in Water: 1,900 mg/L at 20 °C, 8,000 mg/L at 80 °C [11]; 2,000 mg/L at 25 °C [12]; 1,780 mg/L at 15 °C, 0.16 vol% at 20 °C [13]; 2,000 mg/L at 20 °C [14]; 2,090 mg/L at 25 °C [7]; 1,780 mg/L at 15 °C, 2,050 mg/L at 30 °C [15]; 1,930 mg/L at 25 °C [16]; 2,259 mg/L at 35 °C [17]; 0.1 part per 100 parts water at 20 °C, 1.0 per 100 parts water at 100 °C [18]; 0.0311 M at 25 °C [19]; 2,100 ppm at 25 °C [20].

Specific Density: 1.2037 at 20/4 °C [1]; 1.205 at 15/4 °C [21]; 1.205 at 20/4 °C [22]; 1.3440 at 1.5/4 °C, 1.2229 at 0/4 °C, 1.2193 at 4/4 °C, 1.2125 at 10/4 °C, 1.205 at 18/4 °C, 1.1986 at 25/4 °C [18].

Transformation Products: Irradiation of nitrobenzene in the vapor phase produced nitrosobenzene and 4-nitrophenol [23].

Vapor Density: 5.03 g/L at 25 °C, 4.25 (air = 1).

Vapor Pressure: 0.15 mm at 20.0 °C, 0.35 mm at 30.0 °C [11]; 0.27 mm at 20.0 °C [24]; 1 mm at 44.4 °C, 10 mm at 84.9 °C, 40 mm at 115.4 °C, 100 mm at 139.9 °C, 400 mm at 185.8 °C, 760 mm at 210.6 °C [1]; 0.28 mm at 25 °C [12]; 50 mm at 120.2 °C, 200 mm at 160.5 °C, 300 mm at 174.5 °C, 400 mm at 184.5 °C, 500 mm at 192.5 °C, 600 mm at 199.5 °C [18].

FIRE HAZARDS

Flash Point: 88 °C [3].

Lower Explosive Limit (LEL): 1.8% [3].

Upper Explosive Limit (UEL): Unknown [3].

HEALTH HAZARD DATA

Immediately Dangerous to Life or Health (IDLH): 200 ppm [3].

Permissible Exposure Limits (PEL) in Air: 1 ppm ($\approx$5 mg/m^3) [25].

MANUFACTURING

Selected Manufacturers:

American Cyanamid Co.
Organic Chemicals Division
Bound Brook, NJ 08805

E.I. duPont de Nemours and Co.
Explosives Department
Wilmington, DE 19898

First Chemical Corp.
656 North State St.
Jackson, MS 39205

Uses: Solvent for cellulose ethers; modifying esterification of cellulose acetate; ingredient of metal polishes and shoe polishes; manufacture of aniline, benzidine, quinoline, azobenzene, drugs, photographic chemicals.

REFERENCES

1. Weast, R.C., Ed. *CRC Handbook of Chemistry and Physics*, 67th ed. (Boca Raton, FL: CRC Press, Inc., 1986), 2406 p.
2. Jury, W.A., Spencer, W.F., and W.J. Farmer. "Behavior Assessment Model for Trace Organics in Soil: III. Application of Screening Model," *J. Environ. Qual.*, 13(4):573-579 (1984).
3. "NIOSH Pocket Guide to Chemical Hazards," U.S. Department of Health and Human Services, U.S. Government Printing Office (1987), 241 p.
4. Løkke, H. "Sorption of Selected Organic Pollutants in Danish Soils," *Ecotoxicol. Environ. Safety*, 8(5):395-409 (1984).
5. Seip, H.M., Alstad, J., Carlberg, G.E., Matinsen, K., and R. Skaane. "Measurement of Mobility of Organic Compounds in Soils," *Sci. Tot. Environ.*, 50:87-101 (1986).
6. Fujita, T., Iwasa, J., and C. Hansch. "A New Substituent Constant, π, Derived from Partition Coefficients," *J. Am. Chem. Soc.*, 86(23):5175-5180 (1964).
7. Banerjee, S., Yalkowsky, S.H., and S.C. Valvani. "Water Solubility and Octanol/Water Partition Coefficients of Organics. Limitations of the Solubility-Partition Coefficient Correlation," *Environ. Sci.*

Technol., 14(10):1227-1229 (1980).

8. Geyer, H., Politzki, G., and D. Freitag. "Prediction of Ecotoxicological Behaviour of Chemicals: Relationship Between *n*-Octanol/Water Partition Coefficient and Bioaccumulation of Organic Chemicals by Alga *Chlorella*," *Chemosphere*, 13(2):269-284 (1984).

9. Lu, P.-Y., and R.L. Metcalf. "Environmental Fate and Biodegradability of Benzene Derivatives as Studied in a Model Ecosystem," *Environ. Health Perspect.*, 10:269-284 (1975).

10. Leo, A., Hansch, C., and D. Elkins. "Partition Coefficients and Their Uses," *Chem. Rev.*, 71(6):525-616 (1971).

11. Verschueren, K. *Handbook of Environmental Data on Organic Chemicals* (New York: Van Nostrand Reinhold Co., 1983), 1310 p.

12. Warner, H.P., Cohen, J.M., and J.C. Ireland. "Determination of Henry's Law Constants of Selected Priority Pollutants," Office of Science and Development, U.S. EPA Report-600/D-87/229 (1987), 14 p.

13. Meites, L., Ed. *Handbook of Analytical Chemistry*, 1st ed. (New York: McGraw-Hill, Inc., 1963), 1782 p.

14. Haque, R., Ed., *Dynamics, Exposure and Hazardous Assessment of Toxic Chemicals* (Ann Arbor, MI: Ann Arbor Science Publishers, Inc., 1980), 496 p.

15. Gross, P.M., and J.H. Saylor. "The Solubilities of Certain Slightly Soluble Organic Compounds in Water," *J. Am. Chem. Soc.*, 53(5):1744-1751 (1931).

16. Andrews, L.J., and R.M. Keefer. "Cation Complexes of Compounds Containing Carbon-Carbon Double Bonds. VI. The Argentation of Substituted Benzenes," *J. Am. Chem. Soc.*, 72(7):3113-3116 (1950).

17. Hine, J., Haworth, H.W., and O.B. Ramsey. "Polar Effects on Rates and Equilibria. VI. The Effect of Solvent on the Transmission of Polar Effects," *J. Am. Chem. Soc.*, 85(10):1473-1475 (1963).

18. Standen, A., Ed. *Kirk-Othmer Encyclopedia of Chemical Technology, Volume 13*, 2nd ed. (New York: John Wiley and Sons, Inc., 1967), 994 p.

19. Tewari, Y.B., Miller, M.M., Wasik, S.P., and D.E. Martire. "Aqueous Solubility and Octanol/Water Partition Coefficient of Organic Compounds at 25.0 °C," *J. Chem. Eng. Data*, 27(4):451-454 (1982).

20. Amoore, J.E., and E. Hautala. "Odor as an Aide to Chemical Safety: Odor Thresholds Compared with Threshold Limit Values and Volatilities for 214 Industrial Chemicals in Air and Water Dilution," *J. Appl. Toxicol.*, 3(6):272-290 (1983).

21. Sax, N.I. *Dangerous Properties of Industrial Materials* (New York: Van Nostrand Reinhold Co., 1984), 3124 p.

22. *Fluka Catalog 1988/89 - Chemika-Biochemika* (Ronkonkoma, NY: Fluka Chemical Corp., 1988), 1536 p.
23. Hazardous Substances Data Bank. Nitrobenzene, National Library of Medicine, Toxicology Information Program (1989).
24. Lyman, W.J., Reehl, W.F., and D.H. Rosenblatt. *Handbook of Chemical Property Estimation Methods: Environmental Behavior of Organic Compounds* (New York: McGraw-Hill, Inc., 1982).
25. "General Industry Standards for Toxic and Hazardous Substances," U.S. Code of Federal Regulations 1910, Subpart Z Section 1910.1000 (July 1982).

2-NITROPHENOL

Synonyms: 2-Hydroxynitrobenzene; *o*-Hydroxynitrobenzene; 2-Nitro-1-hydroxybenzene; *o*-Nitrophenol; ONP; UN 1663.

Structural Formula:

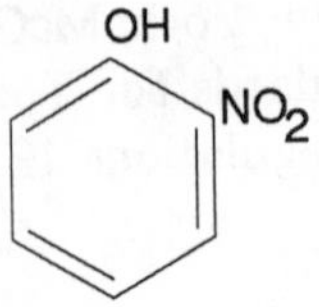

CHEMICAL DESIGNATIONS

CAS Registry Number: 88-75-5

DOT Designation: 1663

Empirical Formula: $C_6H_5NO_3$

Formula Weight: 139.11

RTECS Number: SM 2100000

PHYSICAL AND CHEMICAL PROPERTIES

Appearance and Odor: Pale yellow crystals with an aromatic odor.

Boiling Point: 216 °C [1]; 214.5 °C [2].

Dissociation Constant: 7.23 [3].

Henry's Law Constant: 3.5×10^{-6} atm·m^3/mol [4].

Ionization Potential: No data found.

Log K_{oc}: 1.57 using method of Karickhoff and others [5].

Log K_{ow}: 1.78 [6]; 1.76 [7]; 1.73 [8]; 1.79 [9].

Melting Point: 45-46 °C [1]; 44-45 °C [10].

Solubility in Organics: Soluble in acetone, ethanol, benzene, ether, and chloroform [11].

Solubility in Water: 2,100 mg/L at 20 °C, 10,800 mg/L at 100 °C [12]; 2,000 mg/L at 25 °C [13]; 0.208 wt% at 20 °C, 3.89 g/L at 48 °C [14]; 3,200 mg/L at 38 °C [10]; 1,060 mg/L at 20 °C, 2,500 mg/L at 25 °C [4]; 1.4 g/L at 20 °C [15].

Specific Density: 1.485 at 14/4 °C [1]; 1.495 at 20/4 °C [2]; 1.2942 at 40/4 °C, 1.2712 at 60/4 °C, 1.2482 at 80/4 °C, 1.2323 at 100/4 °C [10].

Transformation Products: No data found.

Vapor Pressure: 20 mm at 105 °C [12]; 1 mm at 49.3 °C, 10 mm at 90.4 °C, 40 mm at 122.1 °C, 100 mm at 146.4 °C, 400 mm at 191.0 °C, 760 mm at 214.5 °C [1]; 0.20 mm at 25 °C [4]; 0.019 mm at 8 °C, 0.12 mm at 25 °C [15].

FIRE HAZARDS

Flash Point: 73.5 °C [16].

Lower Explosive Limit (LEL): Not pertinent [17].

Upper Explosive Limit (UEL): Not pertinent [17].

HEALTH HAZARD DATA

Immediately Dangerous to Life or Health (IDLH): No data found.

Permissible Exposure Limits (PEL) in Air: No standards set.

MANUFACTURING

Selected Manufacturers:

Eastman Organic Chemicals
Rochester, NY 14650

J.T. Baker Chemical Co.
Phillipsburg, NJ 08865

Monsanto Industrial Chemicals Co.
800 North Lindbergh Blvd.
St. Louis, MO 63166

Uses: Indicator; organic synthesis.

REFERENCES

1. Weast, R.C., Ed. *CRC Handbook of Chemistry and Physics*, 67th ed. (Boca Raton, FL: CRC Press, Inc., 1986), 2406 p.
2. Sax, N.I. *Dangerous Properties of Industrial Materials* (New York: Van Nostrand Reinhold Co., 1984), 3124 p.
3. Dean, J.A., Ed., *Lange's Handbook of Chemistry*, 11th ed. (New York: McGraw-Hill, Inc., 1973), 1570 p.
4. Howard, P.H. *Handbook of Environmental Fate and Exposure Data for Organic Chemicals* (Chelsea, MI: Lewis Publishers, Inc., 1989), 574 p.
5. Karickhoff, S.W., Brown, D.S., and T.A. Scott. "Sorption of Hydrophobic Pollutants on Natural Sediments," *Water Res.*, 13:241-248 (1979).
6. Mills, W.B., Porcella, D.B., Ungs, M.J., Gherini, S.A., Summers, K.V., Mok, L., Rupp, G.L., and G.L. Bowie. "Water Quality Assessment: A Screening Procedure for Toxic and Conventional Pollutants in Surface and Groundwater-Part I," Office of Research and Development, U.S. EPA Report-600/6-85-002a (1985), 638 p.
7. Walton, W.C. *Practical Aspects of Ground Water Modeling* (Worthington, OH: National Water Well Association, 1985), 587 p.
8. Leo, A., Hansch, C., and D. Elkins. "Partition Coefficients and Their Uses," *Chem. Rev.*, 71(6):525-616 (1971).
9. Fujita, T., Iwasa, J., and C. Hansch. "A New Substituent Constant, π, Derived from Partition Coefficients," *J. Am. Chem. Soc.*, 86(23):5175-5180 (1964).
10. Standen, A., Ed. *Kirk-Othmer Encyclopedia of Chemical Technology, Volume 13*, 2nd ed. (New York: John Wiley and Sons, Inc., 1967), 994 p.
11. "Chemical, Physical, and Biological Properties of Compounds Present at Hazardous Waste Sites," U.S. EPA Report-530/SW-89-010 (1985), 619 p.
12. Verschueren, K. *Handbook of Environmental Data on Organic Chemicals* (New York: Van Nostrand Reinhold Co., 1983), 1310 p.
13. Morrison, R.T., and R.N. Boyd. *Organic Chemistry* (Boston: Allyn and Bacon, Inc., 1971), 1258 p.

14. Stephen, H., and T. Stephen. *Solubilities of Inorganic and Organic Compounds - Part 1, Volume 1* (London: Pergamon Printing and Art Services, Ltd., 1963), 960 p.

15. Leuenberger, C., Ligocki, M.P., and J.F. Pankow. "Trace Organic Compounds in Rain. 4. Identities, Concentrations, and Scavenging Mechanisms for Phenols in Urban Air and Rain," *Environ. Sci. Technol.*, 19(11):1053-1058 (1985).

16. Sax, N.I., Ed. *Dangerous Properties of Industrial Materials Report* (New York: Van Nostrand Reinhold Co., 1985), 5(3): 88 p.

17. Weiss, G. *Hazardous Chemicals Data Book* (Park Ridge, NJ: Noyes Data Corp., 1986), 1069 p.

4–NITROPHENOL

Synonyms: 4-Hydroxynitrobenzene; *p*-Hydroxynitrobenzene; NCI-C55992; 4-Nitro-1-hydroxybenzene; *p*-Nitrophenol; PNP; RCRA waste number U170; UN 1663.

Structural Formula:

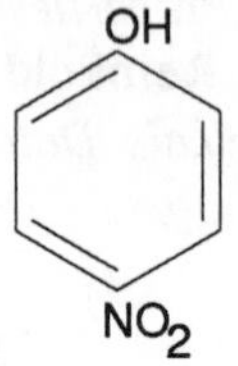

CHEMICAL DESIGNATIONS

CAS Registry Number: 100-02-7

DOT Designation: 1663

Empirical Formula: $C_6H_5NO_3$

Formula Weight: 139.11

RTECS Number: SM 2275000

PHYSICAL AND CHEMICAL PROPERTIES

Appearance and Odor: Colorless to pale yellow, odorless crystals.

Boiling Point: Sublimes and decomposes at 279 °C [1].

Dissociation Constant: 7.15 [2].

Henry's Law Constant: 3.0×10^{-5} atm·m^3/mol at 20 °C (calculated) [3].

Ionization Potential: 9.52 eV [4].

Log K_{oc}: 2.33 [5]; 1.74 [6].

Log K$_{ow}$: 1.90 [7]; 1.91 [8]; 1.85, 1.92 [9]; 2.04 [3]; 1.96 [10].

Melting Point: 114 °C [11].

Solubility in Organics: Soluble in acetone, ethanol, benzene, ether, and chloroform [12].

Solubility in Water: 16,000 mg/L at 25 °C, 269,000 mg/L at 90 °C [13]; 16,900 mg/L at 25 °C [14]; 13,200 mg/L at room temperature [15]; 0.797 wt% at 15 °C, 1.32 wt% at 20 °C, 5.71 wt% at 50 °C [16]; 8.04 g/L at 15 °C, 16 g/L at 25 °C, 29.1 g/L at 90 °C [11]; 11,300 mg/L at 20 °C, 25,000 mg/L at 25 °C [6]; 11.57 g/L at 20 °C in a buffered solution with a pH of 1.5 [3]; 0.1 M at 25 °C [17].

Specific Density: 1.479 at 20/4 °C, 1.270 at 120/4 °C [11].

Transformation Products: Under anaerobic conditions, 4-nitrophenol may undergo nitroreduction to produce *p*-aminophenol [18]. An aqueous solution of 200 ppm exposed to sunlight for one to two months yielded hydroquinone, 4-nitrocatechol and an unidentified polymeric substance [19]. Under artificial sunlight, river water containing 2-5 ppm 4-nitrophenol photodegraded to produce trace amounts of 4-aminophenol [20].

Vapor Pressure: 2.2 mm at 146 °C, 18.7 mm at 186 °C [13]; 10^{-4} mm at 20 °C [3].

FIRE HAZARDS

Flash Point: Not pertinent (combustible solid) [21].

Lower Explosive Limit (LEL): No data found.

Upper Explosive Limit (UEL): No data found.

HEALTH HAZARD DATA

Immediately Dangerous to Life or Health (IDLH): No data found.

Permissible Exposure Limits (PEL) in Air: None [22].

MANUFACTURING

Selected Manufacturers:

Aldrich Chemical Co.
940 West Saint Paul Ave.
Milwaukee, WI 53233

E.I. duPont de Nemours and Co.
Wilmington, DE 19898

Monsanto Industrial Chemicals Co.
800 North Lindbergh Blvd.
St. Louis, MO 63166

Uses: Fungicide for leather; production of parathion; organic synthesis.

REFERENCES

1. Weast, R.C., Ed. *CRC Handbook of Chemistry and Physics*, 67th ed. (Boca Raton, FL: CRC Press, Inc., 1986), 2406 p.
2. Dean, J.A., Ed., *Lange's Handbook of Chemistry*, 11th ed. (New York: McGraw-Hill, Inc., 1973), 1570 p.
3. Schwarzenbach, R., Stierli, R., Folsom, B.R., and J. Zeyer. "Compound Properties Relevant for Assessing the Environmental Partitioning of Nitrophenols," *Environ. Sci. Technol.*, 22(1):83-92 (1988).
4. Gordon, A.J., and R.A. Ford. *The Chemist's Companion* (New York: John Wiley and Sons, Inc., 1972), 551 p.
5. Løkke, H. "Sorption of Selected Organic Pollutants in Danish Soils," *Ecotoxicol. Environ. Safety*, 8(5):395-409 (1984).
6. Howard, P.H. *Handbook of Environmental Fate and Exposure Data for Organic Chemicals* (Chelsea, MI: Lewis Publishers, Inc., 1989), 574 p.
7. Mills, W.B., Porcella, D.B., Ungs, M.J., Gherini, S.A., Summers, K.V., Mok, L., Rupp, G.L., and G.L. Bowie. "Water Quality Assessment: A Screening Procedure for Toxic and Conventional Pollutants in Surface and Groundwater-Part I," Office of Research and Development, U.S. EPA Report-600/6-85-002a (1985), 638 p.
8. Leo, A., Hansch, C., and D. Elkins. "Partition Coefficients and Their Uses," *Chem. Rev.*, 71(6):525-616 (1971).
9. Geyer, H., Politzki, G., and D. Freitag. "Prediction of

Ecotoxicological Behaviour of Chemicals: Relationship Between n-Octanol/Water Partition Coefficient and Bioaccumulation of Organic Chemicals by Alga *Chlorella*," *Chemosphere*, 13(2):269-284 (1984).

10. Fujita, T., Iwasa, J., and C. Hansch. "A New Substituent Constant, π, Derived from Partition Coefficients," *J. Am. Chem. Soc.*, 86(23):5175-5180 (1964).

11. Standen, A., Ed. *Kirk-Othmer Encyclopedia of Chemical Technology, Volume 13*, 2nd ed. (New York: John Wiley and Sons, Inc., 1967), 994 p.

12. "Chemical, Physical, and Biological Properties of Compounds Present at Hazardous Waste Sites," U.S. EPA Report-530/SW-89-010 (1985), 619 p.

13. Verschueren, K. *Handbook of Environmental Data on Organic Chemicals* (New York: Van Nostrand Reinhold Co., 1983), 1310 p.

14. Morrison, R.T., and R.N. Boyd. *Organic Chemistry* (Boston: Allyn and Bacon, Inc., 1971), 1258 p.

15. Meites, L., Ed. *Handbook of Analytical Chemistry*, 1st ed. (New York: McGraw-Hill, Inc., 1963), 1782 p.

16. Stephen, H., and T. Stephen. *Solubilities of Inorganic and Organic Compounds - Part 1, Volume 1* (London: Pergamon Printing and Art Services, Ltd., 1963), 960 p.

17. Caturla, F., Martin-Martinez, J.M., Molina-Sabio, M., Rodriguez-Reinoso, F., and R. Torregrosa. "Adsorption of Substituted Phenols on Activated Carbon," *J. Colloid Interface Sci.*, 124(2):528-534 (1988).

18. Kobayashi, H., and B.E. Rittman. "Microbial Removal of Hazardous Organic Compounds," *Environ. Sci. Technol.*, 16(3):170A-183A (1982).

19. Callahan, M.A., Slimak, M.W., Gable, N.W., May, I.P., Fowler, C.F., Freed, J.R., Jennings, P., Durfee, R.L., Whitmore, F.C., Maestri, B., Mabey, W.R., Holt, B.R., and C. Gould. "Water-Related Environmental Fate of 129 Priority Pollutants Volumes I and II," National Technical Information Service, U.S. EPA Report-440/4-79-029 (1979), 1160 p.

20. Mansour, M., Feicht, E. and P. Méallier. "Improvement of the Photostability of Selected Substances in Aqueous Medium," *Toxicol. Environ. Chem.*, 20-21:139-147 (1989).

21. Weiss, G. *Hazardous Chemicals Data Book* (Park Ridge, NJ: Noyes Data Corp., 1986), 1069 p.

22. *Documentation of the Threshold Limit Values and Biological Exposure Indices* (Cincinnati, OH: American Conference of Governmental Industrial Hygienists, 1986), 744 p.

N–NITROSODIMETHYLAMINE

Synonyms: Dimethylnitrosamine; *n*-Dimethylnitrosamine; *N,N*-Dimethylnitrosamine; Dimethylnitrosomine; DMN; DMNA; **N-Methyl-*n*-nitrosomethanamine;** NDMA; *n*-Nitrosodimethylamine; Nitrous dimethylamide; RCRA waste number P082.

Structural Formula:

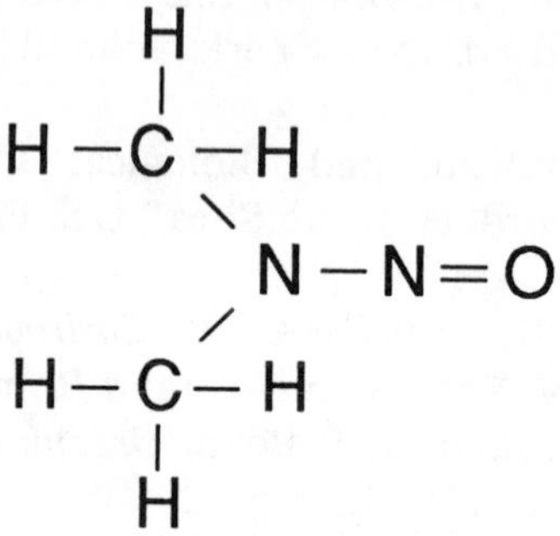

$$
\begin{array}{c}
\text{H} \\
| \\
\text{H} - \text{C} - \text{H} \\
\diagdown \\
\text{N} - \text{N} = \text{O} \\
\diagup \\
\text{H} - \text{C} - \text{H} \\
| \\
\text{H}
\end{array}
$$

CHEMICAL DESIGNATIONS

CAS Registry Number: 62-75-9

DOT Designation: 1955

Empirical Formula: $C_2H_6N_2O$

Formula Weight: 74.09

RTECS Number: IQ 0525000

PHYSICAL AND CHEMICAL PROPERTIES

Appearance and Odor: Yellow liquid with a faint characteristic odor.

Boiling Point: 154 °C [1]; 152 °C [2].

Dissociation Constant: No data found.

Henry's Law Constant: 0.143 atm·m³/mol at 25 °C (estimated using a solubility of 1,000 g/L).

Ionization Potential: No data found.

Log K$_{oc}$: 1.41 using method of Kenaga and Goring [3].

Log K$_{ow}$: 0.06 [4].

Melting Point: No data found.

Solubility in Organics: Soluble in solvents [5] including ethanol and ether [1].

Solubility in Water: Completely miscible [6].

Specific Density: 1.0059 at 20/4 °C [1]; 1.0049 at 18/4 °C [7].

Transformation Products: A Teflon bag containing air and *N*-nitrsodimethylamine was subjected to sunlight on two different days. On a cloudy day, half of the *N*-nitrosodimethylamine was photolyzed in one hour. On a sunny day, half of the *N*-nitrosodimethylamine was photolyzed in 30 minutes. Photolysis products included nitric oxide, carbon monoxide, formaldehyde, and an unidentified compound [8]. In a separate experiment, Tuazon and others [9] irradiated an ozone-rich atmosphere containing *N*-nitrosodimethylamine. Photolysis products they identified included dimethylnitramine, nitromethane, formaldehyde, carbon monoxide, nitrogen dioxide, nitrogen pentoxide, and nitric acid.

Vapor Pressure: 8.1 mm at 25 °C [10]; 2.7 mm at 20 °C [11.

FIRE HAZARDS

Flash Point: 61 °C [12].

Lower Explosive Limit (LEL): No data found.

Upper Explosive Limit (UEL): No data found.

HEALTH HAZARD DATA

Immediately Dangerous to Life or Health (IDLH): Potential human carcinogen [13].

Permissible Exposure Limits (PEL) in Air: No standards set.

MANUFACTURING

Selected Manufacturer:

Aldrich Chemical Co.
940 West Saint Paul Ave.
Milwaukee, WI 53233

Uses: Rubber accelerator; solvent in fiber and plastic industry; rocket fuels; lubricants; condensers to increase dielectric constant; industrial solvent; antioxidant; nematocide; softener for copolymers; research chemical; chemical intermediate for 1,1-dimethylhydrazine.

REFERENCES

1. Weast, R.C., Ed. *CRC Handbook of Chemistry and Physics*, 67th ed. (Boca Raton, FL: CRC Press, Inc., 1986), 2406 p.
2. Sax, N.I. *Dangerous Properties of Industrial Materials* (New York: Van Nostrand Reinhold Co., 1984), 3124 p.
3. Kenaga, E.E., and C.A.I. Goring. "Relationship Between Water Solubility, Soil Sorption, Octanol-Water Partitioning and Concentration of Chemicals in Biota," in *Aquatic Toxicology, ASTM STP 707*, Eaton, J.G., Parrish, P.R., and A.C. Hendricks, Eds. (Philadelphia, PA: American Society for Testing and Materials 1980), pp 78-115.
4. Radding, S.B., Mill, T., Gould, C.W., Lia, D.H., Johnson, H.L., Bomberger, D.S., and C.V. Fojo. "The Environmental Fate of Selected Polynuclear Aromatic Hydrocarbons," Office of Toxic Substances, U.S. EPA Report-560/5-75-009 (1976), 122 p.
5. "Chemical, Physical, and Biological Properties of Compounds Present at Hazardous Waste Sites," U.S. EPA Report-530/SW-89-010 (1985), 619 p.
6. Mirvish, S.S., Issenberg, P., and H.C. Sornson. "Air-Water and Ether-Water Distribution of *n*-Nitroso Compounds: Implications for Laboratory Safety, Analytic Methodology, and Carcinogenicity for the Rat Esophagus, Nose, and Liver," *J. Nat. Cancer Instit.*, 56(6):1125-1129 (1976).
7. Weast, R.C., and M.J. Astle., Eds. *CRC Handbook of Data on Organic Compounds-2 Volumes* (Boca Raton, FL: CRC Press, Inc. 1986).
8. Hanst, P.L., Spence, J.W., and Matthew Miller. "Atmospheric Chemistry of *N*-Nitroso Dimethylamine," *Environ. Sci. Technol.*, 11(4):403-405 (1977).

9. Tuazon, E.C., Carter, W.P.L., Atkinson, R., Winer, A.M., and J.N. Pitts Jr. "Atmospheric Reactions of *N*-Nitrosodimethylamine and Dimethylnitramine," *Environ. Sci. Technol.*, 18(1):49-54 (1984).
10. Mabey, W.R., Smith, J.H., Podoll, R.T., Johnson, H.L., Mill, T., Chou, T.-W., Gates, J., Partridge, I.W., Jaber, H., and D. Vandenberg. "Aquatic Fate Process Data for Organic Priority Pollutants - Final Report," Office of Regulations and Standards, U.S. EPA Report-440/4-81-014 (1982), 407 p.
11. Klein, R.G. "Calculations and Measurements on the Volatility of *N*-Nitrosoamines and Their Aqueous Solutions," *Toxicology*, 23:135-147 (1982).
12. *Catalog Handbook of Fine Chemicals* (Milwaukee, WI: Aldrich Chemical Co., 1988), 2212 p.
13. "NIOSH Pocket Guide to Chemical Hazards," U.S. Department of Health and Human Services, U.S. Government Printing Office (1987), 241 p.

N–NITROSODIPHENYLAMINE

Synonyms: Benzenamine; Curetard A; Delac J; Diphenylnitrosamine; Diphenyl-*n*-nitrosamine; *N,N*-Diphenylnitrosamine; Naugard TJB; NCI-C02880; NDPA; NDPhA; Nitrosodiphenylamine; *n*-Nitroso-*n*-phenylamine; *n*-Nitroso-*n*-phenylbenzenamine; Nitrous diphenylamide; Redax; Retarder J; TJB; Vulcalent A; Vulcatard; Vulcatard A; Vultrol.

Structural Formula:

CHEMICAL DESIGNATIONS

CAS Registry Number: 86-30-6

DOT Designation: None assigned.

Empirical Formula: $C_{12}H_{10}N_2O$

Formula Weight: 198.22

RTECS Number: JJ 9800000

PHYSICAL AND CHEMICAL PROPERTIES

Appearance: Green platy crystals or dark blue crystals.

Boiling Point: No data found.

Dissociation Constant: No data found.

Henry's Law Constant: 2.33 x 10^{-8} atm·m^3/mol at 25 °C (calculated).

Ionization Potential: No data found.

Log K$_{oc}$: 2.76 (estimated) [1].

Log K$_{ow}$: 3.13 [2].

Melting Point: 66.5 °C [3]; 144 °C [4]; 65-66 °C [5].

Solubility in Organics: Soluble in ethanol and benzene [3].

Solubility in Water: 35.1 mg/L at 25 °C [6].

Specific Density: No data found.

Transformation Products: No data found.

Vapor Pressure: No data found however a value of 0.1 mm at 25 °C was assigned by analogy [7].

FIRE HAZARDS

Flash Point: No data found.

Lower Explosive Limit (LEL): No data found.

Upper Explosive Limit (UEL): No data found.

HEALTH HAZARD DATA

Immediately Dangerous to Life or Health (IDLH): No data found.

Permissible Exposure Limits (PEL) in Air: No standards set.

MANUFACTURING
Selected Manufacturers:

Fluka Chemical Corp.
980 South Second St.
Ronkonkoma, NY 11779

Pfaltz & Bauer, Inc.
172 East Aurora St.
Waterbury, CT 06708

Uniroyal Inc.
Uniroyal Chemical Division
Geismar, GA 70734

Uses: Chemical intermediate for *N*-phenyl-*p*-phenylenediamine; rubber processing (vulcanization retarder).

REFERENCES

1. Montgomery, J.H. Unpublished results (1989).
2. Veith, G.D., Macek, K.J., Petrocelli, S.R., and J. Carroll. "An Evaluation of Using Partition Coefficients and Water Solubility to Estimate Bioconcentration Factors for Organic Chemicals in Fish," in *Aquatic Toxicology, ASTM STP 707*, Eaton, J.G., Parrish, P.R., and A.C. Hendricks, Eds. (Philadelphia, PA: American Society for Testing and Materials 1980), pp 116-129.
3. Weast, R.C., Ed. *CRC Handbook of Chemistry and Physics*, 67th ed. (Boca Raton, FL: CRC Press, Inc., 1986), 2406 p.
4. Sax, N.I. *Dangerous Properties of Industrial Materials* (New York: Van Nostrand Reinhold Co., 1984), 3124 p.
5. *Fluka Catalog 1988/89 - Chemika-Biochemika* (Ronkonkoma, NY: Fluka Chemical Corp., 1988), 1536 p.
6. Banerjee, S., Yalkowsky, S.H., and S.C. Valvani. "Water Solubility and Octanol/Water Partition Coefficients of Organics. Limitations of the Solubility-Partition Coefficient Correlation," *Environ. Sci. Technol.*, 14(10):1227-1229 (1980).
7. Mabey, W.R., Smith, J.H., Podoll, R.T., Johnson, H.L., Mill, T., Chou, T.-W., Gates, J., Partridge, I.W., Jaber, H., and D. Vandenberg. "Aquatic Fate Process Data for Organic Priority Pollutants - Final Report," Office of Regulations and Standards, U.S. EPA Report-440/4-81-014 (1982), 407 p.

N–NITROSODI–n–PROPYLAMINE

Synonyms: Dipropylnitrosamine; Di-*n*-propylnitrosamine; DPN; DPNA; NDPA; *n*-Nitrosodipropylamine; **N-Nitroso-*n*-propyl-1-propanamine**; RCRA waste number U111.

Structural Formula:

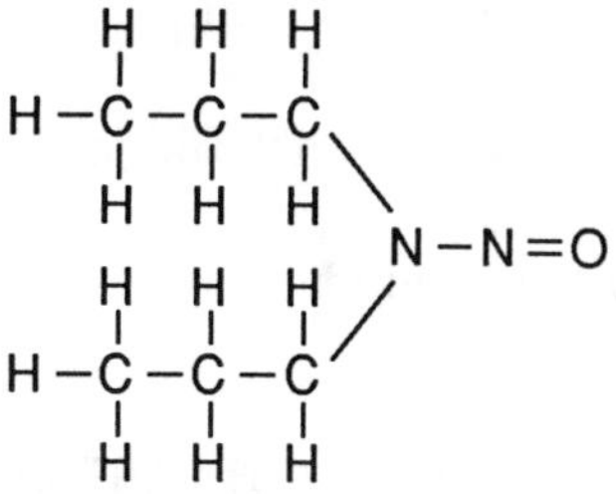

CHEMICAL DESIGNATIONS

CAS Registry Number: 621-64-7

DOT Designation: None assigned.

Empirical Formula: $C_6H_{14}N_2O$

Formula Weight: 130.19

RTECS Number: JL 9700000

PHYSICAL AND CHEMICAL PROPERTIES

Appearance: Yellow to gold colored liquid.

Boiling Point: 205.9 °C [1].

Dissociation Constant: No data found.

Henry's Law Constant: Insufficient vapor pressure data to calculate.

Ionization Potential: No data found.

Log K_{oc}: 1.01 (estimated) [2].

Log K_{ow}: 1.31 (calculated) [3].

Melting Point: No data found.

Solubility in Organics: Soluble in ethanol and ether [4].

Solubility in Water: 9,900 mg/L at 25 °C [5].

Specific Density: 0.9160 at 20/4 °C [6].

Transformation Products: No data found.

Vapor Pressure: No data found.

FIRE HAZARDS

Flash Point: No data found.

Lower Explosive Limit (LEL): No data found.

Upper Explosive Limit (UEL): No data found.

HEALTH HAZARD DATA

Immediately Dangerous to Life or Health (IDLH): No data found.

Permissible Exposure Limits (PEL) in Air: No standards set.

MANUFACTURING

Selected Manufacturer:

> Pfaltz & Bauer, Inc.
> 172 East Aurora St.
> Waterbury, CT 06708

Use: Research chemical.

REFERENCES

1. Dean, J.A., Ed., *Lange's Handbook of Chemistry*, 11th ed. (New York:

McGraw-Hill, Inc., 1973), 1570 p.

2. Montgomery, J.H. Unpublished results (1989).

3. "Treatability Manual - Volume 1: Treatability Data," Office of Research and Development, U.S. EPA Report-600/8-80-042a (1980), 1035 p.

4. Weast, R.C., Ed. *CRC Handbook of Chemistry and Physics*, 67th ed. (Boca Raton, FL: CRC Press, Inc., 1986), 2406 p.

5. Mirvish, S.S., Issenberg, P., and H.C. Sornson. "Air-Water and Ether-Water Distribution of *n*-Nitroso Compounds: Implications for Laboratory Safety, Analytic Methodology, and Carcinogenicity for the Rat Esophagus, Nose, and Liver," *J. Nat. Cancer Instit.*, 56(6):1125-1129 (1976).

6. *IARC Monographs on the Evaluation of Carcinogenic Risk of Chemicals to Man. Some N-Nitroso Compounds Volume 17* (Lyon, France: International Agency for Research on Cancer, 1978), 365 p.

PCB-1016

Synonyms: Arochlor 1016; **Aroclor 1016**; Chlorodiphenyl (41% Cl).

Structural Formula:

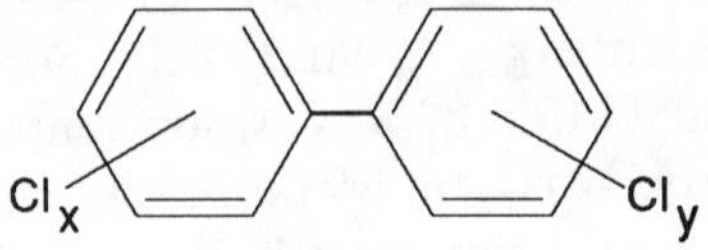

x + y = 0 thru 6

CHEMICAL DESIGNATIONS

CAS Registry Number: 12674-11-2

DOT Designation: 2315

Empirical Formula: Not definitive. PCB-1016 is a mixture of many biphenyls with varying degrees of chlorination. According to Hutzinger [1], the approximate composition of PCB-1016 by weight is as follows: biphenyls (< 0.1%), chlorobiphenyls (1%), dichlorobiphenyls (20%), trichlorobiphenyls (57%), tetrachlorobiphenyls (21%), pentachlorobiphenyls (1%), hexachlorobiphenyls (< 0.1%), and heptachlorobiphenyls (0%).

Formula Weight: 257.9 (average) [1].

RTECS Number: TQ 1351000

PHYSICAL AND CHEMICAL PROPERTIES

Appearance and Odor: Oily light yellow liquid or white powder with a weak odor.

Boiling Point: Distills at 325-356 °C [2].

Henry's Law Constant: 750 atm/mol fraction [3].

Ionization Potential: No data found.

Log K$_{oc}$: 4.70 (estimated) [4].

Log K$_{ow}$: 4.38 [3]; 5.88 [5]; 5.580 for the compounds 2,4,4'-trichlorobiphenyl and 2,2',5-trichlorobiphenyl [6].

Melting Point: No data found.

Solubility in Organics: Soluble in most solvents [7].

Solubility in Water: 0.22-0.25 mg/L [8]; 0.42 mg/L [3]; 0.049 mg/L at 24 °C [9]; 0.085 ppm for the compounds 2,4,4'- and 2,2',5-trichlorobiphenyl [6].

Specific Density: 1.33 at 25/4 °C [2].

Transformation Products: No data found.

Vapor Pressure: 4 x 10^{-4} mm at 25 °C (estimated) [10].

FIRE HAZARDS

Flash Point: Non-flammable [11].

Lower Explosive Limit (LEL): No data found.

Upper Explosive Limit (UEL): No data found.

HEALTH HAZARD DATA

Immediately Dangerous to Life or Health (IDLH): No data found.

Permissible Exposure Limits (PEL) in Air: No standards set.

MANUFACTURING

Selected Manufacturer:

> Monsanto Industrial Chemicals Co.
> 800 North Lindbergh Blvd.
> St. Louis, MO 63166

Uses: Insulator fluid for electric condensers and as an additive in very high pressure lubricants.

REFERENCES

1. Hutzinger, O., Safe, S., and V. Zitko. *The Chemistry of PCB's* (Boca Raton, FL: CRC Press, Inc., 1974), 269 p.
2. Monsanto Industrial Chemicals Co. "PCBs-Aroclors," Technical Bulletin O/PL 306A, (1974).
3. Paris, D.F., Steen, W.C., and G.L. Baughman. "Role of the Physico-Chemical Properties of Aroclors 1016 and 1242 in Determining Their Fate and Transport in Aquatic Environments," *Chemosphere*, 7(4):319-325 (1978).
4. Montgomery, J.H. Unpublished results (1989).
5. Mackay, D. "Correlation of Bioconcentration Factors," *Environ. Sci. Technol.*, 16(5):274-278 (1982).
6. Kenaga, E.E., and C.A.I. Goring. "Relationship Between Water Solubility, Soil Sorption, Octanol-Water Partitioning and Concentration of Chemicals in Biota," in *Aquatic Toxicology, ASTM STP 707*, Eaton, J.G., Parrish, P.R., and A.C. Hendricks, Eds. (Philadelphia, PA: American Society for Testing and Materials 1980), pp 78-115.
7. "Chemical, Physical, and Biological Properties of Compounds Present at Hazardous Waste Sites," U.S. EPA Report-530/SW-89-010 (1985), 619 p.
8. Verschueren, K. *Handbook of Environmental Data on Organic Chemicals* (New York: Van Nostrand Reinhold Co., 1983), 1310 p.
9. Hollifield, H.C. "Rapid Nephelometric Estimate of Water Solubility of Highly Insoluble Organic Chemicals of Environmental Interest," *Bull. Environ. Contam. Toxicol.*, 23(4/5):579-586 (1979).
10. "Treatability Manual - Volume 1: Treatability Data," Office of Research and Development, U.S. EPA Report-600/8-80-042a (1980), 1035 p.
11. Sittig, M. *Handbook of Toxic and Hazardous Chemicals and Carcinogens* (Park Ridge, NJ: Noyes Publications, 1985), 950 p.

PCB-1221

Synonyms: Arochlor 1221; **Aroclor 1221**; Chlorodiphenyl (21% Cl).

Structural Formula:

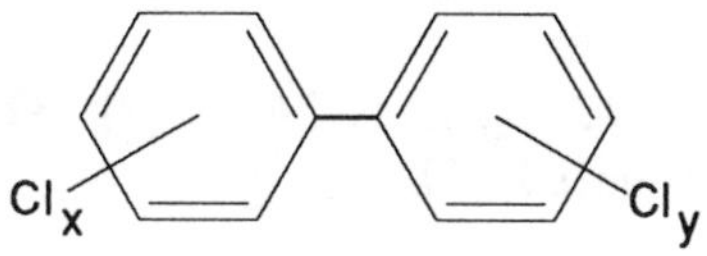

x + y = 0 thru 5

CHEMICAL DESIGNATIONS

CAS Registry Number: 11104-28-2

DOT Designation: 2315

Empirical Formula: Not definitive. PCB-1221 is a mixture of many biphenyls with varying degrees of chlorination. According to Hutzinger [1], the approximate composition of PCB-1221 by weight is as follows: biphenyls (11%), chlorobiphenyls (51%), dichlorobiphenyls (32%), trichlorobiphenyls (4%), tetrachlorobiphenyls (2%), pentachloro-biphenyls (< 0.5%), hexachlorobiphenyls (0%), and heptachlorobiphenyls (0%).

Formula Weight: 192 (average) [1].

RTECS Number: TQ 1352000

PHYSICAL AND CHEMICAL PROPERTIES

Appearance and Odor: Oily, colorless to light yellow liquid with a weak odor.

Boiling Point: Distills at 275-320 °C [2].

Henry's Law Constant: 3.24×10^{-4} atm·m^3/mol (calculated) [3].

Ionization Potential: No data found.

Log K_{oc}: 2.44 (estimated) [4].

Log K_{ow}: 2.8 (estimated) [3].

Melting Point: 1 °C [5].

Solubility in Organics: Soluble in most solvents [6].

Solubility in Water: 0.590 mg/L at 24 °C [7]; 0.200 mg/L at 25 °C [8]; 1.5 mg/L at 25 °C (estimated) [3].

Specific Density: 1.15 at 25/4 °C [1]; 1.182-1.192 at 25/15.5 °C [9].

Transformation Products: No data found.

Vapor Pressure: 0.0067 mm at 25 °C (estimated) [3].

FIRE HAZARDS

Flash Point: 141-150 °C [1].

Lower Explosive Limit (LEL): No data found.

Upper Explosive Limit (UEL): No data found.

HEALTH HAZARD DATA

Immediately Dangerous to Life or Health (IDLH): No data found.

Permissible Exposure Limits (PEL) in Air: No standards set.

MANUFACTURING

Selected Manufacturer:

Monsanto Industrial Chemicals Co.
800 North Lindbergh Blvd.
St. Louis, MO 63166

Uses: In polyvinyl acetate to improve fiber-tear properties; plasticizer for polystyrene; in epoxy resins to improve adhesion and resistance to chemical attack; as an insulator fluid for electric condensers and as an additive in very high pressure lubricants.

REFERENCES

1. Hutzinger, O., Safe, S., and V. Zitko. *The Chemistry of PCB's* (Boca Raton, FL: CRC Press, Inc., 1974), 269 p.
2. Monsanto Industrial Chemicals Co. "PCBs-Aroclors," Technical Bulletin O/PL 306A, (1974).
3. "Treatability Manual - Volume 1: Treatability Data," Office of Research and Development, U.S. EPA Report-600/8-80-042a (1980), 1035 p.
4. Montgomery, J.H. Unpublished results (1989).
5. Hazardous Substances Data Bank. Arochlor 1221, National Library of Medicine, Toxicology Information Program (1989).
6. "Chemical, Physical, and Biological Properties of Compounds Present at Hazardous Waste Sites," U.S. EPA Report-530/SW-89-010 (1985), 619 p.
7. Hollifield, H.C. "Rapid Nephelometric Estimate of Water Solubility of Highly Insoluble Organic Chemicals of Environmental Interest," *Bull. Environ. Contam. Toxicol.*, 23(4/5):579-586 (1979).
8. Sax, N.I., Ed. *Dangerous Properties of Industrial Materials Report* (New York: Van Nostrand Reinhold Co., 1984), 4(6): 105 p.
9. Standen, A., Ed. *Kirk-Othmer Encyclopedia of Chemical Technology, Volume 4*, 2nd ed. (New York: John Wiley and Sons, Inc., 1964), 937 p.

PCB-1232

Synonyms: Arochlor 1232; **Aroclor 1232**; Chlorodiphenyl (32% Cl).

Structural Formula:

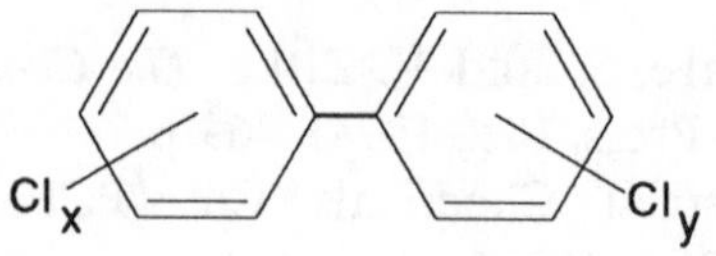

x + y = 0 thru 6

CHEMICAL DESIGNATIONS

CAS Registry Number: 11141-16-5

DOT Designation: 2315

Empirical Formula: Not definitive. PCB-1232 is a mixture of many biphenyls with varying degrees of chlorination. According to Hutzinger [1], the approximate composition of PCB-1232 by weight is as follows: biphenyls (< 0.1%), chlorobiphenyls (31%), dichlorobiphenyls (24%), trichlorobiphenyls (28%), tetrachlorobiphenyls (12%), pentachlorobiphenyls (4%), hexachlorobiphenyls (< 0.1%), and heptachlorobiphenyls (0%).

Formula Weight: 221 (average) [1].

RTECS Number: TQ 1354000

PHYSICAL AND CHEMICAL PROPERTIES

Appearance and Odor: Oily, almost colorless to light yellow liquid with a weak odor.

Boiling Point: Distills at 290-325 °C [2].

Henry's Law Constant: 8.64×10^{-4} atm·m^3/mol (calculated) [3].

Ionization Potential: No data found.

Log K$_{oc}$: 2.83 (estimated) [4].

Log K$_{ow}$: 3.2 (estimated) [3].

Melting Point: -35.5 °C (pour point) [5].

Solubility in Organics: Soluble in most solvents [6].

Solubility in Water: 1.45 mg/L at 25 °C (estimated) [3].

Specific Density: 1.24 at 25/4 °C [2]; 1.270-1.280 at 25/15.5 °C [7].

Transformation Products: No data found.

Vapor Density: ≈9.03 g/L at 25 °C, 7.63 (air = 1).

Vapor Pressure: 0.0046 mm at 25 °C (estimated) [3].

FIRE HAZARDS

Flash Point: 152-154 °C [1].

Lower Explosive Limit (LEL): No data found.

Upper Explosive Limit (UEL): No data found.

HEALTH HAZARD DATA

Immediately Dangerous to Life or Health (IDLH): No data found.

Permissible Exposure Limits (PEL) in Air: No standards set.

MANUFACTURING

Selected Manufacturer:

Monsanto Industrial Chemicals Co.
800 North Lindbergh Blvd.
St. Louis, MO 63166

Uses: In polyvinyl acetate to improve fiber-tear properties; as an insulator fluid for electric condensers and as an additive in very high pressure lubricants.

REFERENCES

1. Hutzinger, O., Safe, S., and V. Zitko. *The Chemistry of PCB's* (Boca Raton, FL: CRC Press, Inc., 1974), 269 p.
2. Monsanto Industrial Chemicals Co. "PCBs-Aroclors," Technical Bulletin O/PL 306A, (1974).
3. "Treatability Manual - Volume 1: Treatability Data," Office of Research and Development, U.S. EPA Report-600/8-80-042a (1980), 1035 p.
4. Montgomery, J.H. Unpublished results (1989).
5. Hazardous Substances Data Bank. Arochlor 1232, National Library of Medicine, Toxicology Information Program (1989).
6. "Chemical, Physical, and Biological Properties of Compounds Present at Hazardous Waste Sites," U.S. EPA Report-530/SW-89-010 (1985), 619 p.
7. Standen, A., Ed. *Kirk-Othmer Encyclopedia of Chemical Technology, Volume 4*, 2nd ed. (New York: John Wiley and Sons, Inc., 1964), 937 p.

PCB-1242

Synonyms: Arochlor 1242; **Aroclor 1242**; Chlorodiphenyl (42% Cl).

Structural Formula:

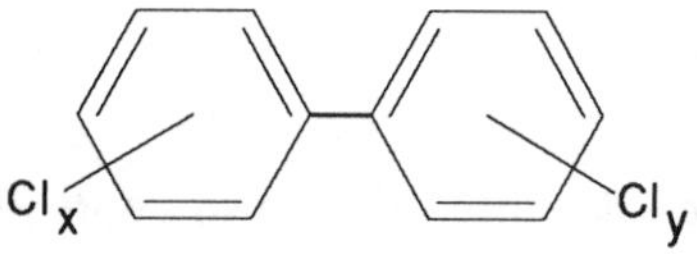

x + y = 0 thru 7

CHEMICAL DESIGNATIONS

CAS Registry Number: 53469-21-9

DOT Designation: 2315

Empirical Formula: Not definitive. PCB-1242 is a mixture of many biphenyls with varying degrees of chlorination. According to Hutzinger [1], the approximate composition of PCB-1242 by weight is as follows: biphenyls (< 0.1%), chlorobiphenyls (1%), dichlorobiphenyls (16%), trichlorobiphenyls (49%), tetrachlorobiphenyls (25%), pentachloro-biphenyls (8%), hexachlorobiphenyls (1%), and heptachlorobiphenyls (< 0.1%).

Formula Weight: Ranges from 154 to 358 [2] with an average value of 261 [1].

RTECS Number: TQ 1356000

PHYSICAL AND CHEMICAL PROPERTIES

Appearance and Odor: Almost colorless to light yellow oily liquid with a weak odor.

Boiling Point: Distills at 325-366 °C [3].

Henry's Law Constant: 5.6×10^{-4} atm·m^3/mol [4]; 421 atm/mol fraction [5].

Ionization Potential: No data found.

Log K_{oc}: 3.71 (estimated) [6].

Log K_{ow}: 5.580 for the compounds 2,4,4'- and 2,2',5-trichlorobiphenyl [7]; 4.11 [5].

Melting Point: -19 °C [8].

Solubility in Organics: Soluble in most solvents [9].

Solubility in Water: 0.100 mg/L at 24 °C [10]; 0.24 mg/L [3]; 0.34 mg/L [5]; 0.24 mg/L at 25 °C [11]; 0.2 mg/L at 20 °C [2]; 0.085 ppm for the compounds 2,4,4'- and 2,2',5-trichlorobiphenyl [7]; 132.9 and 18.6 ppb at 11.5 °C in distilled water and artificial seawater, respectively [12].

Specific Density: 1.392 at 15/4 °C, 1.381-1.392 at 25/15.5 °C [13].

Transformation Products: A strain of *Alcaligenes eutrophus* degraded 81% of the congeners by dechlorination under anaerobic conditions [14].

Vapor Density: $\approx$10.67 g/L at 25 °C, 9.01 (air = 1).

Vapor Pressure: 50 mm at 225 °C [15]; 4.06 x 10^{-4} mm at 25 °C [16]; 0.001 mm at 20 °C [8]; 0.001 mm at 38 °C [2].

FIRE HAZARDS

Flash Point: 176-180 °C [1].

Lower Explosive Limit (LEL): No data found.

Upper Explosive Limit (UEL): No data found.

HEALTH HAZARD DATA

Immediately Dangerous to Life or Health (IDLH): Potential human carcinogen [8].

Permissible Exposure Limits (PEL) in Air: 1 mg/m^3 [17]; 1.0 μg/m^3 10-hour time weighted average [8].

MANUFACTURING

Selected Manufacturer:

Monsanto Industrial Chemicals Co.
800 North Lindbergh Blvd.
St. Louis, MO 63166

Uses: Dielectric liquids; heat-transfer liquid widely used in transformers; swelling agents for transmission seals; ingredient in lubricants, oils, and greases; plasticizers for cellulosics, vinyl, and chlorinated rubbers; in polyvinyl acetate to improve fiber-tear properties.

REFERENCES

1. Hutzinger, O., Safe, S., and V. Zitko. *The Chemistry of PCB's* (Boca Raton, FL: CRC Press, Inc., 1974), 269 p.
2. Nisbet, I.C., and A.F. Sarofim. "Rates and Routes of Transport of PCBs in the Environment," *Environ. Health Perspect.*, (April 1972), p 21-38.
3. Monsanto Industrial Chemicals Co. "PCBs-Aroclors," Technical Bulletin O/PL 306A, (1974).
4. Eisenreich, S.J., Looney, B.B., and J.D. Thornton. "Airborne Organic Contaminants in the Great Lakes Ecosystem," *Environ. Sci. Technol.*, 15(1):30-38 (1981).
5. Paris, D.F., Steen, W.C., and G.L. Baughman. "Role of the Physico-Chemical Properties of Aroclors 1016 and 1242 in Determining Their Fate and Transport in Aquatic Environments," *Chemosphere*, 7(4):319-325 (1978).
6. Montgomery, J.H. Unpublished results (1989).
7. Kenaga, E.E., and C.A.I. Goring. "Relationship Between Water Solubility, Soil Sorption, Octanol-Water Partitioning and Concentration of Chemicals in Biota," in *Aquatic Toxicology, ASTM STP 707*, Eaton, J.G., Parrish, P.R., and A.C. Hendricks, Eds. (Philadelphia, PA: American Society for Testing and Materials 1980), pp 78-115.
8. "NIOSH Pocket Guide to Chemical Hazards," U.S. Department of Health and Human Services, U.S. Government Printing Office (1987), 241 p.
9. "Chemical, Physical, and Biological Properties of Compounds Present at Hazardous Waste Sites," U.S. EPA Report-530/SW-89-010

(1985), 619 p.

10. Hollifield, H.C. "Rapid Nephelometric Estimate of Water Solubility of Highly Insoluble Organic Chemicals of Environmental Interest," *Bull. Environ. Contam. Toxicol.*, 23(4/5):579-586 (1979).

11. "Treatability Manual - Volume 1: Treatability Data," Office of Research and Development, U.S. EPA Report-600/8-80-042a (1980), 1035 p.

12. Dexter, R.N., and S.P. Pavlou. "Mass Solubility and Aqueous Activity Coefficients of Stable Organic Chemicals in the Marine Environment: Polychlorinated Biphenyls," *Mar. Sci.*, 6:41-53 (1978).

13. Standen, A., Ed. *Kirk-Othmer Encyclopedia of Chemical Technology, Volume 4*, 2nd ed. (New York: John Wiley and Sons, Inc., 1964), 937 p.

14. Bedard, D.L., Wagner, R.E., Brennan, M.J., Haberl, M.L., and J.F. Brown Jr. "Extensive Degradation of Aroclors and Environmentally Transformed Polychlorinated Biphenyls by *Alcaligenes eutrophus* H850," *Appl. Environ. Microbiol.*, 53(5):1094-1102 (1987).

15. Verschueren, K. *Handbook of Environmental Data on Organic Chemicals* (New York: Van Nostrand Reinhold Co., 1983), 1310 p.

16. Mackay, D., and A.W. Wolkoff. "Rate of Evaporation of Low-Solubility Contaminants from Water Bodies to Atmosphere," *Environ. Sci. Technol.*, 7(7):611-614 (1973).

17. "General Industry Standards for Toxic and Hazardous Substances," U.S. Code of Federal Regulations 1910, Subpart Z Section 1910.1000 (July 1982).

PCB-1248

Synonyms: Arochlor 1248; **Aroclor 1248**; Chlorodiphenyl (48% Cl).

Structural Formula:

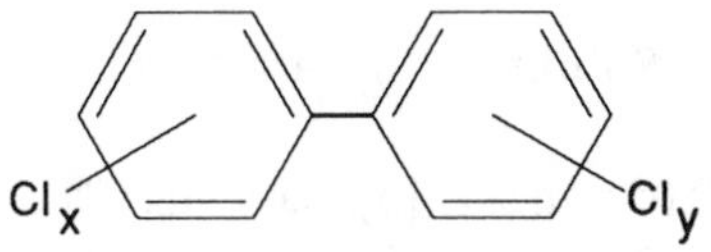

$$x + y = 2 \text{ thru } 6$$

CHEMICAL DESIGNATIONS

CAS Registry Number: 12672-29-6

DOT Designation: 2315

Empirical Formula: Not definitive. PCB-1248 is a mixture of many biphenyls with varying degrees of chlorination. According to Hutzinger [1], the approximate composition of PCB-1248 by weight is as follows: biphenyls (0%), chlorobiphenyls (0%), dichlorobiphenyls (2%), trichlorobiphenyls (18%), tetrachlorobiphenyls (40%), pentachlorobiphenyls (36%), hexachlorobiphenyls (4%), and heptachlorobiphenyls (0%).

Formula Weight: Ranges from 222 to 358 [2] with an average value of 288 [1].

RTECS Number: TQ 1358000

PHYSICAL AND CHEMICAL PROPERTIES

Appearance and Odor: Oily, light yellow liquid with a weak odor.

Boiling Point: Distills at 340-375 °C [3].

Henry's Law Constant: 0.0035 atm·m^3/mol [4].

Ionization Potential: No data found.

Log K_{oc}: 5.64 (estimated) [5].

Log K_{ow}: 6.11 [6]; 6.110 for the compounds 2,2',4,4'- and 2,2',5,5'-tetrachlorobiphenyl [7].

Melting Point: -7 °C (pour point) [8].

Solubility in Organics: Soluble in most solvents [9].

Solubility in Water: 0.054 mg/L [3]; 0.05 mg/L at 20 °C [2]; 0.060 mg/L at 24 °C [9]; 0.017 ppm for the compounds 2,2',4,4'- and 2,2',5,5'-tetrachlorobiphenyl [7].

Specific Density: 1.41 at 25/4 °C [3]; 1.405-1.415 at 65/15.5 °C [11].

Transformation Products: No data found.

Vapor Pressure: 4.94 x 10^{-4} mm at 25 °C [12]; 6 x 10^{-5} mm at 38 °C [2].

FIRE HAZARDS

Flash Point: 193-196 °C [1].

Lower Explosive Limit (LEL): No data found.

Upper Explosive Limit (UEL): No data found.

HEALTH HAZARD DATA

Immediately Dangerous to Life or Health (IDLH): No data found.

Permissible Exposure Limits (PEL) in Air: No standards set.

MANUFACTURING

Selected Manufacturer:

Monsanto Industrial Chemicals Co.
800 North Lindbergh Blvd.
St. Louis, MO 63166

Uses: In epoxy resins to improve adhesion and resistance to chemical attack; as an insulator fluid for electric condensers and as an additive in very high pressure lubricants.

REFERENCES

1. Hutzinger, O., Safe, S., and V. Zitko. *The Chemistry of PCB's* (Boca Raton, FL: CRC Press, Inc., 1974), 269 p.
2. Nisbet, I.C., and A.F. Sarofim. "Rates and Routes of Transport of PCBs in the Environment," *Environ. Health Perspect.*, (April 1972), p 21-38.
3. Monsanto Industrial Chemicals Co. "PCBs-Aroclors," Technical Bulletin O/PL 306A, (1974).
4. Lyman, W.J., Reehl, W.F., and D.H. Rosenblatt. *Handbook of Chemical Property Estimation Methods: Environmental Behavior of Organic Compounds* (New York: McGraw-Hill, Inc., 1982).
5. Montgomery, J.H. Unpublished results (1989).
6. Mackay, D. "Correlation of Bioconcentration Factors," *Environ. Sci. Technol.*, 16(5):274-278 (1982).
7. Kenaga, E.E., and C.A.I. Goring. "Relationship Between Water Solubility, Soil Sorption, Octanol-Water Partitioning and Concentration of Chemicals in Biota," in *Aquatic Toxicology, ASTM STP 707*, Eaton, J.G., Parrish, P.R., and A.C. Hendricks, Eds. (Philadelphia, PA: American Society for Testing and Materials 1980), pp 78-115.
8. Hazardous Substances Data Bank. Arochlor 1248, National Library of Medicine, Toxicology Information Program (1989).
9. "Chemical, Physical, and Biological Properties of Compounds Present at Hazardous Waste Sites," U.S. EPA Report-530/SW-89-010 (1985), 619 p.
10. Hollifield, H.C. "Rapid Nephelometric Estimate of Water Solubility of Highly Insoluble Organic Chemicals of Environmental Interest," *Bull. Environ. Contam. Toxicol.*, 23(4/5):579-586 (1979).
11. Standen, A., Ed. *Kirk-Othmer Encyclopedia of Chemical Technology, Volume 4*, 2nd ed. (New York: John Wiley and Sons, Inc., 1964), 937 p.
12. Mackay, D., and A.W. Wolkoff. "Rate of Evaporation of Low-Solubility Contaminants from Water Bodies to Atmosphere," *Environ. Sci. Technol.*, 7(7):611-614 (1973).

PCB-1254

Synonyms: Arochlor 1254; **Aroclor 1254**; Chlorodiphenyl (54% Cl); NCI-C02664.

Structural Formula:

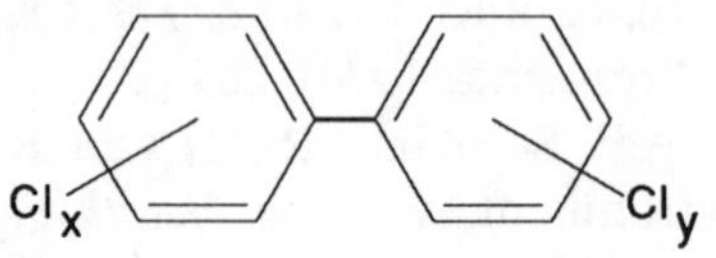

x + y = 0 thru 7

CHEMICAL DESIGNATIONS

CAS Registry Number: 11097-69-1

DOT Designation: 2315

Empirical Formula: Not definitive. PCB-1254 is a mixture of many biphenyls with varying degrees of chlorination. According to Hutzinger [1], the approximate composition of PCB-1254 by weight is as follows: biphenyls (< 0.1%), chlorobiphenyls (< 0.1%), dichlorobiphenyls (0.5%), trichlorobiphenyls (1%), tetrachlorobiphenyls (21%), pentachlorobiphenyls (48%), hexachlorobiphenyls (23%), and heptachlorobiphenyls (6%).

Formula Weight: 327 (average) [1].

RTECS Number: TQ 1360000

PHYSICAL AND CHEMICAL PROPERTIES

Appearance and Odor: Light yellow, viscous liquid with a weak odor.

Boiling Point: Distills at 365-390 °C [2].

Henry's Law Constant: 0.0027 atm·m^3/mol [3]; 0.0023 atm·m^3/mol [4].

Ionization Potential: No data found.

Log K$_{oc}$: 5.61 [5]; 4.628 for the compound 2,2',4,5,5'-pentachlorobiphenyl [6].

Log K$_{ow}$: 6.47 [7]; 5.61 [5]; 6.305 for the compound 2,2',4,5,5'-pentachlorobiphenyl [6].

Melting Point: 10 °C [8].

Solubility in Organics: Soluble in most solvents [9].

Solubility in Water: 0.057 mg/L at 24 °C [10]; 0.012 mg/L at 25 °C [11]; 0.050 mg/L at 20 °C, ≈0.056 mg/L at 26 °C [12]; 0.01 ppm for the compound 2,2',4,5,5'-pentachlorobiphenyl [6]; 24.2 and 4.34 ppb at 11.5 °C in distilled water and artificial seawater, respectively [13].

Specific Density: 1.505 at 15.5/4 °C, 1.495-1.505 at 65/15.5 °C [14].

Transformation Products: A strain of *Alcaligenes eutrophus* degraded 35% of the congeners by dechlorination under anaerobic conditions [15]. Indigenous microbes in the Center Hill Reservoir, TN biooxidized 2-chlorobiphenyl (a congener present in trace quantities) into chlorobenzoic acid and chlorobenzoylformic acid. Biooxidation of the PCB mixture containing 54% wt chlorine was not observed [16].

Vapor Density: ≈13.36 g/L at 25 °C, 11.29 (air = 1).

Vapor Pressure: 7.71 x 10^{-5} mm at 25 °C [17]; 6 x 10^{-5} mm at 20 °C [8].

FIRE HAZARDS

Flash Point: 222 °C [8].

Lower Explosive Limit (LEL): No data found.

Upper Explosive Limit (UEL): No data found.

HEALTH HAZARD DATA

Immediately Dangerous to Life or Health (IDLH): Potential human carcinogen [8].

Permissible Exposure Limits (PEL) in Air: 0.5 mg/m^3 [18]; 1 μg/m^3 10-hour time weighted average [8].

MANUFACTURING

Selected Manufacturer:

> Monsanto Industrial Chemicals Co.
> 800 North Lindbergh Blvd.
> St. Louis, MO 63166

Uses: Secondary plasticizer for polyvinyl chloride; co-polymers of styrene-butadiene and chlorinated rubber to improve chemical resistance to attack.

REFERENCES

1. Hutzinger, O., Safe, S., and V. Zitko. *The Chemistry of PCB's* (Boca Raton, FL: CRC Press, Inc., 1974), 269 p.
2. Monsanto Industrial Chemicals Co. "PCBs-Aroclors," Technical Bulletin O/PL 306A, (1974).
3. Eisenreich, S.J., Looney, B.B., and J.D. Thornton. "Airborne Organic Contaminants in the Great Lakes Ecosystem," *Environ. Sci. Technol.*, 15(1):30-38 (1981).
4. Petrasek, A.C., Kugelman, I.J., Austern, B.M., Pressley, T.A., Winslow, L.A., and R.H. Wise. "Fate of Toxic Organic Compounds in Wastewater Treatment Plants," *J. Water Poll. Control Fed.*, 55(10):1286-1296 (1983).
5. Voice, T.C., and W.J. Weber Jr. "Sorbent Concentration Effects in Liquid/Solid Partitioning," *Environ. Sci. Technol.*, 19(9):789-796 (1985).
6. Kenaga, E.E., and C.A.I. Goring. "Relationship Between Water Solubility, Soil Sorption, Octanol-Water Partitioning and Concentration of Chemicals in Biota," in *Aquatic Toxicology, ASTM STP 707*, Eaton, J.G., Parrish, P.R., and A.C. Hendricks, Eds. (Philadelphia, PA: American Society for Testing and Materials 1980), pp 78-115.
7. Travis, C.C., and A.D. Arms. "Bioconcentration of Organics in Beef, Milk and Vegetation," *Environ. Sci. Technol.*, 22(3):271-274 (1988).
8. "NIOSH Pocket Guide to Chemical Hazards," U.S. Department of Health and Human Services, U.S. Government Printing Office

(1987), 241 p.

9. "Chemical, Physical, and Biological Properties of Compounds Present at Hazardous Waste Sites," U.S. EPA Report-530/SW-89-010 (1985), 619 p.

10. Hollifield, H.C. "Rapid Nephelometric Estimate of Water Solubility of Highly Insoluble Organic Chemicals of Environmental Interest," *Bull. Environ. Contam. Toxicol.*, 23(4/5):579-586 (1979).

11. Callahan, M.A., Slimak, M.W., Gable, N.W., May, I.P., Fowler, C.F., Freed, J.R., Jennings, P., Durfee, R.L., Whitmore, F.C., Maestri, B., Mabey, W.R., Holt, B.R., and C. Gould. "Water-Related Environmental Fate of 129 Priority Pollutants Volumes I and II," National Technical Information Service, U.S. EPA Report-440/4-79-029 (1979), 1160 p.

12. Haque, R., Schmedding, D.W., and V.H. Freed. "Aqueous Solubility, Adsorption, and Vapor Behavior of Polychlorinated Biphenyl Aroclor 1254," *Environ. Sci. Technol.*, 8(2):139-142 (1974).

13. Dexter, R.N., and S.P. Pavlou. "Mass Solubility and Aqueous Activity Coefficients of Stable Organic Chemicals in the Marine Environment: Polychlorinated Biphenyls," *Mar. Sci.*, 6:41-53 (1978).

14. Standen, A., Ed. *Kirk-Othmer Encyclopedia of Chemical Technology, Volume 4*, 2nd ed. (New York: John Wiley and Sons, Inc., 1964), 937 p.

15. Bedard, D.L., Wagner, R.E., Brennan, M.J., Haberl, M.L., and J.F. Brown Jr. "Extensive Degradation of Aroclors and Environmentally Transformed Polychlorinated Biphenyls by *Alcaligenes eutrophus* H850," *Appl. Environ. Microbiol.*, 53(5):1094-1102 (1987).

16. Shiaris, M.P., and G.S. Sayler. "Biotransformation of PCB by Natural Assemblages of Freshwater Microorganisms," *Environ. Sci. Technol.*, 16(6):367-369 (1982).

17. Mackay, D., and A.W. Wolkoff. "Rate of Evaporation of Low-Solubility Contaminants from Water Bodies to Atmosphere," *Environ. Sci. Technol.*, 7(7):611-614 (1973).

18. "General Industry Standards for Toxic and Hazardous Substances," U.S. Code of Federal Regulations 1910, Subpart Z Section 1910.1000 (July 1982).

PCB-1260

Synonyms: Arochlor 1260; **Aroclor 1260**; Chlorodiphenyl (60% Cl); Clophen A60; Kanechlor; Phenoclor DP6.

Structural Formula:

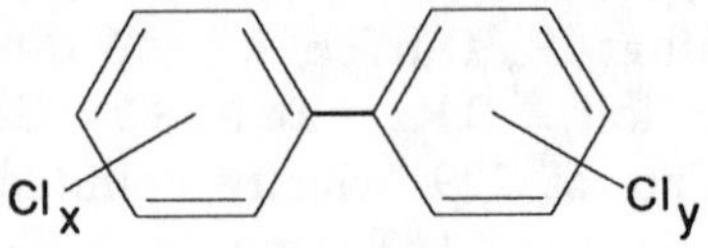

x + y = 4 thru 7

CHEMICAL DESIGNATIONS

CAS Registry Number: 11096-82-5

DOT Designation: 2315

Empirical Formula: Not definitive. PCB-1260 is a mixture of many biphenyls with varying degrees of chlorination. According to Hutzinger [1], the approximate composition of PCB-1260 by weight is as follows: biphenyls (0%), chlorobiphenyls (0%), dichlorobiphenyls (0%), trichloro-biphenyls (0%), tetrachlorobiphenyls (1%), pentachlorobiphenyls (12%), hexachlorobiphenyls (38%), and heptachlorobiphenyls (41%).

Formula Weight: Ranges from 324 to 460 [2] with an average value of 370 [1].

RTECS Number: TQ 1362000

PHYSICAL AND CHEMICAL PROPERTIES

Appearance and Odor: Light yellow sticky, soft resin with a weak odor.

Boiling Point: Distills at 385-420 °C [3].

Henry's Law Constant: 0.0071 atm·m^3/mol [4].

Ionization Potential: No data found.

Log K_{oc}: 6.42 (estimated) [5].

Log K_{ow}: 6.91 [6].

Melting Point: 31 °C [7].

Solubility in Organics: Soluble in most solvents [8].

Solubility in Water: 0.080 mg/L at 24 °C [9]; 0.0027 mg/L [3]; 0.04-0.2 ppm [10].

Specific Density: 1.566 at 15.5/4 °C, 1.555-1.566 at 90/15.5 °C [11].

Transformation Products: No data found.

Vapor Pressure: 4.05×10^{-5} mm at 25 °C [12]; 2×10^{-7} mm at 38 °C [2].

FIRE HAZARDS

Flash Point: None to the boiling point [1].

Lower Explosive Limit (LEL): No data found.

Upper Explosive Limit (UEL): No data found.

HEALTH HAZARD DATA

Immediately Dangerous to Life or Health (IDLH): No data found.

Permissible Exposure Limits (PEL) in Air: No standards set.

MANUFACTURING

Selected Manufacturer:

Monsanto Industrial Chemicals Co.
800 North Lindbergh Blvd.
St. Louis, MO 63166

Uses: Secondary plasticizer for polyvinyl chloride; in polyester resins to increase strength of fiberglass; varnish formulations to improve water and alkali resistance; as an insulator fluid for electric condensers and as an additive in very high pressure lubricants.

REFERENCES

1. Hutzinger, O., Safe, S., and V. Zitko. *The Chemistry of PCB's* (Boca Raton, FL: CRC Press, Inc., 1974), 269 p.
2. Nisbet, I.C., and A.F. Sarofim. "Rates and Routes of Transport of PCBs in the Environment," *Environ. Health Perspect.*, (April 1972), p 21-38.
3. Monsanto Industrial Chemicals Co. "PCBs-Aroclors," Technical Bulletin O/PL 306A, (1974).
4. Lyman, W.J., Reehl, W.F., and D.H. Rosenblatt. *Handbook of Chemical Property Estimation Methods: Environmental Behavior of Organic Compounds* (New York: McGraw-Hill, Inc., 1982).
5. Montgomery, J.H. Unpublished results (1989).
6. Mackay, D. "Correlation of Bioconcentration Factors," *Environ. Sci. Technol.*, 16(5):274-278 (1982).
7. Hazardous Substances Data Bank. Arochlor 1260, National Library of Medicine, Toxicology Information Program (1989).
8. "Chemical, Physical, and Biological Properties of Compounds Present at Hazardous Waste Sites," U.S. EPA Report-530/SW-89-010 (1985), 619 p.
9. Hollifield, H.C. "Rapid Nephelometric Estimate of Water Solubility of Highly Insoluble Organic Chemicals of Environmental Interest," *Bull. Environ. Contam. Toxicol.*, 23(4/5):579-586 (1979).
10. *Drinking Water and Health* (Washington, DC: National Academy of Sciences, 1977), 939 p.
11. Standen, A., Ed. *Kirk-Othmer Encyclopedia of Chemical Technology, Volume 4*, 2nd ed. (New York: John Wiley and Sons, Inc., 1964), 937 p.
12. Mackay, D., and A.W. Wolkoff. "Rate of Evaporation of Low-Solubility Contaminants from Water Bodies to Atmosphere," *Environ. Sci. Technol.*, 7(7):611-614 (1973).

PENTACHLOROPHENOL

Synonyms: Acutox; Chem-penta; Chem-tol; Chlorophen; Cryptogil OL; Dowcide 7; Dowicide 7; Dowicide EC-7; Dowicide G; Dow pentachlorophenol DP-2 antimicrobial; Durotox; EP 30; Fungifen; Fungol; Glazd penta; Grundier arbezol; Lauxtol; Lauxtol A; Liroprem; Monsanto penta; Moosuran; NCI-C54933; NCI-C55378; NCI-C56655; PCP; Penchlorol; Penta; Pentachlorofenol; Pentachlorofenolo; Pentachlorophenate; Pentachlorphenol; 2,3,4,5,6-Pentachlorophenol; Pentacon; Penta-kil; Pentasol; Penwar; Peratox; Permacide; Permaguard; Permasan; Permatox DP-2; Permatox Penta; Permite; Priltox; RCRA waste number U242; Santobrite; Santophen; Santophen 20; Sinituho; Term-i-trol; Thompson's wood fix; Weedone; Witophen P.

Structural Formula:

CHEMICAL DESIGNATIONS

CAS Registry Number: 87-86-5

DOT Designation: 2020

Empirical Formula: C_6HCl_5O

Formula Weight: 266.34

RTECS Number: SM 6300000

PHYSICAL AND CHEMICAL PROPERTIES

Appearance and Odor: White to dark-colored flakes or beads with a characteristic odor.

Boiling Point: 310 °C [1]; 293 °C [2].

Dissociation Constant: 4.74 [3].

Henry's Law Constant: 3.4×10^{-6} atm·m^3/mol [4]; 2.8×10^{-7} atm·m^3/mol [5]; 2.1×10^{-6} atm·m^3/mol [6].

Ionization Potential: No data found.

Log K$_{oc}$: 2.95 [7]; 2.96 [8].

Log K$_{ow}$: 5.01 [9]; 5.86 [10]; 4.84 at pH 1.2, 4.72 at pH 2.4, 4.62 at pH 3.4, 4.57 at pH 4.7, 3.72 at pH 5.9, 3.56 at pH 6.5, 3.32 at pH 7.2, 3.20 at pH 7.8, 3.10 at pH 8.4, 2.75 at pH 8.9, 1.45 at pH 9.3, 1.36 at pH 9.8, 1.30 at pH 10.5, 2.42 at pH 10.5, 1.30 at pH 11.5, 1.67 at pH 12.5, 3.86 at pH 13.5 [11]; 3.69 [12]; 5.24 [13]; 3.807 [14]; 5.0 [15]; 4.16 [16].

Melting Point: 191 °C [17]; 174 °C [18]; 188 °C [19]; 187-189 °C [20].

Solubility in Organics: Very soluble in ethanol and ether; soluble in hot benzene; slightly soluble in solvents and ligroin [21]. Also soluble in acetone, carbitol, and cellosolve [22].

Solubility in Water: 5 mg/L at 0 °C, 35 mg/L at 50 °C, 85 mg/L at 70 °C [23]; 0.002 wt% at 20 °C [24]; 20 mg/L at 20 °C [25]; 20-25 mg/L at 25 °C [2]; 8 mg/100 ml [26]; 14 ppm at 20 °C, 19 ppm at 30 °C [27]; 14 mg/L at 20 °C, 20 mg/L at 30 °C [28]; 2.0 mg/L [29].

Specific Density: 1.978 at 22/4 °C [17]; 1.98 at 15/4 °C [19].

Transformation Products: Under aerobic conditions, microbes in estuarine water partially dechlorinated pentachlorophenol to produce trichlorophenol. In distilled water, pentachlorophenol underwent photolyzed to tetrachlorophenols, trichlorophenols, chlorinated dihydroxybenzenes, and dichloromaleic acid [30]. Under anaerobic conditions, pentachlorophenol may undergo sequential dehalogenation to produce tetra-, tri-, di-, and m-chlorophenol [31]. In aerobic and anaerobic soils, pentachloroanisole was the major metabolite along with 2,3,6-trichlorophenol, 2,3,4,5-, and 2,3,5,6-tetrachlorophenol [32].

Vapor Pressure: 1.7×10^{-4} mm at 20 °C [1]; 1.4×10^{-4} mm at 20 °C [4]; 40 mm at 211.2 °C, 100 mm at 239.6 °C, 400 mm at 285.0 °C, 760 mm at 309.3 °C [17]; 6.7×10^{-7} mm at 19 °C, 1.5×10^{-5} mm at 20 °C [33]; 2.14×10^{-5} mm at 20 °C [34]; 0.12 mm at 100 °C, 1 mm at 135 °C [2]; 1.1×10^{-4} mm at 20 °C, 1.7×10^{-5} mm at 23 °C, 7.6×10^{-6} mm at 30 °C, 2.6×10^{-4} mm at 40 °C [35]; 0.00011 mm at 20 °C [27]; 1.7×10^{-5} at 20.25 °C [29].

FIRE HAZARDS

Flash Point: Not flammable [19].

Lower Explosive Limit (LEL): Not flammable [19].

Upper Explosive Limit (UEL): Not flammable [19].

HEALTH HAZARD DATA

Immediately Dangerous to Life or Health (IDLH): 150 mg/m^3 [24].

Permissible Exposure Limits (PEL) in Air: 0.5 mg/m^3 [36].

MANUFACTURING

Selected Manufacturers:

Dow Chemical Co.
Midland, MI 48640

Reichold Chemicals Inc.
RCI Building
White Plains, NY 10602

Sonford Chemical Co.
Port Neches, TX 77651

Uses: Manufacturing of insecticides, algicides, herbicides, fungicides, and bactericides; wood preservative.

REFERENCES

1. Melnikov, N.N. *Chemistry of Pesticides* (New York: Springer-Verlag, Inc., 1971), 480 p.
2. Bailey, G.W., and J.L. White. "Herbicides: A Compilation of Their Physical, Chemical, and Biological Properties," *Res. Rev.*, 10:97-122 (1965).
3. Mills, W.B., Porcella, D.B., Ungs, M.J., Gherini, S.A., Summers, K.V., Mok, L., Rupp, G.L., and G.L. Bowie. "Water Quality Assessment: A

Screening Procedure for Toxic and Conventional Pollutants in Surface and Groundwater-Part I," Office of Research and Development, U.S. EPA Report-600/6-85-002a (1985), 638 p.

4. Lyman, W.J., Reehl, W.F., and D.H. Rosenblatt. *Handbook of Chemical Property Estimation Methods: Environmental Behavior of Organic Compounds* (New York: McGraw-Hill, Inc., 1982).

5. Eisenreich, S.J., Looney, B.B., and J.D. Thornton. "Airborne Organic Contaminants in the Great Lakes Ecosystem," *Environ. Sci. Technol.*, 15(1):30-38 (1981).

6. Petrasek, A.C., Kugelman, I.J., Austern, B.M., Pressley, T.A., Winslow, L.A., and R.H. Wise. "Fate of Toxic Organic Compounds in Wastewater Treatment Plants," *J. Water Poll. Control Fed.*, 55(10):1286-1296 (1983).

7. Kenaga, E.E. "Predicted Bioconcentration Factors and Soil Sorption Coefficients of Pesticides and other Chemicals," *Ecotoxicol. Environ. Safety*, 4(1):26-38 (1980).

8. Seip, H.M., Alstad, J., Carlberg, G.E., Matinsen, K., and R. Skaane. "Measurement of Mobility of Organic Compounds in Soils," *Sci. Tot. Environ.*, 50:87-101 (1986).

9. Leo, A., Hansch, C., and D. Elkins. "Partition Coefficients and Their Uses," *Chem. Rev.*, 71(6):525-616 (1971).

10. Banerjee, S., Howard, P.H., Rosenburg, A.M., Dombrowski, A.E., Sikka, Harish, and D.L. Tullis. "Development of a General Kinetic Model for Biodegradation and its Application to Chlorophenols and Related Compounds," *Environ. Sci. Technol.*, 18(6):416-422 (1984).

11. Kaiser, K.L.E, and I. Valdmanis. "Apparent Octanol/Water Partition Coefficients of Pentachlorophenol as a Function of pH," *Can. J. Chem.*, 60(16):2104-2106 (1982).

12. Geyer, H., Sheehan, P., Kotzias, D., Freitag, D., and F. Korte. "Prediction of Ecotoxicological Behaviour of Chemicals: Relationship Between Physico-Chemical Properties and Bioaccumulation of Organic Chemicals in the Mussell *Mytilus edulis*," *Chemosphere*, 11(11):1121-1134 (1982).

13. Schellenberg, K., Leuenberger, C., and R.P. Schwarzenbach. "Sorption of Chlorinated Phenols by Natural Sediments and Aquifer Materials," *Environ. Sci. Technol.*, 18(9):652-657 (1984).

14. Lu, P.-Y., and R.L. Metcalf. "Environmental Fate and Biodegradability of Benzene Derivatives as Studied in a Model Ecosystem," *Environ. Health Perspect.*, 10:269-284 (1975).

15. van Gestel, C.A.M., and W.-C. Ma. "Toxicity and Bioaccumulation of Chlorophenols in Earthworms, in Relation to Bioavailability in Soil," *Ecotoxicol. Environ. Safety*, 15(3):289-297 (1988).

16. Rao, P.S.C., and J.M. Davidson. "Estimation of Pesticide Retention

and Transformation Parameters Required in Nonpoint Source Pollution Models," in *Environmental Impact of Nonpoint Source Pollution*, Overcash, M.R., and J.M. Davidson, Eds. (Ann Arbor, MI: Ann Arbor Science Publishers, Inc., 1980), pp 23-67.

17. Weast, R.C., Ed. *CRC Handbook of Chemistry and Physics*, 67th ed. (Boca Raton, FL: CRC Press, Inc., 1986), 2406 p.

18. *IARC Monographs on the Evaluation of Carcinogenic Risk of Chemicals to Man. Some Halogenated Hydrocarbons, Volume 20* (Lyon, France: International Agency for Research on Cancer, 1979), 609 p.

19. Weiss, G. *Hazardous Chemicals Data Book* (Park Ridge, NJ: Noyes Data Corp., 1986), 1069 p.

20. *Fluka Catalog 1988/89 - Chemika-Biochemika* (Ronkonkoma, NY: Fluka Chemical Corp., 1988), 1536 p.

21. "Chemical, Physical, and Biological Properties of Compounds Present at Hazardous Waste Sites," U.S. EPA Report-530/SW-89-010 (1985), 619 p.

22. *Toxic and Hazardous Industrial Chemicals Safety Manual for Handling and Disposal with Toxicity and Hazard Data* (Tokyo, Japan: International Technical Information Institute, 1986), 700 p.

23. Verschueren, K. *Handbook of Environmental Data on Organic Chemicals* (New York: Van Nostrand Reinhold Co., 1983), 1310 p.

24. "NIOSH Pocket Guide to Chemical Hazards," U.S. Department of Health and Human Services, U.S. Government Printing Office (1987), 241 p.

25. Kearney, P.C., and D.D. Kaufman. *Herbicides: Chemistry, Degradation and Mode of Action* (New York: Marcel Dekker, Inc., 1976), 1036 p.

26. Reinbold, K.A., Hassett, J.J., Means, J.C., and W.L. Banwart. "Adsorption of Energy-Related Organic Pollutants: A Literature Review," Office of Research and Development, U.S. EPA Report-600/3-79-086 (1979), 180 p.

27. Bevenue, A., and H. Beckman. "Pentachlorophenol: A Discussion of its Properties and its Occurrence as a Residue in Human and Animal Tissues," *Res. Rev.*, 19:83-134 (1967).

28. Gunther, F.A., Westlake, W.E., and P.S. Jaglan. "Reported Solubilities of 738 Pesticide Chemicals in Water," *Res. Rev.*, 20:1-148 (1968).

29. Gile, J.D., and J.W. Gillett. "Fate of Selected Fungicides in a Terrestrial Laboratory Ecosystem," *J. Agric. Food Chem.*, 27(6):1159-1164 (1979).

30. Hwang, H.-M., Hodson, R.E., and R.F. Lee. "Degradation of Phenol and Chlorophenols by Sunlight and Microbes in Estuarine Water," *Environ. Sci. Technol.*, 20(10):1002-1007 (1986).

31. Kobayashi, H., and B.E. Rittman. "Microbial Removal of Hazardous Organic Compounds," *Environ. Sci. Technol.*, 16(3):170A-183A (1982).

32. Murthy, N.B.K., Kaufman, D.D., and G.F. Fries. "Degradation of Pentachlorophenol (PCP) in Aerobic and Anaerobic Soil," *J. Environ. Sci. Health*, B14(1):1-14 (1979).

33. *Environmental Health Criteria 71: Pentachlorophenol* (Geneva: World Health Organization, 1987), 236 p.

34. Dobbs, A.J., and M.R. Cull. "Volatilisation of Chemicals - Relative Loss Rates and the Estimation of Vapour Pressures," *Environ. Poll. (Series B)*, 3(4):289-298 (1982).

35. Klöpffer, W., Kaufman, G., Rippen, G., and H.-P. Poremski. "A Laboratory Method for Testing the Volatility from Aqueous Solution: First Results and Comparison with Theory," *Ecotoxicol. Environ. Safety*, 6(6):545-559 (1982).

36. "General Industry Standards for Toxic and Hazardous Substances," U.S. Code of Federal Regulations 1910, Subpart Z Section 1910.1000 (July 1982).

PHENANTHRENE

Synonyms: Phenanthren; Phenantrin.

Structural Formula:

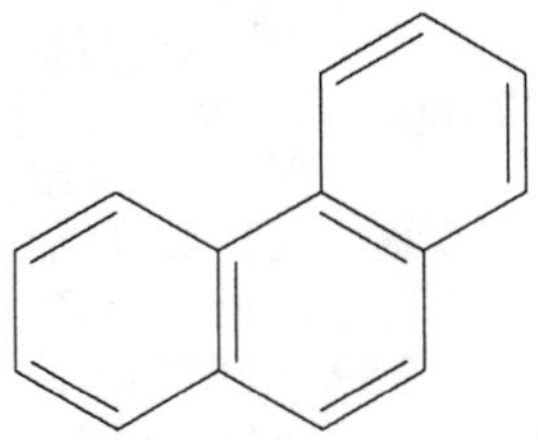

CHEMICAL DESIGNATIONS

CAS Registry Number: 85-01-8

DOT Designation: None assigned.

Empirical Formula: $C_{14}H_{10}$

Formula Weight: 178.24

RTECS Number: SF 7175000

PHYSICAL AND CHEMICAL PROPERTIES

Appearance: Colorless, monoclinic crystals.

Boiling Point: 340 °C [1]; 341.2 °C [2].

Henry's Law Constant: 3.9 x 10^{-5} atm·m^3/mol [3]; 1.3 x 10^{-4} atm·m^3/mol [4]; 2.56 x 10^{-5} atm·m^3/mol at 25 °C [5]; 5.48 x 10^{-5} atm·m^3/mol [6].

Ionization Potential: 8.22 ± 0.29 eV [7].

Log K_{oc}: 3.72 [8]; 4.36 [9]; 4.59 [10].

Log K_{ow}: 4.568 [9]; 4.46 [11]; 4.52 [12]; 4.517 [13]; 4.16 [14].

Melting Point: 100.5 °C [15]; 98-100 °C [16].

Solubility in Organics: Soluble in acetone, ethanol, ether, benzene, acetic acid [17], benzene, and carbon disulfide [18].

Solubility in Water: 0.423 ppm at 8.5 °C, 1.6 mg/L at 15 °C, 0.816 ppm at 21 °C, 1.277 ppm at 30 °C, 0.6 ppm in seawater at 22 °C [19]; 1.18 mg/L at 25 °C [20]; 1.29 mg/L at 25 °C [21]; 0.994 mg/L at 25 °C [22]; 1.002 mg/L at 25 °C, 1.220 mg/L at 29 °C [23]; 1.07 mg/L at 25 °C [24]; 1.600 mg/L [25]; 0.71 mg/L at 25 °C [26]; 423 μg/L at 8.5 °C, 468 μg/L at 10.0 °C, 512 μg/L at 12.5 °C, 601 μg/L at 15 °C, 816 μg/L at 21.0 °C, 995 μg/L at 24.3 °C, 1,277 μg/L at 29.9 °C [27]; 2.81 x 10^{-6} M at 8.4 °C, 3.09 x 10^{-6} M at 11.1 °C, 3.59 x 10^{-6} M at 14.0 °C, 4.40 x 10^{-6} M at 17.5 °C, 4.94 x 10^{-6} M at 20.2 °C, 6.09 x 10^{-6} M at 23.3 °C, 6.46 x 10^{-6} M at 25.0 °C, 7.7 x 10^{-6} M at 29.3 °C, 9.13 x 10^{-6} M at 31.8 °C [28]; 0.235 ppm at 20-25 °C in Narragansett Bay water (pH = 7.7, dissolved organic carbon = 2.9 mg/L, and salinity = 27 wt%) [29].

Specific Density: 0.9800 at 4/4 °C [1]; 1.179 at 25/4 °C [30].

Transformation Products: Catechol is the central metabolite in the bacterial degradation of phenanthrene. Intermediate byproducts include 1-hydroxy-2-napthoic acid, 1,2-dihydroxynaphthalene, and salicylic acid [31]. It was reported that *Beijerinckia*, under aerobic conditions, degraded phenanthrene to *cis*-3,4-dihydroxy-3,4-dihydro-phenanthracene [32].

Vapor Pressure: 6.80 x 10^{-4} mm at 25 °C [33]; 2.1 x 10^{-4} mm at 20 °C [34]; 1 mm at 118.2 °C, 10 mm at 173.0 °C, 40 mm at 215.8 °C, 100 mm at 249.0 °C, 400 mm at 308.0 °C, 760 mm at 340.2 °C [1]; 0.228 mm at 100 °C, 0.309 mm at 105 °C, 0.410 mm at 110 °C, 0.552 mm at 115 °C, 0.720 mm at 120 °C, 0.936 mm at 125 °C, 1.208 mm at 130 °C, 1.554 mm at 135 °C, 1.987 mm at 140 °C, 2.515 mm at 145 °C, 3.156 mm at 150 °C [35]; 0.00349 mm at 51.60 °C, 0.00610 mm at 57.00 °C, 0.00991 mm at 61.85 °C, 0.0159 mm at 67.35 °C, 0.0232 mm at 71.80 °C, 0.0424 mm at 78.90 °C, 0.0667 mm at 83.40 °C, 0.109 mm at 90.3 °C [36]; 0.000064 mm at 36.70 °C, 0.000081 mm at 39.15 °C, 0.000089 mm at 39.85 °C 0.000111 mm at 42.10 °C 0.000120 mm at 42.60 °C, 0.000145 mm at 44.62 °C, 0.000185 mm at 46.70 °C, 0.000223 mm at 48.80 °C, 0.000241 mm at 49.65 °C [37].

FIRE HAZARDS

Flash Point: 171 °C (open cup) [38].

Lower Explosive Limit (LEL): No data found.

Upper Explosive Limit (UEL): No data found.

HEALTH HAZARD DATA

Immediately Dangerous to Life or Health (IDLH): Potential human carcinogen [39].

Permissible Exposure Limits (PEL) in Air: No individual standards have been set, however, as a constituent in coal tar pitch volatiles, the following exposure limits have been established: 0.2 mg/m^3 (benzene-soluble fraction) [40]; 0.1 mg/m^3 10-hour time weighted average (cyclohexane-extractable fraction) [39]; 0.2 mg/m^3 TWA (benzene solubles) [41].

MANUFACTURING

Selected Manufacturers:

Aldrich Chemical Co.
940 West Saint Paul Ave.
Milwaukee, WI 53233

Fluka Chemical Corp.
980 South Second St.
Ronkonkoma, NY 11779

Pfaltz & Bauer, Inc.
172 East Aurora St.
Waterbury, CT 06708

Uses: Explosives; dyestuffs; biochemical research; synthesis of drugs; organic synthesis.

REFERENCES

1.   Weast, R.C., Ed. *CRC Handbook of Chemistry and Physics*, 67th ed. (Boca Raton, FL: CRC Press, Inc., 1986), 2406 p.

2. Zwolinski, B.J., and R.C. Wilhoit. *Handbook of Vapor Pressure and Heats of Vaporization of Hydrocarbons and Related Compounds*, Publication 101 (College Station, TX: Thermodynamics Research Station 971), 329 p.

3. Mackay, D., Shiu, W.Y., and R.P. Sutherland. "Determination of Air-Water Henry's Law Constants for Hydrophobic Pollutants," *Environ. Sci. Technol.*, 13(3):333-337 (1979).

4. Petrasek, A.C., Kugelman, I.J., Austern, B.M., Pressley, T.A., Winslow, L.A., and R.H. Wise. "Fate of Toxic Organic Compounds in Wastewater Treatment Plants," *J. Water Poll. Control Fed.*, 55(10):1286-1296 (1983).

5. Hine, J., and P.K. Mookerjee. "The Intrinsic Hydrophilic Character of Organic Compounds. Correlations in Terms of Structural Contributions," *J. Org. Chem.*, 40(3):292-298 (1975).

6. Southworth, G.R. "The Role Volatilization in Removing Polycyclic Aromatic Hydrocarbons from Aquatic Environments," *Bull. Environ. Contam. Toxicol.*, 21(4/5):507-514 (1979).

7. Franklin, J.L., Dillard, J.G., Rosenstock, H.M., Herron, J.T., Draxl K., and F.H. Field. "Ionization Potentials, Appearance Potentials and Heats of Formation of Gaseous Positive Ions," National Bureau of Standards Report NSRDS-NBS 26, U.S. Government Printing Office (1969), 289 p.

8. Abdul, S.A., Gibson, T.L., and D.N. Rai. "Statistical Correlations for Predicting the Partition Coefficient for Nonpolar Organic Contaminants Between Aquifer Organic Carbon and Water," *Haz. Waste Haz. Mater.*, 4(3):211-222 (1987).

9. Karickhoff, S.W., Brown, D.S., and T.A. Scott. "Sorption of Hydrophobic Pollutants on Natural Sediments," *Water Res.*, 13:241-248 (1979).

10. Socha, S.B., and R. Carpenter. "Factors Affecting the Pore Water Hydrocarbon Concentrations in Puget Sound Sediments," *Geochim. Cosmochim. Acta*, 51(5):1273-1284 (1987).

11. Hansch, C., and T. Fujita. "ρ-σ-π Analysis. A Method for the Correlation of Biological Activity and Chemical Structure," *J. Am. Chem. Soc.*, 86(8):1616-1617 (1964).

12. Yoshida, K., Shigeoka, T., and F. Yamauchi. "Non-Steady State Equilibrium Model for the Preliminary Prediction of the Fate of Chemicals in the Environment," *Ecotoxicol. Environ. Safety*, 7(2):179-190 (1983).

13. Kenaga, E.E., and C.A.I. Goring. "Relationship Between Water Solubility, Soil Sorption, Octanol-Water Partitioning and Concentration of Chemicals in Biota," in *Aquatic Toxicology, ASTM STP 707*, Eaton, J.G., Parrish, P.R., and A.C. Hendricks, Eds.

(Philadelphia, PA: American Society for Testing and Materials 1980), pp 78-115.

14. Landrum, P.F., Nihart, S.R., Eadie, B.J., and W.S. Gardner. "Reverse-Phase Separation Method for Determining Pollutant Binding to Aldrich Humic Acid and Dissolved Organic Carbon of Natural Waters," *Environ. Sci. Technol.*, 19(3):187-192 (1984).

15. Dean, J.A., Ed., *Lange's Handbook of Chemistry*, 11th ed. (New York: McGraw-Hill, Inc., 1973), 1570 p.

16. *Fluka Catalog 1988/89 - Chemika-Biochemika* (Ronkonkoma, NY: Fluka Chemical Corp., 1988), 1536 p.

17. "Chemical, Physical, and Biological Properties of Compounds Present at Hazardous Waste Sites," U.S. EPA Report-530/SW-89-010 (1985), 619 p.

18. *Toxic and Hazardous Industrial Chemicals Safety Manual for Handling and Disposal with Toxicity and Hazard Data* (Tokyo, Japan: International Technical Information Institute, 1986), 700 p.

19. Verschueren, K. *Handbook of Environmental Data on Organic Chemicals* (New York: Van Nostrand Reinhold Co., 1983), 1310 p.

20. Wauchope, R.D., and F.W. Getzen. "Temperature Dependence of Solubilities in Water and Heats of Fusion of Solid Aromatic Hydrocarbons," *J. Chem. Eng. Data*, 17:38-41 (1972).

21. Mackay, D., and W.Y. Shiu. "Aqueous Solubility of Polynuclear Aromatic Hydrocarbons," *J. Chem. Eng. Data*, 22(4):399-402 (1977).

22. Andrews, L.J., and R.M. Keefer. "Cation Complexes of Compounds Containing Carbon-Carbon Double Bonds. IV. The Argentation of Aromatic Hydrocarbons," *J. Am. Chem. Soc.*, 71(11):3644-3647 (1949).

23. May, W.E., Wasik, S.P., and D.H. Freeman. "Determination of the Aqueous Solubility of Polynuclear Aromatic Hydrocarbons by a Coupled Column Liquid Chromatographic Technique," *Anal. Chem.*, 50(1):175-179 (1978).

24. Eganhouse, R.P., and J.A. Calder. "The Solubility of Medium Weight Aromatic Hydrocarbons and the Effect of Hydrocarbon Co-solutes and Salinity," *Geochim. Cosmochim. Acta*, 40(5):555-561 (1976).

25. Davis, W.W., Krahl, M.E., and G.H.A. Clowes. "Solubility of Carcinogenic and Related Hydrocarbons in Water," *J. Am. Chem. Soc.*, 64(1):108-110 (1942).

26. Sahyun, M.R.V. "Binding of Aromatic Compounds to Bovine Serum Albumin," *Nature*, 209(5023):613-614 (1966).

27. May, W.E., Wasik, S.P., and D.H. Freeman. "Determination of the Solubility Behavior of Some Polycyclic Aromatic Hydrocarbons in Water," *Anal. Chem.*, 50(7):997-1000 (1978).

28. Schwarz, F.P. "Determination of Temperature Dependence of Solubilities of Polycyclic Aromatic Hydrocarbons in Aqueous

Solutions by a Fluorescence Method," *J. Chem. Eng. Data*, 22(3):273-277 (1977).

29. Boehm, P.D., and J.G. Quinn. "Solubilization of Hydrocarbons by the Dissolved Organic Matter in Sea Water," *Geochim. Cosmochim. Acta*, 37(11):2459-2477 (1973).

30. Sax, N.I. *Dangerous Properties of Industrial Materials* (New York: Van Nostrand Reinhold Co., 1984), 3124 p.

31. Chapman, P.J. "An Outline of Reaction Sequences Used for the Bacterial Degradation of Phenolic Compounds" in *Degradation of Synthetic Organic Molecules in the Biosphere: Natural, Pesticidal, and Various Other Man-Made Compounds* (Washington, DC: National Academy of Sciences, 1972), pp 17-55.

32. Kobayashi, H., and B.E. Rittman. "Microbial Removal of Hazardous Organic Compounds," *Environ. Sci. Technol.*, 16(3):170A-183A (1982).

33. Radding, S.B., Mill, T., Gould, C.W., Lia, D.H., Johnson, H.L., Bomberger, D.S., and C.V. Fojo. "The Environmental Fate of Selected Polynuclear Aromatic Hydrocarbons," Office of Toxic Substances, U.S. EPA Report-560/5-75-009 (1976), 122 p.

34. Lyman, W.J., Reehl, W.F., and D.H. Rosenblatt. *Handbook of Chemical Property Estimation Methods: Environmental Behavior of Organic Compounds* (New York: McGraw-Hill, Inc., 1982).

35. Osborn, A.G., and D.R. Douslin. "Vapor Pressures and Derived Enthalpies of Vaporization of Some Condensed-Ring Hydrocarbons," *J. Chem. Eng. Data*, 20(3):229-231 (1975).

36. Macknick, A.B., and J.M. Prausnitz. "Vapor Pressures of High Molecular Weight Hydrocarbons," *J. Chem. Eng. Data*, 24(3):175-178 (1979).

37. Bradley, R.S., and T.G. Cleasby. "The Vapour Pressure and Lattice Energy of Some Aromatic Ring Compounds," *J. Chem. Soc. (London)*, pp 1690-1692 (1953).

38. *Fire Protection Guide on Hazardous Materials* (Quincy, MA: National Fire Protection Association, 1984), 443 p.

39. "NIOSH Pocket Guide to Chemical Hazards," U.S. Department of Health and Human Services, U.S. Government Printing Office (1987), 241 p.

40. General Industry Standards for Toxic and Hazardous Substances," U.S. Code of Federal Regulations 1910, Subpart Z Section 1910.1000 (July 1982).

41. *Documentation of the Threshold Limit Values and Biological Exposure Indices* (Cincinnati, OH: American Conference of Governmental Industrial Hygienists, 1986), 744 p.

PHENOL

Synonyms: Baker's P and S liquid and ointment; Benzenol; Carbolic acid; Hydroxybenzene; Monohydroxybenzene; NA 2,821; NCI-C50124; Oxybenzene; Phenic acid; Phenyl hydrate; Phenyl hydroxide; Phenylic acid; Phenylic alcohol; RCRA waste number U188; UN 1671; UN 2312; UN 2821.

Structural Formula:

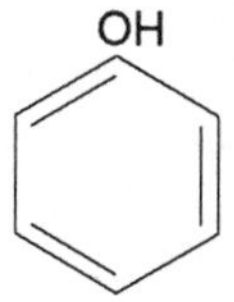

CHEMICAL DESIGNATIONS

CAS Registry Number: 108-95-2

DOT Designation: 1671 (solid); 2312 (molten); 2821 (solution).

Empirical Formula: C_6H_6O

Formula Weight: 94.11

RTECS Number: SJ 3325000

PHYSICAL AND CHEMICAL PROPERTIES

Appearance and Odor: White crystals or light pink liquid which slowly turns brown on exposure to air. Sweet tarry odor.

Boiling Point: 181.7 °C [1]; 183 [2].

Dissociation Constant: 9.99 [3].

Henry's Law Constant: 2.7 x 10^{-7} atm·m^3/mol [4]; 1.71 x 10^{-7} atm·m^3/mol [5]; 3.99 kPa/unit mol at 27.0 °C [6]; 3.97 x 10^{-7} atm·m^3/mol at 25 °C [7].

Ionization Potential: 8.51 eV [8]; 8.47 eV [9].

Log K_{oc}: 1.43 [5]; 1.24 [10].

Log K$_{ow}$: 1.48 [11]; 1.46 [12].

Melting Point: 43 °C [1]; 40.9 °C [13]; 40.5-41.5 °C [14].

Solubility in Organics: Soluble in ethanol, carbon disulfide, and chloroform; very soluble in ether; miscible with carbon tetrachloride and hot benzene [15].

Solubility in Water: 82,000 mg/L at 15 °C [16]; 93,000 mg/L at 25 °C [17]; 8.4 wt% at 20 °C [18]; 67,000 mg/L at 25 °C [19]; 8.2 wt% at 20 °C [20]; 15 parts at 16 °C [2]; 0.866 wt% at 25 °C [21]; 84.12 mg/ml [22]; 2.70 x 10^{-7} M at 12.2 °C, 3.39 x 10^{-7} M at 15.5 °C, 3.91 x 10^{-7} M at 17.4 °C, 4.57 x 10^{-7} M at 20.3 °C, 5.78 x 10^{-7} M at 23.0 °C, 5.82 x 10^{-7} M at 23.3 °C, 6.40 x 10^{-7} M at 25.0 °C, 7.13 x 10^{-7} M at 26.2 °C, 7.18 x 10^{-7} M at 26.7 °C, 8.09 x 10^{-7} M at 28.5 °C, 9.3 x 10^{-7} M at 31.3 °C [23]; 85,000 ppm at 25 °C [24]; 0.90 M at 25 °C [25]; 74 g/L at 8 °C, 82 g/L at 25 °C [26].

Specific Density: 1.0576 at 20/4 °C [1]; 1.058 at 41/4 °C [13]; 1.05760 at 41/4 °C [27]; 1.071 at 25/4 °C [28].

Transformation Products: In an aqueous, oxygenated solution exposed to artificial light (234 nm), phenol photolyzed to hydroquinone, catechol, 2,2'-, 2,4'-, and 4,4'-dihydroxy-biphenyl [29].

Vapor Pressure: 0.2 mm at 20 °C, 1 mm at 40 °C [16]; 1 mm at 40.1 °C, 10 mm at 73.8 °C, 40 mm at 100.1 °C, 100 mm at 121.4 °C, 400 mm at 160.0 °C, 760 mm at 181.9 °C [1]; 0.35 mm at 25 °C [30]; 0.064 mm at 8 °C, 0.34 mm at 25° [26].

FIRE HAZARDS

Flash Point: 79 °C [18].

Lower Explosive Limit (LEL): 1.7% [18].

Upper Explosive Limit (UEL): 8.6% [18].

HEALTH HAZARD DATA

Immediately Dangerous to Life or Health (IDLH): 250 ppm [18].

Permissible Exposure Limits (PEL) in Air: 5 ppm ($\approx$19 mg/m^3) [31]; 20 mg/m^3 ($\approx$5.2 ppm) 10-hour time weighted average, 60 mg/m^3 ($\approx$15.6 ppm) 15-minute ceiling [18].

MANUFACTURING

Selected Manufacturers:

Dow Chemical Co.
Midland, MI 48640

Allied Chemical Corp.
Plastics Division
Morristown, NJ 07960

Monsanto Industrial Chemicals Co.
800 North Lindbergh Blvd.
St. Louis, MO 63166

Uses: Antiseptic and disinfectant; pharmaceuticals; dyes; indicators; slimicide; phenolic resins; epoxy resins (bisphenol-A); nylon-6 (caprolactum); 2,4-D; solvent for refining lubricating oils; preparation of adipic acid, salicylic acid, phenolphthalein, pentachlorophenol, acetophenetidin, picric acid, and many other compounds; germicidal paints; pharmaceuticals; laboratory reagent; dyes and indicators.

REFERENCES

1. Weast, R.C., Ed. *CRC Handbook of Chemistry and Physics*, 67th ed. (Boca Raton, FL: CRC Press, Inc., 1986), 2406 p.
2. Huntress, E.H., and S.P. Mulliken. *Identification of Pure Organic Compounds - Tables of Data on Selected Compounds of Order I* (New York: John Wiley and Sons, Inc., 1941), 691 p.
3. Dean, J.A., Ed., *Lange's Handbook of Chemistry*, 11th ed. (New York: McGraw-Hill, Inc., 1973), 1570 p.
4. Petrasek, A.C., Kugelman, I.J., Austern, B.M., Pressley, T.A., Winslow, L.A., and R.H. Wise. "Fate of Toxic Organic Compounds in Wastewater Treatment Plants," *J. Water Poll. Control Fed.*, 55(10):1286-1296 (1983).
5. Jury, W.A., Spencer, W.F., and W.J. Farmer. "Behavior Assessment Model for Trace Organics in Soil: III. Application of Screening

Model," *J. Environ. Qual.*, 13(4):573-579 (1984).

6. Abd-El-Bary, M.F., Hamoda, M.F., Tanisho, S., and N. Wakao. "Henry's Constants for Phenol over its Diluted Aqueous Solution," *J. Chem. Eng. Data*, 31(2):229-230 (1986).

7. Hine, J., and P.K. Mookerjee. "The Intrinsic Hydrophilic Character of Organic Compounds. Correlations in Terms of Structural Contributions," *J. Org. Chem.*, 40(3):292-298 (1975).

8. Franklin, J.L., Dillard, J.G., Rosenstock, H.M., Herron, J.T., Draxl K., and F.H. Field. "Ionization Potentials, Appearance Potentials and Heats of Formation of Gaseous Positive Ions," National Bureau of Standards Report NSRDS-NBS 26, U.S. Government Printing Office (1969), 289 p.

9. Hazardous Substances Data Bank. Phenol, National Library of Medicine, Toxicology Information Program (1989).

10. Briggs, G.G. "Theoretical and Experimental Relationships Between Soil Adsorption, Octanol-Water Partition Coefficients, Water Solubilities, Bioconcentration Factors, and the Parachor," *J. Agric. Food Chem.*, 29(5):1050-1059 (1981).

11. Leo, A., Hansch, C., and D. Elkins. "Partition Coefficients and Their Uses," *Chem. Rev.*, 71(6):525-616 (1971).

12. Fujita, T., Iwasa, J., and C. Hansch. "A New Substituent Constant, π, Derived from Partition Coefficients," *J. Am. Chem. Soc.*, 86(23):5175-5180 (1964).

13. Weiss, G. *Hazardous Chemicals Data Book* (Park Ridge, NJ: Noyes Data Corp., 1986), 1069 p.

14. *Fluka Catalog 1988/89 - Chemika-Biochemika* (Ronkonkoma, NY: Fluka Chemical Corp., 1988), 1536 p.

15. "Chemical, Physical, and Biological Properties of Compounds Present at Hazardous Waste Sites," U.S. EPA Report-530/SW-89-010 (1985), 619 p.

16. Verschueren, K. *Handbook of Environmental Data on Organic Chemicals* (New York: Van Nostrand Reinhold Co., 1983), 1310 p.

17. Morrison, R.T., and R.N. Boyd. *Organic Chemistry* (Boston: Allyn and Bacon, Inc., 1971), 1258 p.

18. "NIOSH Pocket Guide to Chemical Hazards," U.S. Department of Health and Human Services, U.S. Government Printing Office (1987), 241 p.

19. Warner, H.P., Cohen, J.M., and J.C. Ireland. "Determination of Henry's Law Constants of Selected Priority Pollutants," Office of Science and Development, U.S. EPA Report-600/D-87/229 (1987), 14 p.

20. Stephen, H., and T. Stephen. *Solubilities of Inorganic and Organic Compounds - Part 1, Volume 1* (London: Pergamon Printing and Art

Services, Ltd., 1963), 960 p.

21. Riddick, J.A., Bunger, W.B., and T.K. Sakano. *Organic Solvents - Physical Properties and Methods of Purification. Volume II* (New York: John Wiley and Sons, Inc., 1986), 1325 p.

22. Reinbold, K.A., Hassett, J.J., Means, J.C., and W.L. Banwart. "Adsorption of Energy-Related Organic Pollutants: A Literature Review," Office of Research and Development, U.S. EPA Report-600/3-79-086 (1979), 180 p.

23. Schwarz, F.P. "Determination of Temperature Dependence of Solubilities of Polycyclic Aromatic Hydrocarbons in Aqueous Solutions by a Fluorescence Method," *J. Chem. Eng. Data*, 22(3):273-277 (1977).

24. Amoore, J.E., and E. Hautala. "Odor as an Aide to Chemical Safety: Odor Thresholds Compared with Threshold Limit Values and Volatilities for 214 Industrial Chemicals in Air and Water Dilution," *J. Appl. Toxicol.*, 3(6):272-290 (1983).

25. Caturla, F., Martin-Martinez, J.M., Molina-Sabio, M., Rodriguez-Reinoso, F., and R. Torregrosa. "Adsorption of Substituted Phenols on Activated Carbon," *J. Colloid Interface Sci.*, 124(2):528-534 (1988).

26. Leuenberger, C., Ligocki, M.P., and J.F. Pankow. "Trace Organic Compounds in Rain. 4. Identities, Concentrations, and Scavenging Mechanisms for Phenols in Urban Air and Rain," *Environ. Sci. Technol.*, 19(11):1053-1058 (1985).

27. Standen, A., Ed. *Kirk-Othmer Encyclopedia of Chemical Technology, Volume 15*, 2nd ed. (New York: John Wiley and Sons, Inc., 1968), 923 p.

28. *Criteria for a Recommended Standard. . .Occupational Exposure to Phenol* (Cincinnati, OH: National Institute for Occupational Safety and Health, 1976), 167 p.

29. Callahan, M.A., Slimak, M.W., Gable, N.W., May, I.P., Fowler, C.F., Freed, J.R., Jennings, P., Durfee, R.L., Whitmore, F.C., Maestri, B., Mabey, W.R., Holt, B.R., and C. Gould. "Water-Related Environmental Fate of 129 Priority Pollutants Volumes I and II," National Technical Information Service, U.S. EPA Report-440/4-79-029 (1979), 1160 p.

30. *Documentation of the Threshold Limit Values and Biological Exposure Indices* (Cincinnati, OH: American Conference of Governmental Industrial Hygienists, 1986), 744 p.

31. "General Industry Standards for Toxic and Hazardous Substances," U.S. Code of Federal Regulations 1910, Subpart Z Section 1910.1000 (July 1982).

PYRENE

Synonyms: Benzo[def]phenanthrene; β-Pyrene; β-Pyrine.

Structural Formula:

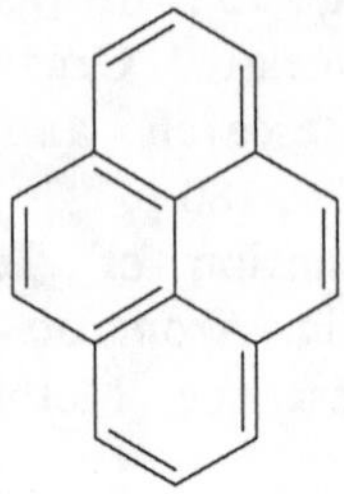

CHEMICAL DESIGNATIONS

CAS Registry Number: 129-00-0

DOT Designation: None assigned.

Empirical Formula: $C_{16}H_{10}$

Formula Weight: 202.26

RTECS Number: UR 2450000

PHYSICAL AND CHEMICAL PROPERTIES

Appearance: Colorless solid (tetracene impurities impart a yellow color). Solutions have a slight blue fluorescence.

Boiling Point: 393 °C [1]; 404 °C [2].

Henry's Law Constant: 1.09×10^{-5} atm·m^3/mol [3]; 1.87×10^{-5} atm·m^3/mol [4].

Ionization Potential: 7.70 ± 0.65 eV [5]; 7.58 eV [6].

Log K_{oc}: 4.66 [7]; 4.92 [8]; 4.88 [9]; 4.81 [10]; 4.80 [11]; 5.13 [12]; 4.67 [13].

Log K_{ow}: 4.88 [14]; 5.18 [15]; 5.32 [16]; 5.09 [10].

Melting Point: 156 °C [1]; 151 °C [17]; 111-114 °C [18]; 149 °C [19].

Solubility in Organics: Soluble in most solvents [20].

Solubility in Water: 0.16 mg/L at 26 °C, 0.032 mg/L at 24 °C (practical grade) [21]; 0.013 mg/L at 25 °C [22]; 0.135 mg/L at 25 °C [23]; 0.135 mg/L at 24 °C [10]; 0.132 mg/L at 25 °C, 0.162 mg/L at 29 °C [24]; 0.171 mg/L at 25 °C [25]; 0.148 mg/L at 25 °C [26]; 0.165 mg/L [27].

Specific Density: 1.271 at 23/4 °C [1].

Transformation Products: Adsorption onto garden soil for 10 days and subjected to ultraviolet radiation produced 1,1'-bipyrene, 1,6-pyrene-dione, 1,8-pyrenedione, and three unidentified compounds [21]. Microbial degradation by *Mycobacterium* sp. yielded the following ring-fission products: 4-phenanthroic acid, 4-hydroxyperinaphthenone, cinnamic acid, and phthalic acids. The compounds pyrenol and the *cis*- and *trans*-4,5-dihydrodiols of pyrene were identified as ring-oxidation products [28].

Vapor Pressure: 6.85×10^{-7} mm at 25 °C [29]; 2.5×10^{-6} mm at 25 °C [30]; 0.0000379 mm at 68.90 °C, 0.000108 mm at 71.75°, 0.000141 mm at 74.15 °C, 0.000167 mm at 75.85 °C, 0.000206 mm at 78.20 °C, 0.000216 mm at 78.90 °C, 0.000292 mm at 81.70 °C, 0.000307 mm at 82.65 °C, 0.000304 mm at 82.70 °C, 0.000379 mm at 85.00 °C, 0.000380 mm at 85.25 °C [31].

FIRE HAZARDS

Flash Point: No data found.

Lower Explosive Limit (LEL): No data found.

Upper Explosive Limit (UEL): No data found.

HEALTH HAZARD DATA

Immediately Dangerous to Life or Health (IDLH): Potential human carcinogen [32].

Permissible Exposure Limits (PEL) in Air: No individual standards have been set, however, as a constituent in coal tar pitch volatiles, the following exposure limits have been established: 0.2 mg/m^3 (benzene-soluble fraction) [33]; 0.1 mg/m^3 10-hour time weighted average (cyclohexane-extractable fraction) [32]; 0.2 mg/m^3 TWA (benzene solubles) [34].

MANUFACTURING

Selected Manufacturers:

Aldrich Chemical Co.
940 West Saint Paul Ave.
Milwaukee, WI 53233

Fluka Chemical Corp.
980 South Second St.
Ronkonkoma, NY 11779

Pfaltz & Bauer, Inc.
172 East Aurora St.
Waterbury, CT 06708

Use: Research chemical. Derived from industrial and experimental coal gasification operations where maximum concentrations detected in gas and coal tar streams were 9.2 mg/m^3 and 24 mg/m^3, respectively [34].

REFERENCES

1. Weast, R.C., Ed. *CRC Handbook of Chemistry and Physics*, 67th ed. (Boca Raton, FL: CRC Press, Inc., 1986), 2406 p.
2. Sax, N.I. *Dangerous Properties of Industrial Materials* (New York: Van Nostrand Reinhold Co., 1984), 3124 p.
3. Mackay, D., and W.Y. Shiu. "A Critical Review of Henry's Law Constants for Chemicals of Environmental Interest," *J. Phys. Chem. Ref. Data*, 10(4):1175-1199 (1981).
4. Southworth, G.R. "The Role Volatilization in Removing Polycyclic Aromatic Hydrocarbons from Aquatic Environments," *Bull. Environ. Contam. Toxicol.*, 21(4/5):507-514 (1979).
5. Franklin, J.L., Dillard, J.G., Rosenstock, H.M., Herron, J.T., Draxl K., and F.H. Field. "Ionization Potentials, Appearance Potentials and Heats of Formation of Gaseous Positive Ions," National Bureau of Standards Report NSRDS-NBS 26, U.S. Government Printing Office (1969), 289 p.
6. Hazardous Substances Data Bank. Pyrene, National Library of

Medicine, Toxicology Information Program (1989).

7. Abdul, S.A., Gibson, T.L., and D.N. Rai. "Statistical Correlations for Predicting the Partition Coefficient for Nonpolar Organic Contaminants Between Aquifer Organic Carbon and Water," *Haz. Waste Haz. Mater.*, 4(3):211-222 (1987).

8. Schwarzenbach, R.P., and J. Westall. "Transport of Nonpolar Organic Compounds from Surface Water to Groundwater. Laboratory Sorption Studies," *Environ. Sci. Technol.*, 15(11):1360-1367 (1981).

9. Chin, Y.-P., Peven, C.S., and W.J. Weber. "Estimating Soil/Sediment Partition Coefficients for Organic Compounds by High Performance Reverse Phase Liquid Chromatography," *Water Res.*, 22(7):873-881 (1988).

10. Means, J.C., Hassett, J.J., Wood, S.G., and W.L. Banwart. Sorption Properties of Energy-Related Pollutants and Sediments, in *Polynuclear Aromatic Hydrocarbons, Third International Symposium on Chemistry, Biology, Carcinogenesis and Mutagenesis*, Jones, P.W., and P. Leber, Eds. (Ann Arbor, MI: Ann Arbor Science Publishers, Inc., 1979), pp 327-340.

11. Hassett, J.J., Means, J.C., Banwart, W.L., and S.G. Wood. "Sorption Properties of Sediments and Energy-Related Pollutants," Office of Research and Development, U.S. EPA Report-600/3-80-041 (1980), 150 p.

12. Hodson, J., and N.A. Williams. "The Estimation of the Adsorption Coefficient (K_{oc}) for Soils by High Performance Liquid Chromatography," *Chemosphere*, 19(1):67-77 (1988).

13. Socha, S.B., and R. Carpenter. "Factors Affecting the Pore Water Hydrocarbon Concentrations in Puget Sound Sediments," *Geochim. Cosmochim. Acta*, 51(5):1273-1284 (1987).

14. Chou, J.T., and P.C. Jurs. "Computer Assisted Computation of Partition Coefficients from Molecular Structures using Fragment Constants," *J. Chem. Info. Comp. Sci.*, 19:172-178 (1979).

15. Karickhoff, S.W., Brown, D.S., and T.A. Scott. "Sorption of Hydrophobic Pollutants on Natural Sediments," *Water Res.*, 13:241-248 (1979).

16. Walton, W.C. *Practical Aspects of Ground Water Modeling* (Worthington, OH: National Water Well Association, 1985), 587 p.

17. Catalog Handbook of Fine Chemicals (Milwaukee, WI: Aldrich Chemical Co., 1988), 2212 p.

18. *Fluka Catalog 1988/89 - Chemika-Biochemika* (Ronkonkoma, NY: Fluka Chemical Corp., 1988), 1536 p.

19. Sims, R.C., Doucette, W.C., McLean, J.E., Grenney, W.J., and R.R. Dupont. "Treatment Potential for 56 EPA Listed Hazardous

Chemicals in Soil," U.S. EPA Report-600/6-88-001 (1988), 105 p.

20. "Chemical, Physical, and Biological Properties of Compounds Present at Hazardous Waste Sites," U.S. EPA Report-530/SW-89-010 (1985), 619 p.

21. Verschueren, K. *Handbook of Environmental Data on Organic Chemicals* (New York: Van Nostrand Reinhold Co., 1983), 1310 p.

22. Miller, M.M., Wasik, S.P., Huang, G.-L., Shiu, W.-Y., and D. Mackay. "Relationships Between Octanol-Water Partition Coefficient and Aqueous Solubility," *Environ. Sci. Technol.*, 19(6):522-529 (1985).

23. Mackay, D., and W.Y. Shiu. "Aqueous Solubility of Polynuclear Aromatic Hydrocarbons," *J. Chem. Eng. Data*, 22(4):399-402 (1977).

24. May, W.E., Wasik, S.P., and D.H. Freeman. "Determination of the Aqueous Solubility of Polynuclear Aromatic Hydrocarbons by a Coupled Column Liquid Chromatographic Technique," *Anal. Chem.*, 50(1):175-179 (1978).

25. Schwarz, F.P., and S.P. Wasik. "Fluorescence Measurements of Benzene, Naphthalene, Anthracene, Pyrene, Fluoranthene, and Benzo[e]pyrene in Water," *Anal. Chem.*, 48(3):524-528 (1976).

26. Wauchope, R.D., and F.W. Getzen. "Temperature Dependence of Solubilities in Water and Heats of Fusion of Solid Aromatic Hydrocarbons," *J. Chem. Eng. Data*, 17:38-41 (1972).

27. Davis, W.W., Krahl, M.E., and G.H.A. Clowes. "Solubility of Carcinogenic and Related Hydrocarbons in Water," *J. Am. Chem. Soc.*, 64(1):108-110 (1942).

28. Heitkamp, M.A., Freeman, J.P., Miller, D.W., and C.C. Cerniglia. "Pyrene Degradation by a Mycobacterium sp.: Identification of Ring Oxidation and Ring Fission Products," *Appl. Environ. Microbiol.*, 54(10):2556-2565 (1988).

29. Radding, S.B., Mill, T., Gould, C.W., Lia, D.H., Johnson, H.L., Bomberger, D.S., and C.V. Fojo. "The Environmental Fate of Selected Polynuclear Aromatic Hydrocarbons," Office of Toxic Substances, U.S. EPA Report-560/5-75-009 (1976), 122 p.

30. Mabey, W.R., Smith, J.H., Podoll, R.T., Johnson, H.L., Mill, T., Chou, T.-W., Gates, J., Partridge, I.W., Jaber, H., and D. Vandenberg. "Aquatic Fate Process Data for Organic Priority Pollutants - Final Report," Office of Regulations and Standards, U.S. EPA Report-440/4-81-014 (1982), 407 p.

31. Bradley, R.S., and T.G. Cleasby. "The Vapour Pressure and Lattice Energy of Some Aromatic Ring Compounds," *J. Chem. Soc. (London)*, pp 1690-1692 (1953).

32. "NIOSH Pocket Guide to Chemical Hazards," U.S. Department of Health and Human Services, U.S. Government Printing Office (1987), 241 p.

33. "General Industry Standards for Toxic and Hazardous Substances," U.S. Code of Federal Regulations 1910, Subpart Z Section 1910.1000 (July 1982).
34. *Documentation of the Threshold Limit Values and Biological Exposure Indices* (Cincinnati, OH: American Conference of Governmental Industrial Hygienists, 1986), 744 p.
35. Cleland, J.G. "Project Summary - Environmental Hazard Rankings of Pollutants Generated in Coal Gasification Processes," Office of Research and Development, U.S. EPA Report-600/S7-81-101 (1981), 19 p.

STYRENE

Synonyms: Cinnamene; Cinnamenol; Cinnamol; Diarex HF77; **Ethyenyl benzene**; NCI-C02200; Phenethylene; Phenylethene; Phenylethylene; Styrene monomer; Styrol; Styrolene; Styron; Styropol; Styropor; UN 2055; Vinyl benzene; Vinyl benzol.

Structural Formula:

CHEMICAL DESIGNATIONS

CAS Registry Number: 100-42-5

DOT Designation: 2055

Empirical Formula: C_8H_8

Formula Weight: 104.15

RTECS Number: WL 3675000

PHYSICAL AND CHEMICAL PROPERTIES

Appearance and Odor: Colorless to light yellow, oily liquid with a sweet, penetrating odor. (Polymerizes readily in the presence of heat, light or a peroxide catalyst. Polymerization is exothermic and may become explosive).

Boiling Point: 145.2 °C [1].

Henry's Law Constant: 0.00261 atm·m^3/mol (calculated) [2].

Ionization Potential: 8.47 eV [3].

Log K_{oc}: 2.87 using method of Chiou and others [4].

Log K_{ow}: 2.95 [5]; 3.16 [6].

Melting Point: -30.6 °C [1]; -33 °C [7].

Solubility in Organics: Soluble in acetone, ethanol, benzene, ether, carbon disulfide, and petroleum ether [1].

Solubility in Water: 280 mg/L at 15 °C, 300 mg/L at 20 °C, 400 mg/L at 40 °C [8]; 160 mg/L at 25 °C [6]; 300 mg/L at 25 °C [9]; 0.031 wt% at 25 °C [10]; 0.032% at 25 °C [11].

Specific Density: 0.9060 at 20/4 °C [1]; 0.9090 at 20/4 °C [7]; 0.9237 at 0/4 °C, 0.9148 at 10/4 °C, 0.9059 at 20/4 °C, 0.8970 at 30/4 °C, 0.8880 at 40/4 °C, 0.8702 at 60/4 °C [11].

Transformation Products: No data found.

Vapor Density: 4.26 g/L at 25 °C, 3.60 (air = 1).

Vapor Pressure: 5 mm at 20 °C, 9.5 mm at 30 °C [8]; 1 mm at -7.0 °C, 10 mm at 30.8 °C, 40 mm at 59.8 °C [1]; 4.3 mm at 15 °C [12]; 6.45 mm at 25 °C [13].

FIRE HAZARDS

Flash Point: 31 °C [3].

Lower Explosive Limit (LEL): 1.1% [3].

Upper Explosive Limit (UEL): 6.1% [3].

HEALTH HAZARD DATA

Immediately Dangerous to Life or Health (IDLH): 5,000 ppm [3].

Permissible Exposure Limits (PEL) in Air: 100 ppm, 200 ppm ceiling, 600 ppm 5 minute/3-hour peak [14]; 50 ppm 10-hour time weighted average, 100 ppm 15-minute ceiling [3]; 50 ppm ($\approx$215 mg/m^3) [12].

MANUFACTURING

Selected Manufacturers:

Dow Chemical Co.
Midland, MI 48640

Amoco Chemical Corp.
130 East Randolph Drive
Chicago, IL 60601

Monsanto Industrial Chemicals Co.
800 North Lindbergh Blvd.
St. Louis, MO 63166

Uses: Preparation of polystyrene, synthetic rubber, resins, protective coatings, and insulators.

REFERENCES

1. Weast, R.C., Ed. *CRC Handbook of Chemistry and Physics*, 67th ed. (Boca Raton, FL: CRC Press, Inc., 1986), 2406 p.
2. "Treatability Manual - Volume 1: Treatability Data," Office of Research and Development, U.S. EPA Report-600/8-80-042a (1980), 1035 p.
3. "NIOSH Pocket Guide to Chemical Hazards," U.S. Department of Health and Human Services, U.S. Government Printing Office (1987), 241 p.
4. Chiou, C.T., Peters, L.J., and V.H. Freed. "A Physical Concept of Soil-Water Equilibria for Nonionic Organic Compounds," *Science*, 206:831-832 (1979).
5. Chou, J.T., and P.C. Jurs. "Computer Assisted Computation of Partition Coefficients from Molecular Structures using Fragment Constants," *J. Chem. Info. Comp. Sci.*, 19:172-178 (1979).
6. Banerjee, S., Yalkowsky, S.H., and S.C. Valvani. "Water Solubility and Octanol/Water Partition Coefficients of Organics. Limitations of the Solubility-Partition Coefficient Correlation," *Environ. Sci. Technol.*, 14(10):1227-1229 (1980).
7. Huntress, E.H., and S.P. Mulliken. *Identification of Pure Organic Compounds - Tables of Data on Selected Compounds of Order I* (New York: John Wiley and Sons, Inc., 1941), 691 p.
8. Verschueren, K. *Handbook of Environmental Data on Organic Chemicals* (New York: Van Nostrand Reinhold Co., 1983), 1310 p.
9. Andrews, L.J., and R.M. Keefer. "Cation Complexes of Compounds Containing Carbon-Carbon Double Bonds. VII. Further Studies on the Argentation of Substituted Benzenes," *J. Am. Chem. Soc.*, 72(11):5034-5037 (1950).
10. Riddick, J.A., Bunger, W.B., and T.K. Sakano. *Organic Solvents - Physical Properties and Methods of Purification. Volume II.* (New

York: John Wiley and Sons, Inc., 1986), 1325 p.

11. Standen, A., Ed. *Kirk-Othmer Encyclopedia of Chemical Technology, Volume 19*, 2nd ed. (New York: John Wiley and Sons, Inc., 1969), 839 p.

12. *Documentation of the Threshold Limit Values and Biological Exposure Indices* (Cincinnati, OH: American Conference of Governmental Industrial Hygienists, 1986), 744 p.

13. *Environmental Health Criteria 26: Styrene* (Geneva: World Health Organization, 1983), 123 p.

14. "General Industry Standards for Toxic and Hazardous Substances," U.S. Code of Federal Regulations 1910, Subpart Z Section 1910.1000 (July 1982).

TCDD

Synonyms: Dioxin; Dioxin (herbicide contaminant); Dioxine; NCI-C03714; TCDBD; 2,3,7,8-TCDD; 2,3,7,8-Tetrachlorodibenzodioxin; 2,3,7,8-Tetrachlorodibenzo[*b,e*][1,4]dioxan; **2,3,7,8-Tetrachlorodibenzo[*b,e*][1,4]dioxin;** 2,3,7,8-Tetrachlorodibenzo-1,4-dioxin; 2,3,7,8-Tetrachlorodibenzo-*p*-dioxin; Tetradioxin.

Structural Formula:

CHEMICAL DESIGNATIONS

CAS Registry Number: 1746-01-6

DOT Designation: None assigned.

Empirical Formula: $C_{12}H_4Cl_4O_2$

Formula Weight: 321.98

RTECS Number: HP 3500000

PHYSICAL AND CHEMICAL PROPERTIES

Appearance: Colorless needles.

Boiling Point: 412.2 °C (estimated) [1]; begins to decompose at 500 °C [2].

Henry's Law Constant: 5.40 x 10^{-23} atm·m³/mol at 18-22 °C (calculated).

Ionization Potential: No data found.

Log K_{oc}: 6.66 [3].

Log K_{ow}: 6.15 [4]; 6.64 [5]; 6.20 [6]; 7.02 [7]; 5.38 [8].

Melting Point: 295 °C [9]; 305 °C [1]; 303-305 °C [10].

Solubility in Organics: Soluble in fats, oils, and other nonpolar solvents [2].

Solubility in Water: 0.2 ppb [10]; 0.0193 ppb at 22 °C [11]; 0.32 ppb [8].

Specific Density: 1.827 (estimated) [1].

Transformation Products: No data found.

Vapor Pressure: 3.46×10^{-9} mm at 30.1 °C [12]; 2.7×10^{-10} mm at 15 °C, 6.4×10^{-10} mm at 20 °C, 1.4×10^{-9} mm at 25 °C, 3.5×10^{-9} mm at 30 °C, 1.63×10^{-8} mm at 40 °C [8]; 7.2×10^{-10} mm at 25 °C [13].

FIRE HAZARDS

Flash Point: No data found.

Lower Explosive Limit (LEL): No data found.

Upper Explosive Limit (UEL): No data found.

HEALTH HAZARD DATA

Immediately Dangerous to Life or Health (IDLH): No data found.

Permissible Exposure Limits (PEL) in Air: No standards set.

MANUFACTURING

Uses: Occurs as an impurity in the herbicide 2,4,5-T (2,4,5-trichlorophenoxyacetic acid).

REFERENCES

1. Schroy, J.M., Hileman, F.D., and S.C. Cheng. "Physical/Chemical Properties of 2,3,7,8-TCDD," *Chemosphere*, 14:877-880 (1985).
2. "Chemical, Physical, and Biological Properties of Compounds

Present at Hazardous Waste Sites," U.S. EPA Report-530/SW-89-010 (1985), 619 p.

3. Walters, R.W., Ostazeski, S.A., and A. Guiseppi-Elie. "Sorption of 2,3,7,8-Tetrachlorodibenzo-*p*-dioxin from Water by Surface Soils," *Environ. Sci. Technol.,* 23(4):480-484 (1989).

4. Travis, C.C., and A.D. Arms. "Bioconcentration of Organics in Beef, Milk and Vegetation," *Environ. Sci. Technol.,* 22(3):271-274 (1988).

5. Marple, L., Berridge, B., and L. Throop. "Measurement of the Water-Octanol Partition Coefficient of 2,3,7,8-Tetrachlorodibenzo-*p*-dioxin," *Environ. Sci. Technol.,* 20(4):397-399 (1986).

6. Doucette, W.J., and A.W. Andren. "Estimation of Octanol/Water Partition Coefficients: Evaluation of Six Methods for Highly Hydrophobic Aromatic Hydrocarbons," *Chemosphere,* 17(2):345-359 (1988).

7. Burkhard, L.P., and D.W. Kuehl. "*n*-Octanol/Water Partition Coefficients by Reverse Phase Liquid Chromatography/Mass Spectrometry for eight Tetrachlorinated Planar Molecules," *Chemosphere,* 15(2):163-167 (1986).

8. Rappe, C., Choudhary, G., and L.H. Keith. *Chlorinated Dioxins and Dibenzofurans in Perspective* (Chelsea, MI: Lewis Publishers, Inc., 1987), 570 p.

9. Weast, R.C., Ed. *CRC Handbook of Chemistry and Physics,* 67th ed. (Boca Raton, FL: CRC Press, Inc., 1986), 2406 p.

10. Crummett, W.B., and R.H. Stehl. "Determination of Chlorinated Dibenzo-*p*-dioxins and Dibenzofurans in Various Materials," *Environ. Health Perspect.,* (September 1973), pp 15-25.

11. Marple, L., Brunck, R., and L. Throop. "Water Solubility of 2,3,7,8-Tetrachlorodibenzo-*p*-dioxin," *Environ. Sci. Technol.,* 20(2):180-182 (1986).

12. Rondorf, B. "Thermal Properties of Dioxins, Furans and Related Compounds," *Chemosphere,* 15:1325-1332 (1986).

13. Podoll, R.T., Jaber, H.M., and T. Mill. "Tetrachlorodioxin: Rates of Volatilization and Photolysis in the Environment," *Environ. Sci. Technol.,* 20(5):490-492 (1986).

1,1,2,2-TETRACHLOROETHANE

Synonyms: Acetosol; Acetylene tetrachloride; Bonoform; Cellon; 1,1-Dichloro-2,2-dichloroethane; Ethane tetrachloride; NCI-C03554; RCRA waste number U208; TCE; Tetrachlorethane; Tetrachloroethane; *sym*-Tetrachloroethane; UN 1702; Westron.

Structural Formula:

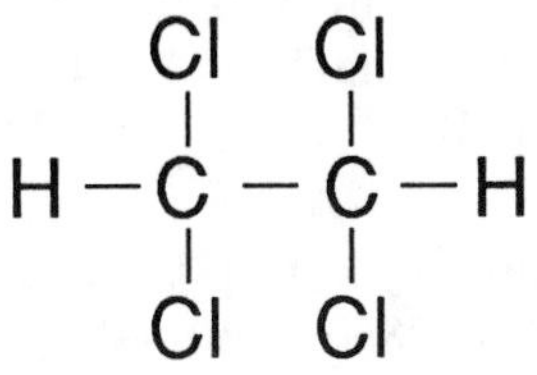

$$\begin{array}{ccc} Cl & & Cl \\ | & & | \\ H-C & - & C-H \\ | & & | \\ Cl & & Cl \end{array}$$

CHEMICAL DESIGNATIONS

CAS Registry Number: 79-34-5

DOT Designation: 1702

Empirical Formula: $C_2H_2Cl_4$

Formula Weight: 167.85

RTECS Number: KI 8575000

PHYSICAL AND CHEMICAL PROPERTIES

Appearance and Odor: Colorless to pale yellow liquid with a sweet chloroform-like odor.

Boiling Point: 146.2 °C [1].

Henry's Law Constant: 3.8 x 10^{-4} atm·m^3/mol [2]; 4.56 x 10^{-4} atm·m^3/mol at 25 °C [3]; 7.1 x 10^{-4} atm·m^3/mol at 37 °C [4].

Ionization Potential: 11.1 eV [5].

Log K$_{oc}$: 2.07 [6]; 1.663 [7].

Log K$_{ow}$: 2.56 [8]; 2.39 [9].

Melting Point: -36 °C [1]]; -42.5 °C [10]; -43 °C [11].

Solubility in Organics: Soluble in acetone, ethanol, ether, benzene, carbon tetrachloride, petroleum ether, chloroform, carbon disulfide, dimethylformamide, and oils [12].

Solubility in Water: 2,900 mg/L at 20 °C [13]; 2,970 mg/L at 25 °C [14]; 0.287 wt% at 25 °C, 0.335 wt% at 55.6 °C [15]; 0.287 wt% at 20 °C [16]; 3,230 ppm at 20 °C [7]; 2,900 mg/L at 25 °C [10]; 0.29 mass% at 20 °C [17]; 0.296 wt% at 23.5 °C [18].

Specific Density: 1.5953 at 20/4 °C [1]; 1.62640 at 0/4 °C, 1.60255 at 15/4 °C, 1.5869 at 25/4 °C, 1.57860 at 30/4 °C [10].

Transformation Product: Monodechlorination by microbes under laboratory conditions produced 1,1,2-trichloroethane [19].

Vapor Density: 5 kg/m^3 at the boiling point [17]; 6.86 g/L at 25 °C, 5.79 (air = 1).

Vapor Pressure: 5 mm at 20 °C, 8.5 mm at 30 °C [13]; 1 mm at -3.8 °C, 10 mm at 33.0 °C, 40 mm at 60.8 °C, 100 mm at 83.2 °C, 400 mm at 124.0 °C, 760 mm at 145.9 °C [1]; 8 mm at 20 °C [5]; 6 mm at 25 °C [20]; 19.5 mm at 45 °C, 62 mm at 70 °C, 140 mm at 91 °C, 220 mm at 104 °C, 350 mm at 118 °C, 470 mm at 128 °C, 620 mm at 138 °C, 760 mm at 146.3 °C [10].

FIRE HAZARDS

Flash Point: Not flammable [5].

Lower Explosive Limit (LEL): Not flammable [21].

Upper Explosive Limit (UEL): Not flammable [21].

HEALTH HAZARD DATA

Immediately Dangerous to Life or Health (IDLH): Potential human carcinogen [5].

Permissible Exposure Limits (PEL) in Air: 5 ppm ($\approx$35 mg/m^3) [22]; lowest detectable limit [5]; 1 ppm ($\approx$7 mg/m^3) [20].

MANUFACTURING

Selected Manufacturers:

Aldrich Chemical Co.
940 West Saint Paul Ave.
Milwaukee, WI 53233

Eastman Organic Chemicals
Rochester, NY 14650

Hooker Chemical Co.
Industrial Chemicals Division
Niagara Falls, NY 14302

Uses: Solvent for chlorinated rubber; insecticide and bleach manufacturing; paint, varnish, and rust remover manufacturing; degreasing, cleansing, and drying of metals; denaturant for ethyl alcohol; preparation of 1,1-dichloroethylene; extractant and solvent for oils and fats; insecticides; weed killer; fumigant; intermediate in the manufacturing of other chlorinated hydrocarbons; herbicide.

REFERENCES

1. Weast, R.C., Ed. *CRC Handbook of Chemistry and Physics*, 67th ed. (Boca Raton, FL: CRC Press, Inc., 1986), 2406 p.
2. Pankow, J.F., and M.E. Rosen. "Determination of Volatile Compounds in Water by Purging Directly to a Capillary Column with Whole Column Cryotrapping," *Environ. Sci. Technol.*, 22(4):398-405 (1988).
3. Hine, J., and P.K. Mookerjee. "The Intrinsic Hydrophilic Character of Organic Compounds. Correlations in Terms of Structural Contributions," *J. Org. Chem.*, 40(3):292-298 (1975).
4. Sato, A., and T. Nakajima. "A Structure-Activity Relationship of Some Chlorinated Hydrocarbons," *Arch. Environ. Health*, 34(2):69-75 (1979).
5. "NIOSH Pocket Guide to Chemical Hazards," U.S. Department of Health and Human Services, U.S. Government Printing Office (1987), 241 p.
6. Schwille, F. *Dense Chlorinated Solvents* (Chelsea, MI: Lewis Publishers, Inc., 1988), 146 p.
7. Chiou, C.T., Peters, L.J., and V.H. Freed. "A Physical Concept of

Soil-Water Equilibria for Nonionic Organic Compounds," *Science*, 206:831-832 (1979).

8. Mills, W.B., Porcella, D.B., Ungs, M.J., Gherini, S.A., Summers, K.V., Mok, L., Rupp, G.L., and G.L. Bowie. "Water Quality Assessment: A Screening Procedure for Toxic and Conventional Pollutants in Surface and Groundwater-Part I," Office of Research and Development, U.S. EPA Report-600/6-85-002a (1985), 638 p.

9. Yoshida, K., Shigeoka, T., and F. Yamauchi. "Non-Steady State Equilibrium Model for the Preliminary Prediction of the Fate of Chemicals in the Environment," *Ecotoxicol. Environ. Safety*, 7(2):179-190 (1983).

10. Standen, A., Ed. *Kirk-Othmer Encyclopedia of Chemical Technology, Volume 4*, 2nd ed. (New York: John Wiley and Sons, Inc., 1964), 937 p.

11. *Catalog Handbook of Fine Chemicals* (Milwaukee, WI: Aldrich Chemical Co., 1988), 2212 p.

12. "Chemical, Physical, and Biological Properties of Compounds Present at Hazardous Waste Sites," U.S. EPA Report-530/SW-89-010 (1985), 619 p.

13. Verschueren, K. *Handbook of Environmental Data on Organic Chemicals* (New York: Van Nostrand Reinhold Co., 1983), 1310 p.

14. Banerjee, S., Yalkowsky, S.H., and S.C. Valvani. "Water Solubility and Octanol/Water Partition Coefficients of Organics. Limitations of the Solubility-Partition Coefficient Correlation," *Environ. Sci. Technol.*, 14(10):1227-1229 (1980).

15. Stephen, H., and T. Stephen. *Solubilities of Inorganic and Organic Compounds - Part 1, Volume 1* (London: Pergamon Printing and Art Services, Ltd., 1963), 960 p.

16. Riddick, J.A., Bunger, W.B., and T.K. Sakano. *Organic Solvents - Physical Properties and Methods of Purification. Volume II* (New York: John Wiley and Sons, Inc., 1986), 1325 p.

17. Konietzko, H. "Chlorinated Ethanes: Sources, Distribution, Environmental Impact, and Health Effects," in *Hazard Assessment of Chemicals, Volume 3*, J. Saxena, Ed. (New York: Academic Press, Inc., 1984), pp 401-448.

18. Schwarz, F.P. "Measurement of the Solubilities of Slightly Soluble Organic Liquids in Water by Elution Chromatography," *Anal. Chem.*, 52(1):10-15 (1980).

19. Smith, L.R., and J. Dragun. "Degradation of Volatile Chlorinated Aliphatic Priority Pollutants in Groundwater," *Environ. Int.*, 19(4):291-298 (1984).

20. *Documentation of the Threshold Limit Values and Biological Exposure Indices* (Cincinnati, OH: American Conference of Governmental

Industrial Hygienists, 1986), 744 p.
21. Weiss, G. *Hazardous Chemicals Data Book* (Park Ridge, NJ: Noyes Data Corp., 1986), 1069 p.
22. "General Industry Standards for Toxic and Hazardous Substances," U.S. Code of Federal Regulations 1910, Subpart Z Section 1910.1000 (July 1982).

TETRACHLOROETHYLENE

Synonyms: Ankilostin; Antisol 1; Carbon bichloride; Carbon dichloride; Dee-Solv; Didakene; Dow-per; ENT 1,860; Ethylene tetrachloride; Fedal-UN; NCI-C04580; Nema; PCE; PER; Perawin; PERC; Perchlor; Perchlorethylene; Perchloroethylene; Perclene; Perclene D; Percosolv; Perk; Perklone; Persec; RCRA waste number U210; Tetlen; Tetracap; Tetrachlorethylene; **Tetrachloroethene;** 1,1,2,2-Tetrachloroethylene; Tetraleno; Tetralex; Tetravec; Tetroguer; Tetropil; UN 1897.

Structural Formula:

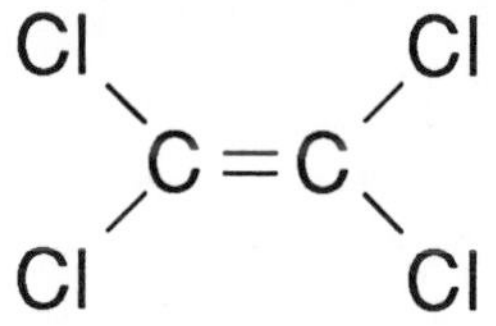

$$\underset{\underset{\displaystyle Cl}{\diagdown}}{\overset{\overset{\displaystyle Cl}{\diagup}}{C}} = \underset{\underset{\displaystyle Cl}{\diagdown}}{\overset{\overset{\displaystyle Cl}{\diagup}}{C}}$$

CHEMICAL DESIGNATIONS

CAS Registry Number: 127-18-4

DOT Designation: 1897

Empirical Formula: C_2Cl_4

Formula Weight: 165.83

RTECS Number: KX 3850000

PHYSICAL AND CHEMICAL PROPERTIES

Appearance and Odor: Colorless liquid with a chloroform or sweet ethereal odor.

Boiling Point: 121.2 °C [1].

Henry's Law Constant: 0.0153 atm·m^3/mol [2]; 0.0131 atm·m^3/mol [3]; 0.00287 atm·m^3/mol [4]; 0.0592 atm·m^3/mol at 37 °C [5].

Ionization Potential: 9.32 eV [6]; 9.71 eV [7].

Log K$_{oc}$: 2.42 [8]; 2.56 [3]; 2.322 [9].

Log K$_{ow}$: 2.60 [10]; 2.88 [11]; 2.10 [12]; 2.53 [13].

Melting Point: -19 °C [14]; -22.4 °C [15].

Solubility in Organics: Soluble in ethanol, benzene, ether [16], and oils [17].

Solubility in Water: 150 mg/L at 25 °C [10]; 150 mg/L at 20 °C [18]; 2,200 mg/L at 20 °C [19]; 400 mg/L at 25 °C [20]; 485 ppm at 25 °C [12]; 149 mg/L at 20 °C [21].

Specific Density: 1.6227 at 20/4 °C [14]; 1.63120 at 10/4 °C, 1.63109 at 15/4 °C, 1.62260 at 20/4 °C, 1.60640 at 30/4 °C, 1.44865 at 120/4 °C [22].

Transformation Products: Sequential dehalogenation by microbes under laboratory conditions produced trichloroethylene, *cis*-1,2-dichloroethylene, *trans*-1,2-dichloroethylene, and vinyl chloride [23].

Vapor Density: 6.78 g/L at 25 °C, 5.72 (air = 1).

Vapor Pressure: 1 mm at -20.6 °C, 5 mm at 2.4 °C, 10 mm at 13.8 °C, 20 mm at 26.3 °C, 41 mm at 40 °C, 67 mm at 50 °C, 104 mm at 60 °C, 155.3 mm at 70 °C, 226.0 mm at 80 °C, 319.2 mm at 90 °C, 438.5 mm at 100 °C, 591.6 mm at 110 °C, 760 mm at mm at 121.2 °C [22]; 24 mm at 30 °C [10]; 100 mm at 61.3 °C, 400 mm at 100.0 °C, 760 mm at 120.8 °C [14]; 19 mm at 25 °C [24]; 20 mm at 25 °C [25]; 15.8 mm at 22 °C [26]; 14 mm at 20 °C [27].

FIRE HAZARDS

Flash Point: Not flammable [6].

Lower Explosive Limit (LEL): Not flammable [15].

Upper Explosive Limit (UEL): Not flammable [15].

HEALTH HAZARD DATA

Immediately Dangerous to Life or Health (IDLH): 500 ppm (carcinogen) [15].

Permissible Exposure Limits (PEL) in Air: 100 ppm, 200 ppm ceiling, 300 ppm 5-minute/3-hour peak [28]; lowest feasible limit [6]; 50 ppm ($\approx$340 mg/m^3), 200 ppm ($\approx$1,340 mg/m^3) STEL [24].

MANUFACTURING

Selected Manufacturers:

> Dow Chemical Co.
> Midland, MI 48640

> PPG Industries, Inc.
> Industrial Chemicals Division
> Barberton, OH 44203

> Vulcan Materials Co.
> Chemicals Division
> Wichita, KS 67201

Uses: Dry cleaning fluid; degreasing and drying metals and other solids; solvent for waxes, greases, fats, oils, gums; manufacturing printing inks and paint removers; preparation of fluorocarbons and trichloroacetic acid; vermifuge; organic synthesis.

REFERENCES

1. Dean, J.A., Ed., *Lange's Handbook of Chemistry*, 11th ed. (New York: McGraw-Hill, Inc., 1973), 1570 p.
2. Pankow, J.F., and M.E. Rosen. "Determination of Volatile Compounds in Water by Purging Directly to a Capillary Column with Whole Column Cryotrapping," *Environ. Sci. Technol.*, 22(4):398-405 (1988).
3. Schwille, F. *Dense Chlorinated Solvents* (Chelsea, MI: Lewis Publishers, Inc., 1988), 146 p.
4. Warner, H.P., Cohen, J.M., and J.C. Ireland, "Determination of Henry's Law Constants of Selected Priority Pollutants," Office of Science and Development, U.S. EPA Report-600/D-87/229 (1987), 14 p.
5. Sato, A. and T. Nakajima. "A Structure-Activity Relationship of Some Chlorinated Hydrocarbons," *Arch. Environ. Health*, 34(2):69-75 (1979).

6. "NIOSH Pocket Guide to Chemical Hazards," U.S. Department of Health and Human Services, U.S. Government Printing Office (1987), 241 p.

7. Yoshida, K., Shigeoka, T., and F. Yamauchi. "Non-Steady State Equilibrium Model for the Preliminary Prediction of the Fate of Chemicals in the Environment," *Ecotoxicol. Environ. Safety*, 7(2):179-190 (1983).

8. Abdul, S.A., Gibson, T.L., and D.N. Rai. "Statistical Correlations for Predicting the Partition Coefficient for Nonpolar Organic Contaminants Between Aquifer Organic Carbon and Water," *Haz. Waste Haz. Mater.*, 4(3):211-222 (1987).

9. Chiou, C.T., Peters, L.J., and V.H. Freed. "A Physical Concept of Soil-Water Equilibria for Nonionic Organic Compounds," *Science*, 206:831-832 (1979).

10. Verschueren, K. *Handbook of Environmental Data on Organic Chemicals* (New York: Van Nostrand Reinhold Co., 1983), 1310 p.

11. Mabey, W.R., Smith, J.H., Podoll, R.T., Johnson, H.L., Mill, T., Chou, T.-W., Gates, J., Partridge, I.W., Jaber, H., and D. Vandenberg. "Aquatic Fate Process Data for Organic Priority Pollutants - Final Report," Office of Regulations and Standards, U.S. EPA Report-440/4-81-014 (1982), 407 p.

12. Banerjee, S., Yalkowsky, S.H., and S.C. Valvani. "Water Solubility and Octanol/Water Partition Coefficients of Organics. Limitations of the Solubility-Partition Coefficient Correlation," *Environ. Sci. Technol.*, 14(10):1227-1229 (1980).

13. Veith, G.D., Macek, K.J., Petrocelli, S.R., and J. Carroll. "An Evaluation of Using Partition Coefficients and Water Solubility to Estimate Bioconcentration Factors for Organic Chemicals in Fish," in *Aquatic Toxicology, ASTM STP 707*, Eaton, J.G., Parrish, P.R., and A.C. Hendricks, Eds. (Philadelphia, PA: American Society for Testing and Materials 1980), pp 116-129.

14. Weast, R.C., Ed. *CRC Handbook of Chemistry and Physics*, 67th ed. (Boca Raton, FL: CRC Press, Inc., 1986), 2406 p.

15. Weiss, G. *Hazardous Chemicals Data Book* (Park Ridge, NJ: Noyes Data Corp., 1986), 1069 p.

16. "Chemical, Physical, and Biological Properties of Compounds Present at Hazardous Waste Sites," U.S. EPA Report-530/SW-89-010 (1985), 619 p.

17. *Toxic and Hazardous Industrial Chemicals Safety Manual for Handling and Disposal with Toxicity and Hazard Data* (Tokyo, Japan: International Technical Information Institute, 1986), 700 p.

18. Pearson, C.R., and G. McConnell. "Chlorinated C_1 and C_2 Hydrocarbons in the Marine Environment," in *Proc. R. Soc. London*,

B189(1096):305-322 (1975).

19. Chiou, C.T., Freed, V.H., Schmedding, D.W., and R.L. Kohnert. "Partition Coefficients and Bioaccumulation of Selected Organic Chemicals," *Environ. Sci. Technol.*, 11(5):475-478 (1977).

20. Kenaga, E.E. *Environmental Dynamics of Pesticides* (New York: Plenum Press, 1975), 243 p.

21. Munz, C., and P.V. Roberts. "Air-Water Phase Equilibria of Volatile Organic Solutes," *J. Am. Water Works Assoc.*, 79(5):62-69 (1987).

22. Standen, A., Ed. *Kirk-Othmer Encyclopedia of Chemical Technology, Volume 4*, 2nd ed. (New York: John Wiley and Sons, Inc., 1964), 937 p.

23. Smith, L.R., and J. Dragun. "Degradation of Volatile Chlorinated Aliphatic Priority Pollutants in Groundwater," *Environ. Int.*, 19(4):291-298 (1984).

24. *Documentation of the Threshold Limit Values and Biological Exposure Indices* (Cincinnati, OH: American Conference of Governmental Industrial Hygienists, 1986), 744 p.

25. Valsaraj, K.T. "On the Physio-Chemical Aspects of Partitioning of Non-Polar Hydrophobic Organics at the Air-Water Interface," *Chemosphere*, 17(5):875-887 (1988).

26. Sax, N.I. *Dangerous Properties of Industrial Materials* (New York: Van Nostrand Reinhold Co., 1984), 3124 p.

27. McConnell, G., Ferguson, D.M., and C.R. Pearson. "Chlorinated Hydrocarbons and the Environment," *Endeavour*, 34(121):13-18 (1975).

28. "General Industry Standards for Toxic and Hazardous Substances," U.S. Code of Federal Regulations 1910, Subpart Z Section 1910.1000 (July 1982).

TOLUENE

Synonyms: Antisal 1a; Methacide; **Methylbenzene**; Methylbenzol; NCI-C07272; Phenylmethane; RCRA waste number U220; Toluol; Tolu-sol; UN 1294.

Structural Formula:

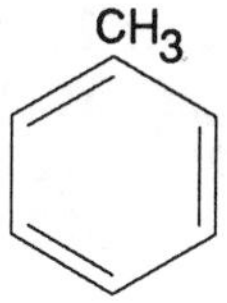

CHEMICAL DESIGNATIONS

CAS Registry Number: 108-88-3

DOT Designation: 1294

Empirical Formula: C_7H_8

Formula Weight: 92.14

RTECS Number: XS 5250000

PHYSICAL AND CHEMICAL PROPERTIES

Appearance and Odor: Colorless, water-white liquid with a pleasant odor similar to benzene.

Boiling Point: 110.6 °C [1].

Henry's Law Constant: 0.0067 atm·m^3/mol [2]; 0.00674 atm·m^3/mol at 25 °C [3].

Ionization Potential: 8.82 eV [4].

Log K_{oc}: 2.06 [5]; 2.18 [6].

Log K_{ow}: 2.65 [7]; 2.69 [8]; 2.21 [9]; 2.63 [10]; 2.50 [11]; 2.11, 2.73, 2.80 [12]; 2.79 [13].

Melting Point: -95 °C [1].

Solubility in Organics: Soluble in acetone, carbon disulfide, and ligroin; miscible with acetic acid, ethanol, benzene, ether, chloroform, and other solvents [14].

Solubility in Water: 515 mg/L at 20 °C [15]; 524 mg/L at 25 °C [16]; 0.05 wt% at 20 °C [4]; 490 mg/L at 25 °C, 0.07 vol% at 20 °C [17]; 535 mg/L at 25 °C [18]; 506.7 mg/L at 25 °C, 418.5 mg/L in natural seawater at 25 °C [19]; 515 at 25 °C [20]; 627 mg/L at 25 °C [21]; 530 mg/L at 25 °C [22]; 570 mg/L at 30 °C [23]; 724 mg/L at 0 °C [24]; 0.0368 wt% at 10 °C, 0.0492 wt% at 25 °C, 0.0344 vol% at 25 °C, 0.627 g/L at 25 °C, 0.057 wt% at 30 °C [25]; 724 ppm at 0 °C, 573 ppm at 25 °C [26]; 534.8 ppm at 25 °C, 379.3 ppm in artificial seawater at 25 °C [27]; 554.0 mg/L at 25 °C, 402.0 mg/L in artificial seawater at 25 °C [28]; 0.00628 M at 25 °C [7]; 580 ppm at 25.00 °C [29]; 220 mg/L in fresh water at 25 °C, 230 mg/L in salt water at 25 °C [30]; 0.0665 wt% at 23.5 °C [31]; 470 mg/L at 30 °C [32]; 0.00669 M at 25 °C [33].

Specific Density: 0.8669 at 20/4 °C [1]; 0.86697 at 20/4 °C, 0.86233 at 25/4 °C [34]; 0.8666 at 20/4 °C, 0.8573 at 30/4 °C, 0.8480 at 40/4 °C [35].

Transformation Products: A mutant of *Pseudomonas putida* oxidized toluene to (+)-*cis*-2,3-dihydroxy-1-methylcyclohexa-1,4-diene [36]. Other metabolites identified in the microbial degradation of toluene include *cis*-2,3-dihydroxy-2,3-dihydrotoluene, 3-methyl catechol, benzyl alcohol, benzaldehyde, benzoic acid, and catechol [15]. In a methanogenic aquifer material, toluene degraded completely to carbon dioxide [37].

Vapor Density: 3.77 g/L at 25 °C, 3.18 (air = 1).

Vapor Pressure: 22 mm at 20 °C, [15]; 1 mm at -26.7 °C, 10 mm at 6.4 °C, 40 mm at 31.8 °C, 100 mm at 51.9 °C, 400 mm at 89.5 °C, 760 mm at 110.6 °C [1]; 36.7 mm at 30 °C [38].

FIRE HAZARDS

Flash Point: 4.4 °C [4].

Lower Explosive Limit (LEL): 1.3% [4].

Upper Explosive Limit (UEL): 7.1% [4].

HEALTH HAZARD DATA

Immediately Dangerous to Life or Health (IDLH): 2,000 ppm [4].

Permissible Exposure Limits (PEL) in Air: 200 ppm, 300 ppm ceiling, 500 ppm 10-minute peak [39]; 100 ppm 10-hour time weighted average, 200 ppm 10-minute ceiling [4]; 100 ppm ($\approx$375 mg/m^3), 150 ppm ($\approx$560 mg/m^3) STEL [40].

MANUFACTURING

Selected Manufacturers:

Exxon Chemical Co.
Houston, TX 77001

Shell Chemical Co.
Petrochemicals Division
Houston, TX 77001

Sun Oil Co.
St. Davids, PA 19087

Uses: Manufacturing of caprolactum, saccharin, medicines, dyes, perfumes, benzoic acid, trinitrotoluene (TNT), and other benzene derivatives; solvent for paints and coatings, gums, resins, rubber, oils, and vinyl compounds; adhesive solvent in plastic toys and model airplanes; diluent and thinner for nitrocellulose lacquers; detergent manufacturing; aviation gasoline and high-octane blending stock; preparation of toluene diisocyanate for polyurethane resins.

REFERENCES

1. Weast, R.C., Ed. *CRC Handbook of Chemistry and Physics*, 67th ed. (Boca Raton, FL: CRC Press, Inc., 1986), 2406 p.
2. Pankow, J.F., and M.E. Rosen. "Determination of Volatile Compounds in Water by Purging Directly to a Capillary Column with Whole Column Cryotrapping," *Environ. Sci. Technol.*,

22(4):398-405 (1988).

3. Hine, J., and P.K. Mookerjee. "The Intrinsic Hydrophilic Character of Organic Compounds. Correlations in Terms of Structural Contributions," *J. Org. Chem.*, 40(3):292-298 (1975).

4. "NIOSH Pocket Guide to Chemical Hazards," U.S. Department of Health and Human Services, U.S. Government Printing Office (1987), 241 p.

5. Abdul, S.A., Gibson, T.L., and D.N. Rai. "Statistical Correlations for Predicting the Partition Coefficient for Nonpolar Organic Contaminants Between Aquifer Organic Carbon and Water," *Haz. Waste Haz. Mater.*, 4(3):211-222 (1987).

6. Garbarini, D.R., and L.W. Lion. "Influence of the Nature of Soil Organics on the Sorption of Toluene and Trichloroethylene," *Environ. Sci. Technol.*, 20(12):1263-1269 (1986).

7. Tewari, Y.B., Miller, M.M., Wasik, S.P., and D.E. Martire. "Aqueous Solubility and Octanol/Water Partition Coefficient of Organic Compounds at 25.0 °C," *J. Chem. Eng. Data*, 27(4):451-454 (1982).

8. Hansch, C., Quinlan, J.E., and G.L. Lawrence. "The Linear Free-Energy Relationship Between Partition Coefficients and the Aqueous Solubility of Organic Liquids," *J. Org. Chem.*, 33(1):347-350 (1968).

9. Veith, G.D., Macek, K.J., Petrocelli, S.R., and J. Carroll. "An Evaluation of Using Partition Coefficients and Water Solubility to Estimate Bioconcentration Factors for Organic Chemicals in Fish," in *Aquatic Toxicology, ASTM STP 707*, Eaton, J.G., Parrish, P.R., and A.C. Hendricks, Eds. (Philadelphia, PA: American Society for Testing and Materials, 1980), pp 116-129.

10. Yalkowsky, S.H., Valvani, S.C., and D. Mackay. "Estimation of the Aqueous Solubility of Some Aromatic Compounds," *Res. Rev.*, 85:43-55 (1983).

11. Walton, B.T., Anderson, T.A., Hendricks, M.S., and S.S. Talmage. "Physicochemical Properties as Predictors of Organic Chemical Effects on Soil Microbial Respiration," *Environ. Toxicol. Chem.*, 8(1):53-63 (1989).

12. Leo, A., Hansch, C., and D. Elkins. "Partition Coefficients and Their Uses," *Chem. Rev.*, 71(6):525-616 (1971).

13. Fujita, T., Iwasa, J., and C. Hansch. "A New Substituent Constant, π, Derived from Partition Coefficients," *J. Am. Chem. Soc.*, 86(23):5175-5180 (1964).

14. "Chemical, Physical, and Biological Properties of Compounds Present at Hazardous Waste Sites," U.S. EPA Report-530/SW-89-010 (1985), 619 p.

15. Verschueren, K. *Handbook of Environmental Data on Organic*

Chemicals (New York: Van Nostrand Reinhold Co., 1983), 1310 p.

16. Banerjee, S. "Solubility of Organic Mixtures in Water," *Environ. Sci. Technol.*, 18(8):587-591 (1984).

17. Meites, L., Ed. *Handbook of Analytical Chemistry*, 1st ed. (New York: McGraw-Hill, Inc., 1963), 1782 p.

18. Walton, W.C. *Practical Aspects of Ground Water Modeling* (Worthington, OH: National Water Well Association, 1985), 587 p.

19. Rossi, S.S., and W.H. Thomas. "Solubility Behavior of Three Aromatic Hydrocarbons in Distilled Water and Natural Seawater," *Environ. Sci. Technol.*, 15(6):715-716 (1981).

20. McAuliffe, C. "Solubility in Water of Paraffin, Cycloparaffin, Olefin, Acetylene, Cycloolefin, and Aromatic Compounds," *J. Phys. Chem.*, 70(4):1267-1275 (1966).

21. Bohon, R.L., and W.F. Claussen. "The Solubility of Aromatic Hydrocarbons in Water," *J. Am. Chem. Soc.*, 73(4):1571-1578 (1951).

22. Andrews, L.J., and R.M. Keefer. "Cation Complexes of Compounds Containing Carbon-Carbon Double Bonds. IV. The Argentation of Aromatic Hydrocarbons," *J. Am. Chem. Soc.*, 71(11):3644-3647 (1949).

23. Gross, P.M., and J.H. Saylor. "The Solubilities of Certain Slightly Soluble Organic Compounds in Water," *J. Am. Chem. Soc.*, 53(5):1744-1751 (1931).

24. Brookman, G.T., Flanagan, M., and J.O. Kebe. "Literature Survey: Hydrocarbon Solubilities and Attenuation Mechanisms," API Publication 4414, (Washington, DC: American Petroleum Institute, 1985), 101 p.

25. Stephen, H., and T. Stephen. *Solubilities of Inorganic and Organic Compounds - Part 1, Volume 1* (London: Pergamon Printing and Art Services, Ltd., 1963), 960 p.

26. Polak, J., and B.C.-Y. Lu. "Mutual Solubilities of Hydrocarbons and Water at 0 and 25 °C," *Can. J. Chem.*, 51(24):4018-4023 (1973).

27. Sutton, C., and J.A. Calder. "Solubility of Alkylbenzenes in Distilled Water and Seawater at 25 °C," *J. Chem. Eng. Data*, 20(3):320-322 (1975).

28. Price, L.C. "Aqueous Solubility of Petroleum as Applied to its Origin and Primary Migration," *Am. Assoc. Pet. Geol. Bull.*, 60(2):213-244 (1976).

29. Keely, D.F., Hoffpauir, M.A., and J.R. Meriwether. "Solubility of Aromatic Hydrocarbons in Water and Sodium Chloride Solutions of Different Ionic Strengths: Benzene and Toluene," *J. Chem. Eng. Data*, 33(2):87-89 (1988).

30. Krasnoshchekova, R.Y., and M. Gubergrits. "Solubility of *n*-Alkylbenzene in Fresh and Salt Waters," [Chemical Abstracts 83(16):136583p]: *Vodn. Resur.*, 2:170-173 (1975).

31. Schwarz, F.P. "Measurement of the Solubilities of Slightly Soluble Organic Liquids in Water by Elution Chromatography," *Anal. Chem.,* 52(1):10-15 (1980).

32. *Drinking Water and Health* (Washington, DC: National Academy of Sciences, 1980), 415 p.

33. Ben-Naim, A., and J. Wilf. "Solubilities and Hydrophobic Interactions in Aqueous Solutions of Monoalkylbenzene Molecules," *J. Phys. Chem.,* 84(6):583-586 (1980).

34. Huntress, E.H., and S.P. Mulliken. *Identification of Pure Organic Compounds - Tables of Data on Selected Compounds of Order I* (New York: John Wiley and Sons, Inc., 1941), 691 p.

35. Sumer, K.M., and A.R. Thompson. "Refraction, Dispersion, and Densities of Benzene, Toluene, and Xylene Mixtures," *J. Chem. Eng. Data,* 13(1):30-34 (1968).

36. Dagley, S. "Microbial Degradation of Stable Chemical Structures: General Features of Metabolic Pathways" in *Degradation of Synthetic Organic Molecules in the Biosphere: Natural, Pesticidal, and Various Other Man-Made Compounds* (Washington, DC: National Academy of Sciences, 1972), pp 1-16.

37. Wilson, B.H., Smith, G.B., and J.F. Rees. "Biotransformations of Selected Alkylbenzenes and Halogenated Aliphatic Hydrocarbons in Methanogenic Aquifer Material: A Microcosm Study," *Environ. Sci. Technol.,* 20(10):997-1002 (1986).

38. Sax, N.I. *Dangerous Properties of Industrial Materials* (New York: Van Nostrand Reinhold Co., 1984), 3124 p.

39. "General Industry Standards for Toxic and Hazardous Substances," U.S. Code of Federal Regulations 1910, Subpart Z Section 1910.1000 (July 1982).

40. *Documentation of the Threshold Limit Values and Biological Exposure Indices* (Cincinnati, OH: American Conference of Governmental Industrial Hygienists, 1986), 744 p.

TOXAPHENE

Synonyms: Agricide maggot killer (F); Alltex; Alltox; Attac 4-2; Attac 4-4; Attac 6; Attac 6-3; Attac 8; Camphechlor; Camphochlor; Camphoclor; Chem-phene M5055; Chlorinated camphene; Chloro-camphene; Clor chem T-590; Compound 3,956; Crestoxo; Crestoxo 90; ENT 9,735; Estonox; Fasco-terpene; Geniphene; Gy-phene; Hercules 3,956; Hercules toxaphene; Huilex; Kamfochlor; M 5,055; Melipax; Motox; NA 2,761; NCI-C00259; Octachlorocamphene; PCC; Penphene; Phenacide; Phenatox; Phenphane; Polychlorcamphene; Polychlorinated camphenes; Polychlorocamphene; RCRA waste number P123; Strobane-T; Strobane T-90; Synthetic 3956; Texadust; Toxakil; Toxon 63; Toxyphen; Vertac 90%; Vertac toxaphene 90.

Structural Formula:

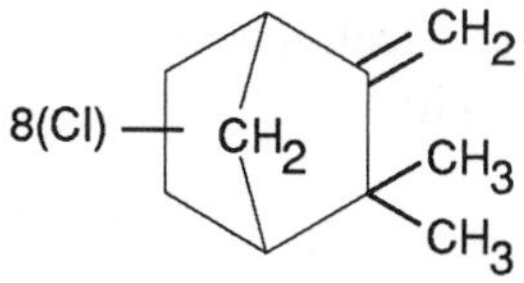

CHEMICAL DESIGNATIONS

CAS Registry Number: 8001-35-2

DOT Designation: 2761

Empirical Formula: $C_{10}H_{10}Cl_8$

Formula Weight: 413.82

RTECS Number: XW 5250000

PHYSICAL AND CHEMICAL PROPERTIES

Appearance and Odor: Yellow, waxy solid with a chlorine-like odor.

Boiling Point: Decomposes above 120 °C [1].

Henry's Law Constant: 0.063 atm·m^3/mol [2].

Ionization Potential: No data found.

Log K$_{oc}$: 3.18 using method of Kenaga and Goring [3].

Log K$_{ow}$: 3.30 [4]; 5.50 [5]; 3.23 [6].

Melting Point: 65-90 °C [7]; 85 °C [8].

Solubility in Organics: Very soluble in most solvents [9].

Solubility in Water: ≈3 ppm at 25 °C [10]; 0.74 ppm at 25 °C [11]; 3 x 10^{-4} wt% at 20 °C [12]; 1.75 mg/L at 25 °C [13]; 0.40 ppm at 20-25 °C [14].

Specific Density: 1.6 at 20/4 °C [15]; 1.519-1.567 at 25/25 °C [16]; 1.6 at 15/4 °C [17].

Transformation Products: Dehydrochlorination will occur after prolonged exposure to sunlight [18].

Vapor Pressure: 0.2-0.4 mm at 25 °C [10]; 3.3 x 10^{-5} mm at 20-25 °C [19]; 1 x 10^{-6} mm [8].

FIRE HAZARDS

Flash Point: 28.9 °C in solution [17]; 34-46 °C [16].

Lower Explosive Limit (LEL): 1.1% (in solvent) [17].

Upper Explosive Limit (UEL): 6.4 (in solvent) [17].

HEALTH HAZARD DATA

Immediately Dangerous to Life or Health (IDLH): 200 mg/m^3 [12].

Permissible Exposure Limits (PEL) in Air: 0.5 mg/m^3 [20]; 1 mg/m^3 STEL [21].

MANUFACTURING

Selected Manufacturers:
Hercules, Inc.
Synthetics Department
Brunswick, GA 31521

Sonford Chemical Co.
Pure-Atlantic Highway
Port Neches, TX 77651

Uses: Pesticide used primarily on cotton, lettuce, tomatoes, corn, peanuts, wheat, and soybean.

REFERENCES

1. "Treatability Manual - Volume 1: Treatability Data," Office of Research and Development, U.S. EPA Report-600/8-80-042a (1980), 1035 p.
2. Petrasek, A.C., Kugelman, I.J., Austern, B.M., Pressley, T.A., Winslow, L.A., and R.H. Wise. "Fate of Toxic Organic Compounds in Wastewater Treatment Plants," *J. Water Poll. Control Fed.*, 55(10):1286-1296 (1983).
3. Kenaga, E.E., and C.A.I. Goring. "Relationship Between Water Solubility, Soil Sorption, Octanol-Water Partitioning and Concentration of Chemicals in Biota," in *Aquatic Toxicology, ASTM STP 707*, Eaton, J.G., Parrish, P.R., and A.C. Hendricks, Eds. (Philadelphia, PA: American Society for Testing and Materials 1980), pp 78-115.
4. Paris, D.F., Lewis, D.L., and J.T. Barnett. "Bioconcentration of Toxaphene by Microorganisms," *Bull. Environ. Contam. Toxicol.*, 17(5):564-572 (1977).
5. Travis, C.C., and A.D. Arms. "Bioconcentration of Organics in Beef, Milk and Vegetation," *Environ. Sci. Technol.*, 22(3):271-274 (1988).
6. Rao, P.S.C., and J.M. Davidson. "Estimation of Pesticide Retention and Transformation Parameters Required in Nonpoint Source Pollution Models," in *Environmental Impact of Nonpoint Source Pollution*, Overcash, M.R., and J.M. Davidson, Eds. (Ann Arbor, MI: Ann Arbor Science Publishers, Inc., 1980), pp 23-67.
7. *IARC Monographs on the Evaluation of Carcinogenic Risk of Chemicals to Man. Some Halogenated Hydrocarbons, Volume 20* (Lyon, France: International Agency for Research on Cancer, 1979), 609 p.
8. Sims, R.C., Doucette, W.C., McLean, J.E., Grenney, W.J., and R.R. Dupont. "Treatment Potential for 56 EPA Listed Hazardous Chemicals in Soil," U.S. EPA Report-600/6-88-001 (1988), 105 p.
9. "Chemical, Physical, and Biological Properties of Compounds Present at Hazardous Waste Sites," U.S. EPA Report-530/SW-89-010 (1985), 619 p.

10. Brooks, G.T. *Chlorinated Insecticides, Volume I, Technology and Applications* (Cleveland, OH: CRC Press, 1974), 249 p.

11. Weil, L., Dure, G., and K.E. Quentin. "Solubility in Water of Insecticide Chlorinated Hydrocarbons and Polychlorinated Biphenyls in View of Water Pollution," *Z. Wasser Forsch.*, 7(6):169-175 (1974).

12. "NIOSH Pocket Guide to Chemical Hazards," U.S. Department of Health and Human Services, U.S. Government Printing Office (1987), 241 p.

13. Warner, H.P., Cohen, J.M., and J.C. Ireland. "Determination of Henry's Law Constants of Selected Priority Pollutants," Office of Science and Development, U.S. EPA Report-600/D-87/229 (1987), 14 p.

14. Weber, J.B. "Interaction of Organic Pesticides with Particulate Matter in Aquatic and Soil Systems," in *Fate of Organic Pesticides in the Aquatic Environment, Advances in Chemistry Series*, R.F. Gould, Ed. (Washington, D.C.: American Chemical Society, 1972), pp 55-120.

15. Melnikov, N.N. *Chemistry of Pesticides* (New York: Springer-Verlag, Inc., 1971), 480 p.

16. Berg, Gordon L., Ed., *The Farm Book* (Willoughby, OH: Meister Publishing Co., 1983), 440 p.

17. Weiss, G. *Hazardous Chemicals Data Book* (Park Ridge, NJ: Noyes Data Corp., 1986), 1069 p.

18. Hazardous Substances Data Bank. Toxaphene, National Library of Medicine, Toxicology Information Program (1989).

19. *Environmental Health Criteria 45: Camphechlor* (Geneva: World Health Organization, 1984), 66 p.

20. "General Industry Standards for Toxic and Hazardous Substances," U.S. Code of Federal Regulations 1910, Subpart Z Section 1910.1000 (July 1982).

21. *Documentation of the Threshold Limit Values and Biological Exposure Indices* (Cincinnati, OH: American Conference of Governmental Industrial Hygienists, 1986), 744 p.

1,2,4–TRICHLOROBENZENE

Synonyms: 1,2,4-TCB; *unsym*-Trichlorobenzene; UN 2321.

Structural Formula:

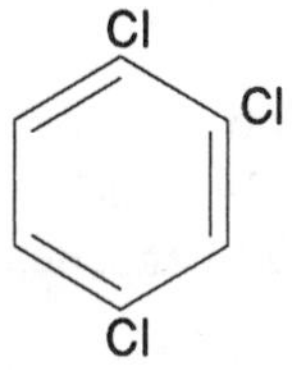

CHEMICAL DESIGNATIONS

CAS Registry Number: 120-82-1

DOT Designation: 2321

Empirical Formula: $C_6H_3Cl_3$

Formula Weight: 181.45

RTECS Number: DC 2100000

PHYSICAL AND CHEMICAL PROPERTIES

Appearance and Odor: Colorless liquid with an odor similar to *o*-dichlorobenzene.

Boiling Point: 213.5 °C [1]; 210 °C [2].

Henry's Law Constant: 0.00232 atm·m^3/mol [3]; 0.0012 atm·m^3/mol [4].

Ionization Potential: No data found.

Log K_{oc}: 2.70 [5]; 3.98 (soils), 4.61 (lacustrine sediments) [6]; 3.09, 3.16 [7].

Log K_{ow}: 4.02 [5]; 3.98 [8]; 4.23 [9]; 4.11 [10]; 3.93 [11]; 4.176 [12]; 4.12 [13].

Melting Point: 17 °C [1].

Solubility in Organics: Soluble in ether [1], other organic solvents, and oils [14].

Solubility in Water: 19 ppm at 22 °C [15]; 31.3 mg/L at 25 °C [16]; 48.8 mg/L at 25 °C [17]; 30 mg/L at 25 °C [18]; 34.57 mg/L at 25 °C [19]; 0.000254 M at 25 °C [20].

Specific Density: 1.4542 at 20/4 °C [1]; 1.4460 at 25/4 °C [2].

Transformation Products: Under aerobic conditions, biodegradation products may include 2,3-dichlorobenzene, 2,4-dichlorobenzene, 2,5-dichlorobenzene, 2,6-dichlorobenzene, and carbon dioxide [21].

Vapor Density: 7.42 g/L at 25 °C, 6.26 (air = 1).

Vapor Pressure: 1 mm at 38.4 °C, 10 mm at 81.7 °C, 40 mm at 114.8 °C, 100 mm at 140.0 °C, 400 mm at 187.7 °C, 760 mm at 213.0 °C [1]; 0.4 mm at 25 °C [17]; 0.29 mm at 25 °C [22].

FIRE HAZARDS

Flash Point: 105 °C [23].

Lower Explosive Limit (LEL): 2.5% at 125 °C [23].

Upper Explosive Limit (UEL): 6.6% at 150 °C [23].

HEALTH HAZARD DATA

Immediately Dangerous to Life or Health (IDLH): No data found.

Permissible Exposure Limits (PEL) in Air: 5 ppm ($\approx$40 mg/m^3) [24].

MANUFACTURING

Selected Manufacturers:

Aldrich Chemical Co.
940 West Saint Paul Ave.
Milwaukee, WI 53233

Fluka Chemical Corp.
980 South Second St.
Ronkonkoma, NY 11779

Pfaltz & Bauer, Inc.
172 East Aurora St.
Waterbury, CT 06708

Uses: Solvent in chemical manufacturing; dyes and intermediates; dielectric fluid; synthetic transformer oils; lubricants; heat-transfer medium; insecticides.

REFERENCES

1. Weast, R.C., Ed. *CRC Handbook of Chemistry and Physics*, 67th ed. (Boca Raton, FL: CRC Press, Inc., 1986), 2406 p.
2. Standen, A., Ed. *Kirk-Othmer Encyclopedia of Chemical Technology, Volume 4*, 2nd ed. (New York: John Wiley and Sons, Inc., 1964), 937 p.
3. Valsaraj, K.T. "On the Physio-Chemical Aspects of Partitioning of Non-Polar Hydrophobic Organics at the Air-Water Interface," *Chemosphere*, 17(5):875-887 (1988).
4. Oliver, B.G. "Desorption of Chlorinated Hydrocarbons from Spiked and Anthropogenically Contaminated Sediments," *Chemosphere*, 14(8):1087-1106 (1985).
5. Abdul, S.A., Gibson, T.L., and D.N. Rai. "Statistical Correlations for Predicting the Partition Coefficient for Nonpolar Organic Contaminants Between Aquifer Organic Carbon and Water," *Haz. Waste Haz. Mater.*, 4(3):211-222 (1987).
6. Chin, Y.-P., Peven, C.S., and W.J. Weber. "Estimating Soil/Sediment Partition Coefficients for Organic Compounds by High Performance Reverse Phase Liquid Chromatography," *Water Res.*, 22(7):873-881 (1988).
7. Banerjee, P., Piwoni, M.D., and K. Ebeid. "Sorption of Organic Contaminants to a Low Carbon Subsurface Core," *Chemosphere*, 14(8):1057-1067 (1985).
8. Chin, Y.-P., Weber, W.J. Jr., and T.C. Voice. "Determination of Partition Coefficients and Aqueous Solubilities by Reverse Phase Chromatography - II. Evaluation of Partitioning and Solubility Models," *Water Res.*, 20(11):1443-1451 (1986).
9. Mackay, D. "Correlation of Bioconcentration Factors," *Environ. Sci. Technol.*, 16(5):274-278 (1982).
10. Hawker, D.W., and D.W. Connell. "Influence of Partition Coefficient of Lipophilic Compounds on Bioconcentration Kinetics with Fish," *Water Res.*, 22(6):701-707 (1988).
11. Könemann, H., Zelle, R., and F. Busser. "Determination of Log P_{oct}

Values of Chloro-Substituted Benzenes, Toluenes, and Anilines by High-Performance Liquid Chromatography on ODS-Silica," *J. Chromatogr.*, 178:559-565 (1979).

12. Kenaga, E.E., and C.A.I. Goring. "Relationship Between Water Solubility, Soil Sorption, Octanol-Water Partitioning and Concentration of Chemicals in Biota," in *Aquatic Toxicology, ASTM STP 707*, Eaton, J.G., Parrish, P.R., and A.C. Hendricks, Eds. (Philadelphia, PA: American Society for Testing and Materials 1980), pp 78-115.

13. Anliker, R., and P. Moser. "The Limits of Bioaccumulation of Organic Pigments in Fish: Their Relation to the Partition Coefficient and the Solubility in Water and Octanol," *Ecotoxicol. Environ. Safety*, 13(1):43-52 (1987).
Toxic and Hazardous Industrial Chemicals Safety Manual for Handling and Disposal with Toxicity and Hazard Data (Tokyo, Japan: International Technical Information Institute, 1986), 700 p.

15. Verschueren, K. *Handbook of Environmental Data on Organic Chemicals* (New York: Van Nostrand Reinhold Co., 1983), 1310 p.

16. Banerjee, S. "Solubility of Organic Mixtures in Water," *Environ. Sci. Technol.*, 18(8):587-591 (1984).

17. Neely, W.B., and G.E. Blau, Eds. *Environmental Exposure from Chemicals. Volume 1* (Boca Raton, FL: CRC Press, Inc. 1985), 245 p.

18. Walton, W.C. *Practical Aspects of Ground Water Modeling* (Worthington, OH: National Water Well Association, 1985), 587 p.

19. Yalkowsky, S.H., Orr, R.J., and S.C. Valvani. "Solubility and Partitioning. 3. The Solubility of Halobenzenes in Water," *Indust. Eng. Chem. Fund.*, 18(4):351-353 (1979).

20. Miller, M.M., Ghodbane, S., Wasik, S.P., Tewari, Y.B., and D.E. Martire. "Aqueous Solubilities, Octanol/Water Partition Coefficients, and Entropies of Melting of Chlorinated Benzenes and Biphenyls," *J. Chem. Eng. Data*, 29(2):184-190 (1984).

21. Kobayashi, H., and B.E. Rittman. "Microbial Removal of Hazardous Organic Compounds," *Environ. Sci. Technol.*, 16(3):170A-183A (1982).

22. Warner, H.P., Cohen, J.M., and J.C. Ireland. "Determination of Henry's Law Constants of Selected Priority Pollutants," Office of Science and Development, U.S. EPA Report-600/D-87/229 (1987), 14 p.

23. *Fire Protection Guide on Hazardous Materials* (Quincy, MA: National Fire Protection Association, 1984), 443 p.

24. *Documentation of the Threshold Limit Values and Biological Exposure Indices* (Cincinnati, OH: American Conference of Governmental Industrial Hygienists, 1986), 744 p.

1,1,1-TRICHLOROETHANE

Synonyms: Aerothene; Aerothene TT; Baltana; Chloroethene; Chloroethene NU; Chlorothane NU; Chlorothene; Chlorothene NU; Chlorothene VG; Chlorten; Genklene; Inhibisol; Methyl chloroform; Methyltrichloromethane; NCI-C04626; RCRA waste number U226; Solvent III; α-T; 1,1,1-TCA; 1,1,1-TCE; α-Trichloroethane; Tri-ethane; UN 2831.

Structural Formula:

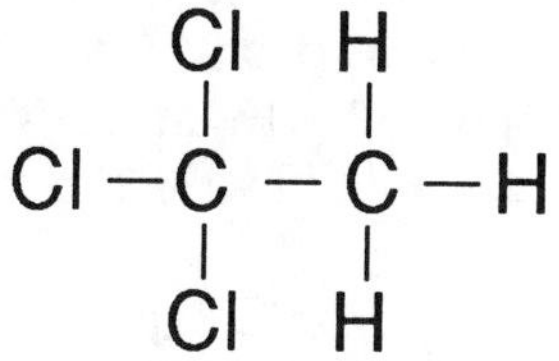

$$
\begin{array}{ccc}
Cl & & H \\
| & & | \\
Cl-C & -\ C & -H \\
| & & | \\
Cl & & H
\end{array}
$$

CHEMICAL DESIGNATIONS

CAS Registry Number: 71-55-6

DOT Designation: 2831

Empirical Formula: $C_2H_3Cl_3$

Formula Weight: 133.40

RTECS Number: KJ 2975000

PHYSICAL AND CHEMICAL PROPERTIES

Appearance and Odor: Colorless, watery liquid with an odor similar to chloroform. Readily corrodes aluminum and aluminum alloys.

Boiling Point: 74.1 °C [1].

Henry's Law Constant: 0.018 atm·m^3/mol [2]; 0.013 atm·m^3/mol [3]; 0.0162 atm·m^3/mol at 25 °C [4]; 0.015 atm·m^3/mol at 20 °C [5]; 0.0274 atm·m^3/mol at 37 °C [6].

Ionization Potential: No data found.

Log K_{oc}: 2.18 [3]; 2.017 [7].

516   1,1,1-Trichloroethane

Log K$_{ow}$: 2.18 [8]; 2.48 [9]; 2.49 [10]; 2.47 [11]; 2.17 [12].

Melting Point: -30.4 °C [1]; < -39 °C [13]; -32.62 °C [14].

Solubility in Organics: Sparingly soluble in ethanol; freely soluble in carbon disulfide and benzene. Also soluble in ether, methanol and carbon tetrachloride [15].

Solubility in Water: 4,400 mg/L at 20 °C [16]; 480 mg/L at 20 °C [17]; 300 mg/L at 25 °C [18]; 950 mg/L at 25 °C [2]; 0.07 wt% at 20 °C [19]; 720 mg/L at 25 °C [3]; 1,334 mg/L at 25 °C [9]; 730 mg/L at 20 °C [20]; 1,550 mg/L at 20 °C [10]; 0.132 wt% at 20 °C [21]; 1,360 ppm at 20 °C [7]; 0.44 mass% at 20 °C [22]; 0.1175 wt% 23.5 °C [23].

Specific Density: 1.3390 at 20/4 °C [1]; 1.31 at 20/4 °C [13]; 1.336 at 20/4 °C [24]; 1.37068 at 0/4 °C, 1.34587 at 15/4 °C, 1.3296 at 30/4 °C [14].

Transformation Products: Microbial degradation by sequential dehalogenation under laboratory conditions produced 1,1-dichloro-ethane, *cis*-1,2-dichloroethylene, *trans*-1,2-dichloroethylene, chloro-ethane, and vinyl chloride; Hydrolysis products included acetic acid and 1,1-dichloroethylene [25]. An anaerobic species of Clostridium biotransformed 1,1,1-trichloroethane yielding 1,1-dichloroethane, acetic acid, and unidentified products [26].

Vapor Density: 5.45 g/L at 25 °C, 4.60 (air = 1).

Vapor Pressure: 37 mm at 0 °C, 62 mm at 10 °C, 100 mm at 20 °C, 150 mm at 30 °C, 240 mm at 40 °C, 340 mm at 50 °C, 470 mm at 60 °C, 660 mm at 70 °C, 900 mm at 80 °C [14]; 1 mm at -52.0 °C, 10 mm at -21.9 °C, 40 mm at 1.6 °C, 100 mm at 20.0 °C, 400 mm at 54.6 °C, 760 mm at 74.1 °C [1]; 124 mm at 25 °C [9]; 96 mm at 20 °C [27].

FIRE HAZARDS

Flash Point: None [28]; ≤ 25 °C [29].

Lower Explosive Limit (LEL): 7.5% [28].

Upper Explosive Limit (UEL): 12.5% [28].

HEALTH HAZARD DATA

Immediately Dangerous to Life or Health (IDLH): 1,000 ppm [13].

Permissible Exposure Limits (PEL) in Air: 350 ppm ($\approx$1,900 mg/m^3) [30]; 350 ppm 15-minute ceiling [19]; 350 ppm ($\approx$1,900 mg/m^3), 450 ppm ($\approx$2,450 mg/m^3) STEL [31].

MANUFACTURING

Selected Manufacturers:

Dow Chemical Co.
Midland, MI 48640

PPG Industries, Inc.
Industrial Chemicals Division
1 Gateway Center
Pittsburgh, PA 15222

Vulcan Materials Co.
Chemicals Division
Wichita, KS 67201

Uses: Organic synthesis; solvent for metal cleaning of precision instruments; textile processing; aerosol propellants; pesticide.

REFERENCES

1. Weast, R.C., Ed. *CRC Handbook of Chemistry and Physics*, 67th ed. (Boca Raton, FL: CRC Press, Inc., 1986), 2406 p.
2. Lyman, W.J., Reehl, W.F., and D.H. Rosenblatt. *Handbook of Chemical Property Estimation Methods: Environmental Behavior of Organic Compounds* (New York: McGraw-Hill, Inc., 1982).
3. Schwille, F. *Dense Chlorinated Solvents* (Chelsea, MI: Lewis Publishers, Inc., 1988), 146 p.
4. Hine, J., and P.K. Mookerjee. "The Intrinsic Hydrophilic Character of Organic Compounds. Correlations in Terms of Structural Contributions," *J. Org. Chem.*, 40(3):292-298 (1975).
5. Roberts, P.V., and P.G. Dändliker. "Mass Transfer of Volatile Organic Contaminants from Aqueous Solution to the Atmosphere

During Surface Aeration," *Environ. Sci. Technol.*, 17(8):484-489 (1983).

6. Sato, A., and T. Nakajima. "A Structure-Activity Relationship of Some Chlorinated Hydrocarbons," *Arch. Environ. Health*, 34(2):69-75 (1979).

7. Chiou, C.T., Peters, L.J., and V.H. Freed. "A Physical Concept of Soil-Water Equilibria for Nonionic Organic Compounds," *Science*, 206:831-832 (1979).

8. Mills, W.B., Porcella, D.B., Ungs, M.J., Gherini, S.A., Summers, K.V., Mok, L., Rupp, G.L., and G.L. Bowie. "Water Quality Assessment: A Screening Procedure for Toxic and Conventional Pollutants in Surface and Groundwater-Part I," Office of Research and Development, U.S. EPA Report-600/6-85-002a (1985), 638 p.

9. Neely, W.B., and G.E. Blau, Eds. *Environmental Exposure from Chemicals. Volume 1* (Boca Raton, FL: CRC Press, Inc. 1985), 245 p.

10. Munz, C., and P.V. Roberts. "Air-Water Phase Equilibria of Volatile Organic Solutes," *J. Am. Water Works Assoc.*, 79(5):62-69 (1987).

11. Veith, G.D., Macek, K.J., Petrocelli, S.R., and J. Carroll. "An Evaluation of Using Partition Coefficients and Water Solubility to Estimate Bioconcentration Factors for Organic Chemicals in Fish," in *Aquatic Toxicology, ASTM STP 707*, Eaton, J.G., Parrish, P.R., and A.C. Hendricks, Eds. (Philadelphia, PA: American Society for Testing and Materials 1980), pp 116-129.

12. Schwarzenbach, R.P., Giger, W., Hoehn, E., and J.K. Schneider. "Behavior of Organic Compounds during Infiltration of River Water to Groundwater. Field Studies," *Environ. Sci. Technol.*, 17(8):472-479 (1983).

13. Weiss, G. *Hazardous Chemicals Data Book* (Park Ridge, NJ: Noyes Data Corp., 1986), 1069 p.

14. Standen, A., Ed. *Kirk-Othmer Encyclopedia of Chemical Technology, Volume 4*, 2nd ed. (New York: John Wiley and Sons, Inc., 1964), 937 p.

15. "Chemical, Physical, and Biological Properties of Compounds Present at Hazardous Waste Sites," U.S. EPA Report-530/SW-89-010 (1985), 619 p.

16. Verschueren, K. *Handbook of Environmental Data on Organic Chemicals* (New York: Van Nostrand Reinhold Co., 1983), 1310 p.

17. Pearson, C.R., and G. McConnell. "Chlorinated C_1 and C_2 Hydrocarbons in the Marine Environment," in *Proc. R. Soc. London*, B189(1096):305-322 (1975).

18. *IARC Monographs on the Evaluation of Carcinogenic Risk of Chemicals to Man. Some Halogenated Hydrocarbons, Volume 20* (Lyon, France: International Agency for Research on Cancer, 1979),

609 p.

19. "NIOSH Pocket Guide to Chemical Hazards," U.S. Department of Health and Human Services, U.S. Government Printing Office (1987), 241 p.

20. Mackay, D., and W.Y. Shiu. "A Critical Review of Henry's Law Constants for Chemicals of Environmental Interest," *J. Phys. Chem. Ref. Data*, 10(4):1175-1199 (1981).

21. Riddick, J.A., Bunger, W.B., and T.K. Sakano. *Organic Solvents - Physical Properties and Methods of Purification. Volume II.* (New York: John Wiley and Sons, Inc., 1986), 1325 p.

22. Konietzko, H. "Chlorinated Ethanes: Sources, Distribution, Environmental Impact, and Health Effects," in *Hazard Assessment of Chemicals, Volume 3*, J. Saxena, Ed. (New York: Academic Press, Inc., 1984), pp 401-448.

23. Schwarz, F.P. "Measurement of the Solubilities of Slightly Soluble Organic Liquids in Water by Elution Chromatography," *Anal. Chem.*, 52(1):10-15 (1980).

24. *Fluka Catalog 1988/89 - Chemika-Biochemika* (Ronkonkoma, NY: Fluka Chemical Corp., 1988), 1536 p.

25. Smith, L.R., and J. Dragun. "Degradation of Volatile Chlorinated Aliphatic Priority Pollutants in Groundwater," *Environ. Int.*, 19(4):291-298 (1984).

26. Gälli, R., and P.L. McCarty. "Biotransformation of 1,1,1-Trichloroethane, Trichloromethane, and Tetrachloromethane by a *Clostridium* sp.," *Appl. Environ. Microbiol.*, 55(4):837-844 (1989).

27. "Treatability Manual - Volume 1: Treatability Data," Office of Research and Development, U.S. EPA Report-600/8-80-042a (1980), 1035 p.

28. *Fire Protection Guide on Hazardous Materials* (Quincy, MA: National Fire Protection Association, 1984), 443 p.

29. Kuchta, J.M., Furno, A.L., Bartkowiak, A., and G.H. Martindill. "Effect of Pressure and Temperature on Flammability Limits of Chlorinated Hydrocarbons in Oxygen-Nitrogen and Nitrogen Tetroxide-Nitrogen Atmospheres," *J. Chem. Eng. Data*, 13(3):421-428 (1968).

30. "General Industry Standards for Toxic and Hazardous Substances," U.S. Code of Federal Regulations 1910, Subpart Z Section 1910.1000 (July 1982).

31. *Documentation of the Threshold Limit Values and Biological Exposure Indices* (Cincinnati, OH: American Conference of Governmental Industrial Hygienists, 1986), 744 p.

1,1,2-TRICHLOROETHANE

Synonyms: Ethane trichloride; NCI-C04579; RCRA waste number U227; β-T; 1,1,2-TCA; 1,2,2-Trichloroethane; β-Trichloroethane; Vinyl trichloride.

Structural Formula:

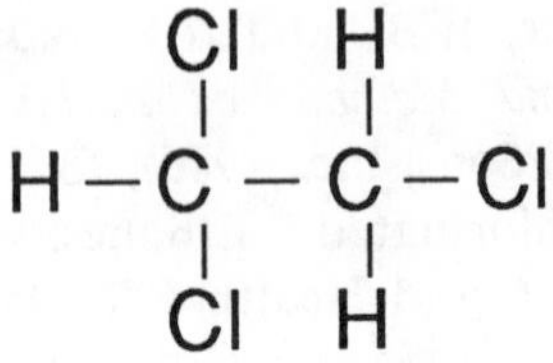

$$\begin{array}{ccc} Cl & & H \\ | & & | \\ H-C & - & C-Cl \\ | & & | \\ Cl & & H \end{array}$$

CHEMICAL DESIGNATIONS

CAS Registry Number: 79-00-5

DOT Designation: 2831

Empirical Formula: $C_2H_3Cl_3$

Formula Weight: 133.40

RTECS Number: KJ 3150000

PHYSICAL AND CHEMICAL PROPERTIES

Appearance and Odor: Colorless liquid with a pleasant odor.

Boiling Point: 113.8 °C [1]; 111-114 °C [2].

Henry's Law Constant: 7.4×10^{-4} atm·m^3/mol [3]; 9.09×10^{-4} atm·m^3/mol at 25 °C [4]; 0.00149 atm·m^3/mol at 37 °C [5].

Ionization Potential: No data found.

Log K_{oc}: 1.75 [6].

Log K_{ow}: 2.18 [7].

Melting Point: -36.5 °C [1]; -37.0 °C [8].

Solubility in Organics: Soluble in ethanol, chloroform, and ether [9].

Solubility in Water: 4,500 mg/L at 20 °C [10]; 4,400 mg/L at 20 °C [11]; 0.44 wt% at 20 °C [12]; 0.45 mass% at 20 °C [13].

Specific Density: 1.4397 at 20/4 °C [1]; 1.434 at 20/4 °C [2]; 1.4410 at 20/4 °C [8].

Transformation Products: No data found.

Vapor Density: 4 kg/m^3 at the boiling point [13]; 5.45 g/L at 25 °C, 4.60 (air = 1).

Vapor Pressure: 19 mm at 20 °C, 32 mm at 30 °C [10]; 1 mm at -24.0 °C, 10 mm at 8.3 °C, 40 mm at 35.2 °C, 100 mm at 55.7 °C, 400 mm at 93.0 °C, 760 mm at 113.9 °C [1].

FIRE HAZARDS

Flash Point: none [14].

Lower Explosive Limit (LEL): 6% [15].

Upper Explosive Limit (UEL): 15.5% [15].

HEALTH HAZARD DATA

Immediately Dangerous to Life or Health (IDLH): Potential human carcinogen [15].

Permissible Exposure Limits (PEL) in Air: 10 ppm ($\approx$45 mg/m^3) [16].

MANUFACTURING

Selected Manufacturers:

Aldrich Chemical Co.
940 West Saint Paul Ave.
Milwaukee, WI 53233

Fluka Chemical Corp.
980 South Second St.
Ronkonkoma, NY 11779

Uses: Solvent for fats, oils, resins, waxes, resins, and other products; organic synthesis.

REFERENCES

1. Weast, R.C., Ed. *CRC Handbook of Chemistry and Physics*, 67th ed. (Boca Raton, FL: CRC Press, Inc., 1986), 2406 p.
2. *Fluka Catalog 1988/89 - Chemika-Biochemika* (Ronkonkoma, NY: Fluka Chemical Corp., 1988), 1536 p.
3. Pankow, J.F., and M.E. Rosen. "Determination of Volatile Compounds in Water by Purging Directly to a Capillary Column with Whole Column Cryotrapping," *Environ. Sci. Technol.*, 22(4):398-405 (1988).
4. Hine, J., and P.K. Mookerjee. "The Intrinsic Hydrophilic Character of Organic Compounds. Correlations in Terms of Structural Contributions," *J. Org. Chem.*, 40(3):292-298 (1975).
5. Sato, A., and T. Nakajima. "A Structure-Activity Relationship of Some Chlorinated Hydrocarbons," *Arch. Environ. Health*, 34(2):69-75 (1979).
6. Schwille, F. *Dense Chlorinated Solvents* (Chelsea, MI: Lewis Publishers, Inc., 1988), 146 p.
7. Mills, W.B., Porcella, D.B., Ungs, M.J., Gherini, S.A., Summers, K.V., Mok, L., Rupp, G.L., and G.L. Bowie. "Water Quality Assessment: A Screening Procedure for Toxic and Conventional Pollutants in Surface and Groundwater-Part I," Office of Research and Development, U.S. EPA Report-600/6-85-002a (1985), 638 p.
8. Standen, A., Ed. *Kirk-Othmer Encyclopedia of Chemical Technology, Volume 4*, 2nd ed. (New York: John Wiley and Sons, Inc., 1964), 937 p.
9. "Chemical, Physical, and Biological Properties of Compounds Present at Hazardous Waste Sites," U.S. EPA Report-530/SW-89-010 (1985), 619 p.
10. Verschueren, K. *Handbook of Environmental Data on Organic Chemicals* (New York: Van Nostrand Reinhold Co., 1983), 1310 p.
11. Dean, J.A. *Handbook of Organic Chemistry* (New York: McGraw-Hill, Inc., 1987), 957 p.
12. Riddick, J.A., Bunger, W.B., and T.K. Sakano. *Organic Solvents - Physical Properties and Methods of Purification. Volume II* (New

York: John Wiley and Sons, Inc., 1986), 1325 p.

13. Konietzko, H. "Chlorinated Ethanes: Sources, Distribution, Environmental Impact, and Health Effects," in *Hazard Assessment of Chemicals, Volume 3*, J. Saxena, Ed. (New York: Academic Press, Inc., 1984), pp 401-448.

14. Dean, J.A., Ed., *Lange's Handbook of Chemistry*, 11th ed. (New York: McGraw-Hill, Inc., 1973), 1570 p.

15. "NIOSH Pocket Guide to Chemical Hazards," U.S. Department of Health and Human Services, U.S. Government Printing Office (1987), 241 p.

16. "General Industry Standards for Toxic and Hazardous Substances," U.S. Code of Federal Regulations 1910, Subpart Z Section 1910.1000 (July 1982).

TRICHLOROETHYLENE

Synonyms: Acetylene trichloride; Algylen; Anamenth; Benzinol; Blacosolv; Blancosolv; Cecolene; Chlorilen; 1-Chloro-2,2-dichloro-ethylene; Chlorylea; Chlorylen; Circosolv; Crawhaspol; Densinfluat; 1,1-Dichloro-2-chloroethylene; Dow-tri; Dukeron; Ethinyl trichloride; Ethylene trichloride; Fleck-flip; Flock-flip; Fluate; Gemalgene; Germalgene; Lanadin; Lethurin; Narcogen; Narkogen; Narkosoid; NCI-C04546; Nialk; Perm-a-chlor; Perm-a-clor; Petzinol; Philex; RCRA waste number U228; TCE; Threthylen; Threthylene; Trethylene; Tri; Triad; Trial; Triasol; Trichloran; Trichloren; **Trichloroethene**; 1,1,2-Trichloroethene; 1,2,2-Trichloroethene; 1,1,2-Trichloroethylene; 1,2,2-Trichloroethylene; Tri-clene; Trielene; Trieline; Triklone; Trilen; Trilene; Triline; Trimar; Triol; Tri-plus; Tri-plus M; UN 1710; Vestrol; Vitran; Westrosol.

Structural Formula:

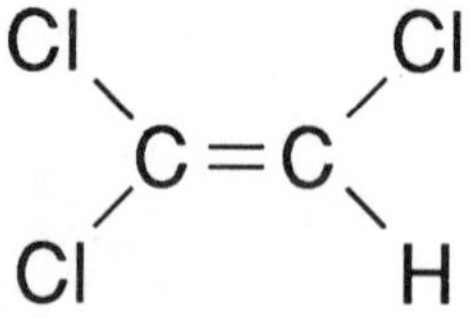

$$Cl_2C=CHCl$$

CHEMICAL DESIGNATIONS

CAS Registry Number: 79-01-6

DOT Designation: 1710

Empirical Formula: C_2HCl_3

Formula Weight: 131.39

RTECS Number: KX 4550000

PHYSICAL AND CHEMICAL PROPERTIES

Appearance and Odor: Clear, colorless, watery-liquid with a chloroform-like odor.

Boiling Point: 87.2 °C [1]; 86.7 °C [2].

Henry's Law Constant: 0.0091 atm·m^3/mol [3]; 0.0117 atm·m^3/mol [4]; 0.0099 atm·m^3/mol at 20 °C [5]; 0.0196 atm·m^3/mol at 37 °C [6].

Ionization Potential: 9.47 eV [7]; 9.94 eV [8].

Log K$_{oc}$: 1.81 [9]; 2.10 [10]; 2.025 [11].

Log K$_{ow}$: 2.53 [12]; 2.29 [13]; 2.42 [14]; 2.60 [15]; 3.24, 3.30 [16]; 2.37 [17]; 3.03 [18].

Melting Point: -73 °C [19]; -84.8 °C [20]; -87.1 °C [2].

Solubility in Organics: Soluble in acetone, ethanol, chloroform, and ether [21].

Solubility in Water: 1,100 mg/L at 25 °C [22]; 1,100 mg/L at 20 °C [23]; 1,470 mg/L at 25 °C [24]; 0.0104 M at 25 °C [18]; 0.1 wt% at 20 °C [7]; 0.07 vol% at 20 °C [25]; 1,080 mg/L at 20 °C [26]; 0.137 wt% at 25 °C [27]; 1,250 mg/L at 60 °C [2].

Specific Density: 1.4642 at 20/4 °C [19]; 1.461 at 20/4 °C [28]; 1.375 at 100/4 °C [2].

Transformation Products: Microbial degradation by sequential dehalogenation may produce cis-1,2-dichloroethylene, trans-1,2-dichloroethylene, and vinyl chloride [29]. In a methanogenic aquifer, biodegradation produced 1,2-dichloroethylene and vinyl chloride [30].

Vapor Density: 5.37 g/L at 25 °C, 4.54 (air = 1).

Vapor Pressure: 5.4 mm at -20 °C, 10.8 mm at -10.8 °C, 20.1 mm at 0 °C, 35.2 mm at 10 °C, 57.8 mm at 20 °C, 94 mm at 30 °C, 146.8 mm at 40 °C, 212 mm at 50 °C, 305.7 mm at 60 °C, 760 mm at 86.7 °C [2]; 74 mm at 25 °C [31]; 1 mm at -43.8 °C, 10 mm at -12.4 °C, 40 mm at 11.9 °C, 100 mm at 31.4 °C, 400 mm at 67.0 °C, 760 mm at 86.7 °C [19]; 58 mm at 20 °C [10]; 57.9 mm at 20 °C [4]; 100 mm at 32 °C [32]; 53.5 mm at 20 °C [26]; 26.5 mm at 5 °C, 56.8 mm at 20 °C, 72.6 mm at 25 °C, 91.5 mm at 30 °C [33].

FIRE HAZARDS

Flash Point: 32.2 °C [34].

Lower Explosive Limit (LEL): 8% [35].

Upper Explosive Limit (UEL): 10.5% [35].

HEALTH HAZARD DATA

Immediately Dangerous to Life or Health (IDLH): 1,000 ppm (carcinogen) [34].

Permissible Exposure Limits (PEL) in Air: 100 ppm, 200 ppm ceiling, 300 ppm 5 minute/2-hour peak [36]; 25 ppm 10-hour time weighted average [7]; 50 ppm ($\approx$270 mg/m^3), 200 ppm ($\approx$1,080 mg/m^3) STEL [37].

MANUFACTURING

Selected Manufacturers:

Dow Chemical Co.
Midland, MI 48640

E.I. duPont de Nemours and Co.
Electrochemicals Department
Wilmington, DE 19898

PPG Industries Inc.
Industrial Chemicals Division
Lake Charles, LA 70601

Uses: Dry cleaning fluid; degreasing and drying metals and electronic parts; extraction solvent for oils, waxes, and fats; solvent for cellulose esters and ethers; removing caffeine from coffee; refrigerant and heat exchange liquid; fumigant; diluent in paints and adhesives; textile processing; aerospace operations (flushing liquid oxygen); anesthetic; medicine; organic synthesis.

REFERENCES

1. Dean, J.A., Ed., *Lange's Handbook of Chemistry*, 11th ed. (New York: McGraw-Hill, Inc., 1973), 1570 p.
2. Standen, A., Ed. *Kirk-Othmer Encyclopedia of Chemical Technology,*

Volume 4, 2nd ed. (New York: John Wiley and Sons, Inc., 1964), 937 p.

3. Pankow, J.F., and M.E. Rosen. "Determination of Volatile Compounds in Water by Purging Directly to a Capillary Column with Whole Column Cryotrapping," *Environ. Sci. Technol.*, 22(4):398-405 (1988).

4. "Treatability Manual - Volume 1: Treatability Data," Office of Research and Development, U.S. EPA Report-600/8-80-042a (1980), 1035 p.

5. Roberts, P.V., and P.G. Dändliker. "Mass Transfer of Volatile Organic Contaminants from Aqueous Solution to the Atmosphere During Surface Aeration," *Environ. Sci. Technol.*, 17(8):484-489 (1983).

6. Sato, A., and T. Nakajima. "A Structure-Activity Relationship of Some Chlorinated Hydrocarbons," *Arch. Environ. Health*, 34(2):69-75 (1979).

7. "NIOSH Pocket Guide to Chemical Hazards," U.S. Department of Health and Human Services, U.S. Government Printing Office (1987), 241 p.

8. Yoshida, K., Shigeoka, T., and F. Yamauchi. "Non-Steady State Equilibrium Model for the Preliminary Prediction of the Fate of Chemicals in the Environment," *Ecotoxicol. Environ. Safety*, 7(2):179-190 (1983).

9. Abdul, S.A., Gibson, T.L., and D.N. Rai. "Statistical Correlations for Predicting the Partition Coefficient for Nonpolar Organic Contaminants Between Aquifer Organic Carbon and Water," *Haz. Waste Haz. Mater.*, 4(3):211-222 (1987).

10. Schwille, F. *Dense Chlorinated Solvents* (Chelsea, MI: Lewis Publishers, Inc., 1988), 146 p.

11. Garbarini, D.R., and L.W. Lion. "Influence of the Nature of Soil Organics on the Sorption of Toluene and Trichloroethylene," *Environ. Sci. Technol.*, 20(12):1263-1269 (1986).

12. Miller, M.M., Wasik, S.P., Huang, G.-L., Shiu, W.-Y., and D. Mackay. "Relationships Between Octanol-Water Partition Coefficient and Aqueous Solubility," *Environ. Sci. Technol.*, 19(6):522-529 (1985).

13. Schwarzenbach, R.P., Giger, W., Hoehn, E., and J.K. Schneider. "Behavior of Organic Compounds during Infiltration of River Water to Groundwater. Field Studies, " *Environ. Sci. Technol.*, 17(8):472-479 (1983).

14. Banerjee, S., Yalkowsky, S.H., and S.C. Valvani. "Water Solubility and Octanol/Water Partition Coefficients of Organics. Limitations of the Solubility-Partition Coefficient Correlation," *Environ. Sci. Technol.*, 14(10):1227-1229 (1980).

15. Hawker, D.W., and D.W. Connell. "Influence of Partition Coefficient of Lipophilic Compounds on Bioconcentration Kinetics with Fish," *Water Res.*, 22(6):701-707 (1988).

16. Geyer, H., Politzki, G., and D. Freitag. "Prediction of Ecotoxicological Behaviour of Chemicals: Relationship Between *n*-Octanol/Water Partition Coefficient and Bioaccumulation of Organic Chemicals by Alga *Chlorella*," *Chemosphere*, 13(2):269-284 (1984).

17. Green, W.J., Lee, G.F., Jones, R.A., and Ted Palit. "Interaction of Clay Soils with Water and Organic Solvents: Implications for the Disposal of Hazardous Wastes," *Environ. Sci. Technol.*, 17(5):278-282 (1983).

18. Tewari, Y.B., Miller, M.M., Wasik, S.P., and D.E. Martire. "Aqueous Solubility and Octanol/Water Partition Coefficient of Organic Compounds at 25.0 °C," *J. Chem. Eng. Data*, 27(4):451-454 (1982).

19. Weast, R.C., Ed. *CRC Handbook of Chemistry and Physics*, 67th ed. (Boca Raton, FL: CRC Press, Inc., 1986), 2406 p.

20. *Catalog Handbook of Fine Chemicals* (Milwaukee, WI: Aldrich Chemical Co., 1988), 2212 p.

21. "Chemical, Physical, and Biological Properties of Compounds Present at Hazardous Waste Sites," U.S. EPA Report-530/SW-89-010 (1985), 619 p.

22. Verschueren, K. *Handbook of Environmental Data on Organic Chemicals* (New York: Van Nostrand Reinhold Co., 1983), 1310 p.

23. Pearson, C.R., and G. McConnell. "Chlorinated C_1 and C_2 Hydrocarbons in the Marine Environment," in *Proc. R. Soc. London*, B189(1096):305-322 (1975).

24. Lyman, W.J., Reehl, W.F., and D.H. Rosenblatt. *Handbook of Chemical Property Estimation Methods: Environmental Behavior of Organic Compounds* (New York: McGraw-Hill, Inc., 1982).

25. Meites, L., Ed. *Handbook of Analytical Chemistry*, 1st ed. (New York: McGraw-Hill, Inc., 1963), 1782 p.

26. Munz, C., and P.V. Roberts. "Air-Water Phase Equilibria of Volatile Organic Solutes," *J. Am. Water Works Assoc.*, 79(5):62-69 (1987).

27. Riddick, J.A., Bunger, W.B., and T.K. Sakano. *Organic Solvents - Physical Properties and Methods of Purification. Volume II* (New York: John Wiley and Sons, Inc., 1986), 1325 p.

28. *Fluka Catalog 1988/89 - Chemika-Biochemika* (Ronkonkoma, NY: Fluka Chemical Corp., 1988), 1536 p.

29. Smith, L.R., and J. Dragun. "Degradation of Volatile Chlorinated Aliphatic Priority Pollutants in Groundwater," *Environ. Int.*, 19(4):291-298 (1984).

30. Wilson, B.H., Smith, G.B., and J.F. Rees. "Biotransformations of

Selected Alkylbenzenes and Halogenated Aliphatic Hydrocarbons in Methanogenic Aquifer Material: A Microcosm Study," *Environ. Sci. Technol.,* 20(10):997-1002 (1986).

31. Mackay, D., and W.Y. Shiu. "A Critical Review of Henry's Law Constants for Chemicals of Environmental Interest," *J. Phys. Chem. Ref. Data,* 10(4):1175-1199 (1981).

32. Sax, N.I. *Dangerous Properties of Industrial Materials* (New York: Van Nostrand Reinhold Co., 1984), 3124 p.

33. Klöpffer, W., Kaufman, G., Rippen, G., and H.-P. Poremski. "A Laboratory Method for Testing the Volatility from Aqueous Solution: First Results and Comparison with Theory," *Ecotoxicol. Environ. Safety,* 6(6):545-559 (1982).

34. Weiss, G. *Hazardous Chemicals Data Book* (Park Ridge, NJ: Noyes Data Corp., 1986), 1069 p.

35. *Fire Protection Guide on Hazardous Materials* (Quincy, MA: National Fire Protection Association, 1984), 443 p.

36. "General Industry Standards for Toxic and Hazardous Substances," U.S. Code of Federal Regulations 1910, Subpart Z Section 1910.1000 (July 1982).

37. *Documentation of the Threshold Limit Values and Biological Exposure Indices* (Cincinnati, OH: American Conference of Governmental Industrial Hygienists, 1986), 744 p.

TRICHLOROFLUOROMETHANE

Synonyms: Algofrene type 1; Arcton 9; Electro-CF 11; Eskimon 11; F 11; FC 11; Fluorocarbon 11; Fluorotrichloromethane; Freon 11; Freon 11A; Freon 11B; Freon HE; Freon MF; Frigen 11; Genetron 11; Halocarbon 11; Isceon 11; Isotron 11; Ledon 11; Monofluorotrichloromethane; NCI-C04637; RCRA waste number U121; Refrigerant 11; Trichloromonofluoromethane; Ucon 11; Ucon fluorocarbon 11; Ucon refrigerant 11.

Structural Formula:

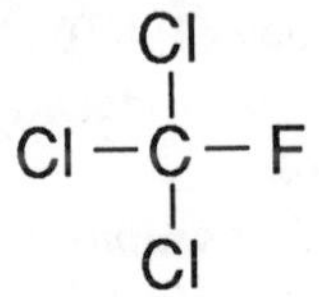

$$Cl-\underset{\underset{\displaystyle Cl}{|}}{\overset{\overset{\displaystyle Cl}{|}}{C}}-F$$

CHEMICAL DESIGNATIONS

CAS Registry Number: 75-69-4

DOT Designation: 1078

Empirical Formula: CCl_3F

Formula Weight: 137.37

RTECS Number: PB 6125000

PHYSICAL AND CHEMICAL PROPERTIES

Appearance and Odor: Colorless, odorless liquid.

Boiling Point: 23.63 °C [1].

Henry's Law Constant: 0.11 atm·m^3/mol [2]; 0.00583 atm·m^3/mol [3]; 1.73 atm·m^3/mol at 25 °C [4].

Ionization Potential: 11.77 eV [5].

Log K_{oc}: 2.20 [6]; 2.13 [7].

Log K_{ow}: 2.53 [8].

Melting Point: -111 °C [9].

Solubility in Organics: Soluble in ethanol, ether, and other solvents [10].

Solubility in Water: 1,100 mg/L at 25 °C [11]; 1,240 mg/L at 25 °C [7]; 1,100 mg/L at 20 °C [12].

Specific Density: 1.476 at 25/4 °C [7]; 1.484 at 17.2/4 °C [13]; 1.487 at 20/4 °C [14].

Transformation Products: No data found.

Vapor Density: 5.85 g/L at 23.77 °C [15]; 5.61 g/L at 25 °C, 4.74 (air = 1).

Vapor Pressure: 687 mm at 20 °C, 980 mm at 30 °C [11]; 667 mm at 20 °C [16]; 760 mm at 23.7 °C, 1,520 mm at 44.1 °C, 3,800 mm at 77.3 °C, 7,600 mm at 108.2 °C, 15,200 mm at 146.7 °C, 30,400 mm at 194.0 °C [17]; 792 mm at 25 °C [18]; 667.4 mm at 20 °C [19].

FIRE HAZARDS

Flash Point: Not flammable [20].

Lower Explosive Limit (LEL): Not flammable [20].

Upper Explosive Limit (UEL): Not flammable [20].

HEALTH HAZARD DATA

Immediately Dangerous to Life or Health (IDLH): 10,000 ppm [5].

Permissible Exposure Limits (PEL) in Air: 1,000 ppm ($\approx$5,600 mg/m^3) [21]; 1,000 ppm ceiling [18].

MANUFACTURING

Selected Manufacturers:

E.I. duPont de Nemours and Co.
Wilmington, DE 19898

Union Carbide Corp.
270 Park Ave.
New York, NY 10017

Uses: Aerosol propellant; refrigerant; solvent; blowing agent for polyurethane foams; fire extinguishing; chemical intermediate; organic synthesis.

REFERENCES

1. Kudchadker, A.P., Kudchadker, S.A., Shukla, R.P., and P.R. Patnaik. "Vapor Pressures and Boiling Points of Selected Halomethanes," *J. Phys. Chem. Ref. Data*, 8(2):499-517 (1979).
2. Pankow, J.F., and M.E. Rosen. "Determination of Volatile Compounds in Water by Purging Directly to a Capillary Column with Whole Column Cryotrapping," *Environ. Sci. Technol.*, 22(4):398-405 (1988).
3. Warner, H.P., Cohen, J.M., and J.C. Ireland. "Determination of Henry's Law Constants of Selected Priority Pollutants," Office of Science and Development, U.S. EPA Report-600/D-87/229 (1987), 14 p.
4. Hine, J., and P.K. Mookerjee. "The Intrinsic Hydrophilic Character of Organic Compounds. Correlations in Terms of Structural Contributions," *J. Org. Chem.*, 40(3):292-298 (1975).
5. "NIOSH Pocket Guide to Chemical Hazards," U.S. Department of Health and Human Services, U.S. Government Printing Office (1987), 241 p.
6. Schwille, F. *Dense Chlorinated Solvents* (Chelsea, MI: Lewis Publishers, Inc., 1988), 146 p.
7. Neely, W.B., and G.E. Blau, Eds. *Environmental Exposure from Chemicals. Volume 1* (Boca Raton, FL: CRC Press, Inc. 1985), 245 p.
8. Hansch, C., Vittoria, A., Silipo, C., and P.Y.C. Jow. "Partition Coefficients and the Structure-Activity Relationship of the Anesthetic Gases," *J. Med. Chem.*, 18(6):546-548 (1975).
9. Windholz, M., Budavari, S., Blumetti, R.F., and E.S. Otterbein, Eds. *The Merck Index*, 10th ed., (Rahway, NJ: Merck and Co., 1983), 1463 p.
10. "Chemical, Physical, and Biological Properties of Compounds Present at Hazardous Waste Sites," U.S. EPA Report-530/SW-89-010 (1985), 619 p.
11. Verschueren, K. *Handbook of Environmental Data on Organic Chemicals* (New York: Van Nostrand Reinhold Co., 1983), 1310 p.

12. Pearson, C.R., and G. McConnell. "Chlorinated C_1 and C_2 Hydrocarbons in the Marine Environment," in *Proc. R. Soc. London*, B189(1096):305-322 (1975).

13. Sax, N.I. *Dangerous Properties of Industrial Materials* (New York: Van Nostrand Reinhold Co., 1984), 3124 p.

14. *Fluka Catalog 1988/89 - Chemika-Biochemika* (Ronkonkoma, NY: Fluka Chemical Corp., 1988), 1536 p.

15. Braker, William, and A.L. Mossman. *Matheson Gas Data Book* (East Rutherford, NJ: Matheson Gas Products, 1971), 574 p.

16. Mills, W.B., Porcella, D.B., Ungs, M.J., Gherini, S.A., Summers, K.V., Mok, L., Rupp, G.L., and G.L. Bowie. "Water Quality Assessment: A Screening Procedure for Toxic and Conventional Pollutants in Surface and Groundwater-Part I," Office of Research and Development, U.S. EPA Report-600/6-85-002a (1985), 638 p.

17. Weast, R.C., Ed. *CRC Handbook of Chemistry and Physics*, 67th ed. (Boca Raton, FL: CRC Press, Inc., 1986), 2406 p.

18. *Documentation of the Threshold Limit Values and Biological Exposure Indices* (Cincinnati, OH: American Conference of Governmental Industrial Hygienists, 1986), 744 p.

19. McConnell, G., Ferguson, D.M., and C.R. Pearson. "Chlorinated Hydrocarbons and the Environment," *Endeavour*, 34(121):13-18 (1975).

20. Weiss, G. *Hazardous Chemicals Data Book* (Park Ridge, NJ: Noyes Data Corp., 1986), 1069 p.

21. "General Industry Standards for Toxic and Hazardous Substances," U.S. Code of Federal Regulations 1910, Subpart Z Section 1910.1000 (July 1982).

2,4,5-TRICHLOROPHENOL

Synonyms: Collunosol; Dowicide 2; Dowicide B; NCI-C61187; Nurelle; Phenachlor; Preventol I; RCRA waste number U230; 2,4,5-TCP; 2,4,5-TCP-Dowicide 2.

Structural Formula:

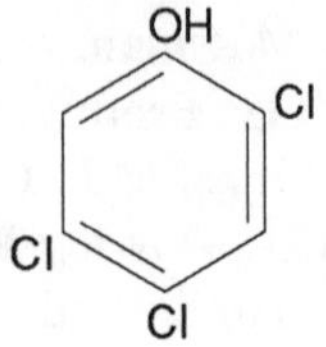

CHEMICAL DESIGNATIONS

CAS Registry Number: 95-95-4

DOT Designation: 2020

Empirical Formula: $C_6H_3Cl_3O$

Formula Weight: 197.45

RTECS Number: SN 1400000

PHYSICAL AND CHEMICAL PROPERTIES

Appearance and Odor: Colorless crystals or yellow to gray flakes with a strong disinfectant or phenolic odor.

Boiling Point: 252 °C [1].

Dissociation Constant: 7.37 [2].

Henry's Law Constant: 1.76 x 10^{-7} atm·m³/mol at 25 °C (estimated) [3].

Ionization Potential: No data found.

Log K_{oc}: 2.85, 3.51 using method of Karickhoff and others [4].

Log K_{ow}: 3.72 [5]; 4.19 [6]; 3.85 [7].

Melting Point: 68-70 °C [1]; 61-63 °C [8]; 57 °C [9]; 64-67 °C [10].

Solubility in Organics: Soluble in ethanol and ligroin [11].

Solubility in Water: 1,190 mg/kg at 25 °C [12]; 1.2 g/L at 25 °C [3].

Specific Density: 1.5 at 75/4 °C [12]; 1.678 at 25/4 °C [8].

Transformation Products: No data found.

Vapor Pressure: 1 mm at 72 °C, 10 mm at 117.3 °C, 40 mm at 151.5 °C, 100 mm at 178.0 °C, 400 mm at 226.5 °C, 760 mm at 251.8 °C [1]; 0.0035 mm at 8 °C, 0.022 mm at 25 °C [3].

FIRE HAZARDS

Flash Point: Not flammable [9].

Lower Explosive Limit (LEL): Not flammable [9].

Upper Explosive Limit (UEL): Not flammable [9].

HEALTH HAZARD DATA

Immediately Dangerous to Life or Health (IDLH): No data found.

Permissible Exposure Limits (PEL) in Air: No standards set.

MANUFACTURING

Selected Manufacturers:

Dow Chemical Co.
Midland, MI 48640

Fike Chemicals, Inc.
Nitro, WV 25143

Hooker Chemical Corp.
Niagara Falls, NY 14302

Uses: Fungicide; bactericide; organic synthesis.

REFERENCES

1. Weast, R.C., Ed. *CRC Handbook of Chemistry and Physics*, 67th ed. (Boca Raton, FL: CRC Press, Inc., 1986), 2406 p.
2. Dean, J.A., Ed., *Lange's Handbook of Chemistry*, 11th ed. (New York: McGraw-Hill, Inc., 1973), 1570 p.
3. Leuenberger, C., Ligocki, M.P., and J.F. Pankow. "Trace Organic Compounds in Rain. 4. Identities, Concentrations, and Scavenging Mechanisms for Phenols in Urban Air and Rain," *Environ. Sci. Technol.*, 19(11):1053-1058 (1985).
4. Karickhoff, S.W., Brown, D.S., and T.A. Scott. "Sorption of Hydrophobic Pollutants on Natural Sediments," *Water Res.*, 13:241-248 (1979).
5. Leo, A., Hansch, C., and D. Elkins. "Partition Coefficients and Their Uses," *Chem. Rev.*, 71(6):525-616 (1971).
6. Schellenberg, K., Leuenberger, C., and R.P. Schwarzenbach. "Sorption of Chlorinated Phenols by Natural Sediments and Aquifer Materials," *Environ. Sci. Technol.*, 18(9):652-657 (1984).
7. Schultz, T.W., Wesley, S.K., and L.L. Baker. "Structure-Activity Relationships for Di and Tri Alkyl and/or Halogen Substituted Phenols," *Bull. Environ. Contam. Toxicol.*, 43(2):192-198 (1989).
8. Sax, N.I. *Dangerous Properties of Industrial Materials* (New York: Van Nostrand Reinhold Co., 1984), 3124 p.
9. Weiss, G. *Hazardous Chemicals Data Book* (Park Ridge, NJ: Noyes Data Corp., 1986), 1069 p.
10. *Fluka Catalog 1988/89 - Chemika-Biochemika* (Ronkonkoma, NY: Fluka Chemical Corp., 1988), 1536 p.
11. "Chemical, Physical, and Biological Properties of Compounds Present at Hazardous Waste Sites," U.S. EPA Report-530/SW-89-010 (1985), 619 p.
12. Verschueren, K. *Handbook of Environmental Data on Organic Chemicals* (New York: Van Nostrand Reinhold Co., 1983), 1310 p.

2,4,6-TRICHLOROPHENOL

Synonyms: Dowicide 2S; NCI-C02904; Omal; Phenachlor; RCRA waste number F027; 2,4,6-TCP; 2,4,6-TCP-Dowicide 25.

Structural Formula:

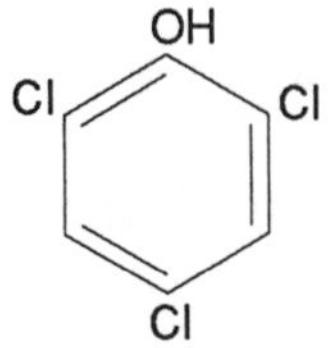

CHEMICAL DESIGNATIONS

CAS Registry Number: 88-06-2

DOT Designation: 2020

Empirical Formula: $C_6H_3Cl_3O$

Formula Weight: 197.45

RTECS Number: SN 1575000

PHYSICAL AND CHEMICAL PROPERTIES

Appearance and Odor: Colorless needles or yellow solid with a strong phenolic odor.

Boiling Point: 246 °C [1].

Dissociation Constant: 7.42 [2].

Henry's Law Constant: 9.07×10^{-8} atm·m^3/mol at 25 °C (estimated) [3].

Ionization Potential: No data found.

Log K_{oc}: 3.03 [2].

Log K_{ow}: 2.80 [4]; 2.97 [5]; 3.72 [6]; 3.06, 3.69 [7].

Melting Point: 69.5 °C [1].

Solubility in Organics: Soluble in ethanol and ether [1].

Solubility in Water: 800 mg/L at 25 °C, 2,430 mg/L at 96 °C [8]; 420 mg/L at 20-25 °C [4]; 0.9 g/L at 20-25 °C [9].

Specific Density: 1.4901 at 75/4 °C [1].

Transformation Products: No data found.

Vapor Pressure: 1 mm at 76.5 °C [10]; 10 mm at 120.2 °C, 40 mm at 152.2 °C, 100 mm at 177.8 °C, 400 mm at 222.5 °C, 760 mm at 246.0 °C [1]; 0.0084 mm at 24 °C [2]; 0.0025 mm at 8 °C, 0.017 mm at 25 °C [3].

FIRE HAZARDS

Flash Point: Non-combustible [11].

Lower Explosive Limit (LEL): No data found.

Upper Explosive Limit (UEL): No data found.

HEALTH HAZARD DATA

Immediately Dangerous to Life or Health (IDLH): No data found.

Permissible Exposure Limits (PEL) in Air: No standards set.

MANUFACTURING

Selected Manufacturers:

Aldrich Chemical Co.
940 West Saint Paul Ave.
Milwaukee, WI 53233

Fluka Chemical Corp.
980 South Second St.
Ronkonkoma, NY 11779

Pfaltz & Bauer, Inc.
172 East Aurora St.
Waterbury, CT 06708

Uses: Manufacturing of fungicides, bactericides, antiseptics, germicides; wood, and glue preservatives; in textiles to prevent mildew; defoliant; disinfectant; organic synthesis.

REFERENCES

1. Weast, R.C., Ed. *CRC Handbook of Chemistry and Physics*, 67th ed. (Boca Raton, FL: CRC Press, Inc., 1986), 2406 p.
2. Howard, P.H. *Handbook of Environmental Fate and Exposure Data for Organic Chemicals* (Chelsea, MI: Lewis Publishers, Inc., 1989), 574 p.
3. Leuenberger, C., Ligocki, M.P., and J.F. Pankow. "Trace Organic Compounds in Rain. 4. Identities, Concentrations, and Scavenging Mechanisms for Phenols in Urban Air and Rain," *Environ. Sci. Technol.*, 19(11):1053-1058 (1985).
4. Geyer, H., Sheehan, P., Kotzias, D., Freitag, D., and F. Korte. "Prediction of Ecotoxicological Behaviour of Chemicals: Relationship Between Physico-Chemical Properties and Bioaccumulation of Organic Chemicals in the Mussell *Mytilus edulis*," *Chemosphere*, 11(11):1121-1134 (1982).
5. Isnard, S., and S. Lambert. "Estimating Bioconcentration Factors from Octanol-Water Partition Coefficient and Aqueous Solubility," *Chemosphere*, 17(1):21-34 (1988).
6. Schellenberg, K., Leuenberger, C., and R.P. Schwarzenbach. "Sorption of Chlorinated Phenols by Natural Sediments and Aquifer Materials," *Environ. Sci. Technol.*, 18(9):652-657 (1984).
7. Leo, A., Hansch, C., and D. Elkins. "Partition Coefficients and Their Uses," *Chem. Rev.*, 71(6):525-616 (1971).
8. Verschueren, K. *Handbook of Environmental Data on Organic Chemicals* (New York: Van Nostrand Reinhold Co., 1983), 1310 p.
9. Kilzer, L., Scheunert, I., Geyer, H., Klein, W., and F. Korte. "Laboratory Screening of the Volatilization Rates of Organic Chemicals from Water and Soil," *Chemosphere*, 8(10):751-761 (1979).
10. Perry, R.H., and C.H. Chilton. *Chemical Engineers Handbook*, 5th ed. (New York, McGraw-Hill, Inc., 1973), p. 3-60.
11. "NIOSH Pocket Guide to Chemical Hazards," U.S. Department of Health and Human Services, U.S. Government Printing Office (1987), 241 p.

VINYL ACETATE

Synonyms: Acetic acid, ethenyl ester; Acetic acid, ethylene ester; Acetic acid, vinyl ester; 1-Acetoxyethylene; Ethenyl acetate; Ethenylethanoate; UN 1301; VAC; VAM; Vinyl acetate H.Q.

Structural Formula:

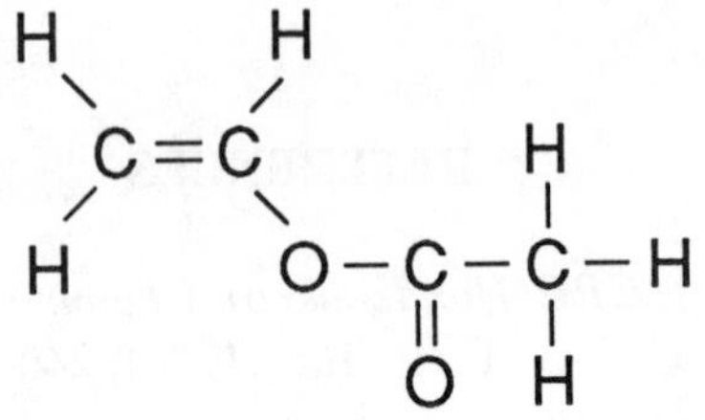

$$
\begin{array}{ccccc}
H & & H & & \\
\backslash & & / & & H \\
C & = & C & & | \\
/ & & \backslash & & H \\
H & & O-C-C-H \\
& & \parallel \;\; | \\
& & O \;\; H
\end{array}
$$

CHEMICAL DESIGNATIONS

CAS Registry Number: 108-05-4

DOT Designation: 1301

Empirical Formula: $C_4H_6O_2$

Formula Weight: 86.09

RTECS Number: AK 0875000

PHYSICAL AND CHEMICAL PROPERTIES

Appearance and Odor: Colorless, watery liquid with a pleasant fruity odor. Slowly polymerizes in light to a colorless, transparent mass.

Boiling Point: 72.2 °C [1].

Henry's Law Constant: 4.81×10^{-4} atm·m³/mol (calculated) [2].

Ionization Potential: 9.19 eV [3].

Log K_{oc}: 0.45 (estimated) [4].

Log K_{ow}: 0.73 [2].

Melting Point: -93.2 °C [1].

Solubility in Organics: Soluble in acetone, ethanol, benzene, chloroform, and ether [1].

Solubility in Water: 25,000 ppm at 25 °C [5]; 20,000 mg/L at 20 °C [6].

Specific Density: 0.9317 at 20/4 °C [1].

Transformation Products: No data found.

Vapor Density: 3.52 g/L at 25 °C, 2.97 (air = 1).

Vapor Pressure: 83 mm at 20 °C, 115 mm at 25 °C, 140 mm at 30 °C [7]; 1 mm at -48 °C, 10 mm at -18.9 °C, 40 mm at 5.3 °C, 100 mm at 23.3 °C, 400 mm at 55.5 °C, 760 mm at 72.5 °C [1]; 100 mm at 21.5 °C [8].

FIRE HAZARDS

Flash Point: -8 °C [9].

Lower Explosive Limit (LEL): 2.6% [10].

Upper Explosive Limit (UEL): 13.4% [10].

HEALTH HAZARD DATA

Immediately Dangerous to Life or Health (IDLH): No data found.

Permissible Exposure Limits (PEL) in Air: 10 ppm ($\approx$30 mg/m^3), 20 ppm ($\approx$60 mg/m^3) STEL [11].

MANUFACTURING

Selected Manufacturers:

Celanese Chemical Co.
245 Park Ave.
New York, NY 10017

National Distiller and Chemical Corp.
U.S. Industrial Chemical Division
Houston, TX 77000

Uses: Manufacturing of polyvinyl acetate, polyvinyl alcohol, polyvinyl chloride-acetate resins; used particularly in latex paint; paper coatings; adhesives; textile finishing; safety glass interlayers.

REFERENCES

1. Weast, R.C., Ed. *CRC Handbook of Chemistry and Physics*, 67th ed. (Boca Raton, FL: CRC Press, Inc., 1986), 2406 p.
2. Howard, P.H. *Handbook of Environmental Fate and Exposure Data for Organic Chemicals* (Chelsea, MI: Lewis Publishers, Inc., 1989), 574 p.
3. *Instruction Manual - Model ISP1 101: Intrinsically Safe Portable Photoionization Analyzer* (Newton, MA: HNU Systems, Inc., 1986), 86 p.
4. Montgomery, J.H. Unpublished results (1989).
5. Amoore, J.E., and E. Hautala. "Odor as an Aide to Chemical Safety: Odor Thresholds Compared with Threshold Limit Values and Volatilities for 214 Industrial Chemicals in Air and Water Dilution," *J. Appl. Toxicol.*, 3(6):272-290 (1983).
6. Dean, J.A., Ed., *Lange's Handbook of Chemistry*, 11th ed. (New York: McGraw-Hill, Inc., 1973), 1570 p.
7. Verschueren, K. *Handbook of Environmental Data on Organic Chemicals* (New York: Van Nostrand Reinhold Co., 1983), 1310 p.
8. Sax, N.I. *Dangerous Properties of Industrial Materials* (New York: Van Nostrand Reinhold Co., 1984), 3124 p.
9. "NIOSH Pocket Guide to Chemical Hazards," U.S. Department of Health and Human Services, U.S. Government Printing Office (1987), 241 p.
10. Weiss, G. *Hazardous Chemicals Data Book* (Park Ridge, NJ: Noyes Data Corp., 1986), 1069 p.
11. *Documentation of the Threshold Limit Values and Biological Exposure Indices* (Cincinnati, OH: American Conference of Governmental Industrial Hygienists, 1986), 744 p.

VINYL CHLORIDE

Synonyms: **Chlorethene;** Chlorethylene; Chloroethene; 1-Chloroethene; Chloroethylene; 1-Chloroethylene; Ethylene monochloride; Monochloroethene; Monochloroethylene; MVC; RCRA waste number U043; Trovidur; UN 1086; VC; VCM; Vinyl C monomer; Vinyl chloride monomer.

Structural Formula:

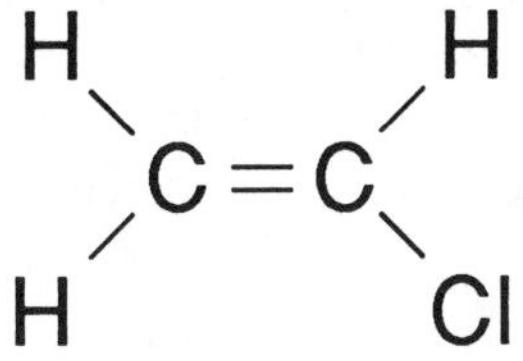

$$\begin{array}{ccc} H & & H \\ \diagdown & & \diagup \\ & C = C & \\ \diagup & & \diagdown \\ H & & Cl \end{array}$$

CHEMICAL DESIGNATIONS

CAS Registry Number: 75-01-4

DOT Designation: 1086

Empirical Formula: C_2H_3Cl

Formula Weight: 62.50

RTECS Number: KU 9625000

PHYSICAL AND CHEMICAL PROPERTIES

Appearance and Odor: Colorless liquified compressed gas with a faint, sweetish odor.

Boiling Point: -13.4 °C [1].

Henry's Law Constant: 2.78 atm·m^3/mol [2]; 0.022 atm·m^3/mol [3]; 2.37 atm·m^3/mol [4]; 1.22 atm·m^3/mol at 10 °C [5]; 0.056 atm·m^3/mol at 25 °C [6].

Ionization Potential: 9.995 eV [7].

Log K$_{oc}$: 0.39 using method of Karickhoff and others [8].

Log K$_{ow}$: 0.60 [9].

Melting Point: -153.8 °C [1].

Solubility in Organics: Soluble in ethanol, carbon tetrachloride, and ether [10].

Solubility in Water: 1,100 mg/L at 25 °C [11]; 60 mg/L at 10 °C [12]; 2,700 mg/L at 25 °C [5]; 0.95 wt% at 15 °C, 0.995 wt% at 16 °C, 0.915 wt% at 20.5 °C, 0.88 wt% at 26 °C, 0.89 wt% at 29.5 °C, 0.94 wt% at 35 °C, 0.89 wt% at 41 °C, 0.88 wt% at 46.5 °C, 0.95 wt% at 55 °C, 0.92 wt% at 65 °C, 0.98 wt% at 72.5 °C, 1.00 wt% at 80 °C, 1.12 wt% at 85 °C [13].

Specific Density: 0.9106 at 20/4 °C [1]; 0.969 at -13/4 °C [14]; 0.94 at 13.9/4 °C, 0.9121 at 20/4 °C [15].

Transformation Products: Irradation of vinyl chloride in the presence of nitrogen dioxide for 160 minutes produced formic acid, hydrochloric acid, carbon monoxide, formaldehyde, and ozone. Trace amounts of formyl chloride and nitric acid were also identified. In the presence of ozone, however, vinyl chloride photooxidized to carbon monoxide, formaldehyde, formic acid, and small amounts of hydrochloric acid [16].

Vapor Density: 2.86 g/L at 0 °C [17]; 2.55 g/L at 25 °C, 2.16 (air = 1).

Vapor Pressure: 240 mm at -40 °C; 580 mm at -20 °C, 2,660 mm at 25 °C [11]; 2,580 mm at 20 °C [18]; 1 mm at -105.6 °C, 10 mm at -83.7 °C, 40 mm at -66.8 °C, 100 mm at -53.2 °C, 400 mm at -28.0 °C, 760 mm at -13.8 °C [1]; 2,530 mm at 20 °C [19]; 2,600 at 25 °C [20]; 0.50 atm at -30 °C, 0.77 atm at -20 °C, 1.17 atm at -10 °C, 1.70 atm at 0 °C, 2.43 atm at 10 °C, 3.33 atm at 20 °C, 4.51 atm at 30 °C, 5.94 atm at 40 °C, 7.80 atm at 50 °C, 9.93 atm at 60 °C [15]; 2,320 mm at 20 °C [21].

FIRE HAZARDS

Flash Point: -78 °C [7].

Lower Explosive Limit (LEL): 3.6% [7].

Upper Explosive Limit (UEL): 33% [7].

HEALTH HAZARD DATA

Immediately Dangerous to Life or Health (IDLH): Potential human carcinogen [7].

Permissible Exposure Limits (PEL) in Air: 1 ppm, 5 ppm 15-minute ceiling [22]; lowest detectable limit [7]; 5 ppm ($\approx$10 mg/m^3) [19].

MANUFACTURING

Selected Manufacturer:

Dow Chemical Co.
Midland, MI 48640

Uses: Manufacturing of polyvinyl chloride and copolymers; adhesives for plastics; refrigerant; extraction solvent; organic synthesis.

REFERENCES

1. Weast, R.C., Ed. *CRC Handbook of Chemistry and Physics*, 67th ed. (Boca Raton, FL: CRC Press, Inc., 1986), 2406 p.
2. Gossett, J.M. "Measurement of Henry's Law Constants for C_1 and C_2 Chlorinated Hydrocarbons," *Environ. Sci. Technol.*, 21(2):202-208 (1987).
3. Pankow, J.F., and M.E. Rosen. "Determination of Volatile Compounds in Water by Purging Directly to a Capillary Column with Whole Column Cryotrapping," *Environ. Sci. Technol.*, 22(4):398-405 (1988).
4. Jury, W.A., Spencer, W.F., and W.J. Farmer. "Behavior Assessment Model for Trace Organics in Soil: III. Application of Screening Model," *J. Environ. Qual.*, 13(4):573-579 (1984).
5. Dilling, W.L. "Interphase Transfer Processes. II. Evaporation Rates of Chloro Methanes, Ethanes, Ethylenes, Propanes, and Propylenes from Dilute Aqueous Solutions. Comparisons with Theoretical Predictions," *Environ. Sci. Technol.*, 11(4):405-409 (1977).
6. Hine, J., and P.K. Mookerjee. "The Intrinsic Hydrophilic Character of Organic Compounds. Correlations in Terms of Structural Contributions," *J. Org. Chem.*, 40(3):292-298 (1975).
7. "NIOSH Pocket Guide to Chemical Hazards," U.S. Department of Health and Human Services, U.S. Government Printing Office

(1987), 241 p.

8. Karickhoff, S.W., Brown, D.S., and T.A. Scott. "Sorption of Hydrophobic Pollutants on Natural Sediments," *Water Res.*, 13:241-248 (1979).

9. Radding, S.B., Mill, T., Gould, C.W., Lia, D.H., Johnson, H.L., Bomberger, D.S., and C.V. Fojo. "The Environmental Fate of Selected Polynuclear Aromatic Hydrocarbons," Office of Toxic Substances, U.S. EPA Report-560/5-75-009 (1976), 122 p.

10. "Chemical, Physical, and Biological Properties of Compounds Present at Hazardous Waste Sites," U.S. EPA Report-530/SW-89-010 (1985), 619 p.

11. Verschueren, K. *Handbook of Environmental Data on Organic Chemicals* (New York: Van Nostrand Reinhold Co., 1983), 1310 p.

12. Pearson, C.R., and G. McConnell. "Chlorinated C_1 and C_2 Hydrocarbons in the Marine Environment," in *Proc. R. Soc. London*, B189(1096):305-322 (1975).

13. DeLassus, P.T., and D.D. Schmidt. "Solubilities of Vinyl Chloride and Vinylidene Chloride in Water," *J. Chem. Eng. Data*, 26(3):274-276 (1981).

14. Weiss, G. *Hazardous Chemicals Data Book* (Park Ridge, NJ: Noyes Data Corp., 1986), 1069 p.

15. Standen, A., Ed. *Kirk-Othmer Encyclopedia of Chemical Technology, Volume 4*, 2nd ed. (New York: John Wiley & Sons, Inc., 1964), 937 p.

16. Gay, B.W. Jr., Hanst, P.L., Bufalini, J.J., and R.C. Noonan. "Atmospheric Oxidation of Chlorinated Ethylenes," *Environ. Sci. Technol.*, 10(1):58-67 (1976).

17. Hayduk, W., and H. Laudie. "Vinyl Chloride Gas Compressibility and Solubility in Water and Aqueous Potassium Laurate Solutions," *J. Chem. Eng. Data*, 19(3):253-257 (1974).

18. Lyman, W.J., Reehl, W.F., and D.H. Rosenblatt. *Handbook of Chemical Property Estimation Methods: Environmental Behavior of Organic Compounds* (New York: McGraw-Hill, Inc., 1982).

19. *Documentation of the Threshold Limit Values and Biological Exposure Indices* (Cincinnati, OH: American Conference of Governmental Industrial Hygienists, 1986), 744 p.

20. Sax, N.I. *Dangerous Properties of Industrial Materials* (New York: Van Nostrand Reinhold Co., 1984), 3124 p.

21. McConnell, G., Ferguson, D.M., and C.R. Pearson. "Chlorinated Hydrocarbons and the Environment," *Endeavour*, 34(121):13-18 (1975).

22. "General Industry Standards for Toxic and Hazardous Substances," U.S. Code of Federal Regulations 1910, Subpart Z Section 1910.1000 (July 1982).

o–XYLENE

Synonyms: 1,2-Dimethylbenzene; *o*-Dimethylbenzene; *o*-Methyltoluene; UN 1307; 1,2-Xylene; ortho-Xylene; *o*-Xylol.

Structural Formula:

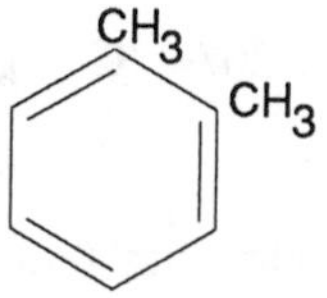

CHEMICAL DESIGNATIONS

CAS Registry Number: 95-47-6

DOT Designation: 1307

Empirical Formula: C_8H_{10}

Formula Weight: 106.17

RTECS Number: ZE 2450000

PHYSICAL AND CHEMICAL PROPERTIES

Appearance: Clear, colorless liquid.

Boiling Point: 144.4 °C [1]; 143-144 °C [2].

Henry's Law Constant: 0.00527 atm·m^3/mol [3]; 0.0050 atm·m^3/mol [4]; 0.00535 atm·m^3/mol at 25 °C [5].

Ionization Potential: 8.56 eV [6].

Log K_{oc}: 2.11 [7].

Log K_{ow}: 2.95 [8]; 3.13 [9]; 2.77 [10]; 3.16 [11]; 3.08 [12].

Melting Point: -25.2 °C [1].

Solubility in Organics: Soluble in acetone, ethanol, benzene, and ether [1].

Solubility in Water: 152 mg/L at 20 °C [13]; 200 mg/L [8]; 204 mg/L at 25 °C [14]; 175 mg/L at 25 °C [3]; 0.00208 M at 25 °C [9]; 142 mg/L at 0 °C, 167 mg/L at 25 °C [15]; 142 ppm at 0 °C, 213 ppm at 25 °C [16]; 170.5 ppm at 25 °C, 129.6 ppm in artificial seawater at 25 °C [17]; 176.2 mg/L at 25 °C [18]; 1,742 mmols/L at 20 °C [12].

Specific Density: 0.8802 at 20/4 °C [1]; 0.88011 at 20/4 °C [19]; 0.87596 at 25/4 °C [20].

Transformation Products: No data found.

Vapor Density: 4.34 g/L at 25 °C, 3.66 (air = 1).

Vapor Pressure: 1 mm at -9.8 °C, 10 mm at 25.9 °C, 40 mm at 52.8 °C, 100 mm at 74.1 °C; 400 mm at 113.8 °C; 760 mm at 136.2 °C [1]; 6.6 mm at 25 °C [3].

FIRE HAZARDS

Flash Point: 17 °C [21]; 46.1 °C (open cup) [22].

Lower Explosive Limit (LEL): 1% [6].

Upper Explosive Limit (UEL): 6% [6].

HEALTH HAZARD DATA

Immediately Dangerous to Life or Health (IDLH): 1,000 ppm (isomeric mixture) [6].

Permissible Exposure Limits (PEL) in Air: For total xylenes (containing *ortho*, *meta* and *para* isomers) - 100 ppm ($\approx$435 mg/m^3) [23]; 100 ppm, 150 ppm ($\approx$655 mg/m^3) STEL [24].

MANUFACTURING

Selected Manufacturers:

ARCO Chemical Co.
260 South Broad St.
Philadelphia, PA 19101

Cities Services Co., Inc.
Petrochemicals Division
60 Wall St.
New York, NY 10005

Shell Chemical Co.
Petrochemicals Division
Houston, TX 77001

Uses: Preparation of phthalic acid, phthalic anhydride, terephthalic acid, isophthalic acid; solvent for alkyd resins, lacquers, enamels, rubber cements; manufacturing of dyes, pharmaceuticals, and insecticides; motor fuels.

REFERENCES

1. Weast, R.C., Ed. *CRC Handbook of Chemistry and Physics*, 67th ed. (Boca Raton, FL: CRC Press, Inc., 1986), 2406 p.
2. *Fluka Catalog 1988/89 - Chemika-Biochemika* (Ronkonkoma, NY: Fluka Chemical Corp., 1988), 1536 p.
3. Mackay, D., and A.W. Wolkoff. "Rate of Evaporation of Low-Solubility Contaminants from Water Bodies to Atmosphere," *Environ. Sci. Technol.*, 7(7):611-614 (1973).
4. Pankow, J.F., and M.E. Rosen. "Determination of Volatile Compounds in Water by Purging Directly to a Capillary Column with Whole Column Cryotrapping," *Environ. Sci. Technol.*, 22(4):398-405 (1988).
5. Hine, J., and P.K. Mookerjee. "The Intrinsic Hydrophilic Character of Organic Compounds. Correlations in Terms of Structural Contributions," *J. Org. Chem.*, 40(3):292-298 (1975).
6. "NIOSH Pocket Guide to Chemical Hazards," U.S. Department of Health and Human Services, U.S. Government Printing Office (1987), 241 p.
7. Abdul, S.A., Gibson, T.L., and D.N. Rai. "Statistical Correlations for Predicting the Partition Coefficient for Nonpolar Organic Contaminants Between Aquifer Organic Carbon and Water," *Haz. Waste Haz. Mater.*, 4(3):211-222 (1987).
8. Isnard, S., and S. Lambert. "Estimating Bioconcentration Factors from Octanol-Water Partition Coefficient and Aqueous Solubility," *Chemosphere*, 17(1):21-34 (1988).
9. Tewari, Y.B., Miller, M.M., Wasik, S.P., and D.E. Martire. "Aqueous Solubility and Octanol/Water Partition Coefficient of Organic

Compounds at 25.0 °C," *J. Chem. Eng. Data*, 27(4):451-454 (1982).

10. Leo, A., Hansch, C., and D. Elkins. "Partition Coefficients and Their Uses," *Chem. Rev.*, 71(6):525-616 (1971).

11. Hodson, J., and N.A. Williams. "The Estimation of the Adsorption Coefficient (K_{oc}) for Soils by High Performance Liquid Chromatography," *Chemosphere*, 19(1):67-77 (1988).

12. Galassi, S., Mingazzini, M., Viganò, L., Cesareo, D., and M.L. Tosato. "Approaches to Modeling Toxic Responses of Aquatic Organisms to Aromatic Hydrocarbons," *Ecotoxicol. Environ. Safety*, 16(2):158-169 (1988).

13. Mackay, D., and W.Y. Shiu. "A Critical Review of Henry's Law Constants for Chemicals of Environmental Interest," *J. Phys. Chem. Ref. Data*, 10(4):1175-1199 (1981).

14. Andrews, L.J., and R.M. Keefer. "Cation Complexes of Compounds Containing Carbon-Carbon Double Bonds. IV. The Argentation of Aromatic Hydrocarbons," *J. Am. Chem. Soc.*, 71(11):3644-3647 (1949).

15. Brookman, G.T., Flanagan, M., and J.O. Kebe. "Literature Survey: Hydrocarbon Solubilities and Attenuation Mechanisms," API Publication 4414, (Washington, DC: American Petroleum Institute, 1985), 101 p.

16. Polak, J., and B.C.-Y. Lu. "Mutual Solubilities of Hydrocarbons and Water at 0 and 25 °C," *Can. J. Chem.*, 51(24):4018-4023 (1973).

17. Sutton, C., and J.A. Calder. "Solubility of Alkylbenzenes in Distilled Water and Seawater at 25 °C," *J. Chem. Eng. Data*, 20(3):320-322 (1975).

18. Hermann, R.B. "Theory of Hydrophobic Bonding. II. The Correlation of Hydrocarbon Solubility in Water with Solvent Cavity Surface Area," *J. Phys. Chem.*, 76(19):2754-2759 (1972).

19. Huntress, E.H., and S.P. Mulliken. *Identification of Pure Organic Compounds - Tables of Data on Selected Compounds of Order I* (New York: John Wiley and Sons, Inc., 1941), 691 p.

20. Standen, A., Ed. *Kirk-Othmer Encyclopedia of Chemical Technology, Volume 22*, 2nd ed. (New York: John Wiley and Sons, Inc., 1970), 702 p.

21. Windholz, M., Budavari, S., Stroumtsos, L.S., and M.N. Fertig, Eds. *The Merck Index*, 9th ed., (Rahway, NJ: Merck and Co., 1976), 1313 p.

22. Hawley, G.G. *The Condensed Chemical Dictionary* (New York: Van Nostrand Reinhold Co., 1981), 1135 p.

23. "General Industry Standards for Toxic and Hazardous Substances," U.S. Code of Federal Regulations 1910, Subpart Z Section 1910.1000 (July 1982).

24. *Documentation of the Threshold Limit Values and Biological Exposure*

Indices (Cincinnati, OH: American Conference of Governmental Industrial Hygienists, 1986), 744 p.

m–XYLENE

Synonyms: 1,3-Dimethylbenzene; *m*-Dimethylbenzene; *m*-Methyltoluene; UN 1307; 1,3-Xylene; *m*-Xylol.

Structural Formula:

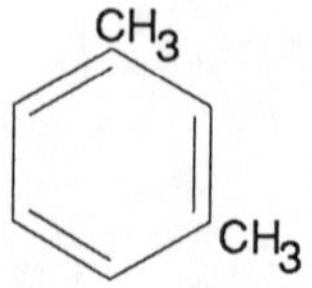

CHEMICAL DESIGNATIONS

CAS Registry Number: 108-38-3

DOT Designation: 1307

Empirical Formula: C_8H_{10}

Formula Weight: 106.17

RTECS Number: ZE 2275000

PHYSICAL AND CHEMICAL PROPERTIES

Appearance and Odor: Clear, colorless, watery liquid with a sweet odor.

Boiling Point: 139.1 °C [1]; 137-138 °C [2].

Henry's Law Constant: 0.0070 atm·m^3/mol [3]; 0.0063 atm·m^3/mol at 25 °C [4].

Ionization Potential: 8.58 eV [5].

Log K_{oc}: 3.20 [6].

Log K_{ow}: 3.20 [7].

Melting Point: -47.9 °C [1].

Solubility in Organics: Soluble in acetone, ethanol, benzene, and ether [1].

Solubility in Water: 158 mg/L [8]; 173 mg/L at 25 °C [9]; 0.00151 M at 25 °C [10]; 0.0146 wt% at 25 °C [11]; 196 ppm at 0 °C, 162 ppm at 25 °C [12]; 146.0 ppm at 25 °C, 106.0 ppm in artificial seawater at 25 °C [13]; 157.0 mg/L at 25 °C [14]; 1,525 mmols/L at 20 °C [15]; 170 ppm at 25 °C [16].

Specific Density: 0.8642 at 20/4 °C [1]; 0.8684 at 25/4 °C [17]; 0.86407 at 20/4 °C, 0.85979 at 25/4 °C [18].

Transformation products: Microbial degradation produced 3-methyl-benzyl alcohol, 3-methylbenzaldehyde, *m*-toluic acid, and 3-methyl catechol [19].

Vapor Density: 4.34 g/L at 25 °C, 3.66 (air = 1).

Vapor Pressure: 1 mm at -6.9 °C, 10 mm at 28.3 °C, 40 mm at 55.3 °C, 100 mm at 76.8 °C; 400 mm at 116.7 °C; 760 mm at 139.1 °C; [1]; 8.287 at 25 °C [20]; 15.2 at 35 °C [21].

FIRE HAZARDS

Flash Point: 25 °C [5].

Lower Explosive Limit (LEL): 1.1% [5].

Upper Explosive Limit (UEL): 7% [5].

HEALTH HAZARD DATA

Immediately Dangerous to Life or Health (IDLH): 1,000 ppm (isomeric mixture) [5].

Permissible Exposure Limits (PEL) in Air: For total xylenes (containing *ortho*, *meta* and *para* isomers) - 100 ppm ($\approx$435 mg/m^3) [22]; 100 ppm, 150 ppm ($\approx$655 mg/m^3) STEL [23].

MANUFACTURING

Selected Manufacturers:

ARCO Chemical Co.
260 South Broad St.
Philadelphia, PA 19101

Chevron Chemical Co.
Industrial Chemicals Division
200 Bush St.
San Francisco, CA 94120

Fallek Chemical Corp.
460 Park Ave.
New York, NY 10022

Uses: Solvent; intermediate for dyes and organic synthesis; insecticides; aviation fuel.

REFERENCES

1. Weast, R.C., Ed. *CRC Handbook of Chemistry and Physics*, 67th ed. (Boca Raton, FL: CRC Press, Inc., 1986), 2406 p.
2. *Fluka Catalog 1988/89 - Chemika-Biochemika* (Ronkonkoma, NY: Fluka Chemical Corp., 1988), 1536 p.
3. Pankow, J.F., and M.E. Rosen. "Determination of Volatile Compounds in Water by Purging Directly to a Capillary Column with Whole Column Cryotrapping," *Environ. Sci. Technol.*, 22(4):398-405 (1988).
4. Hine, J., and P.K. Mookerjee. "The Intrinsic Hydrophilic Character of Organic Compounds. Correlations in Terms of Structural Contributions," *J. Org. Chem.*, 40(3):292-298 (1975).
5. "NIOSH Pocket Guide to Chemical Hazards," U.S. Department of Health and Human Services, U.S. Government Printing Office (1987), 241 p.
6. Abdul, S.A., Gibson, T.L., and D.N. Rai. "Statistical Correlations for Predicting the Partition Coefficient for Nonpolar Organic Contaminants Between Aquifer Organic Carbon and Water," *Haz. Waste Haz. Mater.*, 4(3):211-222 (1987).
7. Leo, A., Hansch, C., and D. Elkins. "Partition Coefficients and Their Uses," *Chem. Rev.*, 71(6):525-616 (1971).
8. Isnard, S., and S. Lambert. "Estimating Bioconcentration Factors from Octanol-Water Partition Coefficient and Aqueous Solubility," *Chemosphere*, 17(1):21-34 (1988).
9. Andrews, L.J., and R.M. Keefer. "Cation Complexes of Compounds Containing Carbon-Carbon Double Bonds. IV. The Argentation of Aromatic Hydrocarbons," *J. Am. Chem. Soc.*, 71(11):3644-3647 (1949).
10. Tewari, Y.B., Miller, M.M., Wasik, S.P., and D.E. Martire. "Aqueous Solubility and Octanol/Water Partition Coefficient of Organic

Compounds at 25.0 °C," *J. Chem. Eng. Data*, 27(4):451-454 (1982).

11. Riddick, J.A., Bunger, W.B., and T.K. Sakano. *Organic Solvents - Physical Properties and Methods of Purification. Volume II* (New York: John Wiley and Sons, Inc., 1986), 1325 p.

12. Polak, J., and B.C.-Y. Lu. "Mutual Solubilities of Hydrocarbons and Water at 0 and 25 °C," *Can. J. Chem.*, 51(24):4018-4023 (1973).

13. Sutton, C., and J.A. Calder. "Solubility of Alkylbenzenes in Distilled Water and Seawater at 25 °C," *J. Chem. Eng. Data*, 20(3):320-322 (1975).

14. Hermann, R.B. "Theory of Hydrophobic Bonding. II. The Correlation of Hydrocarbon Solubility in Water with Solvent Cavity Surface Area," *J. Phys. Chem.*, 76(19):2754-2759 (1972).

15. Galassi, S., Mingazzini, M., Viganò, L., Cesareo, D., and M.L. Tosato. "Approaches to Modeling Toxic Responses of Aquatic Organisms to Aromatic Hydrocarbons," *Ecotoxicol. Environ. Safety*, 16(2):158-169 (1988).

16. Amoore, J.E., and E. Hautala. "Odor as an Aide to Chemical Safety: Odor Thresholds Compared with Threshold Limit Values and Volatilities for 214 Industrial Chemicals in Air and Water Dilution," *J. Appl. Toxicol.*, 3(6):272-290 (1983).

17. Hawley, G.G. *The Condensed Chemical Dictionary* (New York: Van Nostrand Reinhold Co., 1981), 1135 p.

18. Huntress, E.H., and S.P. Mulliken. *Identification of Pure Organic Compounds - Tables of Data on Selected Compounds of Order I* (New York: John Wiley and Sons, Inc., 1941), 691 p.

19. Verschueren, K. *Handbook of Environmental Data on Organic Chemicals* (New York: Van Nostrand Reinhold Co., 1983), 1310 p.

20. Bohon, R.L., and W.F. Claussen. "The Solubility of Aromatic Hydrocarbons in Water," *J. Am. Chem. Soc.*, 73(4):1571-1578 (1951).

21. Hine, J., Haworth, H.W., and O.B. Ramsey. "Polar Effects on Rates and Equilibria. VI. The Effect of Solvent on the Transmission of Polar Effects," *J. Am. Chem. Soc.*, 85(10):1473-1475 (1963).

22. "General Industry Standards for Toxic and Hazardous Substances," U.S. Code of Federal Regulations 1910, Subpart Z Section 1910.1000 (July 1982).

23. *Documentation of the Threshold Limit Values and Biological Exposure Indices* (Cincinnati, OH: American Conference of Governmental Industrial Hygienists, 1986), 744 p.

p–XYLENE

Synonyms: Chromar; 1,4-Dimethylbenzene; *p*-Dimethylbenzene; *p*-Methyltoluene; Scintillar; UN 1307; 1,4-Xylene; *p*-Xylol.

Structural Formula:

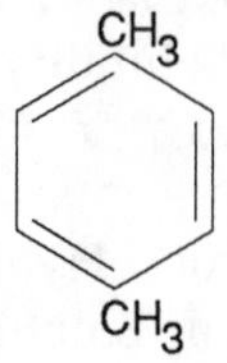

CHEMICAL DESIGNATIONS

CAS Registry Number: 106-42-3

DOT Designation: 1307

Empirical Formula: C_8H_{10}

Formula Weight: 106.17

RTECS Number: ZE 2625000

PHYSICAL AND CHEMICAL PROPERTIES

Appearance and Odor: Clear, colorless, watery liquid with a sweet odor.

Boiling Point: 138.3 °C [1].

Henry's Law Constant: 0.0071 atm·m³/mol [2]; 0.0063 atm·m³/mol at 25 °C [3].

Ionization Potential: 8.44 eV [4].

Log K_{oc}: 2.31 [5].

Log K_{ow}: 3.18 [6]; 3.15 [7].

Melting Point: 13.3 °C [1]; 13-15 °C [8].

Solubility in Organics: Soluble in acetone, ethanol, and benzene [1].

Solubility in Water: 200 mg/L at 25 °C [9]; 0.00202 M at 25 °C [6]; 198 mg/L at 25 °C [10]; 180 mg/L at 25 °C [11]; 0.0156 wt% at 20 °C [12]; 164 ppm at 0 °C, 185 ppm at 25 °C [13]; 156.0 ppm at 25 °C, 110.9 ppm in artificial seawater at 25 °C [14]; 163.3 mg/L at 25 °C [15].

Specific Density: 0.8811 at 20/4 °C [1]; 0.86100 at 20/4 °C, 0.85665 at 25/4 °C [16]; 0.85655 at 25/4 °C [17].

Transformation Products: Microbial degradation of *p*-xylene produced 4-methylbenzyl alcohol, 4-methylbenzaldehyde, *p*-toluic acid, and 4-methyl catechol [18].

Vapor Density: 4.34 g/L at 25 °C, 3.66 (air = 1).

Vapor Pressure: 1 mm at -8.1 °C, 10 mm at 27.3 °C, 40 mm at 54.4 °C, 100 mm at 79.9 °C; 400 mm at 115.9 °C; 760 mm at 138.3 °C [1]; 8.763 mm at 25 °C [10]; 15.8 mm at 35 °C [19].

FIRE HAZARDS

Flash Point: 27.2 °C (open cup) [20].

Lower Explosive Limit (LEL): 1.1% [21].

Upper Explosive Limit (UEL): 7% [21].

HEALTH HAZARD DATA

Immediately Dangerous to Life or Health (IDLH): 1,000 ppm (isomeric mixture) [21].

Permissible Exposure Limits (PEL) in Air: For total xylenes (containing *ortho*, *meta* and *para* isomers) - 100 ppm ($\approx$435 mg/m^3) [22]; 100 ppm, 150 ppm ($\approx$655 mg/m^3) STEL [23].

MANUFACTURING

Selected Manufacturers:

Amoco Chemical Co.
130 East Randolph Drive
Chicago, IL 60601

Chevron Chemical Co.
Industrial Chemicals Division
200 Bush St.
San Francisco, CA 94120

Uses: Preparation of terephthalic acid for polyester resins and fibers (Dacron, Mylar and Terylene), vitamins, pharmaceuticals, and insecticides.

REFERENCES

1. Weast, R.C., Ed. *CRC Handbook of Chemistry and Physics*, 67th ed. (Boca Raton, FL: CRC Press, Inc., 1986), 2406 p.
2. Pankow, J.F., and M.E. Rosen. "Determination of Volatile Compounds in Water by Purging Directly to a Capillary Column with Whole Column Cryotrapping," *Environ. Sci. Technol.*, 22(4):398-405 (1988).
3. Hine, J., and P.K. Mookerjee. "The Intrinsic Hydrophilic Character of Organic Compounds. Correlations in Terms of Structural Contributions," *J. Org. Chem.*, 40(3):292-298 (1975).
4. Franklin, J.L., Dillard, J.G., Rosenstock, H.M., Herron, J.T., Draxl K., and F.H. Field. "Ionization Potentials, Appearance Potentials and Heats of Formation of Gaseous Positive Ions," National Bureau of Standards Report NSRDS-NBS 26, U.S. Government Printing Office (1969), 289 p.
5. Abdul, S.A., Gibson, T.L., and D.N. Rai. "Statistical Correlations for Predicting the Partition Coefficient for Nonpolar Organic Contaminants Between Aquifer Organic Carbon and Water," *Haz. Waste Haz. Mater.*, 4(3):211-222 (1987).
6. Tewari, Y.B., Miller, M.M., Wasik, S.P., and D.E. Martire. "Aqueous Solubility and Octanol/Water Partition Coefficient of Organic Compounds at 25.0 °C," *J. Chem. Eng. Data*, 27(4):451-454 (1982).
7. Leo, A., Hansch, C., and D. Elkins. "Partition Coefficients and Their Uses," *Chem. Rev.*, 71(6):525-616 (1971).
8. *Fluka Catalog 1988/89 - Chemika-Biochemika* (Ronkonkoma, NY: Fluka Chemical Corp., 1988), 1536 p.
9. Andrews, L.J., and R.M. Keefer. "Cation Complexes of Compounds Containing Carbon-Carbon Double Bonds. IV. The Argentation of Aromatic Hydrocarbons," *J. Am. Chem. Soc.*, 71(11):3644-3647 (1949).
10. Bohon, R.L., and W.F. Claussen. "The Solubility of Aromatic Hydrocarbons in Water," *J. Am. Chem. Soc.*, 73(4):1571-1578 (1951).
11. Banerjee, S. "Solubility of Organic Mixtures in Water," *Environ. Sci.*

Technol., 18(8):587-591 (1984).

12. Riddick, J.A., Bunger, W.B., and T.K. Sakano. *Organic Solvents - Physical Properties and Methods of Purification. Volume II* (New York: John Wiley and Sons, Inc., 1986), 1325 p.

13. Polak, J., and B.C.-Y. Lu. "Mutual Solubilities of Hydrocarbons and Water at 0 and 25 °C," *Can. J. Chem.*, 51(24):4018-4023 (1973).

14. Sutton, C., and J.A. Calder. "Solubility of Alkylbenzenes in Distilled Water and Seawater at 25 °C," *J. Chem. Eng. Data*, 20(3):320-322 (1975).

15. Hermann, R.B. "Theory of Hydrophobic Bonding. II. The Correlation of Hydrocarbon Solubility in Water with Solvent Cavity Surface Area," *J. Phys. Chem.*, 76(19):2754-2759 (1972).

16. Huntress, E.H., and S.P. Mulliken. *Identification of Pure Organic Compounds - Tables of Data on Selected Compounds of Order I* (New York: John Wiley and Sons, Inc., 1941), 691 p.

17. Kirchnerová, J., and G.C.B. Cave. "The Solubility of Water in Low-Dielectric Solvents," *Can. J. Chem.*, 54(24):3909-3916 (1976).

18. Verschueren, K. *Handbook of Environmental Data on Organic Chemicals* (New York: Van Nostrand Reinhold Co., 1983), 1310 p.

19. Hine, J., Haworth, H.W., and O.B. Ramsey. "Polar Effects on Rates and Equilibria. VI. The Effect of Solvent on the Transmission of Polar Effects," *J. Am. Chem. Soc.*, 85(10):1473-1475 (1963).

20. Hawley, G.G. *The Condensed Chemical Dictionary* (New York: Van Nostrand Reinhold Co., 1981), 1135 p.

21. "NIOSH Pocket Guide to Chemical Hazards," U.S. Department of Health and Human Services, U.S. Government Printing Office (1987), 241 p.

22. "General Industry Standards for Toxic and Hazardous Substances," U.S. Code of Federal Regulations 1910, Subpart Z Section 1910.1000 (July 1982).

23. *Documentation of the Threshold Limit Values and Biological Exposure Indices* (Cincinnati, OH: American Conference of Governmental Industrial Hygienists, 1986), 744 p.

Conversion Factors

To convert	Into	Multiply by
acre-feet	feet^3	4.356×10^4
	gallons (U.S.)	3.529×10^5
	inches^3	7.527×10^7
	liters	1.233×10^6
	meters^3	1,233
	yards^3	1,613
acre-feet per day	feet^3 per second	0.5042
	gallons (U.S.) per minute	226.3
	liters^3 per second	14.28
	meters^3 per day	1,234
	meters^3 per second	0.01428
acres	feet^2	43,560
	hectares	0.4047
	inches^2	6.273×10^6
	kilometers^2	4.047×10^{-3}
	meters^2	4,047
	miles^2	1.563×10^{-3}
atmospheres	bars	1.01325
	millimeters of Hg	760
	pascals	1.01325×10^5
	torrs	760
centimeters	feet	0.03281
	millimeters	10
	inches	0.3937
	yards	0.01094
centimeters^2	feet^2	0.001076
	inches^2	0.155
	meters^2	1×10^{-4}
	yards^2	1.196×10^{-4}
centimeters^3	fluid ounces	0.03381
	feet^3	3.5314×10^{-5}
	inches^3	0.06102
	liters	0.001
	ounces (U.S., fluid)	0.03381

To convert	Into	Multiply by
centimeters per second	feet per day	2,835
	feet per minute	1.197
	feet per second	0.03281
	liters per meter per second	9.985
	meters per minute	0.6
	miles per minute	3.728×10^{-4}
Darcy	centimeters per second	9.66×10^{-4}
	feet per second	3.173×10^{-5}
	liters per meter per second	8.58×10^{-3}
feet	centimeters	30.48
	inches	12
	kilometers	3.048×10^{-4}
	meters	0.3048
	miles	1.894×10^{-4}
	millimeters	304.8
	yards	0.333
feet^2	acres	2.296×10^{-5}
	hectares	9.29×10^{-9}
	inches^2	144
	kilometers^2	9.29×10^{-8}
	meters^2	9.29×10^{-2}
	miles^2	3.587×10^{-8}
	yards^2	0.1111
feet^3	acre-feet	2.296×10^{-5}
	gallons (U.S.)	7.481
	inches^3	1,728
	liters	28.32
	meters^3	2.832×10^{-3}
	yards^3	3.704×10^{-2}
feet per day	feet per second	1.157×10^{-5}
	meters per second	3.528×10^{-6}
feet^3 per foot per day	gallons per foot per day	7.48052
	liters per meter per day	92.903
	meters^3 per meter per day	0.0929

To convert	Into	Multiply by
feet^3 per foot^2 per day	feet^3 per foot^2 per minute	6.944 x 10^{-4}
	gallons per foot^2 per day	7.4805
	inches^3 per inch^2 per hour	0.5
	liters per meter^2 per day	304.8
	meters^3 per meter^2 per day	0.3048
	millimeters^3 per inch^2 per hour	0.5
	millimeters^3 per millimeter^2 per hour	25.4
feet per second	feet per day	86,400
	feet per hour	3,600
	gallons (U.S.) per foot^2 per day	5.737 x 10^5
	kilometers per hour	1.097
	meters per second	0.3048
	miles per hour	0.6818
feet^3 per second	acre-feet per day	1.983
	feet^3 per minute	60.0
	gallons (U.S.) per minute	448.8
	liters per second	28.32
	meters^3 per second	0.02832
	meters^3 per day	2,447
gallons (U.S.)	acre-feet	3.068 x 10^{-6}
	feet^3	0.1337
	fluid ounces	128.0
	liters	3.785
	meters^3	3.785 x 10^{-3}
	yards^3	4.951 x 10^{-3}
gallons per foot per day	feet^3 per foot per day	0.13368
	liters per meter per day	12.42
	meters^3 per meter per day	0.01242

To convert	Into	Multiply by
gallons (U.S.) per foot^2 per day	centimeters per second	4.717×10^{-5}
	Darcy	0.05494
	feet per day	0.13368
	feet per second	1.547×10^{-6}
	gallons per foot^2 per minute	6.944×10^{-4}
	liters per meter^2 per day	40.7458
	meters per day	0.0407458
	meters per minute	2.83×10^{-5}
	meters per second	4.716×10^{-7}
gallons (U.S.) per foot^2 per minute	meters per day	58.67
	meters per second	0.06791
gallons (U.S.) per minute	acre-feet per day	4.419×10^{-3}
	feet^3 per second	2.228×10^{-3}
	liters per second	6.309×10^{-2}
	meters^3 per day	5.45
	meters^3 per second	6.309×10^{-5}
grams per centimeter^3	kilograms per meter^3	1000
	pounds per feet^3	62.428
	pounds per gallon (U.S.)	8.345
grams	kilograms	0.001
	ounces (avoirdupois)	0.03527
	pounds	0.022046
grams per liter	grains per gallon (U.S.)	58.4178
	grams per centimeter^3	0.001
	kilograms per meter^3	1
	pounds per feet^3	0.0624
	pounds per inch^3	3.61×10^{-5}
	pounds per gallon (U.S.)	8.35×10^{-3}
grams per meter^3	grains per feet^3	0.4370
	milligrams per liter	1.0
	pounds per gallon (U.S.)	8.345×10^{-5}

To convert	Into	Multiply by
hectares	acres	2.471
	feet^2	1.076 x 10^5
	inches^2	1.55 x 10^7
	kilometers^2	0.01
	meters^2	10,000
	miles^2	3.861 x 10^{-3}
	yards^2	11,959.90
inches	centimeters	2.54
	feet	0.8333
	kilometers	2.54 x 10^{-5}
	meters	2.54 x 10^{-2}
	miles	1.578 x 10^{-5}
	millimeters	25.4
inches^2	acres	1.594 x 10^{-8}
	centimeters^2	6.4516
	feet^2	6.944 x 10^{-3}
	hectares	6.452 x 10^{-8}
	kilometers^2	6.452 x 10^{-10}
	meters^2	6.452 x 10^{-4}
	millimeters^2	645.16
inches^3	acre-feet	1.329 x 10^{-8}
	feet^3	5.787 x 10^{-4}
	gallons (U.S.)	4.329 x 10^{-3}
	liters	1.639 x 10^{-2}
	meters^3	1.639 x 10^{-5}
	milliliters	16.387
	yards^3	2.143 x 10^{-5}
kilograms	grams	1000
	ounces (avoirdupois)	35.28
	pounds	2.205
	tons (metric)	0.001
kilograms per meter^3	grams per centimeter^3	0.001
	grams per liter	1.0
	pounds per feet^3	0.0624
	pounds per inch^3	3.6127 x 10^{-5}

To convert	Into	Multiply by
kilometers	feet	3,281
	inches	39,370
	meters	1,000
	miles	0.6214
	millimeters	1×10^6
kilometers^2	acres	247.1
	hectares	100
	inches^2	1.55×10^9
	meters^2	1×10^6
	miles^2	0.3861
kilometers per hour	feet per day	78,740
	feet per second	0.9113
	meters per second	0.2778
	miles per hour	0.6214
liters	acre-feet	8.106×10^{-7}
	decimeters^3	1
	feet^3	3.531×10^{-2}
	fluid ounces	33.814
	gallons (U.S.)	0.2642
	inches^3	61.02
	meters^3	0.001
	yards^3	1.308×10^{-3}
liters per second	acre-feet per day	7.005×10^{-2}
	feet^3 per hour	127.1328
	feet^3 per minute	2.11888
	feet^3 per second	3.531×10^{-2}
	gallons (U.S.) per minute	15.85
	gallons (U.S.) per hour	951.0194
	meters^3 per day	86.4
meters	feet	3.28084
	inches	39.3701
	kilometers	0.001
	miles	6.214×10^{-4}
	millimeters	1,000
	yards	1.0936

To convert	Into	Multiply by
meters^2	acres	2.471×10^{-4}
	feet^2	10.76
	hectares	1×10^{-4}
	inches^2	1.550
	kilometers	1×10^{-6}
	miles^2	3.861×10^{-7}
	yards^2	1.196
meters^3	acre-feet	8.106×10^{-4}
	feet^3	35.31
	gallons (U.S.)	264.2
	inches^3	6.102×10^4
	liters	1,000
	yards^3	1.308
meters per second	feet per day	283,447
	kilometers per hour	3.6
	feet per second	3.281
	miles per hour	2.237
meters^3 per day	acre-feet per day	6.051×10^6
	feet^3 per second	3.051×10^6
	gallons (U.S.) per minute	1.369×10^9
miles	feet	5,280
	kilometers	1.609
	meters	1,609
	millimeters	1.609×10^6
miles^2	acres	640
	feet^2	2.778×10^7
	hectares	259
	inches^2	4.014×10^9
	kilometers^2	2.59
	meters^2	2.59×10^6
miles per hour	feet per day	1.267×10^5
	kilometers per hour	1.609
	meters per second	0.447

To convert	Into	Multiply by
milligrams per liter	grams per meter^3	1.0
	pounds per feet^3	6.2428 x 10^{-5}
milliliters	centimeters^3	1.0
	liters	0.001
millimeters	centimeters	0.1
	feet	3.281 x 10^{-3}
	inches	0.03937
	kilometers	1.0 x 10^{-6}
	meters	1.0 x 10^{-3}
	miles	6.214 x 10^{-7}
millimeters^2	centimeters^2	0.01
	inches^2	0.00155
millimeters^3	centimeters^3	0.001
	inches^3	6.102 x 10^{-5}
	liters	1 x 10^{-6}
millimeters of Hg	atmospheres	1.316 x 10^{-3}
	pascals	133.3224
	torrs	1.0
ounces (avoirdupois)	grams	28.35
	kilograms	0.02835
	pounds	0.0625
pascals	atmospheres	9.869 x 10^{-6}
	millimeters of Hg	7.501 x 10^{-3}
pounds (avoirdupois)	kilograms	0.4535
	ounces (avoirdupois)	16
pounds per centimeter^3	pounds per inch^3	0.0361
pounds per feet^3	grams per centimeter^3	0.016
	grams per liter	27,680
	pounds per inch^3	5.787 x 10^{-4}
	pounds per gallon (U.S.)	0.1337

To convert	Into	Multiply by
pounds per gallon (U.S.)	grams per centimeter^3	0.1198
	grams per liter	119.8
	pounds per feet^3	7.481
	pounds per inch^3	4.329×10^{-3}
pounds per inch^3	pounds per centimeter^3	27.68
	pounds per feet^3	1,728
	pounds per gallon (U.S.)	231
torrs	millimeters of Hg	1.0
	pascals	133.322
yards	centimeters	91.44
	fathoms	0.5
	feet	3.0
	inches	36.0
	meters	0.9144
	miles	5.682×10^{-4}
yards^3	acre-feet	6.198×10^{-4}
	feet^3	27
	gallons (U.S.)	202
	inches^3	4.666×10^4
	liters	764.6
	meters^3	.7646
years (normal calendar)	hours	8,760
	minutes	5.256×10^5
	seconds	3.1536×10^7
	weeks	52.1428

U.S. EPA Approved Test Methods

This table is based on information in "Guidelines Establishing Test Procedures for the Analysis of Pollutants," U.S. Code of Federal Regulations, 40 CFR Ch.1, Part 136, pp 244–453 (July 1986).

Chemical Name	GC	GC/MS
Acenaphthene	610	625, 1625
Acenaphthylene	610	625, 1625
Acetone	----	------
Acrolein	603	624, 1624
Acrylonitrile	603	624, 1624
Aldrin	608	625
Anthracene	610	625, 1625
Benzene	602	624, 1624
Benzidine	----	625, 1625
Benzo[a]anthracene	610	625, 1625
Benzo[b]fluoranthene	610	625, 1625
Benzo[k]fluoranthene	610	625, 1625
Benzoic acid	----	------
Benzo[ghi]perylene	610	625, 1625
Benzo[a]pyrene	610	625, 1625
Benzyl alcohol	----	------
Benzyl butyl phthalate	606	625, 1625
α-BHC	608	625
β-BHC	608	625
δ-BHC	608	625
Bis(2-chloroethoxy)methane	611	625, 1625
Bis(2-chloroethyl)ether	611	625, 1625
Bis(2-chloroisopropyl)ether	611	625, 1625
Bis(2-ethylhexyl)phthalate	606	625, 1625
Bromodichloromethane	601	624, 1624
Bromoform	601	624, 1624
4-Bromophenyl phenyl ether	611	625, 1625
2-Butanone	----	------
Carbon disulfide	----	------
Carbon tetrachloride	601	624, 1624
Chlordane	608	625
cis-Chlordane	----	------
trans-Chlordane	----	------
4-Chloroaniline	----	------
Chlorobenzene	601, 602	624, 1624
p-Chloro-m-cresol	604	625, 1625
Chloroethane	601	624, 1624

Chemical Name	GC	GC/MS
2-Chloroethyl vinyl ether	601	624, 1624
Chloroform	601	624, 1624
2-Chloronaphthalene	612	625, 1625
2-Chlorophenol	604	625, 1625
4-Chlorophenyl phenyl ether	611	625, 1625
Chrysene	610	625, 1625
p,p'-DDD	608	625
p,p'-DDE	608	625
p,p'-DDT	608	625
Dibenz[a,h]anthracene	610	625, 1625
Dibenzofuran	----	------
Dibromochloromethane	601	624, 1624
Di-n-butyl phthalate	606	625, 1625
1,2-Dichlorobenzene	601, 602, 612	624, 625, 1625
1,3-Dichlorobenzene	601, 602, 612	624, 625, 1625
1,4-Dichlorobenzene	601, 602, 612	624, 625, 1625
3,3'-Dichlorobenzidine	----	625, 1625
Dichlorodifluoromethane	601	------
1,1-Dichloroethane	601	624, 1624
1,2-Dichloroethane	601	624, 1624
1,1-Dichloroethylene	601	624, 1624
trans-1,2-Dichloroethylene	601	624, 1624
2,4-Dichlorophenol	604	625, 1625
1,2-Dichloropropane	601	624, 1624
cis-1,3-Dichloropropylene	601	624, 1624
trans-1,3-Dichloropropylene	601	624, 1624
Dieldrin	608	625
Diethyl phthalate	606	625, 1625
2,4-Dimethylphenol	604	625, 1625
Dimethyl phthalate	606	625, 1625
4,6-Dinitro-o-cresol	604	625, 1625
2,4-Dinitrophenol	604	625, 1625
2,4-Dinitrotoluene	609	625, 1625
2,6-Dinitrotoluene	609	625, 1625
Di-n-octyl phthalate	606	625, 1625
1,2-Diphenylhydrazine	----	------
α-Endosulfan	608	625
β-Endosulfan	608	625
Endosulfan sulfate	608	625
Endrin	608	625

Chemical Name	GC	GC/MS
Endrin aldehyde	608	625
Ethylbenzene	602	624, 1624
Fluoranthene	610	625, 1625
Fluorene	610	625, 1625
Heptachlor	608	625
Heptachlor epoxide	608	------
Hexachlorobenzene	612	625, 1625
Hexachlorobutadiene	612	625, 1625
Hexachlorocyclopentadiene	612	625, 1625
Hexachloroethane	612	625, 1625
2-Hexanone	----	------
Indeno[1,2,3-*c,d*]pyrene	610	625, 1625
Isophorone	609	625, 1625
Lindane	608	625
Methoxychlor	----	------
Methyl bromide	601	624, 1624
Methyl chloride	601	624, 1624
Methylene chloride	601	624, 1624
2-Methyl naphthalene	----	------
4-Methyl-2-pentanone	----	------
2-Methylphenol	----	------
4-Methylphenol	----	------
Naphthalene	610	625, 1625
2-Nitroaniline	----	------
3-Nitroaniline	----	------
4-Nitroaniline	----	------
Nitrobenzene	609	625, 1625
2-Nitrophenol	604	625, 1625
4-Nitrophenol	604	625, 1625
N-Nitrosodimethylamine	607	625, 1625
N-Nitrosodiphenylamine	607	625, 1625
N-Nitrosodi-*n*-propylamine	607	625, 1625
PCB-1016	608	625
PCB-1221	608	625
PCB-1232	608	625
PCB-1242	608	625
PCB-1248	608	625
PCB-1254	608	625
PCB-1260	608	625
Pentachlorophenol	604	625, 1625

Chemical Name	GC	GC/MS
Phenanthrene	610	625, 1625
Phenol	604	625, 1625
Pyrene	610	625, 1625
Styrene	----	------
TCDD	----	613
1,1,2,2-Tetrachloroethane	601	624, 1624
Tetrachloroethylene	601	624, 1624
Toluene	602	624, 1624
Toxaphene	608	625
1,2,4-Trichlorobenzene	612	625, 1625
1,1,1-Trichloroethane	601	624, 1624
1,1,2-Trichloroethane	601	624, 1624
Trichloroethylene	601	624, 1624
Trichlorofluoromethane	601	624, 1624
2,4,5-Trichlorophenol	604	625, 1625
2,4,6-Trichlorophenol	----	------
Vinyl acetate	----	------
Vinyl chloride	601	624, 1624

CAS Index

50-29-3	*p,p'*-DDT
50-32-8	Benzo[*a*]pyrene
51-28-5	2,4-Dinitrophenol
53-70-3	Dibenz[*a,h*]anthracene
56-23-5	Carbon tetrachloride
56-55-3	Benzo[*a*]anthracene
57-74-9	Chlordane
58-89-9	Lindane
59-50-7	*p*-Chloro-*m*-cresol
60-57-1	Dieldrin
62-75-9	*N*-Nitrosodimethylamine
65-85-0	Benzoic acid
67-64-1	Acetone
67-66-3	Chloroform
67-72-1	Hexachloroethane
71-43-2	Benzene
71-55-6	1,1,1-Trichloroethane
72-20-8	Endrin
72-43-5	Methoxychlor
72-54-8	*p,p'*-DDD
72-55-9	*p,p'*-DDE
74-83-9	Methyl bromide
74-87-3	Methyl chloride
75-00-3	Chloroethane
75-01-4	Vinyl chloride
75-09-2	Methylene chloride
75-15-0	Carbon disulfide
75-25-2	Bromoform
75-27-4	Bromodichloromethane
75-34-3	1,1-Dichloroethane
75-35-4	1,1-Dichloroethylene
75-69-4	Trichlorofluoromethane
75-71-8	Dichlorodifluoromethane
76-44-8	Heptachlor
77-47-4	Hexachlorocyclopentadiene
78-59-1	Isophorone
78-87-5	1,2-Dichloropropane
78-93-3	2-Butanone
79-00-5	1,1,2-Trichloroethane
79-01-6	Trichloroethylene
79-34-5	1,1,2,2-Tetrachloroethane
83-32-9	Acenaphthene

RTECS Number Index

AB 1254000 Acenaphthylene
AB 1255500 Acenaphthene
AK 0875000 Vinyl acetate
AL 3150000 Acetone
AS 1050000 Acrolein
AT 5250000 Acrylonitrile
BX 0700000 4-Chloroaniline
BY 6650000 2-Nitroaniline
BY 6825000 3-Nitroaniline
BY 7000000 4-Nitroaniline
CA 9350000 Anthracene
CU 1400000 Benzo[b]fluoranthene
CV 9275000 Benzo[a]anthracene
CY 1400000 Benzene
CZ 0175000 Chlorobenzene
CZ 4499000 1,3-Dichlorobenzene
CZ 4500000 1,2-Dichlorobenzene
CZ 4550000 1,4-Dichlorobenzene
DA 0700000 Ethylbenzene
DA 2975000 Hexachlorobenzene
DA 6475000 Nitrobenzene
DC 2100000 1,2,4-Trichlorobenzene
DC 9625000 Benzidine
DD 0525000 3,3'-Dichlorobenzidine
DF 6350000 Benzo[k]fluoranthene
DG 0875000 Benzoic acid
DI 6200500 Benzo[ghi]perylene
DJ 3675000 Benzo[a]pyrene
DN 3150000 Benzyl alcohol
EJ 0700000 Hexachlorobutadiene
EL 6475000 2-Butanone
FF 6650000 Carbon disulfide
FG 4900000 Carbon tetrachloride
FS 9100000 Chloroform
GC 0700000 Chrysene
GO 6300000 2-Methylphenol
GO 6475000 4-Methylphenol
GO 7100000 p-Chloro-m-cresol
GO 9625000 4,6-Dinitro-o-cresol
GV 3500000 α-BHC
GV 4375000 β-BHC
GV 4550000 δ-BHC

GV 4900000   Lindane
GW 7700000   Isophorone
GY 1225000   Hexachlorocyclopentadiene
HN 2625000   Dibenz[a,h]anthracene
HP 3500000   TCDD
IO 1575000   Endrin
IO 1750000   Dieldrin
IO 2100000   Aldrin
IQ 0525000   N-Nitrosodimethylamine
JJ 9800000   N-Nitrosodiphenylamine
JL 9700000   N-Nitrosodi-n-propylamine
KH 7525000   Chloroethane
KI 0175000   1,1-Dichloroethane
KI 0525000   1,2-Dichloroethane
KI 0700000   p,p'-DDD
KI 4025000   Hexachloroethane
KI 8575000   1,1,2,2-Tetrachloroethane
KJ 2975000   1,1,1-Trichloroethane
KJ 3150000   1,1,2-Trichloroethane
KJ 3325000   p,p'-DDT
KJ 3675000   Methoxychlor
KN 0875000   Bis(2-chloroethyl)ether
KN 1750000   Bis(2-chloroisopropyl)ether
KN 6300000   2-Chloroethyl vinyl ether
KU 9625000   Vinyl chloride
KV 9275000   1,1-Dichloroethylene
KV 9400000   trans-1,2-Dichloroethylene
KV 9450000   p,p'-DDE
KX 3850000   Tetrachloroethylene
KX 4550000   Trichloroethylene
LL 4025000   Fluoranthene
LL 5670000   Fluorene
MP 1400000   2-Hexanone
MW 2625000   1,2-Diphenylhydrazine
NK 9300000   Indeno[1,2,3-cd]pyrene
PA 3675000   Bis(2-chloroethoxy)methane
PA 4900000   Methyl bromide
PA 5310000   Bromodichloromethane
PA 6300000   Methyl chloride
PA 6360000   Dibromochloromethane
PA 8050000   Methylene chloride
PA 8200000   Dichlorodifluoromethane
PB 5600000   Bromoform

Empirical Formula Index

CCl_2F_2	Dichlorofluoromethane
CCl_3F	Trichlorofluoromethane
CCl_4	Carbon tetrachloride
$CHBrCl_2$	Bromodichloromethane
$CHBr_2Cl$	Dibromochloromethane
$CHBr_3$	Bromoform
$CHCl_3$	Chloroform
CH_2Cl_2	Methylene chloride
CH_3Br	Methyl bromide
CH_3Cl	Methyl chloride
CS_2	Carbon disulfide
C_2Cl_4	Tetrachloroethylene
C_2Cl_6	Hexachloroethane
C_2HCl_3	Trichloroethylene
$C_2H_2Cl_2$	1,1-Dichloroethylene
	trans-1,2-Dichloroethylene
$C_2H_2Cl_4$	1,1,2,2-Tetrachloroethane
C_2H_3Cl	Vinyl chloride
$C_2H_3Cl_3$	1,1,1-Trichloroethane
	1,1,2-Trichloroethane
$C_2H_4Cl_2$	1,1-Dichloroethane
	1,2-Dichloroethane
C_2H_5Cl	Chloroethane
C_3H_3N	Acrylonitrile
$C_3H_4Cl_2$	cis-1,3-Dichloropropylene
	trans-1,3-Dichloropropylene
C_3H_4O	Acrolein
$C_3H_6Cl_2$	1,2-Dichloropropane
$C_3H_6N_2O$	N-Nitrosodimethylamine
C_3H_6O	Acetone
$C_3H_6O_2$	Vinyl acetate
C_4Cl_6	Hexachlorobutadiene
C_4H_7ClO	2-Chloroethyl vinyl ether
$C_4H_8Cl_2O$	Bis(2-chloroethyl)ether
C_4H_8O	2-Butanone
C_5Cl_6	Hexachlorocyclopentadiene
$C_5H_{10}Cl_2O_2$	Bis(2-chloroethoxy)methane
	Bis(2-chloroisopropyl)ether
C_6Cl_6	Hexachlorobenzene
C_6HCl_5O	Pentachlorophenol

Synonym Index

1,068, *see* Chlordane
Aahepta, *see* Heptachlor
Aalindan, *see* Lindane
Acetosol, *see* 1,1,2,2-Tetrachloroethane
Acenaphthalene, *see* Acenaphthene
Acenaphthene, *see* Acenaphthene
Acenaphthylene, *see* Acenaphthylene
Acetic acid, ethenyl ester, *see* Vinyl acetate
Acetic acid, ethylene ester, *see* Vinyl acetate
Acetic acid, vinyl ester, *see* Vinyl acetate
Acetone, *see* Acetone
1-Acetoxyethylene, *see* Vinyl acetate
Acetylene dichloride, *see* trans-1,2-Dichloroethylene
trans-Acetylene dichloride, *see* trans-1,2-Dichloroethylene
Acetylene tetrachloride, *see* 1,1,2,2-Tetrachloroethane
Acetylene trichloride, *see* Trichloroethylene
Acraldehyde, *see* Acrolein
Acritet, *see* Acrylonitrile
Acrolein, *see* Acrolein
Acrylaldehyde, *see* Acrolein
Acrylic aldehyde, *see* Acrolein
Acrylon, *see* Acrylonitrile
Acrylonitrile, *see* Acrylonitrile
Acrylonitrile monomer, *see* Acrylonitrile
Acutox, *see* Pentachlorophenol
Aerothene, *see* 1,1,1-Trichloroethane
Aerothene MM, *see* Methylene chloride
Aerothene TT, *see* 1,1,1-Trichloroethane
Aethylis, *see* Chloroethane
Aethylis chloridum, *see* Chloroethane
Aficide, *see* Lindane
Agricide maggot killer (F), *see* Toxaphene
Agrisol G-20, *see* Lindane
Agritan, *see* p,p'-DDT
Agroceres, *see* Heptachlor
Agrocide, *see* Lindane
Agrocide 2, *see* Lindane
Agrocide 6G, *see* Lindane
Agrocide 7, *see* Lindane
Agrocide III, *see* Lindane
Agrocide WP, *see* Lindane
Agronexit, *see* Lindane
Aldifen, *see* 2,4-Dinitrophenol

Aldrec, *see* Aldrin
Aldrex, *see* Aldrin
Aldrex 30, *see* Aldrin
Aldrin, *see* Aldrin
Aldrite, *see* Aldrin
Aldrosol, *see* Aldrin
Algofrene type 1, *see* Trichlorofluoromethane
Algofrene type 2, *see* Dichlorodifluoromethane
Algylen, *see* Trichloroethylene
Alltex, *see* Toxaphene
Alltox, *see* Toxaphene
Allyl aldehyde, *see* Acrolein
Altox, *see* Aldrin
Alvit, *see* Dieldrin
Amarthol fast orange R base, *see* 3-Nitroaniline
Amatin, *see* Hexachlorobenzene
Ameisenatod, *see* Lindane
Ameisenmittel merck, *see* Lindane
1-Amino-4-chloroaniline, *see* 4-Chloroaniline
1-Amino-4-chlorobenzene, *see* 4-Chloroaniline
1-Amino-*p*-chlorobenzene, *see* 4-Chloroaniline
4-Aminochlorobenzene, *see* 4-Chloroaniline
p-Aminochlorobenzene, *see* 4-Chloroaniline
3-Aminonitrobenzene, *see* 3-Nitroaniline
m-Aminonitrobenzene, *see* 3-Nitroaniline
4-Aminonitrobenzene, *see* 4-Nitroaniline
p-Aminonitrobenzene, *see* 4-Nitroaniline
1-Amino-2-nitrobenzene, *see* 2-Nitroaniline
1-Amino-3-nitrobenzene, *see* 3-Nitroaniline
1-Amino-4-nitrobenzene, *see* 4-Nitroaniline
An, *see* Acrylonitrile
Anamenth, *see* Trichloroethylene
Ankilostin, *see* Tetrachloroethylene
Annulene, *see* Benzene
Anodynon, *see* Chloroethane
Anofex, *see* *p,p'*-DDT
Anozol, *see* Diethyl phthalate
Anthracene, *see* Anthracene
Anthracin, *see* Anthracene
Anticarie, *see* Hexachlorobenzene
Antinonin, *see* 4,6-Dinitro-*o*-cresol
Antinonnon, *see* 4,6-Dinitro-*o*-cresol
Antisal 1a, *see* Toluene

Antisol 1, *see* Tetrachloroethylene
Aparasin, *see* Lindane
Aphtiria, *see* Lindane
Aplidal, *see* Lindane
Aptal, *see* p-Chloro-m-cresol
Aqualin, *see* Acrolein
Aqualine, *see* Acrolein
Arbitex, *see* Lindane
Arborol, *see* 4,6-Dinitro-o-cresol
Arcton 6, *see* Dichlorodifluoromethane
Arcton 9, *see* Trichlorofluoromethane
Arkotine, *see* p,p'-DDT
Arochlor 1221, *see* PCB-1221
Arochlor 1232, *see* PCB-1232
Arochlor 1242, *see* PCB-1242
Arochlor 1248, *see* PCB-1248
Arochlor 1254, *see* PCB-1254
Arochlor 1260, *see* PCB-1260
Aroclor 1016, *see* PCB-1016
Aroclor 1221, *see* PCB-1221
Aroclor 1232, *see* PCB-1232
Aroclor 1242, *see* PCB-1242
Aroclor 1248, *see* PCB-1248
Aroclor 1254, *see* PCB-1254
Aroclor 1260, *see* PCB-1260
Artic, *see* Methyl chloride
Aspon-chlordane, *see* Chlordane
Attac 4-2, *see* Toxaphene
Attac 4-4, *see* Toxaphene
Attac 6, *see* Toxaphene
Attac 6-3, *see* Toxaphene
Attac 8, *see* Toxaphene
Avlothane, *see* Hexachloroethane
Avolin, *see* Dimethyl phthalate
Azoamine Red ZH, *see* 4-Nitroaniline
Azobase MNA, *see* 3-Nitroaniline
Azoene fast orange GR base, *see* 2-Nitroaniline
Azoene fast orange GR salt, *see* 2-Nitroaniline
Azofix orange GR salt, *see* 2-Nitroaniline
Azofix red GG salt, *see* 4-Nitroaniline
Azogene fast orange GR, *see* 2-Nitroaniline
Azoic diazo component 6, *see* 2-Nitroaniline
Azoic diazo compound 37, *see* 4-Nitroaniline

1,2-Benzenedicarboxylic acid, diethyl ester, *see* Diethyl phthalate
1,2-Benzenedicarboxylic acid, dimethyl ester, *see* Dimethyl phthalate
1,2-Benzenedicarboxylic acid, dioctyl ester, *see* Di-*n*-octyl phthalate
1,2-Benzenedicarboxylic acid, di-*n*-octyl ester, *see* Di-*n*-octyl phthalate
o-Benzenedicarboxylic acid, dioctyl ester, *see* Di-*n*-octyl phthalate
Benzeneformic acid, *see* Benzoic acid
Benzene hexachloride, *see* Lindane
Benzene hexachloride-α-isomer, *see* α-BHC
Benzene-*cis*-hexachloride, *see* β-BHC
Benzene-γ-hexachloride, *see* Lindane
α-Benzene hexachloride, *see* α-BHC
β-Benzene hexachloride, *see* β-BHC
δ-Benzene hexachloride, *see* δ-BHC
γ-Benzene hexachloride, *see* Lindane
trans-α-Benzenehexachloride, *see* β-BHC
Benzene methanoic acid, *see* Benzoic acid
Benzenemethanol, *see* Benzyl alcohol
Benzenol, *see* Phenol
2,3-Benzfluoranthene, *see* Benzo[*b*]fluoranthene
3,4-Benzfluoranthene, *see* Benzo[*b*]fluoranthene
8,9-Benzfluoranthene, *see* Benzo[*k*]fluoranthene
Benzidene base, *see* Benzidine
Benzidine, *see* Benzidine
p-Benzidine, *see* Benzidine
2,3-Benzindene, *see* Fluorene
Benzinoform, *see* Carbon tetrachloride
Benzinol, *see* Trichloroethylene
Benzoanthracene, *see* Benzo[*a*]anthracene
Benzo[*a*]anthracene, *see* Benzo[*a*]anthracene
1,2-Benzoanthracene, *see* Benzo[*a*]anthracene
Benzoate, *see* Benzoic acid
Benzo[*d,e,f*]chrysene, *see* Benzo[*a*]pyrene
Benzoepin, *see* α-Endosulfan, β-Endosulfan
Benzo[*b*]fluoranthene, *see* Benzo[*b*]fluoranthene
2,3-Benzofluoranthene, *see* Benzo[*b*]fluoranthene
3,4-Benzo[*b*]fluoranthene, *see* Benzo[*b*]fluoranthene
11,12-Benzofluoranthene, *see* Benzo[*k*]fluoranthene
Benzo[*e*]fluoranthene, *see* Benzo[*b*]fluoranthene
Benzo[*k*]fluoranthene, *see* Benzo[*k*]fluoranthene
11,12-Benzo[*k*]fluoranthene, *see* Benzo[*k*]fluoranthene
3,4-Benzofluoranthene, *see* Benzo[*b*]fluoranthene
8,9-Benzofluoranthene, *see* Benzo[*k*]fluoranthene
Benzo[*jk*]fluorene, *see* Fluoranthene

Benzoic acid, *see* Benzoic acid
Benzol, *see* Benzene
Benzole, *see* Benzene
Benzolene, *see* Benzene
Benzo[*ghi*]perylene, *see* Benzo[*ghi*]perylene
1,12-Benzoperylene, *see* Benzo[*ghi*]perylene
Benzo[*a*]phenanthrene, *see* Benzo[*a*]anthracene
Benzo[*a*]phenanthrene, *see* Chrysene
Benzo[alpha]phenanthrene, *see* Chrysene
Benzo[*b*]phenanthrene, *see* Benzo[*a*]anthracene
Benzo[*def*]phenanthrene, *see* Pyrene
1,2-Benzophenanthrene, *see* Chrysene
2,3-Benzophenanthrene, *see* Benzo[*a*]anthracene
Benzo[*a*]pyrene, *see* Benzo[*a*]pyrene
Benzo[alpha]pyrene, *see* Benzo[*a*]pyrene
1,2-Benzopyrene, *see* Benzo[*a*]pyrene
3,4-Benzopyrene, *see* Benzo[*a*]pyrene
6,7-Benzopyrene, *see* Benzo[*a*]pyrene
Benzoyl alcohol, *see* Benzyl alcohol
1,12-Benzperylene, *see* Benzo[*ghi*]perylene
1,2-Benzphenanthrene, *see* Chrysene
2,3-Benzphenanthrene, *see* Benzo[*a*]anthracene
Benz[*a*]phenanthrene, *see* Chrysene
Benz[*a*]pyrene, *see* Benzo[*a*]pyrene
1,2-Benzpyrene, *see* Benzo[*a*]pyrene
3,4-Benzpyrene, *see* Benzo[*a*]pyrene
3,4-Benz[*a*]pyrene, *see* Benzo[*a*]pyrene
Benzyl alcohol, *see* Benzyl alcohol
Benzyl butyl phthalate, *see* Benzyl butyl phthalate
Benzyl *n*-butyl phthalate, *see* Benzyl butyl phthalate
3,4-Benzypyrene, *see* Benzo[*a*]pyrene
Beosit, *see* α-Endosulfan, β-Endosulfan
Bexol, *see* Lindane
B(*b*)F, *see* Benzo[*b*]fluoranthene
B(*k*)F, *see* Benzo[*k*]fluoranthene
BHC, *see* Lindane
α-BHC, *see* α-BHC
β-BHC, *see* β-BHC
δ-BHC, *see* δ-BHC
γ-BHC, *see* Lindane
4,4'-Bianiline, *see* Benzidine
N,N'-Bianiline, *see* 1,2-Diphenylhydrazine
p,p'-Bianiline, *see* Benzidine

Bicarburet of hydrogen, *see* Benzene
1,2-Bichloroethane, *see* 1,2-Dichloroethane
2,3,1',8'-Binaphthylene, *see* Benzo[*k*]fluoranthene
Bio 5,462, *see* α-Endosulfan, β-Endosulfan
Biocide, *see* Acrolein
Bioflex 81, *see* Bis(2-ethylhexyl)phthalate
Bioflex DOP, *see* Bis(2-ethylhexyl)phthalate
(1,1'-Biphenyl)-4,4'-diamine, *see* Benzidine
4,4'-Biphenyldiamine, *see* Benzidine
p,p'-Biphenyldiamine, *see* Benzidine
4,4'-Biphenylenediamine, *see* Benzidine
p,p'-Biphenylenediamine, *see* Benzidine
o-Biphenylenemethane, *see* Fluorene
Biphenylene oxide, *see* Dibenzofuran
o-Biphenylmethane, *see* Fluorene
2,2-Bis(*p*-anisyl)-1,1,1-trichloroethane, *see* Methoxychlor
Bis(2-chloroethoxy)methane, *see* Bis(2-chloroethoxy)methane
Bis(β-chloroethyl)acetal ethane, *see* Bis(2-chloroethoxy)methane
Bis(2-chloroethyl)ether, *see* Bis(2-chloroethyl)ether
Bis-(β-chloroethyl)ether, *see* Bis(2-chloroethyl)ether
Bis(2-chloroethyl)formal, *see* Bis(2-chloroethoxy)methane
Bis(β-chloroethyl)formal, *see* Bis(2-chloroethoxy)methane
Bis(2-chloroisopropyl)ether, *see* Bis(2-chloroisopropyl)ether
Bis-(β-chloroisopropyl)ether, *see* Bis(2-chloroisopropyl)ether
Bis(2-chloro-1-methylethyl)ether, *see* Bis(2-chloroisopropyl)ether
1,1-Bis(4-chlorophenyl)-2,2-dichloroethane, *see* *p,p'*-DDD
1,1-Bis(*p*-chlorophenyl)-2,2-dichloroethane, *see* *p,p'*-DDD
2,2-Bis(4-chlorophenyl)-1,1-dichloroethane, *see* *p,p'*-DDD
2,2-Bis(*p*-chlorophenyl)-1,1-dichloroethane, *see* *p,p'*-DDD
2,2-Bis(4-chlorophenyl)-1,1-dichloroethene, *see* *p,p'*-DDE
2,2-Bis(*p*-chlorophenyl)-1,1-dichloroethene, *see* *p,p'*-DDE
1,1-Bis(*p*-chlorophenyl)-2,2,2-trichloroethane, *see* *p,p'*-DDT
2,2-Bis(4-chlorophenyl)-1,1,1-trichloroethane, *see* *p,p'*-DDT
2,2-Bis(*p*-chlorophenyl)-1,1,1-trichloroethane, *see* *p,p'*-DDT
α,α-Bis(*p*-chlorophenyl)-β,β,β-trichloroethane, *see* *p,p'*-DDT
Bis(2-ethylhexyl)phthalate, *see* Bis(2-ethylhexyl)phthalate
Bis(2-ethylhexyl)-1,2-benzenedicarboxylate, *see* Bis(2-ethylhexyl)phthalate
1,1-Bis(*p*-methoxyphenyl)-2,2,2-trichloroethane, *see* Methoxychlor
2,2-Bis(*p*-methoxyphenyl)-1,1,1-trichloroethane, *see* Methoxychlor
Blacosolv, *see* Trichloroethylene
Blancosolv, *see* Trichloroethylene
Bonoform, *see* 1,1,2,2-Tetrachloroethane

Borer sol, *see* 1,2-Dichloroethane
Bosan Supra, *see* *p,p'*-DDT
Bovidermol, *see* *p,p'*-DDT
BP, *see* Benzo[*a*]pyrene
3,4-BP, *see* Benzo[*a*]pyrene
B(*a*)P, *see* Benzo[*a*]pyrene
B(*ghi*)P, *see* Benzo[*ghi*]perylene
Brentamine fast orange GR base, *see* 2-Nitroaniline
Brentamine fast orange GR salt, *see* 2-Nitroaniline
Brocide, *see* 1,2-Dichloroethane
Bromodichloromethane, *see* Bromodichloromethane
4-Bromodiphenyl ether, *see* 4-Bromophenyl phenyl ether
p-Bromodiphenyl ether, *see* 4-Bromophenyl phenyl ether
Bromoform, *see* Bromoform
Brom-o-gaz, *see* Methyl bromide
Brom-o-gaz, *see* Methyl bromide
Bromomethane, *see* Methyl bromide
1-Bromo-4-phenoxybenzene, *see* 4-Bromophenyl phenyl ether
1-Bromo-*p*-phenoxybenzene, *see* 4-Bromophenyl phenyl ether
4-Bromophenyl ether, *see* 4-Bromophenyl phenyl ether
p-Bromophenyl ether, *see* 4-Bromophenyl phenyl ether
4-Bromophenyl phenyl ether, *see* 4-Bromophenyl phenyl ether
p-Bromophenyl phenyl ether, *see* 4-Bromophenyl phenyl ether
Bunt-cure, *see* Hexachlorobenzene
Bunt-no-more, *see* Hexachlorobenzene
Butanone, *see* 2-Butanone
2-Butanone, *see* 2-Butanone
Butyl benzyl phthalate, *see* Benzyl butyl phthalate
n-Butyl benzyl phthalate, *see* Benzyl butyl phthalate
Butyl ketone, *see* 2-Hexanone
Butyl methyl ketone, *see* 2-Hexanone
n-Butyl methyl ketone, *see* 2-Hexanone
Butyl phenylmethyl 1,2-Benzenedicarboxylate, *see* Benzyl butyl phthalate
Butyl phthalate, *see* Di-*n*-butyl phthalate
n-Butyl phthalate, *see* Di-*n*-butyl phthalate
C-56, *see* Hexachlorocyclopentadiene
Camphechlor, *see* Toxaphene
Camphochlor, *see* Toxaphene
Camphoclor, *see* Toxaphene
Camphor tar, *see* Naphthalene
Candaseptic, *see* *p*-Chloro-*m*-cresol
Capsine, *see* 4,6-Dinitro-*o*-cresol

γ-Chlordan, *see* Chlordane
Chlordane, *see* Chlordane
α-Chlordane, *see* cis-Chlordane, *trans*-Chlordane
α(cis)-Chlordane, *see* trans-Chlordane
β-Chlordane, *see* cis-Chlordane
cis-Chlordane, *see* cis-Chlordane
γ-Chlordane, *see* trans-Chlordane
trans-Chlordane, *see* trans-Chlordane
Chloresene, *see* Lindane
Chlorethene, *see* Vinyl chloride
Chlorethyl, *see* Chloroethane
Chlorethylene, *see* Vinyl chloride
2-Chlorethyl vinyl ether, *see* 2-Chloroethyl vinyl ether
Chlorex, *see* Bis(2-chloroethyl)ether
Chloridan, *see* Chlordane
Chloridum, *see* Chloroethane
Chlorilen, *see* Trichloroethylene
Chlorinated camphene, *see* Toxaphene
Chlorinated hydrochloric ether, *see* 1,1-Dichloroethane
Chlorindan, *see* Chlordane
Chlor kil, *see* Chlordane
4-Chloroaniline, *see* 4-Chloroaniline
p-Chloroaniline, *see* 4-Chloroaniline
Chloroben, *see* 1,2-Dichlorobenzene
4-Chlorobenzamine, *see* 4-Chloroaniline
p-Chlorobenzamine, *see* 4-Chloroaniline
Chlorobenzene, *see* Chlorobenzene
Chlorobenzol, *see* Chlorobenzene
Chlorocamphene, *see* Toxaphene
3-Chlorochlordene, *see* Heptachlor
1-Chloro-2-(β-chloroisopropoxy)propane, *see* Bis(2-chloroisopropyl)ether
1-Chloro-2-(β-chloroethoxy)ethane, *see* Bis(2-chloroethyl)ether
Chlorocresol, *see* p-Chloro-m-cresol
4-Chlorocresol, *see* p-Chloro-m-cresol
p-Chlorocresol, *see* p-Chloro-m-cresol
4-Chloro-m-cresol, *see* p-Chloro-m-cresol
6-Chloro-m-cresol, *see* p-Chloro-m-cresol
p-Chloro-m-cresol, *see* p-Chloro-m-cresol
Chlorodane, *see* Chlordane
Chloroden, *see* 1,2-Dichlorobenzene
Chlorodibromomethane, *see* Dibromochloromethane
1-Chloro-2,2-dichloroethylene, *see* Trichloroethylene
Chlorodiphenyl (21% Cl), *see* PCB-1221

Coal naphtha, *see* Benzene
Coal tar naphtha, *see* Benzene
Codechine, *see* Lindane
Collunosol, *see* 2,4,5-Trichlorophenol
Compound 118, *see* Aldrin
Compound 269, *see* Endrin
Compound 497, *see* Dieldrin
Compound 889, *see* Bis(2-ethylhexyl)phthalate
Compound 3,956, *see* Toxaphene
Co-op hexa, *see* Hexachlorobenzene
Corodane, *see* Chlordane
Cortilan-neu, *see* Chlordane
Crawhaspol, *see* Trichloroethylene
2-Cresol, *see* 2-Methylphenol
4-Cresol, *see* 4-Methylphenol
o-Cresol, *see* 2-Methylphenol
p-Cresol, *see* 4-Methylphenol
Crestoxo, *see* Toxaphene
Crestoxo 90, *see* Toxaphene
o-Cresylic acid, *see* 2-Methylphenol
p-Cresylic acid, *see* 4-Methylphenol
Crisulfan, *see* α-Endosulfan, β-Endosulfan
Crolean, *see* Acrolein
Cryptogil OL, *see* Pentachlorophenol
Curetard A, *see* N-Nitrosodiphenylamine
Curithane C126, *see* 3,3'-Dichlorobenzidine
Cyanoethylene, *see* Acrylonitrile
Cyclodan, *see* α-Endosulfan, β-Endosulfan
Cyclohexatriene, *see* Benzene
Cyclopenta[*de*]naphthalene, *see* Acenaphthylene
DAF 68, *see* Bis(2-ethylhexyl)phthalate
Daito orange base R, *see* 3-Nitroaniline
Dawson 100, *see* Methyl bromide
DBA, *see* Dibenz[*a,h*]anthracene
DB[*a,h*]A, *see* Dibenz[*a,h*]anthracene
1,2,5,6-DBA, *see* Dibenz[*a,h*]anthracene
DBH, *see* Lindane
DBP, *see* Di-*n*-butyl phthalate
1,2-DCA, *see* 1,2-Dichloroethane
DCB, *see* 1,2-Dichlorobenzene, 3,3'-Dichlorobenzidine
1,2-DCB, *see* 1,2-Dichlorobenzene
1,3-DCB, *see* 1,3-Dichlorobenzene
1,4-DCB, *see* 1,4-Dichlorobenzene

4,4'-Diamino-3,3'-dichlorobiphenyl, *see* 3,3'-Dichlorobenzidine
4,4'-Diamino-3,3'-dichlorodiphenyl, *see* 3,3'-Dichlorobenzidine
4,4'-Diaminodiphenyl, *see* Benzidine
p-Diaminodiphenyl, *see* Benzidine
p,p'-Diaminodiphenyl, *see* Benzidine
4,4'-Dianiline, *see* Benzidine
p,p'-Dianiline, *see* Benzidine
2,2-Di-*p*-anisyl-1,1,1-trichloroethane, *see* Methoxychlor
Diarex HF77, *see* Styrene
Diazo fast orange GR, *see* 2-Nitroaniline
Diazo fast orange R, *see* 3-Nitroaniline
Diazo fast red GG, *see* 4-Nitroaniline
1,2:5,6-Dibenzanthracene, *see* Dibenz[*a,h*]anthracene
1,2:5,6-Dibenzoanthracene, *see* Dibenz[*a,h*]anthracene
1,2:5,6-Dibenz[*a*]anthracene, *see* Dibenz[*a,h*]anthracene
Dibenz[*a*]anthracene, *see* Dibenz[*a,h*]anthracene
Dibenzo[*a,h*]anthracene, *see* Dibenz[*a,h*]anthracene
Dibenzo[*b,jk*]fluorene, *see* Benzo[*k*]fluoranthene
Dibenzofuran, *see* Dibenzofuran
1,2-Dibenzonaphthalene, *see* Chrysene
1,2,5,6-Dibenzonaphthalene, *see* Chrysene
Dibovan, *see* *p,p'*-DDT
Dibromochloromethane, *see* Dibromochloromethane
Dibutyl-1,2-benzenedicarboxylate, *see* Di-*n*-butyl phthalate
Dibutyl phthalate, *see* Di-*n*-butyl phthalate
Di-*n*-butyl phthalate, *see* Di-*n*-butyl phthalate
1,2-Dichlorbenzene, *see* 1,2-Dichlorobenzene
1,3-Dichlorbenzene, *see* 1,3-Dichlorobenzene
m-Dichlorbenzene, *see* 1,3-Dichlorobenzene
o-Dichlorbenzene, *see* 1,2-Dichlorobenzene
1,2-Dichlorbenzol, *see* 1,2-Dichlorobenzene
1,3-Dichlorbenzol, *see* 1,3-Dichlorobenzene
m-Dichlorbenzol, *see* 1,3-Dichlorobenzene
o-Dichlorbenzol, *see* 1,2-Dichlorobenzene
Dichloremulsion, *see* 1,2-Dichloroethane
1,1-Dichlorethane, *see* 1,1-Dichloroethane
1,2-Dichlorethane, *see* 1,2-Dichloroethane
2,2-Dichlorethyl ether, *see* Bis(2-chloroethyl)ether
Di-chloricide, *see* 1,4-Dichlorobenzene
Di-chlor-mulsion, *see* 1,2-Dichloroethane
1,2-Dichlorobenzene, *see* 1,2-Dichlorobenzene
1,3-Dichlorobenzene, *see* 1,3-Dichlorobenzene
1,4-Dichlorobenzene, *see* 1,4-Dichlorobenzene

cis-1,3-Dichloro-1-propene, *see cis*-1,3-Dichloropropylene
trans-1,3-Dichloro-1-propene, *see trans*-1,3-Dichloropropylene
1,3-Dichloroprop-1-ene, *see cis*-1,3-Dichloropropylene,
  trans-1,3-Dichloropropylene
(*E*)-1,3-Dichloropropene, *see trans*-1,3-Dichloropropylene
(*E*)-1,3-Dichloro-1-propene, *see trans*-1,3-Dichloropropylene
(*Z*)-1,3-Dichloropropene, *see cis*-1,3-Dichloropropylene
(*Z*)-1,3-Dichloro-1-propene, *see cis*-1,3-Dichloropropylene
cis-1,3-Dichloro-1-propylene, *see cis*-1,3-Dichloropropylene
trans-1,3-Dichloro-1-propylene, *see trans*-1,3-Dichloropropylene
cis-1,3-Dichloropropylene, *see cis*-1,3-Dichloropropylene
trans-1,3-Dichloropropylene, *see trans*-1,3-Dichloropropylene
Dicophane, *see p,p'*-DDT
Didakene, *see* Tetrachloroethylene
Didigam, *see p,p'*-DDT
Didimac, *see p,p'*-DDT
Dieldrin, *see* Dieldrin
Dieldrite, *see* Dieldrin
Dieldrix, *see* Dieldrin
Di(2-ethylhexyl)orthophthalate, *see* Bis(2-ethylhexyl)phthalate
Di(2-ethylhexyl)phthalate, *see* Bis(2-ethylhexyl)phthalate
Diethyl phthalate, *see* Diethyl phthalate
Diethyl-*o*-phthalate, *see* Diethyl phthalate
Difluorodichloromethane, *see* Dichlorodifluoromethane
1,2-Dihydroacenaphthylene, *see* Acenaphthene
Dilantin DB, *see* 1,2-Dichlorobenzene
Dilatin DB, *see* 1,2-Dichlorobenzene
Dilene, *see p,p'*-DDD
Dimethoxy-DDT, *see* Methoxychlor
p,p'-Dimethoxydiphenyltrichloroethane, *see* Methoxychlor
Dimethoxy-DT, *see* Methoxychlor
2,2-Di-(*p*-methoxyphenyl)-1,1,1-trichloroethane, *see* Methoxychlor
Di(*p*-methoxyphenyl)trichloromethyl methane, *see* Methoxychlor
1,2-Dimethylbenzene, *see o*-Xylene
1,3-Dimethylbenzene, *see m*-Xylene
1,4-Dimethylbenzene, *see p*-Xylene
o-Dimethylbenzene, *see o*-Xylene
m-Dimethylbenzene, *see m*-Xylene
p-Dimethylbenzene, *see p*-Xylene
Dimethyl-1,2-benzenedicarboxylate, *see* Dimethyl phthalate
Dimethylbenzeneorthodicarboxylate, *see* Dimethyl phthalate
Dimethylformaldehyde, *see* Acetone
Dimethylketal, *see* Acetone

Endosulfan II, *see* β-Endosulfan
α-Endosulfan, *see* α-Endosulfan
β-Endosulfan, *see* β-Endosulfan
Endosulfan sulfate, *see* Endosulfan sulfate
Endosulphan, *see* α-Endosulfan, β-Endosulfan
Endrex, *see* Endrin
Endrin, *see* Endrin
Endrin aldehyde, *see* Endrin aldehyde
ENT 54, *see* Acrylonitrile
ENT 154, *see* 4,6-Dinitro-*o*-cresol
ENT 262, *see* Dimethyl phthalate
ENT 1,506, *see* *p,p'*-DDT
ENT 1,656, *see* 1,2-Dichloroethane
ENT 1,716, *see* Methoxychlor
ENT 1,860, *see* Tetrachloroethylene
ENT 4,225, *see* *p,p'*-DDD
ENT 4,504, *see* Bis(2-chloroethyl)ether
ENT 4,705, *see* Carbon tetrachloride
ENT 7,796, *see* Lindane
ENT 9,232, *see* α-BHC
ENT 9,233, *see* β-BHC
ENT 9,234, *see* δ-BHC
ENT 9,735, *see* Toxaphene
ENT 9,932, *see* Chlordane
ENT 15,152, *see* Heptachlor
ENT 15,406, *see* 1,2-Dichloropropane
ENT 15,949, *see* Aldrin
ENT 16,225, *see* Dieldrin
ENT 17,251, *see* Endrin
ENT 23,979, *see* α-Endosulfan, β-Endosulfan
ENT 25,552-X, *see* Chlordane
ENT 25,584, *see* Heptachlor epoxide
Entomoxan, *see* Lindane
EP 30, *see* Pentachlorophenol
Epoxy heptachlor, *see* Heptachlor epoxide
Ergoplast FDO, *see* Bis(2-ethylhexyl)phthalate
Eskimon 11, *see* Trichlorofluoromethane
Eskimon 12, *see* Dichlorodifluoromethane
Essence of mirbane, *see* Nitrobenzene
Essence of myrbane, *see* Nitrobenzene
Estol 1550, *see* Diethyl phthalate
Estonate, *see* *p,p'*-DDT
Estonox, *see* Toxaphene

Ethane dichloride, *see* 1,2-Dichloroethane
Ethane hexachloride, *see* Hexachloroethane
Ethane tetrachloride, *see* 1,1,2,2-Tetrachloroethane
Ethane trichloride, *see* 1,1,2-Trichloroethane
Ethene dichloride, *see* 1,2-Dichloroethane
Ethenyl acetate, *see* Vinyl acetate
Ethenylethanoate, *see* Vinyl acetate
Ether chloratus, *see* Chloroethane
Ether hydrochloric, *see* Chloroethane
Ether muriatic, *see* Chloroethane
Ethinyl trichloride, *see* Trichloroethylene
Ethyenyl benzene, *see* Styrene
Ethylbenzene, *see* Ethylbenzene
Ethylbenzol, *see* Ethylbenzene
Ethyl chloride, *see* Chloroethane
Ethylene aldehyde, *see* Acrolein
Ethylene chloride, *see* 1,2-Dichloroethane
Ethylene dichloride, *see* 1,2-Dichloroethane
1,2-Ethylene dichloride, *see* 1,2-Dichloroethane
Ethylene hexachloride, *see* Hexachloroethane
Ethylene monochloride, *see* Vinyl chloride
Ethylene naphthalene, *see* Acenaphthene
1,8-Ethylene naphthalene, *see* Acenaphthene
Ethylene tetrachloride, *see* Tetrachloroethylene
Ethylene trichloride, *see* Trichloroethylene
Ethylhexyl phthalate, *see* Bis(2-ethylhexyl)phthalate
2-Ethylhexyl phthalate, *see* Bis(2-ethylhexyl)phthalate
Ethylidene chloride, *see* 1,1-Dichloroethane
Ethylidene dichloride, *see* 1,1-Dichloroethane
1,1-Ethylidene dichloride, *see* 1,1-Dichloroethane
Ethyl methyl ketone, *see* 2-Butanone
Ethyl phthalate, *see* Diethyl phthalate
Eviplast 80, *see* Bis(2-ethylhexyl)phthalate
Eviplast 81, *see* Bis(2-ethylhexyl)phthalate
Evola, *see* 1,4-Dichlorobenzene
Exagama, *see* Lindane
Experimental insecticide no. 269, *see* Endrin
Extrar, *see* 4,6-Dinitro-*o*-cresol
F 11, *see* Trichlorofluoromethane
F 12, *see* Dichlorodifluoromethane
Falkitol, *see* Hexachloroethane
Fasciolin, *see* Carbon tetrachloride and Hexachloroethane
Fasco-terpene, *see* Toxaphene

Forlin, *see* Lindane
Formaldehyde bis(β-chloroethylacetal), *see* Bis(2-chloroethoxy)methane
Formyl trichloride, *see* Chloroform
Freon 10, *see* Carbon tetrachloride
Freon 11, *see* Trichlorofluoromethane
Freon 11A, *see* Trichlorofluoromethane
Freon 11B, *see* Trichlorofluoromethane
Freon 12, *see* Dichlorodifluoromethane
Freon 20, *see* Chloroform
Freon 30, *see* Methylene chloride
Freon 150, *see* 1,2-Dichloroethane
Freon F-12, *see* Dichlorodifluoromethane
Freon HE, *see* Trichlorofluoromethane
Freon MF, *see* Trichlorofluoromethane
Frigen 11, *see* Trichlorofluoromethane
Frigen 12, *see* Dichlorodifluoromethane
Fumigant-1, *see* Methyl bromide
Fumigrain, *see* Acrylonitrile
Fungifen, *see* Pentachlorophenol
Fungol, *see* Pentachlorophenol
Gallogama, *see* Lindane
Gamacid, *see* Lindane
Gamaphex, *see* Lindane
Gamene, *see* Lindane
Gamiso, *see* Lindane
Gammahexa, *see* Lindane
Gammalin, *see* Lindane
Gammexene, *see* Lindane
Gammopaz, *see* Lindane
Gemalgene, *see* Trichloroethylene
Genetron 11, *see* Trichlorofluoromethane
Genetron 12, *see* Dichlorodifluoromethane
Geniphene, *see* Toxaphene
Genitox, *see* p,p'-DDT
Genklene, *see* 1,1,1-Trichloroethane
Germalgene, *see* Trichloroethylene
Gesafid, *see* p,p'-DDT
Gesapon, *see* p,p'-DDT
Gesarex, *see* p,p'-DDT
Gesarol, *see* p,p'-DDT
Gexane, *see* Lindane
Glazd penta, *see* Pentachlorophenol

tetrahydro-4,7-methanoindene, *see* Heptachlor epoxide

2,3,4,5,6,7,7-Heptachloro-1a,1b,5,5a,6,6a-hexahydro-2,5-methano-2*H*-oxireno[*a*]indene, *see* Heptachlor epoxide

1,4,5,6,7,8,8-Heptachloro-3a,4,7,7a-tetrahydro-4,7-methanol-1*H*-indene, *see* Heptachlor

1,4,5,6,7,8,8-Heptachloro-3a,4,7,7a-tetrahydro-4,7-*endo*-methanoindene, *see* Heptachlor

1,4,5,6,7,8,8-Heptachloro-3a,4,7,7a-tetrahydro-4,7-methyleneindene, *see* Heptachlor

1,4,5,6,7,10,10-Heptachloro-4,7,8,9-tetrahydro-4,7-*endo*-methyleneindene, *see* Heptachlor

1(3a),4,5,6,7,8,8-Heptachloro-3a(1),4,7,7a-tetrahydro-4,7-methanoindene, *see* Heptachlor

1,4,5,6,7,8,8-Heptachloro-2,3-epoxy-2,3,3a,4,7,7a-hexahydro-4,7-methanoindene, *see* Heptachlor epoxide

2,3,4,5,6,7,7-Heptachloro-1a,1b,5,5a,6,6a-hexahydro-2,5-methano-2*H*-indeno[*1,2-b*]oxirene, *see* Heptachlor epoxide

1,4,5,6,7,8,8-Heptachloro-3a,4,7,7a-tetrahydro-4,7-methanoindene, *see* Heptachlor

1,4,5,6,7,8,8a-Heptachloro-3a,4,7,7a-tetrahydro-4,7-methanoindene, see Heptachlor

1,4,5,6,7,10,10-Heptachloro-4,7,8,9-tetrahydro-4,7-methanoindene, *see* Heptachlor

Heptadichlorocyclopentadiene, *see* Heptachlor

Heptagran, *see* Heptachlor

Heptagranox, *see* Heptachlor

Heptamak, *see* Heptachlor

Heptamul, *see* Heptachlor

Heptasol, *see* Heptachlor

Heptox, *see* Heptachlor

Hercoflex 260, *see* Bis(2-ethylhexyl)phthalate

Hercules 3,956, *see* Toxaphene

Hercules toxaphene, *see* Toxaphene

Hex, *see* Hexachlorocyclopentadiene

Hexa, *see* Lindane

Hexa C.B., *see* Hexachlorobenzene

γ-Hexachlor, *see* Lindane

Hexachloran, *see* Lindane

α-Hexachloran, *see* α-BHC

γ-Hexachloran, *see* Lindane

Hexachlorane, *see* Lindane

α-Hexachlorane, *see* α-BHC

γ-Hexachlorane, *see* Lindane

Hexachlorbutadiene, *see* Hexachlorobutadiene

α-Hexachlorcyclohexane, *see* α-BHC

Hexachlorobenzene, *see* Hexachlorobenzene

β-Hexachlorobenzene, *see* β-BHC

γ-Hexachlorobenzene, *see* Lindane

1,2,3,7,7-Hexachlorobicyclo[2.2.1]-2-heptene-5,6-bisoxymethylene sulfite,
see α-Endosulfan, β-Endosulfan

α,β-1,2,3,7,7-Hexachlorobicyclo[2.2.1]-2-heptane-5,6-bisoxymethylene
sulfite, *see* α-Endosulfan, β-Endosulfan

Hexachlorobutadiene, *see* Hexachlorobutadiene

1,3-Hexachlorobutadiene, *see* Hexachlorobutadiene

Hexachloro-1,3-butadiene, *see* Hexachlorobutadiene

1,1,2,3,4,4-Hexachlorobutadiene, *see* Hexachlorobutadiene

1,1,2,3,4,4-Hexachloro-1,3-butadiene, *see* Hexachlorobutadiene

1,2,3,4,5,6-Hexachlorocyclohexane, *see* Lindane

1α,2α,3β,4α,5α,6-β-Hexachlorocyclohexane, *see* Lindane

1,2,3,4,5,6-Hexachloro-α-cyclohexane, *see* α-BHC

1,2,3,4,5,6-Hexachloro-β-cyclohexane, *see* β-BHC

1,2,3,4,5,6-Hexachloro-δ-cyclohexane, *see* δ-BHC

1,2,3,4,5,6-Hexachloro-γ-cyclohexane, *see* Lindane

1,2,3,4,5,6-Hexachloro-*trans*-cyclohexane, *see* β-BHC

α-Hexachlorocyclohexane, *see* α-BHC

α-1,2,3,4,5,6-Hexachlorocyclohexane, *see* α-BHC

β-Hexachlorocyclohexane, *see* β-BHC

β-1,2,3,4,5,6-Hexachlorocyclohexane, *see* β-BHC

δ-Hexachlorocyclohexane, *see* δ-BHC

δ-1,2,3,4,5,6-Hexachlorocyclohexane, *see* δ-BHC

δ-(aeeeee)-1,2,3,4,5,6-Hexachlorocyclohexane, *see* δ-BHC

γ-Hexachlorocyclohexane, *see* Lindane

γ-1,2,3,4,5,6-Hexachlorocyclohexane, *see* Lindane

1α,2α,3β,4α,5β,6β-Hexachlorocyclohexane, *see* α-BHC

1α,2β,3α,4β,5α,6β-Hexachlorocyclohexane, *see* β-BHC

1α,2α,3α,4β,5β,6β-Hexachlorocyclohexane, *see* δ-BHC

Hexachlorocyclopentadiene, Hexachlorocyclopentadiene

1,2,3,4,5,5-Hexachloro-1,3-cyclopentadiene, *see*
Hexachlorocyclopentadiene

2,2a,3,3,4,7-Hexachlorodecahydro-1,2,4-methenocyclopenta[*c,d*]pentalene-
5-carboxaldehyde, *see* Endrin aldehyde

1,2,3,4,10,10-Hexachloro-6,7-epoxy-1,4,4a,5,6,7,8,8a-
octahydro-1,4-*endo,exo*-5,8-dimethanonaphthalene, *see* Dieldrin

1,2,3,4,10,10-Hexachloro-6,7-epoxy-1,4,4a,5,6,7,8,8a-
octahydro-*endo,endo*-1,4:5,8-dimethanonaphthalene, *see* Endrin

Hexachloroepoxyoctahydro-*endo-endo*-dimethanonaphthalene, *see*

Endrin

Hexachloroepoxyoctahydro-*endo-exo*-dimethanonaphthalene, *see* Dieldrin

Hexachloroethane, *see* Hexachloroethane

1,1,1,2,2,2-Hexachloroethane, *see* Hexachloroethane

Hexachloroethylene, *see* Hexachloroethane

Hexachlorohexahydro-*endo, exo*-dimethanonaphthalene, *see* Aldrin

1,2,3,4,10,10-Hexachloro-1,4,4a,5,8,8a-hexahydro-1,4:5,8-dimethanonaphthalene, *see* Aldrin

1,2,3,4,10,10-Hexachloro-1,4,4a,5,8,8a-hexahydro-1,4-*endo,exo*-5,8-dimethanonaphthalene, *see* Aldrin

1,2,3,4,10,10-Hexachloro-1,4,4a,5,8,8a-hexahydro-*exo*-1,4-*endo*-5,8-*endo*--dimethanonaphthalene, *see* Aldrin

6,7,8,9,10,10-Hexachloro-1,5,5a,6,9,9a-hexahydro-3,3-dioxide, *see* Endosulfan sulfate

(3α,5aα,6β,9β,9aα)-6,7,8,9,10,10-Hexachloro-1,5,5a,6,9,9a-hexahydro-6,9-methano-2,4,3-benzodioxathiepin-3-oxide, *see* β-Endosulfan

(3α,5aβ,6α,9α,9aβ)-6,7,8,9,10,10-Hexachloro-1,5,5a,6,9,9a-hexahydro-6,9-methano-2,4,3-benzodioxathiepin-3-oxide, *see* α-Endosulfan

Hexachlorohexahydromethano-2,4,3-benzodioxathiepin-3-oxide, *see* α-Endosulfan, β-Endosulfan

1,4,5,6,7,7-Hexachloro-5-norborene-2,3-dimethanol cyclic sulfite, *see* α-Endosulfan, β-Endosulfan

3,4,5,6,9,9-Hexachloro-1a,2,2a,3,6,6a,7,7a-octahydro-2,7:3,6-dimethanonaphth[2,3*b*]oxirene, *see* Dieldrin

Hexadrin, *see* Endrin

1,4,4a,5,8,8a-Hexahydro-1,4-*endo,exo*-5,8-dimethanonaphthalene, *see* Aldrin

Hexanone-2, *see* 2-Hexanone

Hexanone, *see* 4-Methyl-2-pentanone

2-Hexanone, *see* 2-Hexanone

Hexaplas M/B, *see* Di-*n*-butyl phthalate

Hexatox, *see* Lindane

Hexaverm, *see* Lindane

Hexicide, *see* Lindane

Hexone, *see* 4-Methyl-2-pentanone

Hexyclan, *see* Lindane

HGI, *see* Lindane

HHDN, *see* Aldrin

Hildan, *see* α-Endosulfan, β-Endosulfan

Hiltonil fast orange GR base, *see* 2-Nitroaniline

Hiltonil fast orange R base, *see* 3-Nitroaniline

Hiltosal fast orange GR salt, *see* 2-Nitroaniline

Isodrin epoxide, *see* Endrin
Isoforon, *see* Isophorone
Isoforone, *see* Isophorone
β-Isomer, *see* β-BHC
γ-Isomer, *see* Lindane
Isooctaphenone, *see* Isophorone
Isophoron, *see* Isophorone
Isophorone, *see* Isophorone
Isopropylacetone, *see* 4-Methyl-2-pentanone
Isotox, *see* Lindane
Isotron 11, *see* Trichlorofluoromethane
Isotron 2, *see* Dichlorodifluoromethane
Ivoran, *see* p,p'-DDT
Ixodex, *see* p,p'-DDT
Jacutin, *see* Lindane
Julin's carbon chloride, *see* Hexachlorobenzene
K III, *see* 4,6-Dinitro-o-cresol
K IV, *see* 4,6-Dinitro-o-cresol
Kaiser chemicals 12, *see* Dichlorodifluoromethane
Kamfochlor, *see* Toxaphene
Kanechlor, *see* PCB-1260
Kayafume, *see* Methyl bromide
Kelene, *see* Chloroethane
Ketone propane, *see* Acetone
β-Ketopropane, *see* Acetone
Kodaflex DOP, *see* Bis(2-ethylhexyl)phthalate
Kokotine, *see* Lindane
Kopsol, *see* p,p'-DDT
KOP-thiodan, *see* α-Endosulfan, β-Endosulfan
Kresamone, *see* 4,6-Dinitro-o-cresol
p-Kresol, *see* 4-Methylphenol
Krezotol 50, *see* 4,6-Dinitro-o-cresol
Kwell, *see* Lindane
Kypchlor, *see* Chlordane
Lanadin, *see* Trichloroethylene
Lauxtol, *see* Pentachlorophenol
Lauxtol A, *see* Pentachlorophenol
Ledon 11, *see* Trichlorofluoromethane
Ledon 12, *see* Dichlorodifluoromethane
Lendine, *see* Lindane
Lentox, *see* Lindane
Lethurin, *see* Trichloroethylene
Lidenal, *see* Lindane

Lindafor, *see* Lindane
Lindagam, *see* Lindane
Lindagrain, *see* Lindane
Lindagranox, *see* Lindane
Lindane, *see* Lindane
α-Lindane, *see* α-BHC
β-Lindane, *see* β-BHC
δ-Lindane, *see* δ-BHC
γ-Lindane, *see* Lindane
Lindapoudre, *see* Lindane
Lindatox, *see* Lindane
Lindosep, *see* Lindane
Lintox, *see* Lindane
Lipan, *see* 4,6-Dinitro-*o*-cresol
Liroprem, *see* Pentachlorophenol
Lorexane, *see* Lindane
M 140, *see* Chlordane
M 410, *see* Chlordane
M 5,055, *see* Toxaphene
Magnacide, *see* Acrolein
Malix, *see* α-Endosulfan, β-Endosulfan
Maralate, *see* Methoxychlor
Marlate, *see* Methoxychlor
Marlate 50, *see* Methoxychlor
Maroxol-50, *see* 2,4-Dinitrophenol
MB, *see* Methyl bromide
M-B-C fumigant, *see* Methyl bromide
MBK, *see* 2-Hexanone
MBX, *see* Methyl bromide
MCB, *see* Chlorobenzene
ME-1,700, *see* *p,p'*-DDD
MEBR, *see* Methyl bromide
Meetco, *see* 2-Butanone
MEK, *see* 2-Butanone
Melipax, *see* Toxaphene
Mendrin, *see* Endrin
Metafume, *see* Methyl bromide
Methacide, *see* Toluene
Methane dichloride, *see* Methylene chloride
Methane tetrachloride, *see* Carbon tetrachloride
Methane trichloride, *see* Chloroform
6,9-Methano-2,4,3-benzodioxathiepin, *see* Endosulfan sulfate
Methenyl chloride, *see* Chloroform

Methenyl tribromide, *see* Bromoform
Methenyl trichloride, *see* Chloroform
Methogas, *see* Methyl bromide
Methoxcide, *see* Methoxychlor
Methoxo, *see* Methoxychlor
Methoxychlor, *see* Methoxychlor
4,4'-Methoxychlor, *see* Methoxychlor
p,p'-Methoxychlor, *see* Methoxychlor
Methoxy-DDT, *see* Methoxychlor
Methyl acetone, *see* 2-Butanone
Methylbenzene, *see* Toluene
Methylbenzol, *see* Toluene
Methyl bromide, *see* Methyl bromide
Methyl *n*-butyl ketone, *see* 2-Hexanone
Methyl chloride, *see* Methyl chloride
Methyl chloroform, *see* 1,1,1-Trichloroethane
3-Methyl-4-chlorophenol, *see* *p*-Chloro-*m*-cresol
1-Methyl-2,4-dinitrobenzene, *see* 2,4-Dinitrotoluene
2-Methyl-1,3-dinitrobenzene, *see* 2,6-Dinitrotoluene
2-Methyl-4,6-dinitrophenol, *see* 4,6-Dinitro-*o*-cresol
6-Methyl-2,4-dinitrophenol, *see* 4,6-Dinitro-*o*-cresol
Methylene bichloride, *see* Methylene chloride
Methylenebiphenyl, *see* Fluorene
2,2'-Methylenebiphenyl, *see* Fluorene
1,1'-[Methylenebis(oxy)]bis(2-chloroethane), *see* Bis(2-
   chloroethoxy)methane
1,1'-[Methylenebis(oxy)]bis(2-chloroformaldehyde), *see*
   Bis(2-chloroethoxy)methane
Methylene chloride, *see* Methylene chloride
Methylene dichloride, *see* Methylene chloride
Methyl ethyl ketone, *see* 2-Butanone
1-Methyl-4-hydroxybenzene, *see* 4-Methylphenol
2-Methylhydroxybenzene, *see* 2-Methyl phenol
o-Methylhydroxybenzene, *see* 2-Methylphenol
4-Methylhydroxybenzene, *see* 4-Methylphenol
p-Methylhydroxybenzene, *see* 4-Methylphenol
Methyl isobutyl ketone, *see* 4-Methyl-2-pentanone
Methyl ketone, *see* Acetone
2-Methylnaphthalene, *see* 2-Methylnaphthalene
β-Methylnaphthalene, *see* 2-Methylnaphthalene
n-Methyl-*n*-nitrosomethanamine, *see* *N*-Nitrosodimethylamine
2-Methyl-4-pentanone, *see* 4-Methyl-2-pentanone
4-Methyl-2-pentanone, *see* 4-Methyl-2-pentanone

Moxie, *see* Methoxychlor
Mszycol, *see* Lindane
Muriatic ether, *see* Chloroethane
Mutoxin, *see* *p,p'*-DDT
MVC, *see* Vinyl chloride
NA 2,761, *see* Aldrin, Dieldrin, Endrin, Heptachlor, *p,p'*-DDD, Lindane, Toxaphene
NA 2,762, *see* Aldrin, Chlordane
NA 2,821, *see* Phenol
NA 9,037, *see* Hexachloroethane
NA 9,094, *see* Benzoic acid
Naphthalene, *see* Naphthalene
1,2-(1,8-Naphthylene)benzene, *see* Fluoranthene
1,2-(1,8-Naphthalenediyl)benzene, *see* Fluoranthene
Naphthalin, *see* Naphthalene
Naphthaline, *see* Naphthalene
Naphthanthracene, *see* Benzo[*a*]anthracene
Naphthene, *see* Naphthalene
Naphtoelan orange R base, *see* 3-Nitroaniline
Naphtolean red GG base, *see* 4-Nitroaniline
Narcogen, *see* Trichloroethylene
Narcotil, *see* Methylene chloride
Narcotile, *see* Chloroethane
Narkogen, *see* Trichloroethylene
Narkosoid, *see* Trichloroethylene
Natasol fast orange GR salt, *see* 2-Nitroaniline
Naugard TJB, *see* *N*-Nitrosodiphenylamine
NCI-C00044, *see* Aldrin
NCI-C00099, *see* Chlordane
NCI-C00124, *see* Dieldrin
NCI-C00157, *see* Endrin
NCI-C00180, *see* Heptachlor
NCI-C00204, *see* Lindane
NCI-C00259, *see* Toxaphene
NCI-C00464, *see* *p,p'*-DDT
NCI-C00475, *see* *p,p'*-DDD
NCI-C00497, *see* Methoxychlor
NCI-C00511, *see* 1,2-Dichloroethane
NCI-C00555, *see* *p,p'*-DDE
NCI-C00566, *see* α-Endosulfan, β-Endosulfan
NCI-C01854, *see* 1,2-Diphenylhydrazine
NCI-C01865, *see* 2,4-Dinitrotoluene
NCI-C02039, *see* 4-Chloroaniline

NCI-C60082, *see* Nitrobenzene
NCI-C60786, *see* 4-Nitroaniline
NCI-C61187, *see* 2,4,5-Trichlorophenol
NDMA, *see* *N*-Nitrosodimethylamine
NDPA, *see* *N*-Nitrosodiphenylamine
NDPA, *see* *N*-Nitrosodi-*n*-propylamine
NDPhA, *see* *N*-Nitrosodiphenylamine
Neantine, *see* Diethyl phthalate
Necatorina, *see* Carbon tetrachloride
Necatorine, *see* Carbon tetrachloride
Nema, *see* Tetrachloroethylene
Nendrin, *see* Endrin
Neocid, *see* *p,p'*-DDT
Neo-scabicidol, *see* Lindane
Nexen FB, *see* Lindane
Nexit, *see* Lindane
Nexit-stark, *see* Lindane
Nexol-E, *see* Lindane
NIA 5,462, *see* α-Endosulfan, β-Endosulfan
Niagara 5,462, *see* α-Endosulfan, β-Endosulfan
Nialk, *see* Trichloroethylene
Nicochloran, *see* Lindane
Niran, *see* Chlordane
Nitrador, *see* 4,6-Dinitro-*o*-cresol
Nitranilin, *see* 3-Nitroaniline
4-Nitraniline, *see* 4-Nitroaniline
m-Nitraniline, *see* 3-Nitroaniline
o-Nitraniline, *see* 2-Nitroaniline
p-Nitraniline, *see* 4-Nitroaniline
Nitration benzene, *see* Benzene
Nitrazol 2F extra, *see* 4-Nitroaniline
Nitrile, *see* Acrylonitrile
3-Nitroaminobenzene, *see* 3-Nitroaniline
m-Nitroaminobenzene, *see* 3-Nitroaniline
2-Nitroaniline, *see* 2-Nitroaniline
3-Nitroaniline, *see* 3-Nitroaniline
4-Nitroaniline, *see* 4-Nitroaniline
m-Nitroaniline, *see* 3-Nitroaniline
o-Nitroaniline, *see* 2-Nitroaniline
p-Nitroaniline, *see* 4-Nitroaniline
2-Nitrobenzenamine, *see* 2-Nitroaniline
3-Nitrobenzenamine, *see* 3-Nitroaniline
m-Nitrobenzenamine, *see* 3-Nitroaniline

Oxybenzene, *see* Phenol
1,1'-Oxybis(2-chloroethane), *see* Bis(2-chloroethyl)ether
2,2'-Oxybis(1-chloropropane), *see* Bis(2-chloroisopropyl)ether
o-Oxytoluene, *see* 2-Methylphenol
p-Oxytoluene, *see* 4-Methylphenol
Palantinol AH, *see* Bis(2-ethylhexyl)phthalate
Palatinol A, *see* Diethyl phthalate
Palatinol BB, *see* Benzyl butyl phthalate
Palatinol C, *see* Di-*n*-butyl phthalate
Palatinol M, *see* Dimethyl phthalate
Panoram D-31, *see* Dieldrin
Parachlorocidum, *see* *p,p'*-DDT
Paracide, *see* 1,4-Dichlorobenzene
Para-cresol, *see* 4-Methylphenol
Para crystals, *see* 1,4-Dichlorobenzene
Paradi, *see* 1,4-Dichlorobenzene
Paradichlorobenzene, *see* 1,4-Dichlorobenzene
Paradichlorobenzol, *see* 1,4-Dichlorobenzene
Paradow, *see* 1,4-Dichlorobenzene
Paramethylphenol, *see* 4-Methylphenol
Paramoth, *see* 1,4-Dichlorobenzene
Paranaphthalene, *see* Anthracene
Paranuggetts, *see* 1,4-Dichlorobenzene
Parazene, *see* 1,4-Dichlorobenzene
Parmetol, *see* *p*-Chloro-*m*-cresol
Parodi, *see* 1,4-Dichlorobenzene
Parol, *see* *p*-Chloro-*m*-cresol
PCB-1016, *see* PCB-1016
PCB-1221, *see* PCB-1221
PCB-1232, *see* PCB-1232
PCB-1242, *see* PCB-1242
PCB-1248, *see* PCB-1248
PCB-1254, *see* PCB-1254
PCB-1260, *see* PCB-1260
PCC, *see* Toxaphene
PCE, *see* Tetrachloroethylene
PCL, *see* Hexachlorocyclopentadiene
PCMC, *see* *p*-Chloro-*m*-cresol
PCP, *see* Pentachlorophenol
PDB, *see* 1,4-Dichlorobenzene
PDCB, *see* 1,4-Dichlorobenzene
PEB1, *see* *p,p'*-DDT
Pedraczak, *see* Lindane

Penchlorol, *see* Pentachlorophenol
Penphene, *see* Toxaphene
Penta, *see* Pentachlorophenol
Pentachlorin, *see* *p,p'*-DDT
Pentachlorfenol, *see* Pentachlorophenol
Pentachlorofenol, *see* Pentachlorophenol
Pentachlorofenolo, *see* Pentachlorophenol
Pentachlorophenol, *see* Pentachlorophenol
2,3,4,5,6-Pentachlorophenol, *see* Pentachlorophenol
Pentachlorophenate, *see* Pentachlorophenol
Pentachlorophenyl chloride, *see* Hexachlorobenzene
Pentacon, *see* Pentachlorophenol
Penta-kil, *see* Pentachlorophenol
Pentasol, *see* Pentachlorophenol
Pentech, *see* *p,p'*-DDT
Penwar, *see* Pentachlorophenol
PER, *see* Tetrachloroethylene
Peratox, *see* Pentachlorophenol
Perawin, *see* Tetrachloroethylene
PERC, *see* Tetrachloroethylene
Perchlor, *see* Tetrachloroethylene
Perchlorethylene, *see* Tetrachloroethylene
Perchlorobenzene, *see* Hexachlorobenzene
Perchlorobutadiene, *see* Hexachlorobutadiene
Perchlorocyclopentadiene, *see* Hexachlorocyclopentadiene
Perchloroethane, *see* Hexachloroethane
Perchloroethylene, *see* Tetrachloroethylene
Perchloromethane, *see* Carbon tetrachloride
Perclene, *see* Tetrachloroethylene
Perclene D, *see* Tetrachloroethylene
Percosolv, *see* Tetrachloroethylene
Periethylene naphthalene, *see* Acenaphthene
Peritonan, *see* *p*-Chloro-*m*-cresol
Perk, *see* Tetrachloroethylene
Perklone, *see* Tetrachloroethylene
Permacide, *see* Pentachlorophenol
Perm-a-chlor, *see* Trichloroethylene
Perm-a-clor, *see* Trichloroethylene
Permaguard, *see* Pentachlorophenol
Permasan, *see* Pentachlorophenol
Permatox DP-2, *see* Pentachlorophenol
Permatox Penta, *see* Pentachlorophenol
Permite, *see* Pentachlorophenol

Persec, *see* Tetrachloroethylene
Persia-Perazol, *see* 1,4-Dichlorobenzene
Pestmaster, *see* Methyl bromide
Petzinol, *see* Trichloroethylene
Pflanzol, *see* Lindane
Phenachlor, *see* 2,4,5-Trichlorophenol, 2,4,6-Trichlorophenol
Phenacide, *see* Toxaphene
Phenanthren, *see* Phenanthrene
Phenanthrene, *see* Phenanthrene
Phenantrin, *see* Phenanthrene
Phenatox, *see* Toxaphene
Phene, *see* Benzene
Phenethylene, *see* Styrene
Phenic acid, *see* Phenol
Phenoclor DP6, *see* PCB-1260
Phenohep, *see* Hexachloroethane
Phenol, *see* Phenol
Phenol carbinol, *see* Benzyl alcohol
Phenphane, *see* Toxaphene
Phenyl-4-bromophenyl ether, *see* 4-Bromophenyl phenyl
Phenyl-*p*-bromophenyl ether, *see* 4-Bromophenyl phenyl
Phenyl carbinol, *see* Benzyl alcohol
Phenylcarboxylic acid, *see* Benzoic acid
Phenyl chloride, *see* Chlorobenzene
1,10-(1,2-Phenylene)pyrene, *see* Indeno[1,2,3-*cd*]pyrene
1,10-(*o*-Phenylene)pyrene, *see* Indeno[1,2,3-*cd*]pyrene
2,3-Phenylene-*o*-pyrene, *see* Indeno[1,2,3-*cd*]pyrene
2,3-Phenylenepyrene, *see* Indeno[1,2,3-*cd*]pyrene
3,4-(*o*-Phenylene)pyrene, *see* Indeno[1,2,3-*cd*]pyrene
o-Phenylenepyrene, *see* Indeno[1,2,3-*cd*]pyrene
Phenylethane, *see* Ethylbenzene
Phenylethene, *see* Styrene
Phenylethylene, *see* Styrene
Phenylformic acid, *see* Benzoic acid
Phenyl hydrate, *see* Phenol
Phenyl hydride, *see* Benzene
Phenyl hydroxide, *see* Phenol
Phenylic acid, *see* Phenol
Phenylic alcohol, *see* Phenol
Phenylmethane, *see* Toluene
Phenyl methanol, *see* Benzyl alcohol
Phenyl methyl alcohol, *see* Benzyl alcohol
Phenyl perchloryl, *see* Hexachlorobenzene

Pyroacetic acid, *see* Acetone
Pyroacetic ether, *see* Acetone
Pyrobenzol, *see* Benzene
Pyrobenzole, *see* Benzene
Quellada, *see* Lindane
Quintox, *see* Dieldrin
R 10, *see* Carbon tetrachloride
R 12, *see* Dichlorodifluoromethane
R 20, *see* Chloroform
R 20 (refrigerant), *see* Chloroform
Rafex, *see* 4,6-Dinitro-*o*-cresol
Rafex 35, *see* 4,6-Dinitro-*o*-cresol
Raphatox, *see* 4,6-Dinitro-*o*-cresol
Raschit K, *see* *p*-Chloro-*m*-cresol
Raschit, *see* *p*-Chloro-*m*-cresol
Rasenanicon, *see* *p*-Chloro-*m*-cresol
RC plasticizer DOP, *see* Bis(2-ethylhexyl)phthalate
RCRA waste number F027, *see* 2,4,6-Trichlorophenol
RCRA waste number P003, *see* Acrolein
RCRA waste number P004, *see* Aldrin
RCRA waste number P022, *see* Carbon disulfide
RCRA waste number P024, *see* 4-Chloroaniline
RCRA waste number P037, *see* Dieldrin
RCRA waste number P047, *see* 4,6-Dinitro-*o*-cresol
RCRA waste number P048, *see* 2,4-Dinitrophenol
RCRA waste number P050, *see* α-Endosulfan, β-Endosulfan
RCRA waste number P051, *see* Endrin
RCRA waste number P059, *see* Heptachlor
RCRA waste number P077, *see* 4-Nitroaniline
RCRA waste number P082, *see* *N*-Nitrosodimethylamine
RCRA waste number P123, *see* Toxaphene
RCRA waste number U002, *see* Acetone
RCRA waste number U009, *see* Acrylonitrile
RCRA waste number U018, *see* Benzo[*a*]anthracene
RCRA waste number U019, *see* Benzene
RCRA waste number U021, *see* Benzidine
RCRA waste number U022, *see* Benzo[*a*]pyrene
RCRA waste number U024, *see* Bis(2-chloroethoxy)methane
RCRA waste number U025, *see* Bis(2-chloroethyl)ether
RCRA waste number U027, *see* Bis(2-chloroisopropyl)ether
RCRA waste number U028, *see* Bis(2-ethylhexyl)phthalate
RCRA waste number U029, *see* Methyl bromide
RCRA waste number U036, *see* Chlordane

RCRA waste number U161, *see* 4-Methyl-2-pentanone
RCRA waste number U165, *see* Naphthalene
RCRA waste number U169, *see* Nitrobenzene
RCRA waste number U170, *see* 4-Nitrophenol
RCRA waste number U188, *see* Phenol
RCRA waste number U208, *see* 1,1,2,2-Tetrachloroethane
RCRA waste number U210, *see* Tetrachloroethylene
RCRA waste number U211, *see* Carbon tetrachloride
RCRA waste number U220, *see* Toluene
RCRA waste number U225, *see* Bromoform
RCRA waste number U226, *see* 1,1,1-Trichloroethane
RCRA waste number U227, *see* 1,1,2-Trichloroethane
RCRA waste number U228, *see* Trichloroethylene
RCRA waste number U230, *see* 2,4,5-Trichlorophenol
RCRA waste number U242, *see* Pentachlorophenol
RCRA waste number U247, *see* Methoxychlor
Red 2G base, *see* 4-Nitroaniline
Redax, *see* N-Nitrosodiphenylamine
Refrigerant 11, *see* Trichlorofluoromethane
Refrigerant 12, *see* Dichlorodifluoromethane
Reomol D 79P, *see* Bis(2-ethylhexyl)phthalate
Reomol DOP, *see* Bis(2-ethylhexyl)phthalate
Retarder BA, *see* Benzoic acid
Retarder J, *see* N-Nitrosodiphenylamine
Retardex, *see* Benzoic acid
Rhodiachlor, *see* Heptachlor
Rhothane, *see* p,p'-DDD
Rhothane D-3, *see* p,p'-DDD
Rothane, *see* p,p'-DDD
Rotox, *see* Methyl bromide
Rukseam, *see* p,p'-DDT
Salvo liquid, *see* Benzoic acid
Salvo powder, *see* Benzoic acid
Sandolin, *see* 4,6-Dinitro-o-cresol
Sandolin A, *see* 4,6-Dinitro-o-cresol
Sanocide, *see* Hexachlorobenzene
Santicizer 160, *see* Benzyl butyl phthalate
Santobane, *see* p,p'-DDT
Santobrite, *see* Pentachlorophenol
Santochlor, *see* 1,4-Dichlorobenzene
Santophen, *see* Pentachlorophenol
Santophen 20, *see* Pentachlorophenol
Scintillar, *p*-Xylene

Sconatex, *see* 1,1-Dichloroethylene
SD 5,532, *see* Chlordane
Seedrin, *see* Aldrin
Seedrin liquid, *see* Aldrin
Selinon, *see* 4,6-Dinitro-*o*-cresol
Shell MIBK, *see* 4-Methyl-2-pentanone
Shell SD-5532, *see* Chlordane
Shinnippon fast red GG base, *see* 4-Nitroaniline
Sicol 150, *see* Bis(2-ethylhexyl)phthalate
Sicol 160, *see* Benzyl butyl phthalate
Silvanol, *see* Lindane
Sinituho, *see* Pentachlorophenol
Sinox, *see* 4,6-Dinitro-*o*-cresol
Slimicide, *see* Acrolein
Smut-go, *see* Hexachlorobenzene
Snieciotox, *see* Hexachlorobenzene
Solaesthin, *see* Methylene chloride
Soleptax, *see* Heptachlor
Solfo black 2B supra, *see* 2,4-Dinitrophenol
Solfo black B, *see* 2,4-Dinitrophenol
Solfo black BB, *see* 2,4-Dinitrophenol
Solfo black G, *see* 2,4-Dinitrophenol
Solfo black SB, *see* 2,4-Dinitrophenol
Solmethine, *see* Methylene chloride
Solvanol, *see* Diethyl phthalate
Solvanom, *see* Dimethyl phthalate
Solvarone, *see* Dimethyl phthalate
Solvent III, *see* 1,1,1-Trichloroethane
Special termite fluid, *see* 1,2-Dichlorobenzene
Spritz-rapidin, *see* Lindane
Spruehpflanzol, *see* Lindane
Staflex DBP, *see* Di-*n*-butyl phthalate
Staflex DOP, *see* Bis(2-ethylhexyl)phthalate
Strobane-T, *see* Toxaphene
Strobane T-90, *see* Toxaphene
Struenex, *see* Lindane
Styrene, *see* Styrene
Styrene monomer, *see* Styrene
Styrol, *see* Styrene
Styrolen, *see* Styrene
Styron, *see* Styrene
Styropol, *see* Styrene
Styropor, *see* Styrene

Sulphocarbonic anhydride, *see* Carbon disulfide
Synklor, *see* Chlordane
Synthetic 3956, *see* Toxaphene
α-T, *see* 1,1,1-Trichloroethane
β-T, *see* 1,1,2-Trichloroethane
Tap 85, *see* Lindane
Tar camphor, *see* Naphthalene
Tat chlor 4, *see* Chlordane
TBH, *see* Lindane, α-BHC, β-BHC, δ-BHC
1,1,1-TCA, *see* 1,1,1-Trichloroethane
1,1,2-TCA, *see* 1,1,2-Trichloroethane
1,2,4-TCB, *see* 1,2,4-Trichlorobenzene
TCDBD, *see* TCDD
TCDD, *see* TCDD
2,3,7,8-TCDD, *see* TCDD
TCE, *see* 1,1,2,2-Tetrachloroethane, Trichloroethylene
1,1,1-TCE, *see* 1,1,1-Trichloroethane
TCM, *see* Chloroform
2,4,5-TCP, *see* 2,4,5-Trichlorophenol
2,4,5-TCP-Dowicide 2, *see* 2,4,5-Trichlorophenol
2,4,6-TCP, *see* 2,4,6-Trichlorophenol
2,4,6-TCP-Dowicide 25, *see* 2,4,6-Trichlorophenol
TDE, *see* p,p'-DDD
4,4'-TDE, *see* p,p'-DDD
p,p'-TDE, *see* p,p'-DDD
Tenn-plas, *see* Benzoic acid
Terabol, *see* Methyl bromide
Termitkil, *see* 1,2-Dichlorobenzene
Term-i-trol, *see* Pentachlorophenol
Terr-o-gas, *see* Methyl bromide
Tetlen, *see* Tetrachloroethylene
Tetracap, *see* Tetrachloroethylene
Tetrachloormetaan, *see* Carbon tetrachloride
Tetrachlorethane, *see* 1,1,2,2-Tetrachloroethane
Tetrachlorethylene, *see* Tetrachloroethylene
Tetrachlorocarbon, *see* Carbon tetrachloride
2,3,7,8-Tetrachlorodibenzodioxin, *see* TCDD
2,3,7,8-Tetrachlorodibenzo-1,4-dioxin, *see* TCDD
2,3,7,8-Tetrachlorodibenzo-p-dioxin, *see* TCDD
2,3,7,8-Tetrachlorodibenzo[b,e][1,4]dioxan, *see* TCDD
2,3,7,8-Tetrachlorodibenzo[b,e][1,4]dioxin, *see* TCDD
Tetrachlorodiphenylethane, *see* p,p'-DDD
Tetrachloroethane, *see* 1,1,2,2-Tetrachloroethane

1,1,2,2-Tetrachloroethane, *see* 1,1,2,2-Tetrachloroethane
sym-Tetrachloroethane, *see* 1,1,2,2-Tetrachloroethane
Tetrachloroethene, *see* Tetrachloroethylene
Tetrachloroethylene, *see* Tetrachloroethylene
1,1,2,2-Tetrachloroethylene, *see* Tetrachloroethylene
Tetrachloromethane, *see* Carbon tetrachloride
Tetradioxin, *see* TCDD
Tetrafinol, *see* Carbon tetrachloride
Tetraform, *see* Carbon tetrachloride
Tetraleno, *see* Tetrachloroethylene
Tetralex, *see* Tetrachloroethylene
Tetra olive N2G, *see* Anthracene
Tetraphene, *see* Benzo[*a*]anthracene
Tetrasol, *see* Carbon tetrachloride
Tetrasulphur black PB, *see* 2,4-Dinitrophenol
Tetravec, *see* Tetrachloroethylene
Tetroguer, *see* Tetrachloroethylene
Tetropil, *see* Tetrachloroethylene
Tetrosulphur PBR, *see* 2,4-Dinitrophenol
Texadust, *see* Toxaphene
Thifor, *see* α-Endosulfan, β-Endosulfan
Thimul, *see* α-Endosulfan, β-Endosulfan
Thiodan, *see* α-Endosulfan, β-Endosulfan
Thiofor, *see* α-Endosulfan, β-Endosulfan
Thiomul, *see* α-Endosulfan, β-Endosulfan
Thionex, *see* α-Endosulfan, β-Endosulfan
Thiosulfan, *see* α-Endosulfan, β-Endosulfan
Thompson's wood fix, *see* Pentachlorophenol
Threthylen, *see* Trichloroethylene
Threthylene, *see* Trichloroethylene
Tionel, *see* α-Endosulfan, β-Endosulfan
Tiovel, *see* α-Endosulfan, β-Endosulfan
TJB, *see* *N*-Nitrosodiphenylamine
TL 314, *see* Acrylonitrile
Toluene, *see* Toluene
α-Toluenol, *see* Benzyl alcohol
Toluol, *see* Toluene
2-Toluol, *see* 2-Methylphenol
4-Toluol, *see* 4-Methylphenol
o-Toluol, *see* 2-Methylphenol
p-Toluol, *see* 4-Methylphenol
Tolu-sol, *see* Toluene
p-Tolyl alcohol, *see* 4-Methylphenol

Topichlor 20, *see* Chlordane
Topiclor, *see* Chlordane
Topiclor 20, *see* Chlordane
Toxakil, *see* Toxaphene
Toxaphene, *see* Toxaphene
Toxichlor, *see* Chlordane
Toxon 63, *see* Toxaphene
Toxyphen, *see* Toxaphene
Trethylene, *see* Trichloroethylene
Tri, *see* Trichloroethylene
Tri-6, *see* Lindane
Triad, *see* Trichloroethylene
Trial, *see* Trichloroethylene
Triasol, *see* Trichloroethylene
Tribromomethane, *see* Bromoform
Trichloran, *see* Trichloroethylene
Trichloren, *see* Trichloroethylene
1,2,4-Trichlorobenzene, *see* 1,2,4-Trichlorobenzene
unsym-Trichlorobenzene, *see* 1,2,4-Trichlorobenzene
1,1,1-Trichloro-2,2-bis(*p*-anisyl)ethane, *see* Methoxychlor
1,1,1-Trichlorobis(4-chlorophenyl)ethane, *see* *p,p'*-DDT
1,1,1-Trichlorobis(*p*-chlorophenyl)ethane, *see* *p,p'*-DDT
1,1,1-Trichloro-2,2-bis(*p*-chlorophenyl)ethane, *see* *p,p'*-DDT
1,1,1-Trichloro-2,2-bis(*p*-methoxyphenol)ethanol, *see* Methoxychlor
1,1,1-Trichloro-2,2-bis(*p*-methoxyphenyl)ethane, *see* Methoxychlor
1,1,1-Trichloro-2,2-di(4-chlorophenyl)ethane, *see* *p,p'*-DDT
1,1,1-Trichloro-2,2-di(*p*-chlorophenyl)ethane, *see* *p,p'*-DDT
1,1,1-Trichloro-2,2-di(4-methoxyphenyl)ethane, *see* Methoxychlor
Trichloroethene, *see* Trichloroethylene
β-Trichloroethane, *see* 1,1,2-Trichloroethane
1,1,1-Trichloroethane, *see* 1,1,1-Trichloroethane
1,1,2-Trichloroethane, *see* 1,1,2-Trichloroethane
1,2,2-Trichloroethane, *see* 1,1,2-Trichloroethane
α-Trichloroethane, *see* 1,1,1-Trichloroethane
Trichloroethylene, *see* Trichloroethylene
1,1,2-Trichloroethene, *see* Trichloroethylene
1,2,2-Trichloroethene, *see* Trichloroethylene
1,1,2-Trichloroethylene, *see* Trichloroethylene
1,2,2-Trichloroethylene, *see* Trichloroethylene
1,1'-(2,2,2-Trichloroethylidene)bis(4-chlorobenzene), *see* *p,p'*-DDT
1,1'-(2,2,2-Trichloroethylidene)bis(4-methoxybenzene), *see* Methoxychlor
Trichlorofluoromethane, *see* Trichlorofluoromethane
Trichloroform, *see* Chloroform

Trichloromethane, *see* Chloroform
Trichloromonofluoromethane, *see* Trichlorofluoromethane
2,4,5-Trichlorophenol, *see* 2,4,5-Trichlorophenol
2,4,6-Trichlorophenol, *see* 2,4,6-Trichlorophenol
Tri-clene, *see* Trichloroethylene
Trielene, *see* Trichloroethylene
Trieline, *see* Trichloroethylene
Tri-ethane, *see* 1,1,1-Trichloroethane
Trifina, *see* 4,6-Dinitro-*o*-cresol
Trifocide, *see* 4,6-Dinitro-*o*-cresol
Triklone, *see* Trichloroethylene
Trilen, *see* Trichloroethylene
Trilene, *see* Trichloroethylene
Triline, *see* Trichloroethylene
Trimar, *see* Trichloroethylene
1,1,3-Trimethyl-3-cyclohexene-5-one, *see* Isophorone
Trimethylcyclohexenone, *see* Isophorone
3,5,5-Trimethyl-2-cyclohexen-1-one, *see* Isophorone
Triol, *see* Trichloroethylene
Tri-plus, *see* Trichloroethylene
Tri-plus M, *see* Trichloroethylene
Trovidur, *see* Vinyl chloride
Truflex DOP, *see* Bis(2-ethylhexyl)phthalate
Ucon 11, *see* Trichlorofluoromethane
Ucon 12, *see* Dichlorodifluoromethane
Ucon 12/halocarbon 12, *see* Dichlorodifluoromethane
Ucon fluorocarbon 11, *see* Trichlorofluoromethane
Ucon refrigerant 11, *see* Trichlorofluoromethane
UN 1028, *see* Dichlorodifluoromethane
UN 1037, *see* Chloroethane
UN 1062, *see* Methyl bromide
UN 1063, *see* Methyl chloride
UN 1086, *see* Vinyl chloride
UN 1090, *see* Acetone
UN 1092, *see* Acrolein
UN 1093, *see* Acrylonitrile
UN 1114, *see* Benzene
UN 1131, *see* Carbon disulfide
UN 1134, *see* Chlorobenzene
UN 1184, *see* 1,2-Dichloroethane
UN 1193, *see* 2-Butanone
UN 1232, *see* 2-Butanone
UN 1245, *see* 4-Methyl-2-pentanone

VC, *see* Vinyl chloride
VCM, *see* Vinyl chloride
VCN, *see* Acrylonitrile
VDC, *see* 1,1-Dichloroethylene
Velsicol 104, *see* Heptachlor
Velsicol 1,068, *see* Chlordane
Velsicol 53-CS-17, *see* Heptachlor epoxide
Velsicol heptachlor, *see* Heptachlor
Ventox, *see* Acrylonitrile
Vermoestricid, *see* Carbon tetrachloride
Vertac 90%, *see* Toxaphene
Vertac toxaphene 90, *see* Toxaphene
Vestinol 80, *see* Bis(2-ethylhexyl)phthalate
Vestrol, *see* Trichloroethylene
Vinicizer 85, *see* Di-*n*-octyl phthalate
Vinyl acetate, *see* Vinyl acetate
Vinyl acetate H.Q., *see* Vinyl acetate
Vinyl A Monomer, *see* Vinyl acetate
Vinyl benzene, *see* Styrene
Vinyl benzol, *see* Styrene
Vinyl C monomer, *see* Vinyl chloride
Vinyl chloride, *see* Vinyl chloride
Vinyl chloride monomer, *see* Vinyl chloride
Vinyl 2-chloroethyl ether, *see* 2-Chloroethyl vinyl ether
Vinyl *β*-chloroethyl ether, *see* 2-Chloroethyl vinyl ether
Vinyl cyanide, *see* Acrylonitrile
Vinyl trichloride, *see* 1,1,2-Trichloroethane
Vinylidene chloride (II), *see* 1,1-Dichloroethylene
Vinylidene chloride, *see* 1,1-Dichloroethylene
Vinylidene dichloride, *see* 1,1-Dichloroethylene
Vinylidine chloride, *see* 1,1-Dichloroethylene
Viton, *see* Lindane
Vitran, *see* Trichloroethylene
Vulcalent A, *see* *N*-Nitrosodiphenylamine
Vulcatard, *see* *N*-Nitrosodiphenylamine
Vulcatard A, *see* *N*-Nitrosodiphenylamine
Vultrol, *see* *N*-Nitrosodiphenylamine
VyAc, *see* Vinyl acetate
Weedone, *see* Pentachlorophenol
Weeviltox, *see* Carbon disulfide
Westron, *see* 1,1,2,2-Tetrachloroethane
Westrosol, *see* Trichloroethylene
White tar, *see* Naphthalene